ELEMENTS
OF
ENGINEERING
ELECTROMAGNETICS

ELEMENTS OF ENGINEERING ELECTROMAGNETICS

N. NARAYANA RAO

Professor of Electrical Engineering
University of Illinois at Urbana–Champaign

PRENTICE-HALL, INC.
Englewood Cliffs, New Jersey 07632

Library of Congress Cataloging in Publication Data

NARAYANA RAO, NANNAPANENI.
  Elements of engineering electromagnetics.

  Bibliography: p.
  Includes index.
  1. Electromagnetic theory. I. Title.
QC670.N3   530.1′41   76-44234
ISBN 0-13-264150-X

10 9 8 7 6

Printed in the United States of America

PRENTICE-HALL INTERNATIONAL, INC., *London*
PRENTICE-HALL OF AUSTRALIA PTY. LIMITED, *Sydney*
PRENTICE-HALL OF CANADA, LTD., *Toronto*
PRENTICE-HALL OF INDIA PRIVATE LIMITED, *New Delhi*
PRENTICE-HALL OF JAPAN, INC., *Tokyo*
PRENTICE-HALL OF SOUTHEAST ASIA PTE. LTD., *Singapore*
WHITEHALL BOOKS LIMITED, *Wellington, New Zealand*

To the memory
of
MY FATHER

CONTENTS

PREFACE

Traditionally, the first undergraduate course in engineering electromagnetics has been based upon developing static fields in a historical manner, and culminating in Maxwell's equations, with perhaps a brief discussion of uniform plane waves. This is then followed by one or more courses dealing with transmission lines and wave propagation. Due to the pressure of increasing areas of interest and fewer required courses, there has been in recent years a growing trend in electrical engineering curricula toward limiting the requirement in electromagnetics to a one-semester course or its equivalent. Consequently, and in view of the student's earlier exposure in engineering physics to static fields and Maxwell's equations, it has become increasingly expedient to deviate from the historical approach and to base the first course in electromagnetics upon dynamic fields and their engineering applications.

There are many texts, including one by the author, which fulfill the requirements of the traditional approach. There are also several books devoted to wave propagation and related topics; these, however, rely upon a first course of the traditional type or a variation of it to provide the required background. Thus a need has arisen for a one-semester text in which the basic material is built up on time-varying fields and their engineering applications so as to enhance its utility for the one-semester student of engineering electromagnetics, while enabling the student who will continue to take futher (elective) courses in electromagnetics to learn many of the same field concepts and mathematical tools and techniques provided by the traditional treatment. This book represents an attempt to satisfy this need.

The thread of development of the material is evident from a reading of the table of contents. Some of the salient features of the first nine chapters consist of introducing:

1. the bulk of the material through the use of the Cartesian coordinate system to keep the geometry simple and yet sufficient to learn the physical concepts and mathematical tools, while employing the other coordinate systems where necessary;
2. Maxwell's equations for time-varying fields first in integral form and then in differential form very early in the book;
3. uniform plane wave propagation by obtaining the field solution to the infinite plane current sheet of uniform sinusoidally time-varying density;
4. material media by considering their interaction with uniform plane wave fields;
5. transmission lines by first considering uniform plane waves guided by two parallel, plane perfect conductors and then extending to a line of arbitrary cross section through graphical field mapping;
6. waveguides by considering the superposition of two obliquely propagating uniform plane waves and then placing perfect conductors in appropriate planes so as to satisfy the boundary conditions;
7. antennas by obtaining the complete field solution to the Hertzian dipole through a successive extension of the quasistatic field solution so as to satisfy simultaneously the two Maxwell's curl equations; and
8. Maxwell's equations for static fields as specializations of Maxwell's equations for time-varying fields and then proceeding with the discussion of the more important topics of static and quasistatic fields.

The final chapter is devoted to seven independent special topics, each based upon one or more of the previous six chapters. It is intended that the instructor will choose one (or more) of these topics for discussion following the corresponding previous chapter(s). Material on cylindrical and spherical coordinate systems is presented as appendices so that it can be studied either immediately following the discussion of the corresponding material on the Cartesian coordinate system or only when necessary.

From considerations of varying degrees of background preparation at different schools, a greater amount of material than can be covered in an average class of three semester-hour credits is included in the book. Since it has been found that nearly eight chapters can be completed during the semester, the first six chapters plus an equivalent of about two chapters from the remaining four is suggested to be typical of coverage. When the background preparation permits an accelerated discussion of the first three chapters, it is possible to cover a greater amount of material. Worked-out examples are distributed throughout the text to illustrate and, in some cases, extend the various concepts. Summary of the material and a number of questions are included for each chapter to facilitate review of the chapters. Problems are arranged in the same order as the text material, and answers are provided for the odd-numbered problems.

This text is based primarily on lecture notes for classes taught by the author at the University of Illinois at Urbana–Champaign. The author wishes to express his appreciation to Patricia Sammann for the excellent typing work. Finally, although great care has been exercised, some errors are inevitable. The author earnestly requests readers to inform him of any errors that they may find and to contribute suggestions for improvement.

Urbana, Illinois   N. NARAYANA RAO

ELEMENTS
OF
ENGINEERING
ELECTROMAGNETICS

1. VECTORS AND FIELDS

Electromagnetics deals with the study of electric and magnetic "fields." It is at once apparent that we need to familiarize ourselves with the concept of a "field," and in particular with "electric" and "magnetic" fields. These fields are vector quantities and their behavior is governed by a set of laws known as "Maxwell's equations." The mathematical formulation of Maxwell's equations and their subsequent application in our study of the elements of engineering electromagnetics require that we first learn the basic rules pertinent to mathematical manipulations involving vector quantities. With this goal in mind, we shall devote this chapter to vectors and fields.

We shall first study certain simple rules of vector algebra without the implication of a coordinate system and then introduce the Cartesian coordinate system, which is the coordinate system employed for the most part of our study in this book. After learning the vector algebraic rules, we shall turn our attention to a discussion of scalar and vector fields, static as well as time-varying, by means of some familiar examples. We shall devote particular attention to sinusoidally time-varying fields, scalar as well as vector, and to the phasor technique of dealing with sinusoidally time-varying quantities. With this general introduction to vectors and fields, we shall then devote the remainder of the chapter to an introduction of the electric and magnetic field concepts, from considerations of the experimental laws of Coulomb and Ampere.

1.1 VECTOR ALGEBRA

In the study of elementary physics we come across several quantities such as mass, temperature, velocity, acceleration, force, and charge. Some of these quantities have associated with them not only a magnitude but also a direction in space whereas others are characterized by magnitude only. The former class of quantities are known as "vectors" and the latter class of quantities are known as "scalars." Mass, temperature, and charge are scalars whereas velocity, acceleration, and force are vectors. Other examples are voltage and current for scalars and electric and magnetic fields for vectors.

Vector quantities are represented by boldface roman type symbols, e.g., $\mathbf{A}$, in order to distinguish them from scalar quantities which are represented by lightface italic type symbols, e.g., A. Graphically, a vector, say $\mathbf{A}$, is represented by a straight line with an arrowhead pointing in the direction of $\mathbf{A}$ and having a length proportional to the magnitude of $\mathbf{A}$, denoted $|\mathbf{A}|$ or simply A. Figures 1.1(a)–(d) show four vectors drawn to the same scale. If

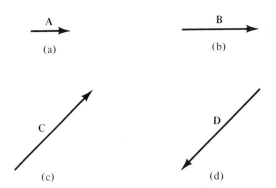

Figure 1.1. Graphical representation of vectors.

the top of the page represents north, then vectors $\mathbf{A}$ and $\mathbf{B}$ are directed eastward with the magnitude of $\mathbf{B}$ being twice that of $\mathbf{A}$. Vector $\mathbf{C}$ is directed toward the northeast and has a magnitude three times that of $\mathbf{A}$. Vector $\mathbf{D}$ is directed toward the southwest and has a magnitude equal to that of $\mathbf{C}$. Since $\mathbf{C}$ and $\mathbf{D}$ are equal in magnitude but opposite in direction, one is the negative of the other.

Since a vector may have in general an arbitrary orientation in three dimensions, we need to define a set of three reference directions at each and every point in space in terms of which we can describe vectors drawn at that point. It is convenient to choose these three reference directions to be mutually orthogonal as, for example, east, north and upward or the three contiguous edges of a rectangular room. Thus let us consider three mutually orthogonal reference directions and direct "unit vectors" along the three directions as shown, for example, in Fig. 1.2(a). A unit vector has magnitude unity. We shall represent a unit vector by the symbol i and use a subscript to denote its direction. We shall denote the three directions by subscripts 1, 2, and 3. We note that for a fixed orientation of i_1, two combinations are possible for the orientations of i_2 and i_3, as shown in Figs. 1.2(a) and (b). If we take a right-hand screw and turn it from i_1 to i_2 through the $90°$-angle, it progresses in the direction of i_3 in Fig. 1.2(a) but opposite to the direction of i_3 in Fig. 1.2(b). Alternatively, a left-hand screw when turned from i_1 to i_2 in Fig. 1.2(b) will progress in the direction of i_3. Hence the set of unit vectors in Fig. 1.2(a) corresponds to a right-handed system whereas the set in Fig. 1.2(b) corresponds to a left-handed system. We shall work consistently with the right-handed system.

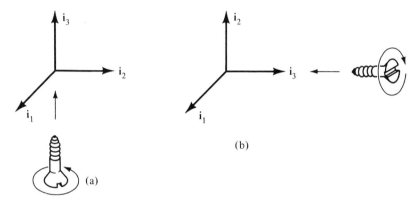

Figure 1.2. (a) Set of three orthogonal unit vectors in a right-handed system. (b) Set of three orthogonal unit vectors in a left-handed system.

A vector of magnitude different from unity along any of the reference directions can be represented in terms of the unit vector along that direction. Thus $4i_1$ represents a vector of magnitude 4 units in the direction of i_1, $6i_2$ represents a vector of magnitude 6 units in the direction of i_2, and $-2i_3$ represents a vector of magnitude 2 units in the direction opposite to that of i_3, as shown in Fig. 1.3. Two vectors are added by placing the beginning of the second vector at the tip of the first vector and then drawing the sum vector from the beginning of the first vector to the tip of the second vector. Thus to add $4i_1$ and $6i_2$, we simply slide $6i_2$ without changing its direction until its beginning coincides with the tip of $4i_1$ and then draw the vector $4i_1 + 6i_2$ from the beginning of $4i_1$ to the tip of $6i_2$, as shown in Fig. 1.3. By adding $-2i_3$ to this vector $4i_1 + 6i_2$ in a similar manner, we obtain the vector

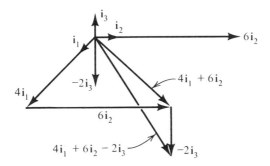

Figure 1.3. Graphical addition of vectors.

$(4\mathbf{i}_1 + 6\mathbf{i}_2 - 2\mathbf{i}_3)$, as shown in Fig. 1.3. We note that the magnitude of $(4\mathbf{i}_1 + 6\mathbf{i}_2)$ is $\sqrt{4^2 + 6^2}$ or 7.211 and that the magnitude of $(4\mathbf{i}_1 + 6\mathbf{i}_2 - 2\mathbf{i}_3)$ is $\sqrt{4^2 + 6^2 + 2^2}$ or 7.483. Conversely to the foregoing discussion, a vector **A** at a given point is simply the superposition of three vectors $A_1\mathbf{i}_1$, $A_2\mathbf{i}_2$, and $A_3\mathbf{i}_3$ which are the projections of **A** onto the reference directions at that point. A_1, A_2, and A_3 are known as the components of **A** along the 1, 2, and 3 directions, respectively. Thus

$$\mathbf{A} = A_1\mathbf{i}_1 + A_2\mathbf{i}_2 + A_3\mathbf{i}_3 \tag{1.1}$$

We now consider three vectors **A**, **B**, and **C** given by

$$\mathbf{A} = A_1\mathbf{i}_1 + A_2\mathbf{i}_2 + A_3\mathbf{i}_3 \tag{1.2a}$$

$$\mathbf{B} = B_1\mathbf{i}_1 + B_2\mathbf{i}_2 + B_3\mathbf{i}_3 \tag{1.2b}$$

$$\mathbf{C} = C_1\mathbf{i}_1 + C_2\mathbf{i}_2 + C_3\mathbf{i}_3 \tag{1.2c}$$

at a point and discuss several algebraic operations involving vectors as follows.

VECTOR ADDITION AND SUBTRACTION: Since a given pair of like components of two vectors are parallel, addition of two vectors consists simply of adding the three pairs of like components of the vectors. Thus

$$\mathbf{A} + \mathbf{B} = (A_1\mathbf{i}_1 + A_2\mathbf{i}_2 + A_3\mathbf{i}_3) + (B_1\mathbf{i}_1 + B_2\mathbf{i}_2 + B_3\mathbf{i}_3)$$
$$= (A_1 + B_1)\mathbf{i}_1 + (A_2 + B_2)\mathbf{i}_2 + (A_3 + B_3)\mathbf{i}_3 \tag{1.3}$$

Vector subtraction is a special case of addition. Thus

$$\mathbf{B} - \mathbf{C} = \mathbf{B} + (-\mathbf{C}) = (B_1\mathbf{i}_1 + B_2\mathbf{i}_2 + B_3\mathbf{i}_3) + (-C_1\mathbf{i}_1 - C_2\mathbf{i}_2 - C_3\mathbf{i}_3)$$
$$= (B_1 - C_1)\mathbf{i}_1 + (B_2 - C_2)\mathbf{i}_2 + (B_3 - C_3)\mathbf{i}_3 \tag{1.4}$$

MULTIPLICATION AND DIVISION BY A SCALAR: Multiplication of a vector **A** by a scalar m is the same as repeated addition of the vector. Thus

$$m\mathbf{A} = m(A_1\mathbf{i}_1 + A_2\mathbf{i}_2 + A_3\mathbf{i}_3) = mA_1\mathbf{i}_1 + mA_2\mathbf{i}_2 + mA_3\mathbf{i}_3 \qquad (1.5)$$

Division by a scalar is a special case of multiplication by a scalar. Thus

$$\frac{\mathbf{B}}{n} = \frac{1}{n}(\mathbf{B}) = \frac{B_1}{n}\mathbf{i}_1 + \frac{B_2}{n}\mathbf{i}_2 + \frac{B_3}{n}\mathbf{i}_3 \qquad (1.6)$$

MAGNITUDE OF A VECTOR: From the construction of Fig. 1.3 and the associated discussion, we have

$$|\mathbf{A}| = |A_1\mathbf{i}_1 + A_2\mathbf{i}_2 + A_3\mathbf{i}_3| = \sqrt{A_1^2 + A_2^2 + A_3^2} \qquad (1.7)$$

UNIT VECTOR ALONG **A**:   The unit vector $\mathbf{i}_A$ has a magnitude equal to unity but its direction is the same as that of **A**. Hence

$$\mathbf{i}_A = \frac{\mathbf{A}}{|\mathbf{A}|} = \frac{A_1\mathbf{i}_1 + A_2\mathbf{i}_2 + A_3\mathbf{i}_3}{\sqrt{A_1^2 + A_2^2 + A_3^2}}$$

$$= \frac{A_1}{\sqrt{A_1^2 + A_2^2 + A_3^2}}\mathbf{i}_1 + \frac{A_2}{\sqrt{A_1^2 + A_2^2 + A_3^2}}\mathbf{i}_2 + \frac{A_3}{\sqrt{A_1^2 + A_2^2 + A_3^2}}\mathbf{i}_3 \quad (1.8)$$

SCALAR OR DOT PRODUCT OF TWO VECTORS: The scalar or dot product of two vectors **A** and **B** is a scalar quantity equal to the product of the magnitudes of **A** and **B** and the cosine of the angle between **A** and **B**. It is represented by a dot between **A** and **B**. Thus if α is the angle between **A** and **B**, then

$$\mathbf{A} \cdot \mathbf{B} = |\mathbf{A}||\mathbf{B}|\cos\alpha = AB\cos\alpha \qquad (1.9)$$

For the unit vectors $\mathbf{i}_1, \mathbf{i}_2, \mathbf{i}_3$, we have

$$\mathbf{i}_1 \cdot \mathbf{i}_1 = 1 \qquad \mathbf{i}_1 \cdot \mathbf{i}_2 = 0 \qquad \mathbf{i}_1 \cdot \mathbf{i}_3 = 0 \qquad (1.10a)$$

$$\mathbf{i}_2 \cdot \mathbf{i}_1 = 0 \qquad \mathbf{i}_2 \cdot \mathbf{i}_2 = 1 \qquad \mathbf{i}_2 \cdot \mathbf{i}_3 = 0 \qquad (1.10b)$$

$$\mathbf{i}_3 \cdot \mathbf{i}_1 = 0 \qquad \mathbf{i}_3 \cdot \mathbf{i}_2 = 0 \qquad \mathbf{i}_3 \cdot \mathbf{i}_3 = 1 \qquad (1.10c)$$

By noting that $\mathbf{A} \cdot \mathbf{B} = A(B\cos\alpha) = B(A\cos\alpha)$, we observe that the dot product operation consists of multiplying the magnitude of one vector by the scalar obtained by projecting the second vector onto the first vector as shown in Figs. 1.4(a) and (b). The dot product operation is commutative since

$$\mathbf{B} \cdot \mathbf{A} = BA\cos\alpha = AB\cos\alpha = \mathbf{A} \cdot \mathbf{B} \qquad (1.11)$$

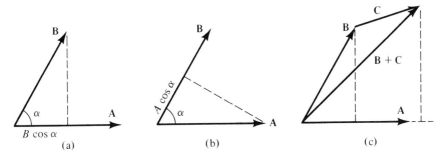

Figure 1.4. (a) and (b) For showing that the dot product of two vectors **A** and **B** is the product of the magnitude of one vector and the projection of the second vector onto the first vector. (c) For proving the distributive property of the dot product operation.

The distributive property also holds for the dot product as can be seen from the construction of Fig. 1.4(c), which illustrates that the projection of **B** + **C** onto **A** is equal to the sum of the projections of **B** and **C** onto **A**. Thus

$$\mathbf{A} \cdot (\mathbf{B} + \mathbf{C}) = \mathbf{A} \cdot \mathbf{B} + \mathbf{A} \cdot \mathbf{C} \qquad (1.12)$$

Using this property, and the relationships (1.10a)–(1.10c), we have

$$
\begin{aligned}
\mathbf{A} \cdot \mathbf{B} &= (A_1 \mathbf{i}_1 + A_2 \mathbf{i}_2 + A_3 \mathbf{i}_3) \cdot (B_1 \mathbf{i}_1 + B_2 \mathbf{i}_2 + B_3 \mathbf{i}_3) \\
&= A_1 \mathbf{i}_1 \cdot B_1 \mathbf{i}_1 + A_1 \mathbf{i}_1 \cdot B_2 \mathbf{i}_2 + A_1 \mathbf{i}_1 \cdot B_3 \mathbf{i}_3 \\
&\quad + A_2 \mathbf{i}_2 \cdot B_1 \mathbf{i}_1 + A_2 \mathbf{i}_2 \cdot B_2 \mathbf{i}_2 + A_2 \mathbf{i}_2 \cdot B_3 \mathbf{i}_3 \\
&\quad + A_3 \mathbf{i}_3 \cdot B_1 \mathbf{i}_1 + A_3 \mathbf{i}_3 \cdot B_2 \mathbf{i}_2 + A_3 \mathbf{i}_3 \cdot B_3 \mathbf{i}_3 \\
&= A_1 B_1 + A_2 B_2 + A_3 B_3 \qquad (1.13)
\end{aligned}
$$

Thus the dot product of two vectors is the sum of the products of the like components of the two vectors.

VECTOR OR CROSS PRODUCT OF TWO VECTORS: The vector or cross product of two vectors **A** and **B** is a vector quantity whose magnitude is equal to the product of the magnitudes of **A** and **B** and the sine of the angle α between **A** and **B** and whose direction is the direction of advance of a right-hand screw as it is turned from **A** to **B** through the angle α, as shown in Fig. 1.5. It is represented by a cross between **A** and **B**. Thus if $\mathbf{i}_N$ is the unit vector in the direction of advance of the right-hand screw, then

$$\mathbf{A} \times \mathbf{B} = |\mathbf{A}| |\mathbf{B}| \sin \alpha \, \mathbf{i}_N = AB \sin \alpha \, \mathbf{i}_N \qquad (1.14)$$

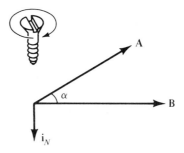

Figure 1.5. The cross product operation $\mathbf{A} \times \mathbf{B}$.

For the unit vectors $\mathbf{i}_1$, $\mathbf{i}_2$, $\mathbf{i}_3$, we have

$$\mathbf{i}_1 \times \mathbf{i}_1 = 0 \qquad \mathbf{i}_1 \times \mathbf{i}_2 = \mathbf{i}_3 \qquad \mathbf{i}_1 \times \mathbf{i}_3 = -\mathbf{i}_2 \qquad (1.15a)$$

$$\mathbf{i}_2 \times \mathbf{i}_1 = -\mathbf{i}_3 \qquad \mathbf{i}_2 \times \mathbf{i}_2 = 0 \qquad \mathbf{i}_2 \times \mathbf{i}_3 = \mathbf{i}_1 \qquad (1.15b)$$

$$\mathbf{i}_3 \times \mathbf{i}_1 = \mathbf{i}_2 \qquad \mathbf{i}_3 \times \mathbf{i}_2 = -\mathbf{i}_1 \qquad \mathbf{i}_3 \times \mathbf{i}_3 = 0 \qquad (1.15c)$$

Note that the cross product of identical vectors is zero. If we arrange the unit vectors in the manner $\mathbf{i}_1\mathbf{i}_2\mathbf{i}_3\mathbf{i}_1\mathbf{i}_2$ and then go forward, the cross product of any two successive unit vectors is equal to the following unit vector, but if we go backward, the cross product of any two successive unit vectors is the negative of the following unit vector.

The cross product operation is not commutative since

$$\mathbf{B} \times \mathbf{A} = |\mathbf{B}||\mathbf{A}| \sin \alpha \, (-\mathbf{i}_N) = -AB \sin \alpha \, \mathbf{i}_N = -\mathbf{A} \times \mathbf{B} \qquad (1.16)$$

The distributive property holds for the cross product (we shall prove this later in this section) so that

$$\mathbf{A} \times (\mathbf{B} + \mathbf{C}) = \mathbf{A} \times \mathbf{B} + \mathbf{A} \times \mathbf{C} \qquad (1.17)$$

Using this property and the relationships (1.15a)–(1.15c), we obtain

$$\begin{aligned}
\mathbf{A} \times \mathbf{B} &= (A_1\mathbf{i}_1 + A_2\mathbf{i}_2 + A_3\mathbf{i}_3) \times (B_1\mathbf{i}_1 + B_2\mathbf{i}_2 + B_3\mathbf{i}_3) \\
&= A_1\mathbf{i}_1 \times B_1\mathbf{i}_1 + A_1\mathbf{i}_1 \times B_2\mathbf{i}_2 + A_1\mathbf{i}_1 \times B_3\mathbf{i}_3 \\
&\quad + A_2\mathbf{i}_2 \times B_1\mathbf{i}_1 + A_2\mathbf{i}_2 \times B_2\mathbf{i}_2 + A_2\mathbf{i}_2 \times B_3\mathbf{i}_3 \\
&\quad + A_3\mathbf{i}_3 \times B_1\mathbf{i}_1 + A_3\mathbf{i}_3 \times B_2\mathbf{i}_2 + A_3\mathbf{i}_3 \times B_3\mathbf{i}_3 \\
&= A_1B_2\mathbf{i}_3 - A_1B_3\mathbf{i}_2 - A_2B_1\mathbf{i}_3 + A_2B_3\mathbf{i}_1 \\
&\quad + A_3B_1\mathbf{i}_2 - A_3B_2\mathbf{i}_1 \\
&= (A_2B_3 - A_3B_2)\mathbf{i}_1 + (A_3B_1 - A_1B_3)\mathbf{i}_2 \\
&\quad + (A_1B_2 - A_2B_1)\mathbf{i}_3 \qquad (1.18)
\end{aligned}$$

This can be expressed in determinant form in the manner

$$\mathbf{A} \times \mathbf{B} = \begin{vmatrix} \mathbf{i}_1 & \mathbf{i}_2 & \mathbf{i}_3 \\ A_1 & A_2 & A_3 \\ B_1 & B_2 & B_3 \end{vmatrix} \tag{1.19}$$

A triple cross product involves three vectors in two cross product operations. Caution must be exercised in evaluating a triple cross product since the order of evaluation is important, that is, $\mathbf{A} \times (\mathbf{B} \times \mathbf{C})$ is not equal to $(\mathbf{A} \times \mathbf{B}) \times \mathbf{C}$. This can be illustrated by means of a simple example involving unit vectors. Thus if $\mathbf{A} = \mathbf{i}_1$, $\mathbf{B} = \mathbf{i}_1$, and $\mathbf{C} = \mathbf{i}_2$, then

$$\mathbf{A} \times (\mathbf{B} \times \mathbf{C}) = \mathbf{i}_1 \times (\mathbf{i}_1 \times \mathbf{i}_2) = \mathbf{i}_1 \times \mathbf{i}_3 = -\mathbf{i}_2$$

whereas

$$(\mathbf{A} \times \mathbf{B}) \times \mathbf{C} = (\mathbf{i}_1 \times \mathbf{i}_1) \times \mathbf{i}_2 = 0 \times \mathbf{i}_2 = 0$$

SCALAR TRIPLE PRODUCT: The scalar triple product involves three vectors in a dot product operation and a cross product operation as, for example, $\mathbf{A} \cdot \mathbf{B} \times \mathbf{C}$. It is not necessary to include parentheses since this quantity can be evaluated in only one manner, that is, by evaluating $\mathbf{B} \times \mathbf{C}$ first and then dotting the resulting vector with $\mathbf{A}$. It is meaningless to try to evaluate the dot product first since it results in a scalar quantity and hence we cannot proceed any further. From (1.13) and (1.19), we have

$$\mathbf{A} \cdot \mathbf{B} \times \mathbf{C} = (A_1 \mathbf{i}_1 + A_2 \mathbf{i}_2 + A_3 \mathbf{i}_3) \cdot \begin{vmatrix} \mathbf{i}_1 & \mathbf{i}_2 & \mathbf{i}_3 \\ B_1 & B_2 & B_3 \\ C_1 & C_2 & C_3 \end{vmatrix} = \begin{vmatrix} A_1 & A_2 & A_3 \\ B_1 & B_2 & B_3 \\ C_1 & C_2 & C_3 \end{vmatrix}$$
$$\tag{1.20}$$

Since the value of the determinant on the right side of (1.20) remains unchanged if the rows are interchanged in a cyclical manner,

$$\mathbf{A} \cdot \mathbf{B} \times \mathbf{C} = \mathbf{B} \cdot \mathbf{C} \times \mathbf{A} = \mathbf{C} \cdot \mathbf{A} \times \mathbf{B} \tag{1.21}$$

We shall now show that the distributive law holds for the cross product operation by using (1.21). Thus let us consider $\mathbf{A} \times (\mathbf{B} + \mathbf{C})$. Then if $\mathbf{D}$ is any arbitrary vector, we have

$$\mathbf{D} \cdot \mathbf{A} \times (\mathbf{B} + \mathbf{C}) = (\mathbf{B} + \mathbf{C}) \cdot (\mathbf{D} \times \mathbf{A}) = \mathbf{B} \cdot (\mathbf{D} \times \mathbf{A}) + \mathbf{C} \cdot (\mathbf{D} \times \mathbf{A})$$
$$= \mathbf{D} \cdot \mathbf{A} \times \mathbf{B} + \mathbf{D} \cdot \mathbf{A} \times \mathbf{C} = \mathbf{D} \cdot (\mathbf{A} \times \mathbf{B} + \mathbf{A} \times \mathbf{C})$$
$$\tag{1.22}$$

where we have used the distributive property of the dot product operation. Since (1.22) holds for any $\mathbf{D}$, it follows that

$$\mathbf{A} \times (\mathbf{B} + \mathbf{C}) = \mathbf{A} \times \mathbf{B} + \mathbf{A} \times \mathbf{C}.$$

Example 1.1.   Given three vectors

$$\mathbf{A} = \mathbf{i}_1 + \mathbf{i}_2$$
$$\mathbf{B} = \mathbf{i}_1 + 2\mathbf{i}_2 - 2\mathbf{i}_3$$
$$\mathbf{C} = \mathbf{i}_2 + 2\mathbf{i}_3$$

let us carry out several of the vector algebraic operations.

(a) $\mathbf{A} + \mathbf{B} = (\mathbf{i}_1 + \mathbf{i}_2) + (\mathbf{i}_1 + 2\mathbf{i}_2 - 2\mathbf{i}_3) = 2\mathbf{i}_1 + 3\mathbf{i}_2 - 2\mathbf{i}_3$

(b) $\mathbf{B} - \mathbf{C} = (\mathbf{i}_1 + 2\mathbf{i}_2 - 2\mathbf{i}_3) - (\mathbf{i}_2 + 2\mathbf{i}_3) = \mathbf{i}_1 + \mathbf{i}_2 - 4\mathbf{i}_3$

(c) $4\mathbf{C} = 4(\mathbf{i}_2 + 2\mathbf{i}_3) = 4\mathbf{i}_2 + 8\mathbf{i}_3$

(d) $|\mathbf{B}| = |\mathbf{i}_1 + 2\mathbf{i}_2 - 2\mathbf{i}_3| = \sqrt{(1)^2 + (2)^2 + (-2)^2} = 3$

(e) $\mathbf{i}_B = \dfrac{\mathbf{B}}{|\mathbf{B}|} = \dfrac{\mathbf{i}_1 + 2\mathbf{i}_2 - 2\mathbf{i}_3}{3} = \dfrac{1}{3}\mathbf{i}_1 + \dfrac{2}{3}\mathbf{i}_2 - \dfrac{2}{3}\mathbf{i}_3$

(f) $\mathbf{A} \cdot \mathbf{B} = (\mathbf{i}_1 + \mathbf{i}_2) \cdot (\mathbf{i}_1 + 2\mathbf{i}_2 - 2\mathbf{i}_3) = (1)(1) + (1)(2) + (0)(-2) = 3$

(g) $\mathbf{A} \times \mathbf{B} = \begin{vmatrix} \mathbf{i}_1 & \mathbf{i}_2 & \mathbf{i}_3 \\ 1 & 1 & 0 \\ 1 & 2 & -2 \end{vmatrix} = (-2 - 0)\mathbf{i}_1 + (0 + 2)\mathbf{i}_2 + (2 - 1)\mathbf{i}_3$

$$= -2\mathbf{i}_1 + 2\mathbf{i}_2 + \mathbf{i}_3$$

(h) $(\mathbf{A} \times \mathbf{B}) \times \mathbf{C} = \begin{vmatrix} \mathbf{i}_1 & \mathbf{i}_2 & \mathbf{i}_3 \\ -2 & 2 & 1 \\ 0 & 1 & 2 \end{vmatrix} = 3\mathbf{i}_1 + 4\mathbf{i}_2 - 2\mathbf{i}_3$

(i) $\mathbf{A} \cdot \mathbf{B} \times \mathbf{C} = \begin{vmatrix} 1 & 1 & 0 \\ 1 & 2 & -2 \\ 0 & 1 & 2 \end{vmatrix} = (1)(6) + (1)(-2) + (0)(1) = 4$ ∎

1.2   CARTESIAN COORDINATE SYSTEM

In the previous section we introduced the technique of expressing a vector at a point in space in terms of its component vectors along a set of three mutually orthogonal directions defined by three mutually orthogonal unit vectors at that point. Now in order to relate vectors at one point in space to vectors at another point in space, we must define the set of three reference directions at each and every point in space. To do this in a systematic manner, we need to use a coordinate system. Although there are several different

coordinate systems, we shall use for the most part of our study the simplest
of these, namely, the Cartesian coordinate system, also known as the "rectangular coordinate system," to keep the geometry simple and yet sufficient to
learn many of the elements of engineering electromagnetics. We shall, however, find it necessary in a few cases to resort to the use of cylindrical and
spherical coordinate systems. Hence a discussion of these coordinate systems
is included in Appendix A. In this section we introduce the Cartesian coordinate system.

The Cartesian coordinate system is defined by a set of three mutually
orthogonal planes as shown in Fig. 1.6(a). The point at which the three planes
intersect is known as the origin O. The origin is the reference point relative
to which we locate any other point in space. Each pair of planes intersects
in a straight line. Hence the three planes define a set of three straight lines
which form the coordinate axes. These coordinate axes are denoted as the
x, y, and z axes. Values of x, y, and z are measured from the origin and hence
the coordinates of the origin are $(0, 0, 0)$, that is, $x = 0$, $y = 0$, and $z = 0$.
Directions in which values of x, y, and z increase along the respective coordinate axes are indicated by arrowheads. The same set of three directions is
used to erect a set of three unit vectors, denoted $\mathbf{i}_x$, $\mathbf{i}_y$, and $\mathbf{i}_z$, as shown in
Fig. 1.6(a), for the purpose of describing vectors drawn at the origin. Note
that the positive x, y, and z directions are chosen such that they form a
right-handed system, that is, a system for which $\mathbf{i}_x \times \mathbf{i}_y = \mathbf{i}_z$.

On one of the three planes, namely, the yz plane, the value of x is constant
and equal to zero, its value at the origin, since movement on this plane does
not require any movement in the x direction. Similarly, on the zx plane the
value of y is constant and equal to zero, and on the xy plane the value of z
is constant and equal to zero. Any point other than the origin is now given
by the intersection of three planes obtained by incrementing the values of
the coordinates by appropriate amounts. For example, by displacing the
$x = 0$ plane by 2 units in the positive x direction, the $y = 0$ plane by 5 units
in the positive y direction, and the $z = 0$ plane by 4 units in the positive z
direction, we obtain the planes $x = 2$, $y = 5$, and $z = 4$, respectively, which
intersect at the point $(2, 5, 4)$ as shown in Fig. 1.6(b). The intersections of
pairs of these planes define three straight lines along which we can erect the
unit vectors $\mathbf{i}_x$, $\mathbf{i}_y$, and $\mathbf{i}_z$ toward the directions of increasing values of x, y,
and z, respectively, for the purpose of describing vectors drawn at that point.
These unit vectors are parallel to the corresponding unit vectors drawn at
the origin, as can be seen from Fig. 1.6(b). The same is true for any point in
space in the Cartesian coordinate system. Thus each one of the three unit
vectors in the Cartesian coordinate system has the same direction at all points
and hence it is uniform. This behavior does not, however, hold for all unit
vectors in the cylindrical and spherical coordinate systems.

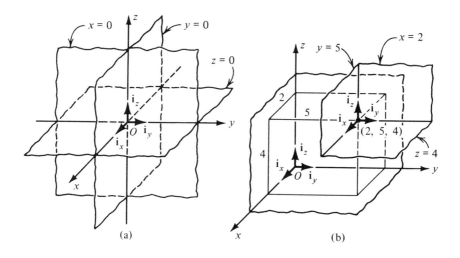

(a)

(b)

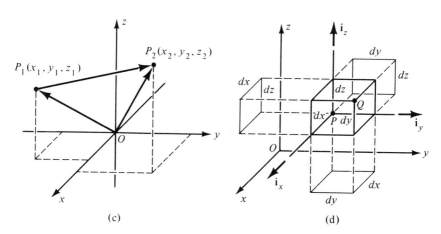

(c)

(d)

Figure 1.6. Cartesian coordinate system. (a) The three orthogonal planes defining the coordinate system. (b) Unit vectors at an arbitrary point. (c) Vector from one arbitrary point to another arbitrary point. (d) Differential lengths, surfaces, and volume formed by incrementing the coordinates.

It is now a simple matter to apply what we have learned in Sec. 1.1 to vectors in Cartesian coordinates. All we need to do is to replace the subscripts 1, 2, and 3 for the unit vectors and the components along the unit vectors by the subscripts x, y, and z, respectively, and also utilize the property that $\mathbf{i}_x$, $\mathbf{i}_y$, and $\mathbf{i}_z$ are uniform vectors. Thus let us, for example, obtain the expression for the vector drawn from point $P_1(x_1, y_1, z_1)$ to point $P_2(x_2, y_2, z_2)$, as shown in Fig. 1.6(c). To do this, we note that the vector drawn from the origin to the point P_1, that is, $\mathbf{OP}_1$ is given by

$$\mathbf{OP}_1 = x_1 \mathbf{i}_x + y_1 \mathbf{i}_y + z_1 \mathbf{i}_z \qquad (1.23)$$

and that the vector drawn from the origin to the point P_2, that is, $\mathbf{OP}_2$ is given by

$$\mathbf{OP}_2 = x_2 \mathbf{i}_x + y_2 \mathbf{i}_y + z_2 \mathbf{i}_z \qquad (1.24)$$

Since, from the rule for vector addition, $\mathbf{OP}_1 + \mathbf{P}_1\mathbf{P}_2 = \mathbf{OP}_2$, we obtain

$$\mathbf{P}_1\mathbf{P}_2 = \mathbf{OP}_2 - \mathbf{OP}_1 = (x_2 - x_1)\mathbf{i}_x + (y_2 - y_1)\mathbf{i}_y + (z_2 - z_1)\mathbf{i}_z \qquad (1.25)$$

In our study of electromagnetic fields, we have to work with line integrals, surface integrals, and volume integrals. As in elementary calculus, these involve differential lengths, surfaces, and volumes obtained by incrementing the coordinates by infinitesimal amounts. Since in the Cartesian coordinate system the three coordinates represent lengths, the differential length elements obtained by incrementing one coordinate at a time keeping the other two coordinates constant are $dx\,\mathbf{i}_x$, $dy\,\mathbf{i}_y$, and $dz\,\mathbf{i}_z$ for the x, y, and z coordinates, respectively, as shown in Fig. 1.6(d), at an arbitrary point $P(x, y, z)$. The three differential length elements form the contiguous edges of a rectangular box in which the corner Q diagonally opposite to P has the coordinates $(x + dx, y + dy, z + dz)$. The differential length vector $d\mathbf{l}$ from P to Q is simply the vector sum of the three differential length elements. Thus

$$d\mathbf{l} = dx\,\mathbf{i}_x + dy\,\mathbf{i}_y + dz\,\mathbf{i}_z \qquad (1.26)$$

The box has six differential surfaces with each surface defined by two of the three length elements, as shown by the projections onto the coordinate planes in Fig. 1.6(d). The orientation of a differential surface dS is specified by a unit vector normal to it, that is, a unit vector perpendicular to any two vectors tangential to the surface. Unless specified, the normal vector can be drawn toward any one of the two sides of a given surface. Thus the differential surfaces formed by the pairs of differential length elements are

$$\pm dS\,\mathbf{i}_z = \pm dx\,dy\,\mathbf{i}_z = \pm dx\,\mathbf{i}_x \times dy\,\mathbf{i}_y \qquad (1.27\text{a})$$

$$\pm dS\,\mathbf{i}_x = \pm dy\,dz\,\mathbf{i}_x = \pm dy\,\mathbf{i}_y \times dz\,\mathbf{i}_z \qquad (1.27\text{b})$$

$$\pm dS\,\mathbf{i}_y = \pm dz\,dx\,\mathbf{i}_y = \pm dz\,\mathbf{i}_z \times dx\,\mathbf{i}_x \qquad (1.27\text{c})$$

Finally, the differential volume dv formed by the three differential lengths is simply the volume of the box, that is,

$$dv = dx \, dy \, dz \tag{1.28}$$

We shall now briefly review some elementary analytic geometrical details that will be useful in our study of electromagnetics. An arbitrary surface is defined by an equation of the form

$$f(x, y, z) = 0 \tag{1.29}$$

In particular, the equation for a plane surface making intercepts a, b, and c on the x, y, and z axes, respectively, is given by

$$\frac{x}{a} + \frac{y}{b} + \frac{z}{c} = 1 \tag{1.30}$$

Since a curve is the intersection of two surfaces, an arbitrary curve is defined by a pair of equations

$$f(x, y, z) = 0 \quad \text{and} \quad g(x, y, z) = 0 \tag{1.31}$$

Alternatively, a curve is specified by a set of three parametric equations

$$x = x(t), \qquad y = y(t), \qquad z = z(t) \tag{1.32}$$

where t is an independent parameter. For example, a straight line passing through the origin and making equal angles with the positive x, y, and z axes is given by the pair of equations $y = x$ and $z = x$, or by the set of three parametric equations $x = t$, $y = t$, and $z = t$.

Example 1.2.   Let us find a unit vector normal to the plane

$$5x + 2y + 4z = 20$$

By writing the given equation for the plane in the form

$$\frac{x}{4} + \frac{y}{10} + \frac{z}{5} = 1$$

we identify the intercepts made by the plane on the x, y, and z axes to be 4, 10, and 5, respectively. The portion of the plane lying in the first octant of the coordinate system is shown in Fig. 1.7.

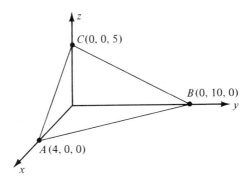

Figure 1.7. The plane surface $5x + 2y + 4z = 20$.

To find a unit vector normal to the plane, we consider two vectors lying in the plane and evaluate their cross product. Thus considering the vectors **AB** and **AC**, we have from (1.25),

$$\mathbf{AB} = (0 - 4)\mathbf{i}_x + (10 - 0)\mathbf{i}_y + (0 - 0)\mathbf{i}_z = -4\mathbf{i}_x + 10\mathbf{i}_y$$
$$\mathbf{AC} = (0 - 4)\mathbf{i}_x + (0 - 0)\mathbf{i}_y + (5 - 0)\mathbf{i}_z = -4\mathbf{i}_x + 5\mathbf{i}_z$$

The cross product of **AB** and **AC** is then given by

$$\mathbf{AB} \times \mathbf{AC} = \begin{vmatrix} \mathbf{i}_x & \mathbf{i}_y & \mathbf{i}_z \\ -4 & 10 & 0 \\ -4 & 0 & 5 \end{vmatrix} = 50\mathbf{i}_x + 20\mathbf{i}_y + 40\mathbf{i}_z$$

This vector is perpendicular to both **AB** and **AC** and hence to the plane. Finally, the required unit vector is obtained by dividing **AB** × **AC** by its magnitude. Thus it is equal to

$$\frac{50\mathbf{i}_x + 20\mathbf{i}_y + 40\mathbf{i}_z}{|50\mathbf{i}_x + 20\mathbf{i}_y + 40\mathbf{i}_z|} = \frac{5\mathbf{i}_x + 2\mathbf{i}_y + 4\mathbf{i}_z}{\sqrt{25 + 4 + 16}} = \frac{1}{3\sqrt{5}}(5\mathbf{i}_x + 2\mathbf{i}_y + 4\mathbf{i}_z) \quad \blacksquare$$

1.3 SCALAR AND VECTOR FIELDS

Before we take up the task of studying electromagnetic fields, we must understand what is meant by a "field." A field is associated with a region in space and we say that a field exists in the region if there is a physcial phenomenon associated with points in that region. For example, in everyday life we are familiar with the earth's gravitational field. We do not "see" the field in the same manner as we see light rays but we know of its existence in the sense that objects are acted upon by the gravitational force of the earth. In a broader context, we can talk of the field of any physical quantity as being a description, mathematical or graphical, of how the quantity varies from one point to another in the region of the field and with time. We can talk of scalar or vector fields depending on whether the quantity of interest is a scalar or a vector. We can talk of static or time-varying fields depending on whether the quantity of interest is independent of or changing with time.

We shall begin our discussion of fields with some simple examples of scalar fields. Thus let us consider the case of the conical pyramid shown in Fig. 1.8(a). A description of the height of the pyramidal surface versus position on its base is an example of a scalar field involving two variables. Choosing the origin to be the projection of the vertex of the cone onto the base and setting up an xy coordinate system to locate points on the base, we obtain the height field as a function of x and y to be

$$h(x, y) = 6 - 2\sqrt{x^2 + y^2} \tag{1.33}$$

Although we are able to depict the height variation of points on the conical surface graphically by using the third coordinate for h, we will have to be content with the visualization of the height field by a set of constant-height contours on the xy plane if only two coordinates were available, as in the case of a two-dimensional space. For the field under consideration, the constant-height contours are circles in the xy plane centered at the origin and equally spaced for equal increments of the height value as shown in Fig. 1.8(a).

For an example of a scalar field in three dimensions, let us consider a rectangular room and the distance field of points in the room from one corner of the room as shown in Fig. 1.8(b). For convenience, we choose this corner to be the origin O and set up a Cartesian coordinate system with the three contiguous edges meeting at that point as the coordinate axes. Each point in the room is defined by a set of values for the three coordinates x, y, and z. The distance r from the origin to that point is $\sqrt{x^2 + y^2 + z^2}$. Thus the distance field of points in the room from the origin is given by

$$r(x, y, z) = \sqrt{x^2 + y^2 + z^2} \tag{1.34}$$

Since the three coordinates are already used up for defining the points in the field region, we have to visualize the distance field by means of a set of constant-distance surfaces. A constant-distance surface is a surface for which points on it correspond to a particular constant value of r. For the case under consideration, the constant-distance surfaces are spherical surfaces centered at the origin and are equally spaced for equal increments in the value of the distance as shown in Fig. 1.8(b).

The fields we have discussed thus far are static fields. A simple example of a time-varying scalar field is provided by the temperature field associated with points in a room, especially when it is being heated or cooled. Just as in the case of the distance field of Fig. 1.8(b), we set up a three-dimensional coordinate system and to each set of three coordinates corresponding to the

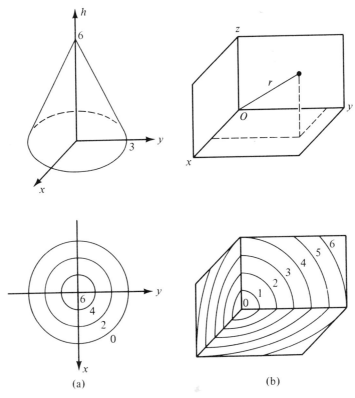

(a)   (b)

Figure 1.8. (a) A conical pyramid lying above the xy plane, and a set of constant-height contours for the conical surface. (b) A rectangular room, and a set of constant-distance surfaces depicting the distance field of points in the room from one corner of the room.

location of a point in the room, we assign a number to represent the temperature T at that point. Since the temperature at that point, however, varies with time t, this number is a function of time. Thus we describe mathematically the time-varying temperature field in the room by a function $T(x, y, z, t)$.

For any given instant of time, we can visualize a set of constant-temperature or isothermal surfaces corresponding to particular values of T as representing the temperature field for that value of time. For a different instant of time, we will have a different set of isothermal surfaces for the same values of T. Thus we can visualize the time-varying temperature field in the room by a set of isothermal surfaces continuously changing their shapes as though in a motion picture.

The foregoing discussion of scalar fields may now be extended to vector fields by recalling that a vector quantity has associated with it a direction in space in addition to magnitude. Hence in order to describe a vector field we attribute to each point in the field region a vector that represents the magnitude and direction of the physical quantity under consideration at that point. Since a vector at a given point can be expressed as the sum of its components along the set of unit vectors at that point, a mathematical description of the vector field involves simply the descriptions of the three component scalar fields. Thus for a vector field $\mathbf{F}$ in the Cartesian coordinate system, we have

$$\mathbf{F}(x, y, z, t) = F_x(x, y, z, t)\mathbf{i}_x + F_y(x, y, z, t)\mathbf{i}_y + F_z(x, y, z, t)\mathbf{i}_z \quad (1.35)$$

Similar expressions hold in the cylindrical and spherical coordinate systems. We should, however, note that two of the unit vectors in the cylindrical coordinate system and all the unit vectors in the spherical coordinate system are themselves functions of the coordinates.

To illustrate the graphical description of a vector field, let us consider the linear velocity vector field associated with points on a circular disk rotating about its center with a constant angular velocity ω rad/s. We know that the magnitude of the linear velocity of a point on the disk is then equal to the product of the angular velocity ω and the radial distance r of the point from the center of the disk. The direction of the linear velocity is tangential to the circle drawn through that point and concentric with the disk. Hence we may depict the linear velocity field by drawing at several points on the disk vectors that are tangential to the circles concentric with the disk and passing through those points, and whose lengths are proportional to the radii of the circles, as shown in Fig. 1.9(a), where the points are carefully selected in order to reveal the circular symmetry of the field with respect to the center of the disk. We, however, find that this method of representation of the vector field results in a congested sketch of vectors. Hence we may simplify the sketch by omitting the vectors and simply placing arrowheads along the circles, giving us a set of "direction lines," also known as "stream lines" and "flux lines," which simply represent the direction of the field at points on them. We note that for the field under consideration the direction lines are also contours of constant magnitude of the velocity, and hence by

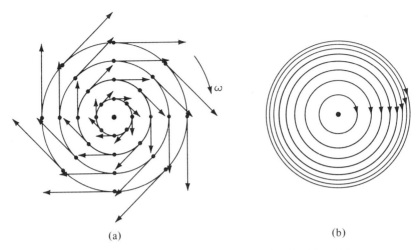

(a)   (b)

Figure 1.9. (a) Linear velocity vector field associated with points on a rotating disk. (b) Same as (a) except that the vectors are omitted, and the density of direction lines is used to indicate the magnitude variation.

increasing the density of the direction lines as r increases, we can indicate the magnitude variation, as shown in Fig. 1.9(b).

1.4   SINUSOIDALLY TIME-VARYING FIELDS

In our study of electromagnetic fields we will be particularly interested in fields that vary sinusoidally with time. Hence we shall devote this section to a discussion of sinusoidally time-varying fields. Let us first consider a scalar sinusoidal function of time. Such a function is given by an expression of the form $A \cos (\omega t + \phi)$ where A is the peak amplitude of the sinusoidal variation, $\omega = 2\pi f$ is the radian frequency, f is the linear frequency, and $(\omega t + \phi)$ is the phase. In particular, the phase of the function for $t = 0$ is ϕ. A plot of this function versus t shown in Fig. 1.10 illustrates how the function changes periodically between positive and negative values. If we now have a sinusoidally time-varying scalar field, we can visualize the field quantity varying sinusoidally with time at each point in the field region with the amplitude and phase governed by the spatial dependence of the field quantity. Thus, for example, the field $A e^{-\alpha z} \cos (\omega t - \beta z)$ where A, α, and β are positive constants is characterized by sinusoidal time variations with amplitude decreasing exponentially with z and the phase at any given time decreasing linearly with z.

For a sinusoidally time-varying vector field, the behavior of each component of the field may be visualized in the manner just discussed. If we now

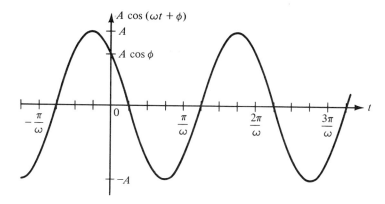

Figure 1.10. Sinusoidally time-varying scalar function $A \cos (\omega t + \phi)$.

fix our attention on a particular point in the field region, we can visualize the sinusoidal variation with time of a particular component at that point by a vector changing its magnitude and direction as shown, for example, for the x component in Fig. 1.11(a). Since the tip of the vector simply moves back and forth along a line, which in this case is parallel to the x axis, the component vector is said to be "linearly polarized" in the x direction. Similarly, the sinusoidal variation with time of the y component of the field can be visualized by a vector changing its magnitude and direction as shown in Fig. 1.11(b), not necessarily with the same amplitude and phase as those of the x component. Since the tip of the vector moves back and forth parallel

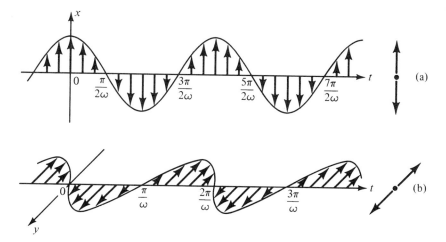

Figure 1.11. (a) Time variation of a linearly polarized vector in the x direction. (b) Time variation of a linearly polarized vector in the y direction.

to the y axis, the y component is said to be linearly polarized in the y direction. In the same manner, the z component is linearly polarized in the z direction.

If two components sinusoidally time-varying vectors have arbitrary amplitudes but are in phase as, for example,

$$\mathbf{F}_1 = F_1 \cos(\omega t + \phi)\,\mathbf{i}_x \qquad (1.36a)$$

$$\mathbf{F}_2 = F_2 \cos(\omega t + \phi)\,\mathbf{i}_y \qquad (1.36b)$$

then the sum vector $\mathbf{F} = \mathbf{F}_1 + \mathbf{F}_2$ is linearly polarized in a direction making an angle

$$\alpha = \tan^{-1}\frac{F_y}{F_x} = \tan^{-1}\frac{F_2}{F_1}$$

with the x direction as shown in the series of sketches in Fig. 1.12 illustrating the time history of the magnitude and direction of $\mathbf{F}$ over an interval of one period.

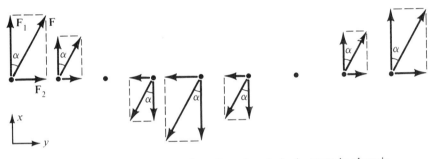

Figure 1.12. The sum vector of two linearly polarized vectors in phase is a linearly polarized vector.

If two component sinusoidally time-varying vectors have equal amplitudes, differ in direction by 90°, and differ in phase by $\pi/2$, as, for example,

$$\mathbf{F}_1 = F_0 \cos(\omega t + \phi)\,\mathbf{i}_x \qquad (1.37a)$$

$$\mathbf{F}_2 = F_0 \sin(\omega t + \phi)\,\mathbf{i}_y \qquad (1.37b)$$

then, to determine the "polarization" of the sum vector $\mathbf{F} = \mathbf{F}_1 + \mathbf{F}_2$, we note that the magnitude of $\mathbf{F}$ is given by

$$|\mathbf{F}| = |F_0 \cos(\omega t + \phi)\,\mathbf{i}_x + F_0 \sin(\omega t + \phi)\,\mathbf{i}_y| = F_0 \qquad (1.38)$$

and that the angle α which $\mathbf{F}$ makes with $\mathbf{i}_x$ is given by

$$\alpha = \tan^{-1}\frac{F_y}{F_x} = \tan^{-1}\left[\frac{F_0 \sin(\omega t + \phi)}{F_0 \cos(\omega t + \phi)}\right] = \omega t + \phi \qquad (1.39)$$

Thus the sum vector rotates with constant magnitude F_0 and at a rate of ω rad/s so that its tip describes a circle. The sum vector is then said to be "circularly polarized." The series of sketches in Fig. 1.13 illustrates the time history of the magnitude and direction of $\mathbf{F}$ over an interval of one period.

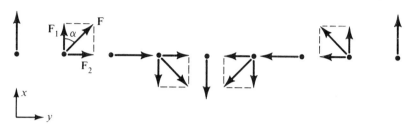

Figure 1.13. Circular polarization.

For the general case in which two component sinusoidally time-varying vectors differ in amplitude, direction, and phase by arbitrary amounts, the sum vector is "elliptically polarized," that is, its tip describes an ellipse.

Example 1.3.   Given two vectors $\mathbf{F}_1 = (3\mathbf{i}_x - 4\mathbf{i}_z)\cos\omega t$ and $\mathbf{F}_2 = 5\mathbf{i}_y \sin\omega t$, we wish to determine the polarization of the vector $\mathbf{F} = \mathbf{F}_1 + \mathbf{F}_2$.

We note that the vector $\mathbf{F}_1$, consisting of two component vectors in phase, is linearly polarized with amplitude $\sqrt{3^2 + (-4)^2}$ or 5 which is equal to that of $\mathbf{F}_2$. Since $\mathbf{F}_1$ varies as $\cos\omega t$ and $\mathbf{F}_2$ varies as $\sin\omega t$, they differ in phase by $\pi/2$. Also,

$$\mathbf{F}_1 \cdot \mathbf{F}_2 = (3\mathbf{i}_x - 4\mathbf{i}_z) \cdot 5\mathbf{i}_y = 0$$

so that $\mathbf{F}_1$ and $\mathbf{F}_2$ are perpendicular. Thus $\mathbf{F}_1$ and $\mathbf{F}_2$ are two linearly polarized vectors having equal amplitudes but differing in direction by 90° and differing in phase by $\pi/2$. Hence $\mathbf{F} = \mathbf{F}_1 + \mathbf{F}_2$ is circularly polarized. ∎

In the remainder of this section we shall briefly review the phasor technique which, as the student may have already learned in sinusoidal steady-state circuit analysis, is very useful in carrying out mathematical manipulations involving sinusoidally time-varying quantities. Let us consider the simple problem of adding the two quantities $10 \cos\omega t$ and $10 \sin(\omega t - 30°)$. To illustrate the basis behind the phasor technique, we carry out the following steps:

$$10 \cos \omega t + 10 \sin (\omega t - 30°) = 10 \cos \omega t + 10 \cos (\omega t - 120°)$$
$$= \text{Re}[10e^{j\omega t}] + \text{Re}[10e^{j(\omega t - 2\pi/3)}]$$
$$= \text{Re}[10e^{j0}e^{j\omega t}] + \text{Re}[10e^{-j2\pi/3}e^{j\omega t}]$$
$$= \text{Re}[(10e^{j0} + 10e^{-j2\pi/3})e^{j\omega t}]$$
$$= \text{Re}[10e^{-j\pi/3}e^{j\omega t}]$$
$$= \text{Re}[10e^{j(\omega t - \pi/3)}]$$
$$= 10 \cos (\omega t - 60°) \tag{1.40}$$

where Re stands for "real part of," and the addition of the two complex numbers $10e^{j0}$ and $10e^{-j2\pi/3}$ is performed by locating them in the complex plane and then using the parallelogram law of addition of complex numbers, as shown in Fig. 1.14. Alternatively, the complex numbers may be expressed in terms of their real and imaginary parts and then added up for conversion into exponential form in the manner

$$10e^{j0} + 10e^{-j2\pi/3} = (10 + j0) + (-5 - j8.66)$$
$$= 5 - j8.66 = \sqrt{5^2 + 8.66^2}\, e^{-j \tan^{-1} 8.66/5}$$
$$= 10e^{-j\pi/3} \tag{1.41}$$

In practice, we do not write all of the steps shown in (1.40). First, we express all functions in their cosine forms and then recognize the phasor corresponding to each cosine function as the complex number having the magnitude equal to the amplitude of the cosine function and phase angle equal to the phase angle of the cosine function for $t = 0$. For the above exam-

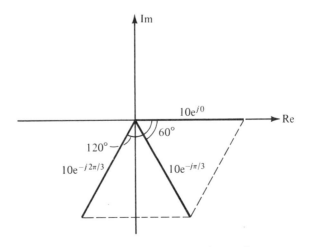

Figure 1.14. Addition of two complex numbers.

ple, the complex numbers $10e^{j0}$ and $10e^{-j2\pi/3}$ are the phasors corresponding to $10 \cos \omega t$ and $10 \sin (\omega t - 30°)$, respectively. Then we add the phasors and from the sum phasor write down the required cosine function. Thus the steps involved are as shown in Fig. 1.15.

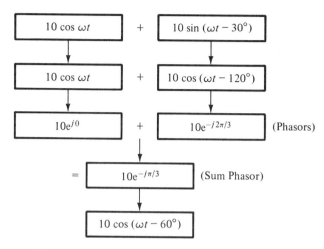

Figure 1.15. Block diagram of steps involved in the application of phasor technique to the addition of two sinusoidally time-varying functions.

The same technique is adopted for solving differential equations by recognizing, for example, that

$$\frac{d}{dt} [A \cos (\omega t + \theta)] = -A\omega \sin (\omega t + \theta) = A\omega \cos (\omega t + \theta + \pi/2)$$

and hence the phasor for $\frac{d}{dt} [A \cos (\omega t + \theta)]$ is

$$A\omega e^{j(\theta + \pi/2)} = A\omega e^{j\pi/2} e^{j\theta} = j\omega A e^{j\theta}$$

or $j\omega$ times the phasor for $A \cos (\omega t + \theta)$. Thus the differentiation operation is replaced by $j\omega$ for converting the differential equation into an algebraic equation involving phasors. To illustrate this, let us consider the differential equation

$$10^{-3} \frac{di}{dt} + i = 10 \cos 1000t \tag{1.42}$$

The solution for this is of the form $i = I_0 \cos (\omega t + \theta)$. Recognizing that $\omega = 1000$ and replacing d/dt by $j1000$ and all time functions by their phasors, we obtain the corresponding algebraic equation as

$$10^{-3}j1000\bar{I} + \bar{I} = 10e^{j0} \tag{1.43}$$

or

$$\bar{I}(1 + j1) = 10e^{j0} \qquad (1.44)$$

where the overbar above I indicates the complex nature of the quantity. Solving (1.44) for $\bar{I}$, we obtain

$$\bar{I} = \frac{10e^{j0}}{1 + j1} = \frac{10e^{j0}}{\sqrt{2}\, e^{j\pi/4}} = 7.07e^{-j\pi/4} \qquad (1.45)$$

and finally

$$i = 7.07 \cos\left(1000t - \frac{\pi}{4}\right) \qquad (1.46)$$

1.5   THE ELECTRIC FIELD

Basic to our study of the elements of engineering electromagnetics is an understanding of the concepts of electric and magnetic fields. Hence we shall devote this and the following section to an introduction of the electric and magnetic fields. From our study of Newton's law of gravitation in elementary physics, we are familiar with the gravitational force field associated with material bodies by virtue of their physical property known as "mass." Newton's experiments showed that the gravitational force of attraction between two bodies of masses m_1 and m_2 separated by a distance R, which is very large compared to their sizes, is equal to $m_1 m_2 G/R^2$ where G is the constant of universal gravitation. In a similar manner, a force field known as the "electric field" is associated with bodies that are "charged." A material body may be charged positively or negatively or may possess no net charge. In the International System of Units which we shall use throughout this book, the unit of charge is coulomb, abbreviated C. The charge of an electron is -1.60219×10^{-19} C. Alternatively, approximately 6.24×10^{18} electrons represent a charge of one negative coulomb.

Experiments conducted by Coulomb showed that the following hold for two charged bodies that are very small in size compared to their separation so that they can be considered as "point charges":

1. The magnitude of the force is proportional to the product of the magnitudes of the charges.
2. The magnitude of the force is inversely proportional to the square of the distance between the charges.
3. The magnitude of the force depends on the medium.
4. The direction of the force is along the line joining the charges.
5. Like charges repel; unlike charges attract.

For free space, the constant of proportionality is $1/4\pi\epsilon_0$ where ϵ_0 is known as the permittivity of free space, having a value 8.854×10^{-12} or approximately equal to $10^{-9}/36\pi$. Thus if we consider two point charges Q_1 C and Q_2 C separated R m in free space, as shown in Fig. 1.16, then the forces F_1 and F_2 experienced by Q_1 and Q_2, respectively, are given by

$$\mathbf{F}_1 = \frac{Q_1 Q_2}{4\pi\epsilon_0 R^2}\mathbf{i}_{21} \tag{1.47a}$$

and

$$\mathbf{F}_2 = \frac{Q_2 Q_1}{4\pi\epsilon_0 R^2}\mathbf{i}_{12} \tag{1.47b}$$

where $\mathbf{i}_{21}$ and $\mathbf{i}_{12}$ are unit vectors along the line joining Q_1 and Q_2 as shown in Fig. 1.16. Equations (1.47a) and (1.47b) represent Coulomb's law. Since the units of force are newtons, we note that ϵ_0 has the units (coulomb)^2 per (newton-meter^2). These are commonly known as "farads per meter" where a farad is (coulomb)^2 per newton-meter.

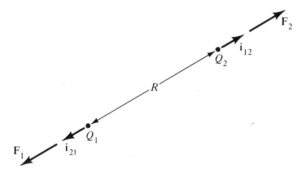

Figure 1.16. Forces experienced by two point charges Q_1 and Q_2.

In the case of the gravitational field of a material body, we define the gravitational field intensity as the force per unit mass experienced by a small test mass placed in that field. In a similar manner, the force per unit charge experienced by a small test charge placed in an electric field is known as the "electric field intensity," denoted by the symbol $\mathbf{E}$. Alternatively, if in a region of space, a test charge q experiences a force $\mathbf{F}$, then the region is said to be characterized by an electric field of intensity $\mathbf{E}$ given by

$$\mathbf{E} = \frac{\mathbf{F}}{q} \tag{1.48}$$

The unit of electric field intensity is newton per coulomb, or more commonly volt per meter, where a volt is newton-meter per coulomb. The test charge should be so small that it does not alter the electric field in which it is placed. Ideally, $\mathbf{E}$ is defined in the limit that q tends to zero, that is,

$$\mathbf{E} = \operatorname*{Lim}_{q \to 0} \frac{\mathbf{F}}{q} \qquad (1.49)$$

Equation (1.49) is the defining equation for the electric field intensity irrespective of the source of the electric field. Just as one body by virtue of its mass is the source of a gravitational field acting upon other bodies by virtue of their masses, a charged body is the source of an electric field acting upon other charged bodies. We will, however, learn in Chap. 2 that there exists another source for the electric field, namely, a time-varying magnetic field.

Returning now to Coulomb's law and letting one of the two charges in Fig. 1.16, say Q_2, be a small test charge q, we have

$$\mathbf{F}_2 = \frac{Q_1 q}{4\pi\epsilon_0 R^2} \mathbf{i}_{12} \qquad (1.50)$$

The electric field intensity $\mathbf{E}_2$ at the test charge due to the point charge Q_1 is then given by

$$\mathbf{E}_2 = \frac{\mathbf{F}_2}{q} = \frac{Q_1}{4\pi\epsilon_0 R^2} \mathbf{i}_{12} \qquad (1.51)$$

Generalizing this result by making R a variable, that is, by moving the test charge around in the medium, writing the expression for the force experienced by it, and dividing the force by the test charge, we obtain the electric field intensity $\mathbf{E}$ of a point charge Q to be

$$\mathbf{E} = \frac{Q}{4\pi\epsilon_0 R^2} \mathbf{i}_R \qquad (1.52)$$

where R is the distance from the point charge to the point at which the field intensity is to be computed and $\mathbf{i}_R$ is the unit vector along the line joining the two points under consideration and directed away from the point charge. The electric field intensity due to a point charge is thus directed everywhere radially away from the point charge and its constant-magnitude surfaces are spherical surfaces centered at the point charge, as shown in Fig. 1.17.

If we now have several point charges $Q_1, Q_2, \ldots$, as shown in Fig. 1.18, the force experienced by a test charge situated at a point P is the vector sum of the forces experienced by the test charge due to the individual charges. It then follows that the electric field intensity at point P is the superposition of the electric field intensities due to the individual charges, that is,

$$\mathbf{E} = \frac{Q_1}{4\pi\epsilon_0 R_1^2} \mathbf{i}_{R_1} + \frac{Q_2}{4\pi\epsilon_0 R_2^2} \mathbf{i}_{R_2} + \cdots + \frac{Q_n}{4\pi\epsilon_0 R_n^2} \mathbf{i}_{R_n} \qquad (1.53)$$

Let us now consider an example.

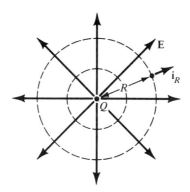

Figure 1.17. Direction lines and constant-magnitude surfaces of electric field due to a point charge.

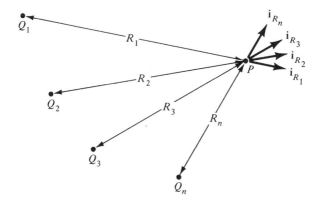

Figure 1.18. A collection of point charges and unit vectors along the directions of their electric fields at a point P.

Example 1.4.   Figure 1.19 shows eight point charges situated at the corners of a cube. We wish to find the electric field intensity at each point charge, due to the remaining seven point charges.

First we note from (1.52) that the electric field intensity at a point $B(x_2, y_2, z_2)$ due to a point charge Q at point $A(x_1, y_1, z_1)$ is given by

$$\mathbf{E}_B = \frac{Q}{4\pi\epsilon_0(AB)^2}\mathbf{i}_{AB} = \frac{Q}{4\pi\epsilon_0(AB)^2}\frac{\mathbf{AB}}{(AB)} = \frac{Q(\mathbf{AB})}{4\pi\epsilon_0(AB)^3}$$

$$= \frac{Q}{4\pi\epsilon_0}\frac{(x_2 - x_1)\mathbf{i}_x + (y_2 - y_1)\mathbf{i}_y + (z_2 - z_1)\mathbf{i}_z}{[(x_2 - x_1)^2 + (y_2 - y_1)^2 + (z_2 - z_1)^2]^{3/2}} \qquad (1.54)$$

Let us now consider the point $(1, 1, 1)$. Applying (1.54) to each of the charges at the seven other points and using (1.53), we obtain the electric field intensity

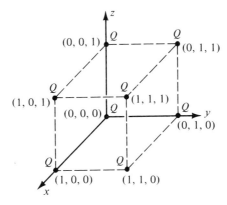

Figure 1.19. A cubical arrangement of point charges.

at the point $(1, 1, 1)$ to be

$$
\mathbf{E}_{(1,1,1)} = \frac{Q}{4\pi\epsilon_0}\Bigg[\frac{\mathbf{i}_x}{(1)^{3/2}} + \frac{\mathbf{i}_y}{(1)^{3/2}} + \frac{\mathbf{i}_z}{(1)^{3/2}} + \frac{\mathbf{i}_y + \mathbf{i}_z}{(2)^{3/2}} + \frac{\mathbf{i}_z + \mathbf{i}_x}{(2)^{3/2}}
$$

$$
+ \frac{\mathbf{i}_x + \mathbf{i}_y}{(2)^{3/2}} + \frac{\mathbf{i}_x + \mathbf{i}_y + \mathbf{i}_z}{(3)^{3/2}}\Bigg]
$$

$$
= \frac{Q}{4\pi\epsilon_0}\left(1 + \frac{1}{\sqrt{2}} + \frac{1}{3\sqrt{3}}\right)(\mathbf{i}_x + \mathbf{i}_y + \mathbf{i}_z)
$$

$$
= \frac{3.29Q}{4\pi\epsilon_0}\left(\frac{\mathbf{i}_x + \mathbf{i}_y + \mathbf{i}_z}{\sqrt{3}}\right)
$$

Noting that $(\mathbf{i}_x + \mathbf{i}_y + \mathbf{i}_z)/\sqrt{3}$ is the unit vector directed from $(0, 0, 0)$ to $(1, 1, 1)$, we find the electric field intensity at $(1, 1, 1)$ to be directed diagonally away from $(0, 0, 0)$, with a magnitude equal to $\dfrac{3.29Q}{4\pi\epsilon_0}$ N/C. From symmetry considerations, if then follows that the electric field intensity at each point charge, due to the remaining seven point charges, has a magnitude $\dfrac{3.29Q}{4\pi\epsilon_0}$ N/C, and it is directed away from the corner opposite to that charge. ■

The foregoing illustration of the computation of the electric field intensity due to a multitude of point charges may be extended to the computation of the field intensity for a continuous charge distribution by dividing the region in which the charge exists into elemental lengths, surfaces, or volumes depending on whether the charge is distributed along a line, over a surface, or in a volume, and treating the charge in each elemental length, surface, or volume as a point charge and then applying superposition. We shall include some of the simpler cases in the problems for the interested reader.

Let us now consider the motion of a cloud of electrons, distributed uniformly with density N, under the influence of a time-varying electric field of intensity

$$\mathbf{E} = E_0 \cos \omega t \; \mathbf{i}_x \tag{1.55}$$

Each electron experiences a force given by

$$\mathbf{F} = e\mathbf{E} = eE_0 \cos \omega t \; \mathbf{i}_x \tag{1.56}$$

where e is the charge of the electron. The equation of motion of the electron is then given by

$$m\frac{d\mathbf{v}}{dt} = eE_0 \cos \omega t \; \mathbf{i}_x \tag{1.57}$$

where m is the mass of the electron and $\mathbf{v}$ is its velocity. Solving (1.57) for $\mathbf{v}$, we obtain

$$\mathbf{v} = \frac{eE_0}{m\omega} \sin \omega t \; \mathbf{i}_x + \mathbf{C} \tag{1.58}$$

where $\mathbf{C}$ is the constant of integration. Assuming an initial condition of $\mathbf{v} = 0$ for $t = 0$ gives us $\mathbf{C} = 0$, reducing (1.58) to

$$\mathbf{v} = \frac{eE_0}{m\omega} \sin \omega t \; \mathbf{i}_x = -\frac{|e| E_0}{m\omega} \sin \omega t \; \mathbf{i}_x \tag{1.59}$$

The motion of the electron cloud gives rise to current flow. To find the current crossing an infinitesimal surface of area ΔS oriented such that the normal vector to the surface makes an angle α with the x direction as shown in Fig. 1.20, let us for instance consider an infinitesimal time interval Δt

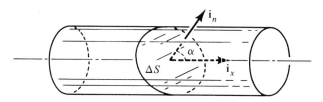

Figure 1.20. For finding the current crossing an infinitesimal area in a moving cloud of electrons.

when v_x is negative. The number of electrons crossing the area ΔS from its right side to its left side in this time interval is the same as that which exists in a column of length $|v_x| \, \Delta t$ and cross-sectional area $\Delta S \cos \alpha$ to the right of the area under consideration. Thus the negative charge ΔQ crossing the area ΔS in time Δt to its left side is given by

$$\Delta Q = (\Delta S \cos \alpha)(|v_x|\Delta t)\, Ne$$
$$= Ne|v_x|\Delta S \cos \alpha\, \Delta t \qquad (1.60)$$

The current ΔI flowing through the area ΔS from its left side to its right side is then given by

$$\Delta I = \frac{|\Delta Q|}{\Delta t} = N|e|\,|v_x|\Delta S \cos \alpha$$
$$= \frac{N|e|^2}{m\omega} E_0 \sin \omega t\, \Delta S \cos \alpha$$
$$= \frac{Ne^2}{m\omega} E_0 \sin \omega t\, \mathbf{i}_x \cdot \Delta S\, \mathbf{i}_n \qquad (1.61)$$

where $\mathbf{i}_n$ is the unit vector normal to the area ΔS as shown in Fig. 1.20.

We can now talk of a current density vector $\mathbf{J}$, associated with the current flow. The current density vector has a magnitude equal to the current per unit area and a direction normal to the area when the area is oriented in order to maximize the current crossing it. The current crossing ΔS is maximized when $\alpha = 0$, that is, when the area is oriented such that $\mathbf{i}_n = \mathbf{i}_x$. The current per unit area is then equal to $\dfrac{Ne^2}{m\omega} E_0 \sin \omega t$. Thus the current density vector is given by

$$\mathbf{J} = \frac{Ne^2}{m\omega} E_0 \sin \omega t\, \mathbf{i}_x$$
$$= Ne\mathbf{v} \qquad (1.62)$$

Finally, by substituting (1.62) back into (1.61), we note that the current crossing any area $\Delta \mathbf{S} = \Delta S\, \mathbf{i}_n$ is simply equal to $\mathbf{J} \cdot \Delta \mathbf{S}$.

1.6   THE MAGNETIC FIELD

In the preceding section we presented an experimental law known as Coulomb's law having to do with the electric force associated with two charged bodies, and we introduced the electric field intensity vector as the force per unit charge experienced by a test charge placed in the electric field. In this section we present another experimental law known as "Ampere's law of force," analogous to Coulomb's law, and use it to introduce the magnetic field concept.

Ampere's law of force is concerned with "magnetic" forces associated with two loops of wire carrying currents by virtue of motion of charges in the loops. Figure 1.21 shows two loops of wire carrying currents I_1 and I_2

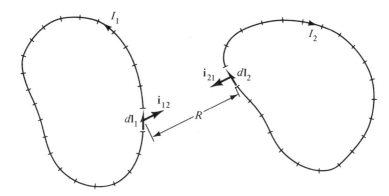

Figure 1.21. Two loops of wire carrying currents I_1 and I_2.

and each of which is divided into a large number of elements having infinitesimal lengths. The total force experienced by a loop is the vector sum of forces experienced by the infinitesimal current elements comprising the loop. The force experienced by each of these current elements is the vector sum of the forces exerted on it by the infinitesimal current elements comprising the second loop. If the number of elements in loop 1 is m and the number of elements in loop 2 is n, then there are $m \times n$ pairs of elements. A pair of magnetic forces is associated with each pair of these elements just as a pair of electric forces is associated with a pair of point charges. Thus if we consider an element dl_1 in loop 1 and an element dl_2 in loop 2, then the forces $d\mathbf{F}_1$ and $d\mathbf{F}_2$ experienced by the elements dl_1 and dl_2, respectively, are given by

$$d\mathbf{F}_1 = I_1 \, dl_1 \times \left(\frac{k I_2 \, dl_2 \times \mathbf{i}_{21}}{R^2} \right) \tag{1.63a}$$

$$d\mathbf{F}_2 = I_2 \, dl_2 \times \left(\frac{k I_1 \, dl_1 \times \mathbf{i}_{12}}{R^2} \right) \tag{1.63b}$$

where $\mathbf{i}_{21}$ and $\mathbf{i}_{12}$ are unit vectors along the line joining the two current elements, R is the distance between them, and k is a constant of proportionality that depends on the medium. For free space, k is equal to $\mu_0/4\pi$ where μ_0 is known as the permeability of free space, having a value $4\pi \times 10^{-7}$. From (1.63a) or (1.63b), we note that the units of μ_0 are newtons per ampere squared. These are commonly known as "henrys per meter" where a henry is a newton-meter per ampere squared.

Equations (1.63a) and (1.63b) represent Ampere's force law as applied to a pair of current elements. Some of the features evident from these equations are as follows:

1. The magnitude of the force is proportional to the product of the two currents and to the product of the lengths of the two current elements.

2. The magnitude of the force is inversely proportional to the square of the distance between the current elements.

3. To determine the direction of the force acting on the current element $d\mathbf{l}_1$, we first find the cross product $d\mathbf{l}_2 \times \mathbf{i}_{21}$ and then cross $d\mathbf{l}_1$ into the resulting vector. Similarly, to determine the direction of the force acting on the current element $d\mathbf{l}_2$, we first find the cross product $d\mathbf{l}_1 \times \mathbf{i}_{12}$ and then cross $d\mathbf{l}_2$ into the resulting vector. For the general case of arbitrary orientations of $d\mathbf{l}_1$ and $d\mathbf{l}_2$, these operations yield $d\mathbf{F}_{12}$ and $d\mathbf{F}_{21}$ which are not equal and opposite. This is not a violation of Newton's third law since isolated current elements do not exist without sources and sinks of charges at their ends. Newton's third law, however, must and does hold for complete current loops.

The forms of (1.63a) and (1.63b) suggest that each current element is acted upon by a field which is due to the other current element. By definition, this field is the magnetic field and is characterized by a quantity known as the "magnetic flux density vector," denoted by the symbol $\mathbf{B}$. Thus we note from (1.63b) that the magnetic flux density at the element $d\mathbf{l}_2$ due to the element $d\mathbf{l}_1$ is given by

$$\mathbf{B}_1 = \frac{\mu_0}{4\pi} \frac{I_1 \, d\mathbf{l}_1 \times \mathbf{i}_{12}}{R^2} \qquad (1.64)$$

and that this flux density acting upon $d\mathbf{l}_2$ results in a force on it given by

$$d\mathbf{F}_2 = I_2 \, d\mathbf{l}_2 \times \mathbf{B}_1 \qquad (1.65)$$

Similarly, we note from (1.63a) that the magnetic flux density at the element $d\mathbf{l}_1$ due to the element $d\mathbf{l}_2$ is given by

$$\mathbf{B}_2 = \frac{\mu_0}{4\pi} \frac{I_2 \, d\mathbf{l}_2 \times \mathbf{i}_{21}}{R^2} \qquad (1.66)$$

and that this flux density acting upon $d\mathbf{l}_1$ results in a force on it given by

$$d\mathbf{F}_1 = I_1 \, d\mathbf{l}_1 \times \mathbf{B}_2 \qquad (1.67)$$

From (1.65) and (1.67), we see that the units of $\mathbf{B}$ are newtons per ampere-meter, commonly known as "webers/meter^2" where a weber is a newton-meter per ampere. The units of webers per unit area give the character of flux density to the quantity $\mathbf{B}$.

Generalizing (1.64) and (1.66), we obtain the magnetic flux density due to an infinitesimal current element of length $d\mathbf{l}$ and carrying current I to be

$$\mathbf{B} = \frac{\mu_0}{4\pi} \frac{I \, d\mathbf{l} \times \mathbf{i}_R}{R^2} \qquad (1.68)$$

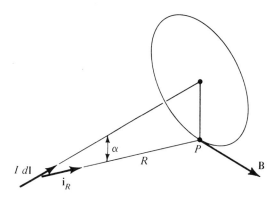

Figure 1.22. Magnetic flux density due to an infinitesimal current element.

where R is the distance from the current element to the point at which the flux density is to be computed and i_R is the unit vector along the line joining the current element and the point under consideration and directed away from the current element as shown in Fig. 1.22. Equation (1.68) is known as the "Biot-Savart law" and is analogous to the expression for the electric field intensity due to a point charge. The Biot-Savart law tells us that the magnitude of **B** at a point P is proportional to the current I, the element length dl, and the sine of the angle α between the current element and the line joining it to the point P and is inversely proportional to the square of the distance from the current element to the point P. Hence the magnetic flux density is zero at points along the axis of the current element. The direction of **B** at point P is normal to the plane containing the current element and the line joining the current element to P as given by the cross product operation $dl \times i_R$, that is, right circular to the axis of the wire. As a numerical example, for a current element $0.01i_z$ m situated at the origin and carrying current 2 amperes, the magnetic flux density at the point $(0, 1, 1)$ has a magnitude $10^{-9}/\sqrt{2}$ Wb/m² and is directed in the $-i_x$ direction. The magnetic field due to a given current distribution can be found by dividing the current distribution into a number of infinitesimal current elements, applying the Biot-Savart law to find the magnetic field due to each current element, and then using superposition. We shall include some simple cases in the problems for the interested reader.

Turning our attention now to (1.65) and (1.67) and generalizing, we say that an infinitesimal current element of length dl and current I placed in a magnetic field of flux density **B** experiences a force $d\mathbf{F}$ given by

$$d\mathbf{F} = I\,dl \times \mathbf{B} \tag{1.69}$$

Alternatively, if a current element experiences a force in a region of space, then the region is said to be characterized by a magnetic field. Since current is due to flow of charges, (1.69) can be formulated in terms of the moving charge causing the flow of current. Thus if the time taken by the charge dq contained in the length $d\mathbf{l}$ of the current element to flow with a velocity $\mathbf{v}$ across the cross-sectional area of the wire is dt, then $I = dq/dt$, and $d\mathbf{l} = \mathbf{v}\,dt$ so that

$$d\mathbf{F} = \frac{dq}{dt}\mathbf{v}\,dt \times \mathbf{B} = dq\,\mathbf{v} \times \mathbf{B} \qquad (1.70)$$

It then follows that the force $\mathbf{F}$ experienced by a test charge q moving with a velocity $\mathbf{v}$ in a magnetic field of flux density $\mathbf{B}$ is given by

$$\mathbf{F} = q\mathbf{v} \times \mathbf{B} \qquad (1.71)$$

We may now obtain a defining equation for $\mathbf{B}$ in terms of the moving test charge. To do this, we note from (1.71) that the magnetic force is directed normally to both $\mathbf{v}$ and $\mathbf{B}$ as shown in Fig. 1.23, and that its magnitude is

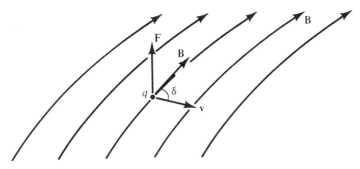

Figure 1.23. Force experienced by a test charge q moving with a velocity $\mathbf{v}$ in a magnetic field $\mathbf{B}$.

equal to $qvB \sin \delta$ where δ is the angle between $\mathbf{v}$ and $\mathbf{B}$. A knowledge of the force $\mathbf{F}$ acting on a test charge moving with an arbitrary velocity $\mathbf{v}$ provides only the value of $B \sin \delta$. To find $\mathbf{B}$, we must determine the maximum force qvB that occurs for δ equal to 90° by trying out several directions of $\mathbf{v}$, keeping its magnitude constant. Thus if this maximum force is $\mathbf{F}_m$ and it occurs for a velocity $v\mathbf{i}_m$, then

$$\mathbf{B} = \frac{\mathbf{F}_m \times \mathbf{i}_m}{qv} \qquad (1.72)$$

As in the case of defining the electric field intensity, we assume that the test charge does not alter the magnetic field in which it is placed. Ideally, $\mathbf{B}$ is defined in the limit that qv tends to zero, that is,

$$\mathbf{B} = \lim_{qv \to 0} \frac{\mathbf{F}_m \times \mathbf{i}_m}{qv} \qquad (1.73)$$

Equation (1.73) is the defining equation for the magnetic flux density irrespective of the source of the magnetic field. We have learned in this section that an electric current or a charge in motion is a source of the magnetic field. We will learn in Chap. 2 that there exists another source for the magnetic field, namely, a time-varying electric field.

We can now combine (1.48) and (1.71) to write the expression for the total force acting on a test charge q moving with a velocity $\mathbf{v}$ in a region characterized by an electric field of intensity $\mathbf{E}$ and a magnetic field of flux density $\mathbf{B}$ as

$$\mathbf{F} = q\mathbf{E} + q\mathbf{v} \times \mathbf{B} = q(\mathbf{E} + \mathbf{v} \times \mathbf{B}) \tag{1.74}$$

Equation (1.74) is known as the "Lorentz force equation." We shall now consider an example.

Example 1.5. The forces experienced by a test charge q for three different velocities at a point in a region characterized by electric and magnetic fields are given by

$$\mathbf{F}_1 = q[E_0\mathbf{i}_x + (E_0 - v_0B_0)\mathbf{i}_y] \quad \text{for } \mathbf{v}_1 = v_0\mathbf{i}_x$$

$$\mathbf{F}_2 = q[(E_0 + v_0B_0)\mathbf{i}_x + E_0\mathbf{i}_y] \quad \text{for } \mathbf{v}_2 = v_0\mathbf{i}_y$$

$$\mathbf{F}_3 = q[E_0\mathbf{i}_x + E_0\mathbf{i}_y] \quad \text{for } \mathbf{v}_3 = v_0\mathbf{i}_z$$

where v_0, E_0, and B_0 are constants. Find $\mathbf{E}$ and $\mathbf{B}$ at the point.

From Lorentz force equation, we have

$$q\mathbf{E} + qv_0\mathbf{i}_x \times \mathbf{B} = q[E_0\mathbf{i}_x + (E_0 - v_0B_0)\mathbf{i}_y] \tag{1.75a}$$

$$q\mathbf{E} + qv_0\mathbf{i}_y \times \mathbf{B} = q[(E_0 + v_0B_0)\mathbf{i}_x + E_0\mathbf{i}_y] \tag{1.75b}$$

$$q\mathbf{E} + qv_0\mathbf{i}_z \times \mathbf{B} = q[E_0\mathbf{i}_x + E_0\mathbf{i}_y] \tag{1.75c}$$

Eliminating $\mathbf{E}$ by subtracting (1.75a) from (1.75b) and (1.75c) from (1.75b), we obtain

$$(\mathbf{i}_y - \mathbf{i}_x) \times \mathbf{B} = B_0(\mathbf{i}_x + \mathbf{i}_y) \tag{1.76a}$$

$$(\mathbf{i}_y - \mathbf{i}_z) \times \mathbf{B} = B_0\mathbf{i}_x \tag{1.76b}$$

It follows from these two equations that $\mathbf{B}$ is perpendicular to both $(\mathbf{i}_x + \mathbf{i}_y)$ and $\mathbf{i}_x$. Hence it is equal to $C(\mathbf{i}_x + \mathbf{i}_y) \times \mathbf{i}_x$ or $-C\mathbf{i}_z$ where C is to be determined. To do this, we substitute $\mathbf{B} = -C\mathbf{i}_z$ in (1.76a) to obtain

$$(\mathbf{i}_y - \mathbf{i}_x) \times (-C\mathbf{i}_z) = B_0(\mathbf{i}_x + \mathbf{i}_y)$$

$$-C(\mathbf{i}_x + \mathbf{i}_y) = B_0(\mathbf{i}_x + \mathbf{i}_y)$$

or $C = -B_0$. Thus we get

$$\mathbf{B} = B_0\mathbf{i}_z$$

Substituting this result in (1.75c), we obtain

$$\mathbf{E} = E_0(\mathbf{i}_x + \mathbf{i}_y) \qquad \blacksquare$$

1.7 SUMMARY

We first learned in this chapter several rules of vector algebra that are necessary for our study of the elements of engineering electromagnetics by considering vectors expressed in terms of their components along three mutually orthogonal directions. To carry out the manipulations involving vectors at different points in space in a systematic manner, we then introduced the Cartesian coordinate system and discussed the application of the vector algebraic rules to vectors in the Cartesian coordinate system. To summarize these rules, we consider three vectors

$$\mathbf{A} = A_x\mathbf{i}_x + A_y\mathbf{i}_y + A_z\mathbf{i}_z$$
$$\mathbf{B} = B_x\mathbf{i}_x + B_y\mathbf{i}_y + B_z\mathbf{i}_z$$
$$\mathbf{C} = C_x\mathbf{i}_x + C_y\mathbf{i}_y + C_z\mathbf{i}_z$$

in a right-handed Cartesian coordinate system, that is, with $\mathbf{i}_x \times \mathbf{i}_y = \mathbf{i}_z$. We then have

$$\mathbf{A} + \mathbf{B} = (A_x + B_x)\mathbf{i}_x + (A_y + B_y)\mathbf{i}_y + (A_z + B_z)\mathbf{i}_z$$
$$\mathbf{B} - \mathbf{C} = (B_x - C_x)\mathbf{i}_x + (B_y - C_y)\mathbf{i}_y + (B_z - C_z)\mathbf{i}_z$$
$$m\mathbf{A} = mA_x\mathbf{i}_x + mA_y\mathbf{i}_y + mA_z\mathbf{i}_z$$
$$\frac{\mathbf{B}}{n} = \frac{B_x}{n}\mathbf{i}_x + \frac{B_y}{n}\mathbf{i}_y + \frac{B_z}{n}\mathbf{i}_z$$
$$|\mathbf{A}| = \sqrt{A_x^2 + A_y^2 + A_z^2}$$
$$\mathbf{i}_A = \frac{A_x}{\sqrt{A_x^2 + A_y^2 + A_z^2}}\mathbf{i}_x + \frac{A_y}{\sqrt{A_x^2 + A_y^2 + A_z^2}}\mathbf{i}_y + \frac{A_z}{\sqrt{A_x^2 + A_y^2 + A_z^2}}\mathbf{i}_z$$
$$\mathbf{A} \cdot \mathbf{B} = A_xB_x + A_yB_y + A_zB_z$$
$$\mathbf{A} \times \mathbf{B} = \begin{vmatrix} \mathbf{i}_x & \mathbf{i}_y & \mathbf{i}_z \\ A_x & A_y & A_z \\ B_x & B_y & B_z \end{vmatrix}$$
$$\mathbf{A} \cdot \mathbf{B} \times \mathbf{C} = \begin{vmatrix} A_x & A_y & A_z \\ B_x & B_y & B_z \\ C_x & C_y & C_z \end{vmatrix}$$

Other useful expressions are

$$dl = dx\,\mathbf{i}_x + dy\,\mathbf{i}_y + dz\,\mathbf{i}_z$$
$$d\mathbf{S} = \pm dx\,dy\,\mathbf{i}_z, \qquad \pm dy\,dz\,\mathbf{i}_x, \qquad \pm dz\,dx\,\mathbf{i}_y$$
$$dv = dx\,dy\,dz$$

As a prelude to the introduction of electric and magnetic fields, we discussed the concepts of scalar and vector fields, static and time-varying, by means of some simple examples such as the height of points on a conical surface above its base, the temperature field of points in a room, and the velocity vector field associated with points on a disk rotating about its center. We learned about the visualization of fields by means of constant-magnitude contours or surfaces and in addition by means of direction lines in the case of vector fields. Particular attention was devoted to sinusoidally time-varying fields. Polarization of vector fields as a means of describing how the orientation of a vector at a point changes with time was discussed. The phasor technique as a means of facilitating mathematical operations involving sinusoidally time-varying quantities was reviewed.

Having obtained the necessary background vector algebraic tools and physical field concepts, we then introduced the electric and magnetic fields from considerations of experimental laws known as Coulomb's law and Ampere's force law, having to do with the electric forces between two point charges and the magnetic forces between two current elements, respectively. From these laws, we deduced the expressions for the electric field intensity $\mathbf{E}$ due to a point charge Q and the magnetic flux density $\mathbf{B}$ due to a current element $I\,d\mathbf{l}$. These expressions are

$$\mathbf{E} = \frac{Q}{4\pi\epsilon_0 R^2}\mathbf{i}_R$$

$$\mathbf{B} = \frac{\mu_0 I\,d\mathbf{l} \times \mathbf{i}_R}{4\pi R^2}$$

where ϵ_0 and μ_0 are the permittivity and the permeability, respectively, of free space, R is the distance from the source to the point, say P, at which the field is to be computed, and $\mathbf{i}_R$ is the unit vector directed from the source toward the point P. We learned that the electric field is a force field acting on charges merely by virtue of the property of charge. The electric force is given simply by

$$\mathbf{F} = q\mathbf{E}$$

On the other hand, the magnetic field exerts forces only on moving charges, or current elements, as given by

$$\mathbf{F} = dq\,\mathbf{v} \times \mathbf{B} = I\,d\mathbf{l} \times \mathbf{B}$$

Combining the electric and magnetic field concepts, we finally introduced the Lorentz force equation for the force exerted on a charge q moving with a velocity $\mathbf{v}$ in a region of electric and magnetic fields $\mathbf{E}$ and $\mathbf{B}$, respectively, as

$$\mathbf{F} = q(\mathbf{E} + \mathbf{v} \times \mathbf{B})$$

REVIEW QUESTIONS

1.1. Give some examples of scalars.

1.2. Give some examples of vectors.

1.3. State all conditions for which $\mathbf{A} \cdot \mathbf{B} = 0$.

1.4. State all conditions for which $\mathbf{A} \times \mathbf{B} = 0$.

1.5. What is the significance of $\mathbf{A} \cdot \mathbf{B} \times \mathbf{C} = 0$?

1.6. Is it necessary for the reference vectors $\mathbf{i}_1$, $\mathbf{i}_2$, and $\mathbf{i}_3$ to be an orthogonal set?

1.7. State whether $\mathbf{i}_1$, $\mathbf{i}_2$, and $\mathbf{i}_3$ directed westward, northward, and downward, respectively, is a right-handed or a left-handed set.

1.8. What is the particular advantageous characteristic associated with the unit vectors in the Cartesian coordinate system?

1.9. How do you find a vector perpendicular to a plane?

1.10. How do you find the perpendicular distance from a point to a plane?

1.11. What is the total distance around the circumference of a circle of radius 1 m? What is the total vector distance around the circle?

1.12. What is the total surface area of a cube of sides 1 m? Assuming the normals to the surfaces to be directed outward of the cubical volume, what is the total vector surface area of the cube?

1.13. Describe briefly your concept of a scalar field and illustrate with an example.

1.14. Describe briefly your concept of a vector field and illustrate with an example.

1.15. How do you depict pictorially the gravitational field of the earth?

1.16. A sinusoidally time-varying vector is expressed in terms of its components along the x, y, and z axes. What is the polarization of each of the components?

1.17. What are the conditions for the sum of two linearly polarized sinusoidally time-varying vectors to be circularly polarized?

1.18. What is the polarization for the general case of the sum of two sinusoidally time-varying linearly polarized vectors having arbitrary amplitudes, phase angles, and directions?

1.19. Considering the second hand on your watch to be a vector, state its polarization. What is the frequency?

1.20. What is a phasor?

1.21. Is there any relationship between a phasor and a vector? Explain.

1.22. Describe the phasor technique of adding two sinusoidal functions of time.

1.23. Describe the phasor technique of solving a differential equation for the sinusoidal steady-state solution.

1.24. State Coulomb's law. To what law in mechanics is Coulomb's law analogous?

1.25. What is the definition of the electric field intensity?

1.26. What are the units of the electric field intensity?

1.27. What is the permittivity of free space? What are its units?

1.28. Describe the electric field due to a point charge.

1.29. How do you find the electric field intensity due to a continuous charge distribution?

1.30. How is current density defined? What are its units?

1.31. For a current flowing on a sheet, how would you define the current density at a point on the sheet? What are the units?

1.32. State Ampere's force law as applied to current elements.

1.33. Why is it not necessary for Newton's third law to hold for current elements?

1.34. What is the permeability of free space? What are its units?

1.35. Describe the magnetic field due to a current element.

1.36. How is the magnetic flux density defined in terms of force on a current element?

1.37. How is the magnetic flux density defined in terms of force on a moving charge?

1.38. What are the units of the magnetic flux density?

1.39. State Lorentz force equation.

1.40. If it is assumed that there is no electric field, the magnetic field at a point can be found from the knowledge of forces exerted on a moving test charge for two different velocities. Explain.

PROBLEMS

1.1. A bug starts at a point and travels 1 m northward, $\frac{1}{2}$ m eastward, $\frac{1}{4}$ m southward, $\frac{1}{8}$ m westward, $\frac{1}{16}$ m northward, and so on, making a 90°-turn to the right and halving the distance each time. (a) What is the total distance traveled by the bug? (b) Find the final position of the bug relative to its starting location. (c) Find the straight-line distance from the starting location to the final position.

1.2. Solve the following equations for $\mathbf{A}$, $\mathbf{B}$, and $\mathbf{C}$:

$$\mathbf{A} + \mathbf{B} + \mathbf{C} = 2\mathbf{i}_1 + 3\mathbf{i}_2 + 2\mathbf{i}_3$$
$$2\mathbf{A} + \mathbf{B} - \mathbf{C} = \mathbf{i}_1 + 3\mathbf{i}_2$$
$$\mathbf{A} - 2\mathbf{B} + 3\mathbf{C} = 4\mathbf{i}_1 + 5\mathbf{i}_2 + \mathbf{i}_3$$

1.3. Show that $(\mathbf{A} + \mathbf{B}) \cdot (\mathbf{A} - \mathbf{B}) = A^2 - B^2$ and that $(\mathbf{A} + \mathbf{B}) \times (\mathbf{A} - \mathbf{B}) = 2\mathbf{B} \times \mathbf{A}$. Verify the above for $\mathbf{A} = 3\mathbf{i}_1 - 5\mathbf{i}_2 + 4\mathbf{i}_3$ and $\mathbf{B} = \mathbf{i}_1 + \mathbf{i}_2 - 2\mathbf{i}_3$.

1.4. Given $\mathbf{A} = -2\mathbf{i}_1 + \mathbf{i}_2$, $\mathbf{B} = \mathbf{i}_1 - 2\mathbf{i}_2 + \mathbf{i}_3$, and $\mathbf{C} = 3\mathbf{i}_1 + 2\mathbf{i}_2 + \mathbf{i}_3$, find $\mathbf{A} \times (\mathbf{B} \times \mathbf{C}) + \mathbf{B} \times (\mathbf{C} \times \mathbf{A}) + \mathbf{C} \times (\mathbf{A} \times \mathbf{B})$.

1.5. Show that $\frac{1}{2}|\mathbf{A} \times \mathbf{B}|$ is equal to the area of the triangle having $\mathbf{A}$ and $\mathbf{B}$ as two of its sides. Then find the area of the triangle formed by the points $(1, 2, 1)$, $(-3, -4, 5)$, and $(2, -1, -3)$.

1.6. Show that $\mathbf{A} \cdot \mathbf{B} \times \mathbf{C}$ is the volume of the parallelepiped having $\mathbf{A}$, $\mathbf{B}$, and $\mathbf{C}$ as three of its contiguous edges. Then find the volume if $\mathbf{A} = 4\mathbf{i}_x$, $\mathbf{B} = 2\mathbf{i}_x + \mathbf{i}_y + 3\mathbf{i}_z$, and $\mathbf{C} = 2\mathbf{i}_y + 6\mathbf{i}_z$. Comment on your result.

1.7. Given $\mathbf{i}_x \times \mathbf{A} = -\mathbf{i}_y + 2\mathbf{i}_z$ and $\mathbf{i}_y \times \mathbf{A} = \mathbf{i}_x - 2\mathbf{i}_z$, find $\mathbf{A}$.

1.8. Find the component of the vector drawn from $(5, 0, 3)$ to $(3, 3, 2)$ along the direction of the vector drawn from $(6, 2, 4)$ to $(3, 3, 6)$.

1.9. Find the unit vector normal to the plane $4x - 5y + 3z = 60$. Then find the distance from the origin to the plane.

1.10. Write the expression for the differential length vector $d\mathbf{l}$ at the point $(1, 2, 8)$ on the straight line $y = 2x$, $z = 4y$, and having the projection dx on the x axis.

1.11. Write the expression for the differential length vector $d\mathbf{l}$ at the point $(4, 4, 2)$ on the curve $x = y = z^2$ and having the projection dz on the z axis.

1.12. Write the expression for the differential surface vector $d\mathbf{S}$ at the point $(1, 1, \frac{1}{2})$ on the plane $x + 2z = 2$ and having the projection $dx\, dy$ on the xy plane.

1.13. Find two differential length vectors tangential to the surface $y = x^2$ at the point $(2, 4, 1)$ and then find a unit vector normal to the surface at that point.

1.14. A hemispherical bowl of radius 2 m lies with its base on the xy plane and with its center at the origin. Write the expression for the scalar field, describing the height of points on the bowl as a function of x and y.

1.15. A number equal to the sum of its coordinates is assigned to each point in a rectangular room having three of its contiguous edges as the coordinate axes. Draw a sketch of the constant-magnitude surfaces for the number field generated in this manner.

1.16. Write the expression for the vector distance of a point in a rectangular room from one corner of the room, choosing the three edges meeting at that point as the coordinate axes. Describe the vector distance field associated with the points in the room.

1.17. For the rotating disk of Fig. 1.9, write the expression for the linear velocity vector field associated with the points on the disk; use an xy coordinate system with the origin at the center of the disk.

1.18. Given $f(z, t) = 10 \cos (2\pi \times 10^7 t - 0.1\pi z)$, (a) draw sketches of f versus z for $t = 0$, $\frac{1}{8} \times 10^{-7}$, $\frac{1}{4} \times 10^{-7}$, $\frac{3}{8} \times 10^{-7}$, and $\frac{1}{2} \times 10^{-7}$ s, and (b) draw sketches of f versus t for $z = 0$, 2.5, 5, 7.5, and 10 m. From your sketches of part (a), what can you say about the function $f(z, t)$?

1.19. Repeat Problem 1.18 for $f(z, t) = 10 \cos (2\pi \times 10^7 t + 0.1\pi z)$.

1.20. Repeat Problem 1.18 for $f(z, t) = 10 \cos 2\pi \times 10^7 t \cos 0.1\pi z$.

1.21. For each of the following vector fields, find the polarization:
(a) $1 \cos (\omega t + 30°) \mathbf{i}_x + \sqrt{2} \cos (\omega t + 30°) \mathbf{i}_y$
(b) $1 \cos (\omega t + 30°) \mathbf{i}_x + 1 \cos (\omega t - 60°) \mathbf{i}_y$
(c) $1 \cos (\omega t + 30°) \mathbf{i}_x + \sqrt{2} \cos (\omega t - 60°) \mathbf{i}_y$

1.22. Determine the polarization of the sum vector obtained by adding the two vector fields

$$\mathbf{F}_1 = (-\sqrt{3}\,\mathbf{i}_x + \mathbf{i}_y) \cos \omega t$$

$$\mathbf{F}_2 = \left(\frac{1}{2}\mathbf{i}_x + \frac{\sqrt{3}}{2}\mathbf{i}_y - \sqrt{3}\,\mathbf{i}_z\right) \sin \omega t$$

1.23. For the vector field $1 \cos \omega t\, \mathbf{i}_x + \sqrt{2} \sin \omega t\, \mathbf{i}_y$, draw sketches similar to those of Figs. 1.12 and 1.13 and describe the polarization.

1.24. Find $10 \cos (\omega t - 30°) + 10 \cos (\omega t + 210°)$ by using the phasor technique.

1.25. Find $3 \cos (\omega t + 60°) - 4 \cos (\omega t + 150°)$ by using the phasor technique.

1.26. Solve the differential equation $5 \times 10^{-6} \dfrac{di}{dt} + 12i = 13 \cos 10^6 t$ by using the phasor technique.

1.27. Two point charges each of mass m and charge q are suspended by strings of length l from a common point. Find the value of q for which the angle made by the strings at the common point is 90°.

1.28. Point charges Q and $-Q$ are situated at $(0, 0, 1)$ and $(0, 0, -1)$, respectively. Find the electric field intensity at (a) $(0, 0, 100)$, and (b) $(100, 0, 0)$.

1.29. For the point charge configuration of Example 1.4, find **E** at the point $(2, 2, 2)$.

1.30. A line charge consists of charge distributed along a line just as graphite in a pencil lead. We then talk of line charge density, or charge per unit length, having the units C/m. Obtain a series expression for the electric field intensity at $(0, 1, 0)$ for a line charge situated along the z axis between $(0, 0, -1)$ and $(0, 0, 1)$ with uniform density 10^{-3} C/m by dividing the line into 100 equal segments. Consider the charge in each segment to be a point charge located at the center of the segment, and use superposition.

1.31. Repeat Problem 1.30, but assume the line charge density to be $10^{-3} |z|$ C/m.

1.32. Charge is distributed uniformly with density 10^{-3} C/m on a circular ring of radius 2 m lying in the xy plane and centered at the origin. Obtain the electric field intensity at the point (0, 0, 1) by using the procedure described in Problem 1.30.

1.33. A surface charge consists of charge distributed on a surface just as paint on a table top. We then talk of surface charge density or, charge per unit area, having the units C/m^2. Obtain a series expression for the electric field intensity at $(0, 0, 1)$ for a surface charge of uniform density 10^{-3} C/m^2 situated within the square on the xy plane having the corners $(1, 1, 0)$, $(-1, 1, 0)$, $(-1, -1, 0)$, and $(1, -1, 0)$ by dividing the square into 10,000 equal areas. Consider the charge in each area as a point charge located at the center of the area, and use superposition.

1.34. Repeat Problem 1.33, but assume the surface charge density to be 10^{-3} $|xy^2|$ C/m^2.

1.35. For an electron cloud of uniform density $N = 10^{12}$ m^{-3} oscillating under the influence of an electric field $E = 10^{-3} \cos 2\pi \times 10^7 t\, i_x$ V/m, find (a) the current density, and (b) the current crossing the surface $0.01(i_x + i_y)$ m^2.

1.36. An object of mass m and charge q, suspended by a spring of spring constant k is acted upon by the earth's gravitational field and an electric field $E_0 \cos \omega t$ parallel to the gravitational field. Obtain the steady-state solution for the velocity of the object.

1.37. Find dF_1 and dF_2 for $I_1\, dl_1 = I_1\, dx\, i_x$ located at the origin and for $I_2\, dl_2 = I_2\, dy\, i_y$ located at $(0, 1, 0)$.

1.38. For an infinitesimal current element $I\, dx\, (i_x + 2i_y + 2i_z)$ located at the point $(1, 0, 0)$, find the magnetic flux density at (a) the point $(0, 1, 1)$ and (b) the point $(2, 2, 2)$.

1.39. A square loop of wire of sides 0.01 m lies in the xy plane, with its sides parallel to the x and y axes and with its center at the origin. It carries a current of 1 ampere in the clockwise sense as seen along the positive z axis. Find the magnetic flux density at (a) $(0, 0, 1)$ and (b) $(0, 1, 0)$.

1.40. A straight wire along the z axis carries current I amperes in the positive z direction. Consider the portion of the wire lying between $(0, 0, -1)$ and $(0, 0, 1)$. By dividing this portion into 100 equal segments and using superposition, obtain a series expression for B at $(0, 1, 0)$.

1.41. A circular loop of wire of radius 2 m is situated in the xy plane and with its center at the origin. It carries a current of 1 ampere in the clockwise sense as seen along the positive z axis. Find B at $(0, 0, 1)$ by dividing the loop into a large number of equal infinitesimal segments and by using superposition.

1.42. Obtain the expression for the orbital frequency for an electron moving in a circular orbit normal to a uniform magnetic field of flux density B_0 Wb/m². Compute its value for B_0 equal to 5×10^{-5}.

1.43. A magnetic field $B = B_0(i_x + 2i_y - 4i_z)$ exists at a point. What should be the electric field at that point if the force experienced by a test charge moving with a velocity $v = v_0(3i_x - i_y + 2i_z)$ is to be zero?

1.44. The forces experienced by a test charge q at a point in a region of electric and magnetic fields are given as follows for three different velocities of the test charge:

$$\mathbf{F}_1 = 0 \qquad \text{for } \mathbf{v} = v_0 \mathbf{i}_x$$
$$\mathbf{F}_2 = 0 \qquad \text{for } \mathbf{v} = v_0 \mathbf{i}_y$$
$$\mathbf{F}_3 = -qE_0 \mathbf{i}_z \qquad \text{for } \mathbf{v} = v_0(\mathbf{i}_x + \mathbf{i}_y)$$

where v_0 and E_0 are constants. (a) Find $\mathbf{E}$ and $\mathbf{B}$ at that point. (b) Find the force experienced by the test charge for $\mathbf{v} = v_0(\mathbf{i}_x - \mathbf{i}_y)$.

2. MAXWELL'S EQUATIONS IN INTEGRAL FORM

In Chap. 1 we learned the simple rules of vector algebra and familiarized ourselves with the basic concepts of fields, particularly those associated with electric and magnetic fields. We now have the necessary background to introduce the additional tools required for the understanding of the various quantities associated with Maxwell's equations and then discuss Maxwell's equations. In particular, our goal in this chapter is to learn Maxwell's equations in integral form as a prerequisite to the derivation of their differential forms in the next chapter. Maxwell's equations in integral form govern the interdependence of certain field and source quantities associated with regions in space, that is, contours, surfaces, and volumes. The differential forms of Maxwell's equations, however, relate the characteristics of the field vectors at a given point to one another and to the source densities at that point.

Maxwell's equations in integral form are a set of four laws resulting from several experimental findings and a purely mathematical contribution. We shall, however, consider them as postulates and learn to understand their physical significance as well as their mathematical formulation. The source quantities involved in their formulation are charges and currents. The field quantities have to do with the line and surface integrals of the electric and magnetic field vectors. We shall therefore first introduce line and surface integrals and then consider successively the four Maxwell's equations in integral form.

2.1 THE LINE INTEGRAL

Let us consider in a region of electric field **E** the movement of a test charge q from the point A to the point B along the path C as shown in Fig. 2.1(a). At each and every point along the path the electric field exerts a force on the test charge and hence does a certain amount of work in moving the charge to another point an infinitesimal distance away. To find the total amount of work done from A to B, we divide the path into a number of infinitesimal segments $\Delta\mathbf{l}_1$, $\Delta\mathbf{l}_2$, $\Delta\mathbf{l}_3$, . . ., $\Delta\mathbf{l}_n$, as shown in Fig. 2.1(b), find the infinitesimal amount of work done for each segment and then add up the contributions from all the segments. Since the segments are infinitesimal in length, we can consider each of them to be straight and the electric field at all points within a segment to be the same and equal to its value at the start of the segment.

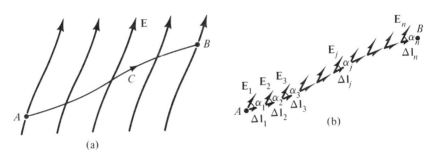

Figure 2.1. For evaluating the total amount of work done in moving a test charge along a path C from point A to point B in a region of electric field.

If we now consider one segment, say the jth segment, and take the component of the electric field for that segment along the length of that segment, we obtain the result $E_j \cos \alpha_j$ where α_j is the angle between the direction of the electric field vector $\mathbf{E}_j$ at the start of that segment and the direction of that segment. Since the electric field intensity has the meaning of force per unit charge, the electric force along the direction of the jth segment is then equal to $qE_j \cos \alpha_j$ where q is the value of the test charge. To obtain the work done in carrying the test charge along the length of the jth segment, we then multiply this electric force component by the length Δl_j of that segment. Thus for the jth segment, we obtain the result for the work done by the electric field as

$$\Delta W_j = qE_j \cos \alpha_j \, \Delta l_j \qquad (2.1)$$

If we do this for all the infinitesimal segments and add up all the contributions, we get the total work done by the electric field in moving the test charge from A to B as

$$
\begin{aligned}
W_A^B &= \Delta W_1 + \Delta W_2 + \Delta W_3 + \ldots + \Delta W_n \\
&= qE_1 \cos \alpha_1 \, \Delta l_1 + qE_2 \cos \alpha_2 \, \Delta l_2 + qE_3 \cos \alpha_3 \, \Delta l_3 \\
&\quad + \ldots + qE_n \cos \alpha_n \, \Delta l_n \\
&= q \sum_{j=1}^{n} E_j \cos \alpha_j \, \Delta l_j
\end{aligned}
\tag{2.2}
$$

In vector notation we make use of the dot product operation between two vectors to write this quantity as

$$
W_A^B = q \sum_{j=1}^{n} \mathbf{E}_j \cdot \Delta \mathbf{l}_j
\tag{2.3}
$$

Example 2.1. Let us consider the electric field given by

$$
\mathbf{E} = y\mathbf{i}_y
$$

and determine the work done by the field in carrying 3 μC of charge from the point $A(0, 0, 0)$ to the point $B(1, 1, 0)$ along the parabolic path $y = x^2, z = 0$ shown in Fig. 2.2(a).

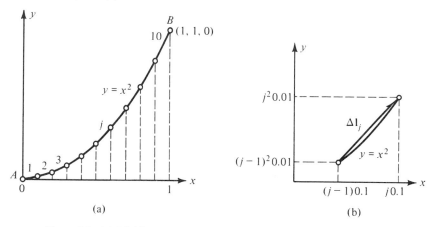

Figure 2.2. (a) Division of the path $y = x^2$ from A $(0,0,0)$ to B $(1,1,0)$ into ten segments. (b) The length vector corresponding to the jth segment of part (a) approximated as a straight line.

For convenience, we shall divide the path into ten segments having equal widths along the x direction, as shown in Fig. 2.2(a). We shall number the segments 1, 2, 3, . . ., 10. The coordinates of the starting and ending points of the jth segment are as shown in Fig. 2.2(b). The electric field at the start of the

jth segment is given by

$$\mathbf{E}_j = (j - 1)^2 \, 0.01 \mathbf{i}_y$$

The length vector corresponding to the jth segment, approximated as a straight line connecting its starting and ending points is

$$\Delta \mathbf{l}_j = 0.1 \mathbf{i}_x + [j^2 - (j - 1)^2] \, 0.01 \mathbf{i}_y$$
$$= 0.1 \mathbf{i}_x + (2j - 1) \, 0.01 \mathbf{i}_y$$

The required work is then given by

$$W_A^B = 3 \times 10^{-6} \sum_{j=1}^{10} \mathbf{E}_j \cdot \Delta \mathbf{l}_j$$

$$= 3 \times 10^{-6} \sum_{j=1}^{10} [(j - 1)^2 0.01 \mathbf{i}_y] \cdot [0.1 \mathbf{i}_x + (2j - 1) 0.01 \mathbf{i}_y]$$

$$= 3 \times 10^{-10} \sum_{j=1}^{10} (j - 1)^2 (2j - 1)$$

$$= 3 \times 10^{-10}[0 + 3 + 20 + 63 + 144 + 275 + 468 + 735$$
$$+ 1088 + 1539]$$

$$= 3 \times 10^{-10} \times 4335 = 1.3005 \, \mu\text{J} \qquad \blacksquare$$

The result that we have obtained in Example 2.1, for W_A^B, is approximate since we divided the path from A to B into a finite number of segments. By dividing it into larger and larger numbers of segments, we can obtain more and more accurate results. In fact, the problem can be conveniently formulated for a computer solution and by varying the number of segments from a small value to a large value, the convergence of the result can be verified. The value to which the result converges is that for which $n = \infty$. The summation in (2.3) then becomes an integral, which represents exactly the work done by the field and is given by

$$W_A^B = q \int_A^B \mathbf{E} \cdot d\mathbf{l} \qquad (2.4)$$

The integral on the right side of (2.4) is known as the "line integral of $\mathbf{E}$ from A to B."

Example 2.2. We shall illustrate the evaluation of the line integral by computing the exact value of the work done by the electric field in Example 2.1.

To do this, we note that at any arbitrary point $(x, y, 0)$ on the curve $y = x^2, z = 0$, the infinitesimal length vector tangential to the curve is given by

$$d\mathbf{l} = dx\,\mathbf{i}_x + dy\,\mathbf{i}_y$$
$$= dx\,\mathbf{i}_x + d(x^2)\,\mathbf{i}_y$$
$$= dx\,\mathbf{i}_x + 2x\,dx\,\mathbf{i}_y$$

The value of $\mathbf{E} \cdot d\mathbf{l}$ at the point $(x, y, 0)$ is

$$\mathbf{E} \cdot d\mathbf{l} = y\mathbf{i}_y \cdot (dx\,\mathbf{i}_x + dy\,\mathbf{i}_y)$$
$$= x^2\mathbf{i}_y \cdot (dx\,\mathbf{i}_x + 2x\,dx\,\mathbf{i}_y)$$
$$= 2x^3\,dx$$

Thus the required work is given by

$$W_A^B = q \int_A^B \mathbf{E} \cdot d\mathbf{l} = 3 \times 10^{-6} \int_{(0,0,0)}^{(1,1,0)} 2x^3\,dx$$
$$= 3 \times 10^{-6}\left[\frac{2x^4}{4}\right]_{x=0}^{x=1} = 1.5\ \mu\mathrm{J} \qquad\blacksquare$$

Dividing both sides of (2.4) by q, we note that the line integral of $\mathbf{E}$ from A to B has the physical meaning of work per unit charge done by the field in moving the test charge from A to B. This quantity is known as the "voltage between A and B" and is denoted by the symbol $[V]_A^B$, having the units of volts. Thus

$$[V]_A^B = \int_A^B \mathbf{E} \cdot d\mathbf{l} \tag{2.5}$$

When the path under consideration is a closed path, as shown in Fig. 2.3, the line integral is written with a circle associated with the integral sign in the manner $\oint_C \mathbf{E} \cdot d\mathbf{l}$. The line integral of a vector around a closed path is known as the "circulation" of that vector. In particular, the line integral of $\mathbf{E}$ around a closed path is the work per unit charge done by the field in moving a test charge around the closed path. It is the voltage around the closed path and is

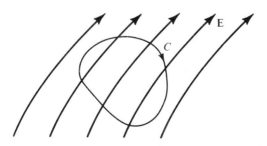

Figure 2.3. Closed path C in a region of electric field.

also known as the "electromotive force." We shall now consider an example of evaluating the line integral of a vector around a closed path.

Example 2.3. Let us consider the force field

$$\mathbf{F} = x\mathbf{i}_y$$

and evaluate $\oint_C \mathbf{F} \cdot d\mathbf{l}$ where C is the closed path $ABCDA$ shown in Fig. 2.4.

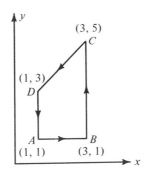

Figure 2.4. For evaluating the line integral of a vector field around a closed path.

Noting that

$$\oint_{ABCDA} \mathbf{F} \cdot d\mathbf{l} = \int_A^B \mathbf{F} \cdot d\mathbf{l} + \int_B^C \mathbf{F} \cdot d\mathbf{l} + \int_C^D \mathbf{F} \cdot d\mathbf{l} + \int_D^A \mathbf{F} \cdot d\mathbf{l} \quad (2.6)$$

we simply evaluate each of the line integrals on the right side of (2.6) and add them up to obtain the required quantity. Thus for the side AB,

$$y = 1, \quad dy = 0, \quad d\mathbf{l} = dx\,\mathbf{i}_x + (0)\mathbf{i}_y = dx\,\mathbf{i}_x$$
$$\mathbf{F} \cdot d\mathbf{l} = (x\mathbf{i}_y) \cdot (dx\,\mathbf{i}_x) = 0$$
$$\int_A^B \mathbf{F} \cdot d\mathbf{l} = 0$$

For the side BC,

$$x = 3, \quad dx = 0, \quad d\mathbf{l} = (0)\mathbf{i}_x + dy\,\mathbf{i}_y = dy\,\mathbf{i}_y$$
$$\mathbf{F} \cdot d\mathbf{l} = (3\mathbf{i}_y) \cdot (dy\,\mathbf{i}_y) = 3\,dy$$
$$\int_B^C \mathbf{F} \cdot d\mathbf{l} = \int_1^5 3\,dy = 12$$

For the side *CD*,

$$y = 2 + x, \qquad dy = dx, \qquad d\mathbf{l} = dx\,\mathbf{i}_x + dx\,\mathbf{i}_y$$
$$\mathbf{F} \cdot d\mathbf{l} = (x\mathbf{i}_y) \cdot (dx\,\mathbf{i}_x + dx\,\mathbf{i}_y) = x\,dx$$
$$\int_C^D \mathbf{F} \cdot d\mathbf{l} = \int_3^1 x\,dx = -4$$

For the side *DA*,

$$x = 1, \qquad dx = 0, \qquad d\mathbf{l} = (0)\mathbf{i}_x + dy\,\mathbf{i}_y$$
$$\mathbf{F} \cdot d\mathbf{l} = (\mathbf{i}_y) \cdot (dy\,\mathbf{i}_y) = dy$$
$$\int_D^A \mathbf{F} \cdot d\mathbf{l} = \int_3^1 dy = -2$$

Finally,

$$\oint_{ABCDA} \mathbf{F} \cdot d\mathbf{l} = 0 + 12 - 4 - 2 = 6 \qquad \blacksquare$$

2.2   THE SURFACE INTEGRAL

Let us consider a region of magnetic field and an infinitesimal surface at a point in that region. Since the surface is infinitesimal, we can assume the magnetic flux density to be uniform on the surface, although it may be non-uniform over a wider region. If the surface is oriented normal to the magnetic field lines, as shown in Fig. 2.5(a), then the magnetic flux crossing the surface is simply given by the product of the surface area and the magnetic flux density on the surface, that is, $B\,\Delta S$. If, however, the surface is oriented parallel to the magnetic field lines, as shown in Fig. 2.5(b), there is no

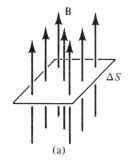
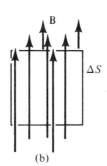
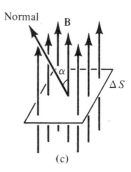

Figure 2.5. An infinitesimal surface ΔS in a magnetic field **B** oriented (a) normal to the field, (b) parallel to the field, and (c) with its normal making an angle α to the field.

magnetic flux crossing the surface. If the surface is oriented in such a manner that the normal to the surface makes an angle α with the magnetic field lines as shown in Fig. 2.5(c), then the amount of magnetic flux crossing the surface can be determined by considering that the component of **B** normal to the surface is $B \cos \alpha$ and the component tangential to the surface is $B \sin \alpha$. The component of **B** normal to the surface results in a flux of $(B \cos \alpha) \Delta S$ crossing the surface whereas the component tangential to the surface does not contribute at all to the flux crossing the surface. Thus the magnetic flux crossing the surface in this case is $(B \cos \alpha) \Delta S$. We can obtain this result alternatively by noting that the projection of the surface onto the plane normal to the magnetic field lines is $\Delta S \cos \alpha$.

Let us now consider a large surface S in the magnetic field region, as shown in Fig. 2.6. The magnetic flux crossing this surface can be found by

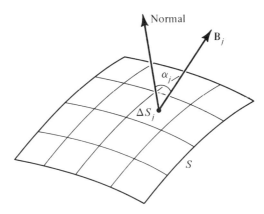

Figure 2.6. Division of a large surface S in a magnetic field region into a number of infinitesimal surfaces.

dividing the surface into a number of infinitesimal surfaces $\Delta S_1, \Delta S_2, \Delta S_3,$ $\ldots, \Delta S_n$ and applying the result obtained above for each infinitesimal surface and adding up the contributions from all the surfaces. To obtain the contribution from the jth surface, we draw the normal vector to that surface and find the angle α_j between the normal vector and the magnetic flux density vector $\mathbf{B}_j$ associated with that surface. Since the surface is infinitesimal, we can assume $\mathbf{B}_j$ to be the value of **B** at the centroid of the surface and we can also erect the normal vector at that point. The contribution to the total magnetic flux from the jth infinitesimal surface is then given by

$$\Delta\psi_j = B_j \cos \alpha_j \, \Delta S_j \qquad (2.7)$$

where the symbol ψ represents magnetic flux. The total magnetic flux crossing the surface S is then given by

$$
\begin{aligned}
[\psi]_S &= \Delta\psi_1 + \Delta\psi_2 + \Delta\psi_3 + \ldots + \Delta\psi_n \\
&= B_1 \cos\alpha_1 \, \Delta S_1 + B_2 \cos\alpha_2 \, \Delta S_2 + B_3 \cos\alpha_3 \, \Delta S_3 \\
&\quad + \ldots + B_n \cos\alpha_n \, \Delta S_n \\
&= \sum_{j=1}^{n} B_j \cos\alpha_j \, \Delta S_j
\end{aligned} \tag{2.8}
$$

In vector notation we make use of the dot product operation between two vectors to write this quantity as

$$
[\psi]_S = \sum_{j=1}^{n} \mathbf{B}_j \cdot \Delta S_j \, \mathbf{i}_{nj} \tag{2.9}
$$

where $\mathbf{i}_{nj}$ is the unit vector normal to the surface ΔS_j. In fact, by recalling that the infinitesimal surface can be considered as a vector quantity having magnitude equal to the area of the surface and direction normal to the surface, that is,

$$
\Delta\mathbf{S}_j = \Delta S_j \, \mathbf{i}_{nj} \tag{2.10}
$$

we can write (2.9) as

$$
[\psi]_S = \sum_{j=1}^{n} \mathbf{B}_j \cdot \Delta\mathbf{S}_j \tag{2.11}
$$

Example 2.4. Let us consider the magnetic field given by

$$
\mathbf{B} = 3xy^2 \mathbf{i}_z \ \text{Wb/m}^2
$$

and determine the magnetic flux crossing the portion of the xy plane lying between $x = 0$, $x = 1$, $y = 0$, and $y = 1$.

For convenience, we shall divide the surface into 25 equal areas as shown in Fig. 2.7(a). We shall designate the squares as 11, 12, $\ldots$, 15, 21, 22, $\ldots$, 55 where the first digit represents the number of the square in the x direction and the second digit represents the number of the square in the y direction. The x and y coordinates of the midpoint of the ijth square are $(2i - 1)0.1$ and $(2j - 1)0.1$, respectively, as shown in Fig. 2.7(b). The magnetic field at the center of the ijth square is then given by

$$
\mathbf{B}_{ij} = 3(2i - 1)(2j - 1)^2 0.001 \mathbf{i}_z
$$

Since we have divided the surface into equal areas and since all areas are in the xy plane,

$$
\Delta\mathbf{S}_{ij} = 0.04\mathbf{i}_z \qquad \text{for all } i \text{ and } j
$$

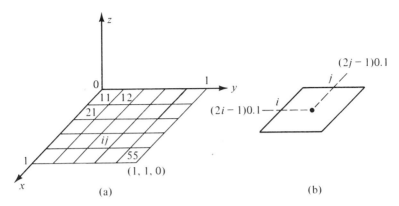

Figure 2.7. (a) Division of the portion of the xy plane lying between $x = 0$, $x = 1, y = 0$, and $y = 1$ into 25 squares. (b) The area corresponding to the ijth square.

The required magnetic flux is then given by

$$[\psi]_s = \sum_{i=1}^{5} \sum_{j=1}^{5} \mathbf{B}_{ij} \cdot \Delta \mathbf{S}_{ij}$$

$$= \sum_{i=1}^{5} \sum_{j=1}^{5} 3(2i - 1)(2j - 1)^2 0.001 \mathbf{i}_z \cdot 0.04 \mathbf{i}_z$$

$$= 0.00012 \sum_{i=1}^{5} \sum_{j=1}^{5} (2i - 1)(2j - 1)^2$$

$$= 0.00012(1 + 3 + 5 + 7 + 9)(1 + 9 + 25 + 49 + 81)$$

$$= 0.495 \text{ Wb} \quad \blacksquare$$

The result that we have obtained for $[\psi]_s$ in Example 2.4 is approximate since we have divided the surface S into a finite number of areas. By dividing it into larger and larger numbers of squares, we can obtain more and more accurate results. In fact, the problem can be conveniently formulated for a computer solution, and by varying the number of squares from a small value to a large value, the convergence of the result can be verified. The value to which the result converges is that for which the number of squares in each direction is infinity. The summation in (2.11) then becomes an integral that represents exactly the magnetic flux crossing the surface and is given by

$$[\psi]_s = \int_S \mathbf{B} \cdot d\mathbf{S} \quad (2.12)$$

where the symbol S associated with the integral sign denotes that the integration is performed over the surface S. The integral on the right side of (2.12) is known as the "surface integral of $\mathbf{B}$ over S." The surface integral is a double integral since dS is equal to the product of two differential lengths. In fact, the work in Example 2.4 indicates that as i and j tend to infinity, the double summation becomes a double integral involving the variables of integration x and y.

Example 2.5.   We shall illustrate the evaluation of the surface integral by computing the exact value of the magnetic flux in Example 2.4.

To do this, we note that at any arbitrary point (x, y) on the surface, the infinitesimal surface vector is given by

$$dS = dx\,dy\,\mathbf{i}_z$$

The value of $\mathbf{B} \cdot dS$ at the point (x, y) is

$$\mathbf{B} \cdot dS = 3xy^2\mathbf{i}_z \cdot dx\,dy\,\mathbf{i}_z$$
$$= 3xy^2\,dx\,dy$$

Thus the required magnetic flux is given by

$$[\psi]_S = \int_S \mathbf{B} \cdot dS$$
$$= \int_{x=0}^{1} \int_{y=0}^{1} 3xy^2\,dx\,dy = 0.5\,\text{Wb} \qquad \blacksquare$$

When the surface under consideration is a closed surface, the surface integral is written with a circle associated with the integral sign in the manner $\oint_S \mathbf{B} \cdot dS$. The surface integral of $\mathbf{B}$ over the closed surface S is simply the magnetic flux emanating from the volume bounded by the surface. We shall now consider an example of evaluating $\oint_S \mathbf{B} \cdot dS$.

Example 2.6.   Let us consider the magnetic field

$$\mathbf{B} = (x + 2)\mathbf{i}_x + (1 - 3y)\mathbf{i}_y + 2z\mathbf{i}_z$$

and evaluate $\oint_S \mathbf{B} \cdot dS$ where S is the surface of the cubical box bounded by the planes

$$x = 0, \qquad x = 1$$
$$y = 0, \qquad y = 1$$
$$z = 0, \qquad z = 1$$

as shown in Fig. 2.8.

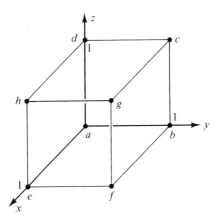

Figure 2.8. For evaluating the surface integral of a vector field over a closed surface.

Noting that

$$\oint_S \mathbf{B} \cdot d\mathbf{S} = \int_{abcd} \mathbf{B} \cdot d\mathbf{S} + \int_{efgh} \mathbf{B} \cdot d\mathbf{S} + \int_{aehd} \mathbf{B} \cdot d\mathbf{S} + \int_{bfgc} \mathbf{B} \cdot d\mathbf{S}$$

$$+ \int_{aefb} \mathbf{B} \cdot d\mathbf{S} + \int_{dhgc} \mathbf{B} \cdot d\mathbf{S} \tag{2.13}$$

we simply evaluate each of the surface integrals on the right side of (2.13) and add them up to obtain the required quantity. In doing so, we recognize that since the quantity we want is the magnetic flux out of the box, we should direct the normal vectors toward the outside of the box. Thus for the surface *abcd*,

$$x = 0, \quad \mathbf{B} = 2\mathbf{i}_x + (1 - 3y)\mathbf{i}_y + 2z\mathbf{i}_z, \quad d\mathbf{S} = -dy\,dz\,\mathbf{i}_x$$

$$\mathbf{B} \cdot d\mathbf{S} = -2\,dy\,dz$$

$$\int_{abcd} \mathbf{B} \cdot d\mathbf{S} = \int_{z=0}^{1} \int_{y=0}^{1} (-2)\,dy\,dz = -2$$

For the surface *efgh*,

$$x = 1, \quad \mathbf{B} = 3\mathbf{i}_x + (1 - 3y)\mathbf{i}_y + 2z\mathbf{i}_z, \quad d\mathbf{S} = dy\,dz\,\mathbf{i}_x$$

$$\mathbf{B} \cdot d\mathbf{S} = 3\,dy\,dz$$

$$\int_{efgh} \mathbf{B} \cdot d\mathbf{S} = \int_{z=0}^{1} \int_{y=0}^{1} 3\,dy\,dz = 3$$

For the surface *aehd*,

$$y = 0, \quad \mathbf{B} = (x + 2)\mathbf{i}_x + 1\mathbf{i}_y + 2z\mathbf{i}_z, \quad d\mathbf{S} = -dz\,dx\,\mathbf{i}_y$$

$$\mathbf{B} \cdot d\mathbf{S} = -dz\,dx$$

$$\int_{aehd} \mathbf{B} \cdot d\mathbf{S} = \int_{x=0}^{1} \int_{z=0}^{1} (-1)\,dz\,dx = -1$$

For the surface *bfgc*,

$$y = 1, \quad \mathbf{B} = (x + 2)\mathbf{i}_x - 2\mathbf{i}_y + 2z\mathbf{i}_z, \quad d\mathbf{S} = dz\,dx\,\mathbf{i}_y$$

$$\mathbf{B} \cdot d\mathbf{S} = -2\,dz\,dx$$

$$\int_{bfgc} \mathbf{B} \cdot d\mathbf{S} = \int_{x=0}^{1} \int_{z=0}^{1} (-2)\,dz\,dx = -2$$

For the surface *aefb*,

$$z = 0, \quad \mathbf{B} = (x + 2)\mathbf{i}_x + (1 - 3y)\mathbf{i}_y + 0\mathbf{i}_z, \quad d\mathbf{S} = -dx\,dy\,\mathbf{i}_z$$

$$\mathbf{B} \cdot d\mathbf{S} = 0$$

$$\int_{aefb} \mathbf{B} \cdot d\mathbf{S} = 0$$

For the surface *dhgc*,

$$z = 1, \quad \mathbf{B} = (x + 2)\mathbf{i}_x + (1 - 3y)\mathbf{i}_y + 2\mathbf{i}_z, \quad d\mathbf{S} = dx\,dy\,\mathbf{i}_z$$

$$\mathbf{B} \cdot d\mathbf{S} = 2\,dx\,dy$$

$$\int_{dhgc} \mathbf{B} \cdot d\mathbf{S} = \int_{y=0}^{1} \int_{x=0}^{1} 2\,dx\,dy = 2$$

Finally,

$$\oint_{S} \mathbf{B} \cdot d\mathbf{S} = -2 + 3 - 1 - 2 + 0 + 2 = 0 \qquad \blacksquare$$

2.3 FARADAY'S LAW

In the previous sections we introduced the line and surface integrals. We are now ready to consider Maxwell's equations in integral form. The first equation, which we shall discuss in this section, is a consequence of an experimental finding by Michael Faraday in 1831 that time-varying magnetic fields give rise to electric fields and hence it is known as "Faraday's law." Faraday discovered that when the magnetic flux enclosed by a loop of wire changes with time, a current is produced in the loop, indicating that a voltage or an "electromotive force," abbreviated as emf, is induced around the loop.

The variation of the magnetic flux can result from the time variation of the magnetic flux enclosed by a fixed loop or from a moving loop in a static magnetic field or from a combination of the two, that is, a moving loop in a time-varying magnetic field.

Thus far we have merely stated Faraday's finding without regard to the polarity of the induced emf around the loop or that of the magnetic flux enclosed by the loop. To clarify the point, let us consider a planar circular loop in the plane of the paper as shown in Fig. 2.9. Then we can talk of emf induced in the clockwise sense or in the counterclockwise sense. The emf induced in the clockwise sense is the line integral of $\mathbf{E}$ ($\oint \mathbf{E} \cdot d\mathbf{l}$) evaluated by traversing the loop in the clockwise direction, as shown in Figs. 2.9(a) and 2.9(b). The emf induced in the counterclockwise sense is the line integral of $\mathbf{E}$($\oint \mathbf{E} \cdot d\mathbf{l}$) evaluated by traversing the loop in the counterclockwise direction, as shown in Figs. 2.9(c) and 2.9(d). One is, of course, the negative of the other. Similarly, we can talk of enclosed magnetic flux directed into the paper or out of the paper. The enclosed magnetic flux into the paper is the surface integral of $\mathbf{B}$ ($\int \mathbf{B} \cdot d\mathbf{S}$) evaluated over the plane surface bounded by the loop and with the normal to the surface directed into the paper, as shown in Figs. 2.9(a) and 2.9(c). The enclosed magnetic flux out of the paper is the surface integral of $\mathbf{B}$ ($\int \mathbf{B} \cdot d\mathbf{S}$) evaluated over the plane surface bounded by the loop and with the normal to the surface directed out of the paper, as shown in Figs. 2.9(b) and 2.9(d). One is, of course, the negative of the other.

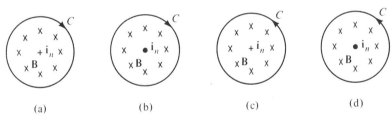

(a)　　　　　　　(b)　　　　　　　(c)　　　　　　　(d)

Figure 2.9. Four possible pairs of directions of traversal around a planar circular loop and normal to the surface bounded by the loop.

If we do not pay any attention to the polarities, we can write four equations relating the emf around the loop to the magnetic flux enclosed by the loop. These are

$$[\text{emf}]_{\text{clockwise}} = \frac{d}{dt}[\text{magnetic flux}]_{\text{into the paper}} \tag{2.14a}$$

$$[\text{emf}]_{\text{clockwise}} = \frac{d}{dt}[\text{magnetic flux}]_{\text{out of the paper}} \tag{2.14b}$$

$$[\text{emf}]_{\text{counterclockwise}} = \frac{d}{dt}[\text{magnetic flux}]_{\text{into the paper}} \tag{2.14c}$$

$$[\text{emf}]_{\text{counterclockwise}} = \frac{d}{dt}[\text{magnetic flux}]_{\text{out of the paper}} \tag{2.14d}$$

The fourth equation is, however, consistent with the first and the third equation is consistent with the second. Thus we are left with a choice between the first and the second. Only one of them can be correct since they provide contradictory results for the emf. Faraday's experiments showed that the second equation is the one that should be used. Alternatively, if we wish to work with clockwise induced emf and magnetic flux into the paper (or with counterclockwise induced emf and magnetic flux out of the paper), we must include a minus sign in front of the time derivative. This is, in fact, what is done conventionally. The convention is to use that normal to the surface which is directed toward the advancing direction of a right-hand screw when it is turned in the sense in which the loop is traversed, as shown in Figs. 2.10(a) and 2.10(b). This is known as the "right-hand screw rule" and is applied consistently for all electromagnetic field laws. Hence, it is well worthwhile digesting it at this early stage.

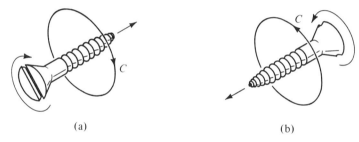

(a)   (b)

Figure 2.10. Right-hand screw rule convention employed in the formulation of electromagnetic field laws.

We can now express Faraday's law mathematically as

$$\oint_c \mathbf{E} \cdot d\mathbf{l} = -\frac{d}{dt} \int_s \mathbf{B} \cdot d\mathbf{S} \qquad (2.15)$$

where S is a surface bounded by C. For the law to be unique, the surface S need not be a plane surface and can be any curved surface bounded by C. This tells us that the magnetic flux through all possible surfaces bounded by C must be the same. We shall make use of this later. In fact, if C is not a planar loop, we cannot have a plane surface bounded by C. A further point of interest is that C need not represent a loop of wire but can be an imaginary closed path. It means that the time-varying magnetic flux induces an electric field in the region and this results in an emf around the closed path. If a wire is placed in the position occupied by the closed path, the emf will produce a current in the loop simply because the charges in the wire are constrained to move along the wire. Let us now consider some examples.

Example 2.7.   A rectangular loop of wire with three sides fixed and the fourth side movable is situated in a plane perpendicular to a uniform magnetic field

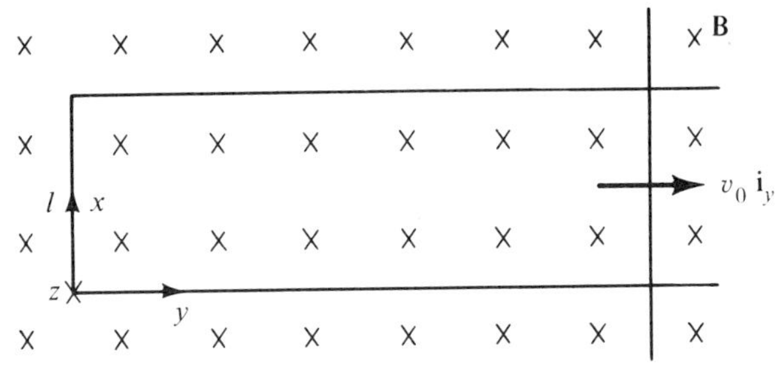

Figure 2.11. A rectangular loop of wire with a movable side situated in a uniform magnetic field.

$\mathbf{B} = B_0 \mathbf{i}_z$, as illustrated in Fig. 2.11. The movable side consists of a conducting bar moving with a velocity v_0 in the y direction. It is desired to find the emf induced in the loop.

Letting the position of the movable side at any time t be $y_0 + v_0 t$, we obtain the magnetic flux enclosed by the loop and directed into the paper as

$$\psi = (\text{area of the loop})B_0$$
$$= l(y_0 + v_0 t)B_0$$

The emf induced in the loop in the clockwise sense is then given by

$$\oint \mathbf{E} \cdot d\mathbf{l} = -\frac{d}{dt}\psi$$
$$= -\frac{d}{dt}[l(y_0 + v_0 t)B_0]$$
$$= -B_0 l v_0$$

Thus if the bar is moving to the right, the induced emf produces a current in the counterclockwise sense. Note that this polarity of the current is such that it gives rise to a magnetic field directed out of the paper inside the loop. The flux of this magnetic field is in opposition to the flux of the original magnetic field and hence tends to decrease it. This observation is in accordance with "Lenz's law," which states that the induced emf is such that it acts to oppose the change in the magnetic flux producing it. The minus sign on the right side of Faraday's law ensures that Lenz's law is always satisfied.

It is also of interest to note that the induced emf can also be interpreted as due to the electric field induced in the moving bar by virtue of its motion perpendicular to the magnetic field. Thus a charge Q in the bar experiences a force $\mathbf{F} = Q\mathbf{v} \times \mathbf{B}$ or $Qv_0\mathbf{i}_y \times B_0\mathbf{i}_z = Qv_0 B_0\mathbf{i}_x$. To an observer moving with the bar, this force appears as an electric force due to an electric field $\mathbf{F}/Q = v_0 B_0\mathbf{i}_x$. Viewed from inside the loop, this electric field is in the counterclockwise direction and hence the induced emf is $v_0 B_0 l$ in that sense as deduced above from Faraday's law. This concept of induced emf is known as the "motional emf concept," which is employed widely in the study of electromechanics.   ■

Example 2.8.   A time-varying magnetic field is given by

$$\mathbf{B} = B_0 \cos \omega t \, \mathbf{i}_y$$

where B_0 is a constant. It is desired to find the induced emf around a rectangular loop in the xz plane as shown in Fig. 2.12.

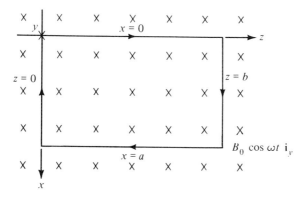

Figure 2.12. A rectangular loop in the xz plane situated in a time-varying magnetic field.

The magnetic flux enclosed by the loop and directed into the paper is given by

$$\psi = \int_S \mathbf{B} \cdot d\mathbf{S} = \int_{z=0}^{b} \int_{x=0}^{a} B_0 \cos \omega t \, \mathbf{i}_y \cdot dx \, dz \, \mathbf{i}_y$$

$$= B_0 \cos \omega t \int_{z=0}^{b} \int_{x=0}^{a} dx \, dz = ab B_0 \cos \omega t$$

The induced emf in the clockwise sense is then given by

$$\oint_C \mathbf{E} \cdot d\mathbf{l} = -\frac{d}{dt} \int_S \mathbf{B} \cdot d\mathbf{S}$$

$$= -\frac{d}{dt} [ab B_0 \cos \omega t] = ab B_0 \omega \sin \omega t$$

The time variations of the magnetic flux enclosed by the loop and the induced emf around the loop are shown in Fig. 2.13. It can be seen that when the magnetic flux enclosed by the loop is decreasing with time, the induced

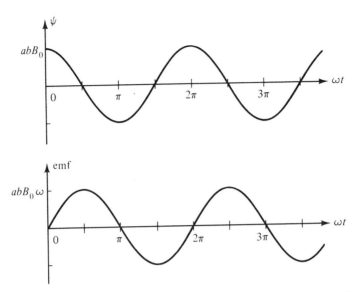

Figure 2.13. Time variations of magnetic flux ψ enclosed by the loop of Fig. 2.12, and the resulting induced emf around the loop.

emf is positive, thereby producing a clockwise current if the loop were a wire. This polarity of the current gives rise to a magnetic field directed into the paper inside the loop and hence acts to increase the magnetic flux enclosed by the loop. When the magnetic flux enclosed by the loop is increasing with time, the induced emf is negative, thereby producing a counterclockwise current around the loop. This polarity of the current gives rise to a magnetic field directed out of the paper inside the loop and hence acts to decrease the magnetic flux enclosed by the loop. These observations are once again consistent with Lenz's law.   ■

2.4 AMPERE'S CIRCUITAL LAW

In the previous section we introduced Faraday's law, one of Maxwell's equations, in integral form. In this section we introduce another of Maxwell's equations in integral form. This equation, known as "Ampere's circuital law," is a combination of an experiemental finding of Oersted that electric currents generate magnetic fields and a mathematical contribution of Maxwell that time-varying electric fields give rise to magnetic fields. It is this contribution of Maxwell that led to the prediction of electromagnetic wave propagation even before the phenomenon was discovered experimentally. In mathematical form, Ampere's circuital law is analogous to Faraday's law and is given by

$$\oint_c \frac{\mathbf{B}}{\mu_0} \cdot d\mathbf{l} = \int_S \mathbf{J} \cdot d\mathbf{S} + \frac{d}{dt} \int_S \epsilon_0 \mathbf{E} \cdot d\mathbf{S} \qquad (2.16)$$

where S is any surface bounded by C. Here again, in order to evaluate the surface integrals on the right side of (2.16), we choose that normal to the surface which is directed toward the advancing direction of a right-hand screw when it is turned in the sense of C, just as in the case of Faraday's law. Also, both integrals on the right side of (2.16) must be evaluated on the same surface, whatever be the surface chosen.

The quantity $\mathbf{J}$ on the right side of (2.16) is the volume current density vector having the magnitude equal to the maximum value of current per unit area (amp/m^2) at the point under consideration, as discussed in Sec. 1.5. Thus the quantity $\int_S \mathbf{J} \cdot d\mathbf{S}$, being the surface integral of $\mathbf{J}$ over S, has the meaning of current due to flow of charges crossing the surface S bounded by C. It also includes line currents, that is, currents flowing along thin filamentary wires enclosed by C, and surface currents, that is, currents flowing along ribbon-like wires enclosed by C. Thus $\int_S \mathbf{J} \cdot d\mathbf{S}$, although formulated in terms of the volume current density vector $\mathbf{J}$, represents the algebraic sum of all the currents due to flow of charges across the surface S.

The quantity $\int_S \epsilon_0 \mathbf{E} \cdot d\mathbf{S}$ on the right side of (2.16) is the flux of the vector field $\epsilon_0 \mathbf{E}$ crossing the surface S. The vector $\epsilon_0 \mathbf{E}$ is known as the "displacement vector" or the "displacement flux density vector" and is denoted by the symbol $\mathbf{D}$. By recalling from (1.52) that $\mathbf{E}$ has the units of (charge) per [(permittivity)(distance)^2], we note that the quantity $\mathbf{D}$ has the units of charge per unit area or C/m^2. Hence the quantity $\int_S \epsilon_0 \mathbf{E} \cdot d\mathbf{S}$, that is, the displacement flux has the units of charge, and the quantity $\frac{d}{dt} \int_S \epsilon_0 \mathbf{E} \cdot d\mathbf{S}$ has the units of $\frac{d}{dt}$(charge) or current and is known as the "displacement current." Physically, it is not a current in the sense that it does not represent the flow of charges, but mathematically it is equivalent to a current crossing the surface S.

The quantity $\oint_C \frac{\mathbf{B}}{\mu_0} \cdot d\mathbf{l}$ on the left side of (2.16) is the line integral of the vector field $\mathbf{B}/\mu_0$ around the closed path C. We learned in Sec. 2.1 that the quantity $\oint_C \mathbf{E} \cdot d\mathbf{l}$ has the physical meaning of work per unit charge associated with the movement of a test charge around the closed path C. The quantity $\oint_C \frac{\mathbf{B}}{\mu_0} \cdot d\mathbf{l}$ does not have a similar physical meaning. This is because magnetic force on a moving charge is directed perpendicular to the direction of motion of the charge as well as to the direction of the magnetic field and hence does not do work in the movement of the charge. The vector $\mathbf{B}/\mu_0$ is known as the "magnetic field intensity vector" and is denoted by the symbol $\mathbf{H}$. By recalling from (1.68) that $\mathbf{B}$ has the units of [(permeability)(current)(length)] per [(distance)^2], we note that the quantity **H** has the units of current per unit distance or amp/m. This gives the units of current or amp to $\oint_C \mathbf{H} \cdot d\mathbf{l}$. In analogy with the name "electromotive force" for $\oint_C \mathbf{E} \cdot d\mathbf{l}$, the quantity $\oint_C \mathbf{H} \cdot d\mathbf{l}$ is known as the "magnetomotive force," abbreviated as mmf.

Replacing $\mathbf{B}/\mu_0$ and $\epsilon_0 \mathbf{E}$ in (2.16) by **H** and **D**, respectively, we rewrite Ampere's circuital law as

$$\oint_C \mathbf{H} \cdot d\mathbf{l} = \int_S \mathbf{J} \cdot d\mathbf{S} + \frac{d}{dt} \int_S \mathbf{D} \cdot d\mathbf{S} \qquad (2.17)$$

In words, (2.17) states that "the magnetomotive force around a closed path C is equal to the total current, that is, the current due to actual flow of charges plus the displacement current bounded by C." When we say "the total current bounded by C," we mean "the total current crossing any given surface S bounded by C." This implies that the total current crossing all possible surfaces bounded by C must be the same since for a given C, $\oint_C \mathbf{H} \cdot d\mathbf{l}$ must have a unique value.

Example 2.9. An infinitely long, thin, straight wire situated along the z axis carries a current I amperes in the z direction. It is desired to find $\oint_C \mathbf{H} \cdot d\mathbf{l}$ around a circle of radius a lying on the xy plane and centered at the origin as shown in Fig. 2.14.

Let us consider the plane surface enclosed by C. The total current crossing the surface consists entirely of the current I carried by the wire. In fact, since the wire is infinitely long, the total current crossing any of the infinite number of surfaces bounded by C is equal to I. The situation is illustrated in Fig. 2.14(a) for a few of the infinite number of surfaces. Thus, noting that the current I is bounded by C in the right-hand sense, and that it is uniquely given, we obtain

$$\oint_C \mathbf{H} \cdot d\mathbf{l} = I \qquad (2.18)$$

We can proceed further and evaluate **H** at points on the circular path from symmetry considerations. In order for $\oint_C \mathbf{H} \cdot d\mathbf{l}$ to be nonzero, **H** must be directed (or have a component) tangential to the circular path and then from symmetry considerations, it must have the same magnitude at all points on the circle since the circle is centered at the wire. We, however, know from elementary considerations of the magnetic field due to a current element that **H** must be directed entirely tangential to the circular path. Thus let us divide

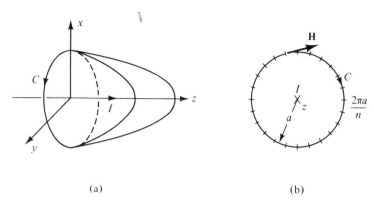

(a)   (b)

Figure 2.14. (a) For illustrating the uniqueness of a wire current enclosed by a closed path for an infinitely long, straight wire. (b) For finding the magnetic field due to the wire.

the circle into a large number of equal segments, say n, as shown in Fig. 2.14(b). Since the length of each segment is $2\pi a/n$ and since **H** is parallel to the segment, $\mathbf{H} \cdot d\mathbf{l}$ for the segment is $(2\pi a/n)H$ and

$$\oint_c \mathbf{H} \cdot d\mathbf{l} = \frac{2\pi a}{n} H \text{ (number of segments)}$$

$$= \frac{2\pi a}{n} H \cdot n = 2\pi a H$$

From (2.18), we then have

$$2\pi a H = I$$

or

$$H = \frac{I}{2\pi a}$$

Thus the magnetic field intensity due to the infinitely long wire is directed circular to the wire in the right-hand sense and has a magnitude $I/2\pi a$ where a is the distance of the point from the wire. The method we have discussed here is a standard procedure for the determination of the static magnetic field due to current distributions possessing certain symmetries. We shall include some cases in the problems for the interested reader. ∎

If the wire of Example 2.9 is finitely long, say, extending from $-d$ to $+d$ on the z axis, then the construction of Fig. 2.15 illustrates that for some surfaces the wire pierces through the surface whereas for some other surfaces it does not. Thus for this case there is no unique value of the wire current alone that is enclosed by C. Hence there must be a displacement current

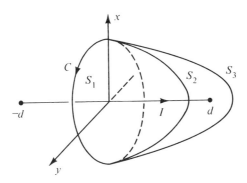

Figure 2.15. For illustrating that the wire current enclosed by a closed path is not unique for a finitely long wire.

through the surfaces in addition to the wire current so that the total current enclosed by C is uniquely given. In fact, this displacement current is provided by the time-varying electric field due to charges accumulating at one end and depleting at the other end of the current-carrying wire. Thus considering, for example, the surfaces S_1 and S_3 and setting the total currents through S_1 and S_3 to be equal, we have

$$\int_{S_1} \mathbf{J} \cdot d\mathbf{S} + \frac{d}{dt} \int_{S_1} \mathbf{D} \cdot d\mathbf{S} = \int_{S_3} \mathbf{J} \cdot d\mathbf{S} + \frac{d}{dt} \int_{S_3} \mathbf{D} \cdot d\mathbf{S} \qquad (2.19)$$

Now, since the wire pierces through S_1 in the right-hand sense,

$$\int_{S_1} \mathbf{J} \cdot d\mathbf{S} = I \qquad (2.20)$$

The wire does not pierce through S_3. Hence

$$\int_{S_3} \mathbf{J} \cdot d\mathbf{S} = 0 \qquad (2.21)$$

Substituting (2.20) and (2.21) into (2.19), we get

$$I + \frac{d}{dt} \int_{S_1} \mathbf{D} \cdot d\mathbf{S} = 0 + \frac{d}{dt} \int_{S_3} \mathbf{D} \cdot d\mathbf{S} \qquad (2.22)$$

or

$$\frac{d}{dt} \int_{S_3} \mathbf{D} \cdot d\mathbf{S} - \frac{d}{dt} \int_{S_1} \mathbf{D} \cdot d\mathbf{S} = I \qquad (2.23)$$

Reversing the sense of evaluation of the surface integral of $\mathbf{D}$ over S_1 and changing the minus sign to a plus sign, we obtain

$$\frac{d}{dt} \oint_{S_3+S_1} \mathbf{D} \cdot d\mathbf{S} = I \tag{2.24}$$

Thus the displacement current emanating from the closed surface $S_1 + S_3$ is equal to I.

Another example in which the wire current enclosed by C is not uniquely defined is shown in Fig. 2.16 which is that of a simple circuit consisting of a capacitor driven by an alternating voltage source. Considering two surfaces S_1 and S_2 where S_1 cuts through the wire and S_2 passes between the plates of the capacitor, we have

$$\int_{S_1} \mathbf{J} \cdot d\mathbf{S} = I \tag{2.25}$$

and

$$\int_{S_2} \mathbf{J} \cdot d\mathbf{S} = 0 \tag{2.26}$$

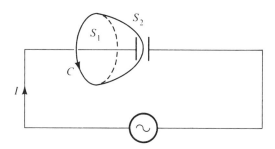

Figure 2.16. A capacitor circuit illustrating that the wire current enclosed by a closed path is not unique.

If we neglect fringing and assume that the electric field in the capacitor is contained entirely within the region between the plates, then

$$\int_{S_1} \mathbf{D} \cdot d\mathbf{S} = 0 \tag{2.27}$$

For $\oint_C \mathbf{H} \cdot d\mathbf{l}$ to be unique,

$$\int_{S_1} \mathbf{J} \cdot d\mathbf{S} + \frac{d}{dt} \int_{S_1} \mathbf{D} \cdot d\mathbf{S} = \int_{S_2} \mathbf{J} \cdot d\mathbf{S} + \frac{d}{dt} \int_{S_2} \mathbf{D} \cdot d\mathbf{S} \tag{2.28}$$

Substituting (2.25), (2.26), and (2.27) into (2.28), we obtain

$$\frac{d}{dt} \int_{S_2} \mathbf{D} \cdot d\mathbf{S} = I \tag{2.29}$$

Thus the displacement current, that is, the time rate of change of the displacement flux between the capacitor plates, is equal to the wire current.

Example 2.10. A time-varying electric field is given by

$$\mathbf{E} = E_0 z \sin \omega t \, \mathbf{i}_x$$

where E_0 is a constant. It is desired to find the induced mmf around a rectangular loop in the yz plane, as shown in Fig. 2.17.

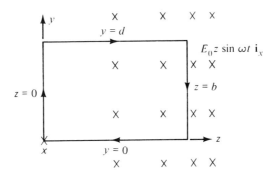

Figure 2.17. A rectangular loop in a time-varying electric field.

The total current here is composed entirely of displacement current. The displacement flux enclosed by the loop and directed into the paper is given by

$$\int_S \mathbf{D} \cdot d\mathbf{S} = \int_{z=0}^{b} \int_{y=0}^{d} \epsilon_0 E_0 z \sin \omega t \, \mathbf{i}_x \cdot dy \, dz \, \mathbf{i}_x$$

$$= \epsilon_0 E_0 \sin \omega t \int_{z=0}^{b} \int_{y=0}^{d} z \, dy \, dz$$

$$= \epsilon_0 \frac{b^2 d}{2} E_0 \sin \omega t$$

The induced mmf around C is then given by

$$\oint_C \mathbf{H} \cdot d\mathbf{l} = \frac{d}{dt} \int_S \mathbf{D} \cdot d\mathbf{S}$$

$$= \frac{d}{dt} \left(\epsilon_0 \frac{b^2 d}{2} E_0 \sin \omega t \right)$$

$$= \epsilon_0 \frac{b^2 d}{2} E_0 \omega \cos \omega t$$

2.5 GAUSS' LAW FOR THE MAGNETIC FIELD

In the previous two sections we learned two of the four Maxwell's equations. These two equations have to do with the line integrals of the electric and magnetic fields around closed paths. The remaining two Maxwell's equations are pertinent to the surface integrals of the electric and magnetic fields over closed surfaces. These are known as Gauss' laws. The Gauss' law for the magnetic field states that "the total magnetic flux emanating from a closed surface S is equal to zero." In mathematical form, this is given by

$$\oint_S \mathbf{B} \cdot d\mathbf{S} = 0 \tag{2.30}$$

Equation (2.30) is not independent of Faraday's law. This can be shown by considering a closed path C and two surfaces S_1 and S_2, both of which are bounded by C as shown in Fig. 2.18. Applying Faraday's law to C and S_1, we have

$$\oint_C \mathbf{E} \cdot d\mathbf{l} = -\frac{d}{dt} \int_{S_1} \mathbf{B} \cdot d\mathbf{S}_1 \tag{2.31}$$

where $d\mathbf{S}_1$ is directed out of the volume bounded by the closed surface $S_1 + S_2$. Applying Faraday's law to C and S_2, we have

$$\oint_C \mathbf{E} \cdot d\mathbf{l} = \frac{d}{dt} \int_{S_2} \mathbf{B} \cdot d\mathbf{S}_2 \tag{2.32}$$

where $d\mathbf{S}_2$ is directed out of the volume bounded by $S_1 + S_2$. Combining (2.31) and (2.32), we obtain

$$-\frac{d}{dt} \int_{S_1} \mathbf{B} \cdot d\mathbf{S}_1 = \frac{d}{dt} \int_{S_2} \mathbf{B} \cdot d\mathbf{S}_2 \tag{2.33}$$

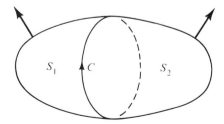

Figure 2.18. A closed path C, and two surfaces S_1 and S_2 bounded by C.

or

$$\frac{d}{dt} \oint_{S_1+S_2} \mathbf{B} \cdot d\mathbf{S} = 0 \qquad (2.34)$$

or

$$\oint_{S_1+S_2} \mathbf{B} \cdot d\mathbf{S} = \text{constant with time} \qquad (2.35)$$

Since there is no experimental evidence that the right side of (2.35) is nonzero, it follows that

$$\oint_{S} \mathbf{B} \cdot d\mathbf{S} = 0$$

where we have replaced $S_1 + S_2$ by S.

In physical terms, (2.30) signifies that magnetic charges do not exist and magnetic flux lines are closed. Whatever magnetic flux enters (or leaves) a certain part of a closed surface must leave (or enter) through the remainder of the closed surface.

2.6 GAUSS' LAW FOR THE ELECTRIC FIELD

Gauss' law for the electric field states that "the total displacement flux emanating from a closed surface S is equal to the total charge contained within the volume V bounded by that surface." This statement, although familiarly known as Gauss' law, has its origin in experiments conducted by Faraday. In mathematical form, Gauss' law for the electric field is given by

$$\oint_{S} \mathbf{D} \cdot d\mathbf{S} = \int_{V} \rho \, dv \qquad (2.36)$$

where ρ is the volume charge density associated with points in the volume V. The volume charge density at a point is defined as the charge per unit volume (C/m^3) at that point in the limit that the volume shrinks to zero. Thus

$$\rho = \lim_{\Delta v \to 0} \frac{\Delta Q}{\Delta v} \qquad (2.37)$$

As an illustration of the computation of the charge contained in a given volume for a specified charge density, let us consider

$$\rho = x + y + z \ C/m^3$$

and the cubical volume V bounded by the planes $x = 0$, $x = 1$, $y = 0$, $y = 1$, $z = 0$, and $z = 1$. Then the charge Q contained within the cubical volume is given by

$$Q = \int_V \rho \, dv = \int_{x=0}^{1} \int_{y=0}^{1} \int_{z=0}^{1} (x + y + z) \, dx \, dy \, dz$$

$$= \int_{x=0}^{1} \int_{y=0}^{1} \left[xz + yz + \frac{z^2}{2} \right]_{z=0}^{1} dx \, dy$$

$$= \int_{x=0}^{1} \int_{y=0}^{1} \left(x + y + \frac{1}{2} \right) dx \, dy$$

$$= \int_{x=0}^{1} \left[xy + \frac{y^2}{2} + \frac{y}{2} \right]_{y=0}^{1} dx$$

$$= \int_{x=0}^{1} (x + 1) \, dx$$

$$= \left[\frac{x^2}{2} + x \right]_{x=0}^{1}$$

$$= \frac{3}{2} \, C$$

Although the quantity on the right side of (2.36), that is, the charge contained within the volume V bounded by the surface S associated with the quantity on the left side of (2.36) is formulated in terms of the volume charge density, it includes surface charges, line charges, and point charges enclosed by S. Thus it represents the algebraic sum of all the charges contained in the volume V. Let us now consider an example.

Example 2.11. A point charge Q is situated at the origin. It is desired to find $\oint_S \mathbf{D} \cdot d\mathbf{S}$ and $\mathbf{D}$ over the surface of a sphere of radius a centered at the origin.

According to Gauss' law for the electric field, the required displacement flux is given by

$$\oint_S \mathbf{D} \cdot d\mathbf{S} = Q \tag{2.38}$$

To evaluate $\mathbf{D}$ on the surface of the sphere, we note that in order for $\oint_S \mathbf{D} \cdot d\mathbf{S}$ to be nonzero, $\mathbf{D}$ must be directed normal to the spherical surface. From symmetry considerations, it must have the same magnitude at all points on the spherical surface since the surface is centered at the origin. Thus let us divide the spherical surface into a large number of infinitesimal areas, as shown in Fig. 2.19. Since $\mathbf{D}$ is normal to each area, $\mathbf{D} \cdot d\mathbf{S}$ for each area is simply equal to $D \, dS$. Hence

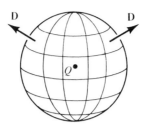

Figure 2.19. For evaluating the displacement flux density over the surface of a sphere centered at a point charge.

$$\oint_S \mathbf{D} \cdot d\mathbf{S} = D \int_S dS$$
$$= D \text{ (surface area of the sphere)}$$
$$= 4\pi a^2 D$$

From (2.38), we then have

$$4\pi a^2 D = Q$$

or

$$D = \frac{Q}{4\pi a^2}$$

Thus the displacement flux density due to the point charge is directed away from the charge and has a magnitude $Q/4\pi a^2$ where a is the distance of the point from the charge. The method we have discussed here is a standard procedure for the determination of the static electric field due to charge distributions possessing certain symmetries. We shall include some cases in the problems for the interested reader.   ■

Gauss' law for the electric field is not independent of Ampere's circuital law if we recognize that, in view of conservation of electric charge, "the total current due to flow of charges emanating from a closed surface S is equal to the time rate of decrease of the charge within the volume V bounded by S," that is,

$$\oint_S \mathbf{J} \cdot d\mathbf{S} = -\frac{d}{dt} \int_V \rho \, dv$$

or

$$\oint_S \mathbf{J} \cdot d\mathbf{S} + \frac{d}{dt} \int_V \rho \, dv = 0 \qquad (2.39)$$

This statement is known as the "law of conservation of charge." In fact, it is this consideration that led to the mathematical contribution of Maxwell to Ampere's circuital law. Ampere's circuital law in its original form did not include the displacement current term which resulted in an inconsistency with (2.39) for time-varying fields.

Returning to the discussion of the dependency of Gauss' law on Ampere's circuital law through (2.39), let us consider the geometry of Fig. 2.18. Applying Ampere's circuital law to C and S_1 and to C and S_2, we get

$$\oint_C \mathbf{H} \cdot d\mathbf{l} = \int_{S_1} \mathbf{J} \cdot d\mathbf{S}_1 + \frac{d}{dt} \int_{S_1} \mathbf{D} \cdot d\mathbf{S}_1 \qquad (2.40a)$$

and

$$\oint_C \mathbf{H} \cdot d\mathbf{l} = -\int_{S_2} \mathbf{J} \cdot d\mathbf{S}_2 - \frac{d}{dt} \int_{S_2} \mathbf{D} \cdot d\mathbf{S}_2 \qquad (2.40b)$$

respectively. Combining (2.40a) and (2.40b), we obtain

$$\oint_{S_1+S_2} \mathbf{J} \cdot d\mathbf{S} + \frac{d}{dt} \oint_{S_1+S_2} \mathbf{D} \cdot d\mathbf{S} = 0 \qquad (2.41)$$

Now, using (2.39), we have

$$-\frac{d}{dt} \int_V \rho \, dv + \frac{d}{dt} \oint_S \mathbf{D} \cdot d\mathbf{S} = 0$$

or

$$\frac{d}{dt} \left[\oint_S \mathbf{D} \cdot d\mathbf{S} - \int_V \rho \, dv \right] = 0 \qquad (2.42)$$

where we have replaced $S_1 + S_2$ by S and where V is the volume enclosed by $S_1 + S_2$. Thus from (2.42), we get

$$\oint_S \mathbf{D} \cdot d\mathbf{S} - \int_V \rho \, dv = \text{constant with time} \qquad (2.43)$$

Since there is no experimental evidence that the right side of (2.43) is nonzero, it follows that

$$\oint_S \mathbf{D} \cdot d\mathbf{S} = \int_V \rho \, dv$$

thereby giving Gauss' law for the electric field.

2.7   SUMMARY

We first learned in this chapter how to evaluate line and surface integrals of vector quantities and then we introduced Maxwell's equations in integral form. These equations, which form the basis of electromagnetic field theory, are given as follows in words and in mathematical form and are illustrated in Figs. 2.20 through 2.23.

Faraday's Law: The electromotive force around a closed path C is equal to the negative of the time rate of change of the magnetic flux enclosed by that path, that is,

$$\oint_C \mathbf{E} \cdot d\mathbf{l} = -\frac{d}{dt} \int_S \mathbf{B} \cdot d\mathbf{S} \qquad (2.44)$$

Ampere's Circuital Law: The magnetomotive force around a closed path C is equal to the sum of the current enclosed by that path due to the actual flow of charges and the displacement current due to the time rate of change of the displacement flux enclosed by that path, that is,

$$\oint_C \mathbf{H} \cdot d\mathbf{l} = \int_S \mathbf{J} \cdot d\mathbf{S} + \frac{d}{dt} \int_S \mathbf{D} \cdot d\mathbf{S} \qquad (2.45)$$

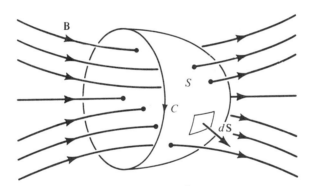

Figure 2.20. For illustrating Faraday's law.

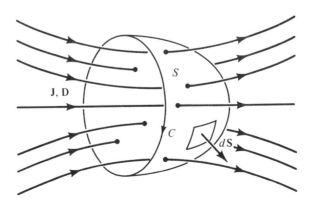

Figure 2.21. For illustrating Ampere's circuital law.

GAUSS' LAW FOR THE MAGNETIC FIELD: The magnetic flux emanating from a closed surface S is equal to zero, that is,

$$\oint_S \mathbf{B} \cdot d\mathbf{S} = 0 \qquad (2.46)$$

GAUSS' LAW FOR THE ELECTRIC FIELD: The displacement flux emanating from a closed surface S is equal to the charge enclosed by that surface, that is,

$$\oint_S \mathbf{D} \cdot d\mathbf{S} = \int_V \rho \, dv \qquad (2.47)$$

The vectors $\mathbf{D}$ and $\mathbf{H}$, known as the displacement flux density and the magnetic field intensity vectors, respectively, are related to $\mathbf{E}$ and $\mathbf{B}$, known as the electric field intensity and the magnetic flux density vectors, respec-

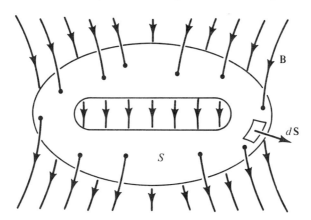

Figure 2.22. For illustrating Gauss' law for the magnetic field.

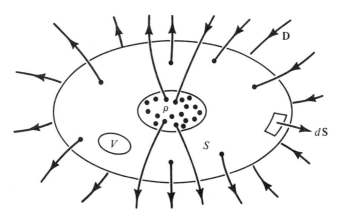

Figure 2.23. For illustrating Gauss' law for the electric field.

tively, in the manner

$$\mathbf{D} = \epsilon_0 \mathbf{E} \tag{2.48}$$

$$\mathbf{H} = \frac{\mathbf{B}}{\mu_0} \tag{2.49}$$

where ϵ_0 and μ_0 are the permittivity and the permeability of free space, respectively. In evaluating the right sides of (2.44) and (2.45), the normal vectors to the surfaces must be chosen such that they are directed in the right-hand sense, that is, toward the side of advance of a right-hand screw as it is turned around C, as shown in Figs. 2.20 and 2.21. We have also learned that (2.46) is not independent of (2.44) and that (2.47) follows from (2.45) with the aid of the law of conservation of charge given by

$$\oint_S \mathbf{J} \cdot d\mathbf{S} + \frac{d}{dt} \int_V \rho \, dv = 0 \tag{2.50}$$

In words, (2.50) states that the sum of the current due to the flow of charges across a closed surface S and the time rate of increase of the charge within the volume V bounded by S is equal to zero. In (2.46), (2.47), and (2.50) the surface integrals must be evaluated in order to find the flux outward from the volume bounded by the surface.

REVIEW QUESTIONS

2.1. How do you find the work done in moving a test charge by an infinitesimal distance in an electric field?

2.2. What is the amount of work involved in moving a test charge normal to the electric field?

2.3. What is the physical interpretation of the line integral of $\mathbf{E}$ between two points A and B?

2.4. How do you find the approximate value of the line integral of a vector along a given path?

2.5. How do you find the exact value of the line integral?

2.6. What is the physical significance of the line integral of the earth's gravitational field intensity?

2.7. What is the value of the line integral of the earth's gravitational field intensity around a closed path?

2.8. How do you find the magnetic flux crossing an infinitesimal surface?

2.9. What is the magnetic flux crossing an infinitesimal surface oriented parallel to the magnetic flux density vector?

2.10. For what orientation of the infinitesimal surface relative to the magnetic flux density vector is the magnetic flux crossing the surface a maximum?

2.11. How do you find the approximate value of the surface integral over a given surface?

2.12. How do you find the exact value of the surface integral?

2.13. Provide physical interpretations for the closed surface integrals of any two vectors of your choice.

2.14. State Faraday's law.

2.15. Why is it necessary to have the minus sign associated with the time rate of increase of magnetic flux on the right side of Faraday's law?

2.16. What is electromotive force?

2.17. What are the different ways in which an emf is induced around a loop?

2.18. To find the induced emf around a planar loop, is it necessary to consider the magnetic flux crossing the plane surface bounded by the loop?

2.19. Discuss briefly the motional emf concept.

2.20. What is Lenz's law?

2.21. How would you orient a loop antenna in order to obtain maximum signal from an incident electromagnetic wave which has its magnetic field linearly polarized in the north–south direction?

2.22. State three applications of Faraday's law.

2.23. State Ampere's circuital law.

2.24. What are the units of the magnetic field intensity vector?

2.25. What are the units of the displacement flux density vector?

2.26. What is displacement current? Give an example involving displacement current.

2.27. Why is it necessary to have the displacement current term on the right side of Ampere's circuital law?

2.28. When can you say that the current in a wire enclosed by a closed path is uniquely defined? Give two examples.

2.29. Give an example in which the current in a wire enclosed by a closed path is not uniquely defined.

2.30. Is it meaningful to consider two different surfaces bounded by a closed path to compute the two different currents on the right side of Ampere's circuital law to find $\oint \mathbf{H} \cdot d\mathbf{l}$ around the closed path?

2.31. Discuss briefly the application of Ampere's circuital law to determine the magnetic field due to current distributions.

2.32. State Gauss' law for the magnetic field. How is it derived from Faraday's law?

2.33. What is the physical interpretation of Gauss' law for the magnetic field?

2.34. State Gauss' law for the electric field.

2.35. How is volume charge density defined?

2.36. State the law of conservation of charge.

2.37. How is Gauss' law for the electric field derived from Ampere's circuital law?

2.38. Discuss briefly the application of Gauss' law for the electric field to determine the electric field due to charge distributions.

2.39. Summarize Maxwell's equations in integral form.

2.40. Which two of the Maxwell's equations are independent?

PROBLEMS

2.1. For the force field $\mathbf{F} = x^2 \mathbf{i}_y$, find the approximate value of the line integral of $\mathbf{F}$ from the origin to the point $(1, 3, 0)$ along a straight line path by dividing the path into ten equal segments.

2.2. For the force field $\mathbf{F} = x^2 \mathbf{i}_y$, obtain a series expression for the line integral of $\mathbf{F}$ from the origin to the point $(1, 3, 0)$ along a straight line path by dividing the path into n equal segments. Express the sum of the series in closed form and compute its value for values of n equal to 5, 10, 100, and ∞.

2.3. For the force field $\mathbf{F} = x^2 \mathbf{i}_y$, find the exact value of the line integral of $\mathbf{F}$ from the origin to the point $(1, 3, 0)$ along a straight line path.

2.4. Given $\mathbf{E} = y\mathbf{i}_x + x\mathbf{i}_y$, find $\int_{(0,0,0)}^{(1,1,0)} \mathbf{E} \cdot d\mathbf{l}$ along the following paths: (a) straight line path $y = x, z = 0$, (b) straight line path from $(0, 0, 0)$ to $(1, 0, 0)$ and then straight line path from $(1, 0, 0)$ to $(1, 1, 0)$, and (c) any path of your choice.

2.5. Show that for any closed path C, $\oint_C d\mathbf{l} = 0$ and hence show that for a uniform field $\mathbf{F}$, $\oint_C \mathbf{F} \cdot d\mathbf{l} = 0$.

2.6. Given $\mathbf{F} = y\mathbf{i}_x - x\mathbf{i}_y$, find $\oint_C \mathbf{F} \cdot d\mathbf{l}$ where C is the closed path in the xy plane consisting of the following: the straight line path from $(0, 0, 0)$ to $(-1, 1, 0)$, the straight line path from $(-1, 1, 0)$ to $(0, \sqrt{2}, 0)$, the straight line path from $(0, \sqrt{2}, 0)$ to $(0, 1, 0)$, the circular path from $(0, 1, 0)$ to $(1, 0, 0)$ having its center at $(0, 0, 0)$, and the straight line path from $(1, 0, 0)$ to $(0, 0, 0)$.

2.7. Given $\mathbf{F} = xy\mathbf{i}_x + yz\mathbf{i}_y + zx\mathbf{i}_z$, find $\oint_C \mathbf{F} \cdot d\mathbf{l}$ where C is the closed path comprising the straight lines from $(0, 0, 0)$ to $(1, 1, 1)$, from $(1, 1, 1)$ to $(1, 1, 0)$, and from $(1, 1, 0)$ to $(0, 0, 0)$.

2.8. For the magnetic flux density vector $\mathbf{B} = x^2 e^{-y} \mathbf{i}_z$ Wb/m^2, find the approximate value of the magnetic flux crossing the portion of the xy plane lying between $x = 0$, $x = 1$, $y = 0$, and $y = 1$, by dividing the area into 100 equal parts.

2.9. For the magnetic flux density vector $\mathbf{B} = x^2 e^{-y}\mathbf{i}_z$ Wb/m^2, obtain a series expression for the magnetic flux crossing the portion of the xy plane lying between $x = 0$, $x = 1$, $y = 0$, and $y = 1$, by dividing the area into n^2 equal parts. Express the sum of the series in closed form and compute its value for values of n equal to 5, 10, 100, and ∞.

2.10. For the magnetic flux density vector $\mathbf{B} = x^2 e^{-y}\mathbf{i}_z$ Wb/m^2, find the exact value of the magnetic flux crossing the portion of the xy plane lying between $x = 0$, $x = 1$, $y = 0$, and $y = 1$, by evaluating the surface integral of $\mathbf{B}$.

2.11. Given $\mathbf{A} = x\mathbf{i}_x + y\mathbf{i}_y + z\mathbf{i}_z$, find $\int_S \mathbf{A} \cdot d\mathbf{S}$ where S is the hemispherical surface of radius 2 m lying above the xy plane and having its center at the origin.

2.12. Show that for any closed surface S, $\oint_S d\mathbf{S} = 0$ and hence show that for a uniform field $\mathbf{A}$, $\oint_S \mathbf{A} \cdot d\mathbf{S} = 0$.

2.13. Given $\mathbf{J} = 3x\mathbf{i}_x + (y - 3)\mathbf{i}_y + (2 + z)\mathbf{i}_z$ amp/m^2, find $\oint_S \mathbf{J} \cdot d\mathbf{S}$, that is, the current flowing out of the surface S of the rectangular box bounded by the planes $x = 0$, $x = 1$, $y = 0$, $y = 2$, $z = 0$, and $z = 3$.

2.14. Given $\mathbf{E} = (y\mathbf{i}_x - x\mathbf{i}_y) \cos \omega t$ V/m, find the time rate of decrease of the magnetic flux crossing toward the positive z side and enclosed by the path in the xy plane from $(0, 0, 0)$ to $(1, 0, 0)$ along $y = 0$, from $(1, 0, 0)$ to $(1, 1, 0)$ along $x = 1$, and from $(1, 1, 0)$ to $(0, 0, 0)$ along $y = x^3$.

2.15. A magnetic field is given in the xz plane by $\mathbf{B} = \dfrac{B_0}{x}\mathbf{i}_y$ Wb/m^2, where B_0 is a constant. A rigid rectangular loop is situated in the xz plane and with its corners at the points (x_0, z_0), $(x_0, z_0 + b)$, $(x_0 + a, z_0 + b)$, and $(x_0 + a, z_0)$. If the loop is moving in that plane with a velocity $\mathbf{v} = v_0\mathbf{i}_x$ m/s, where v_0 is a constant, find by using Faraday's law the induced emf around the loop in the sense defined by connecting the above specified points in succession. Discuss your result by using the motional emf concept.

2.16. Assuming the rectangular loop of Problem 2.15 to be stationary, find the induced emf around the loop if $\mathbf{B} = \dfrac{B_0}{x} \cos \omega t\, \mathbf{i}_y$ Wb/m^2.

2.17. Assuming the rectangular loop of Problem 2.15 to be moving with the velocity $\mathbf{v} = v_0\mathbf{i}_x$ m/s, find the induced emf around the loop if $\mathbf{B} = \dfrac{B_0}{x} \cos \omega t\, \mathbf{i}_y$ Wb/m^2.

2.18. For $\mathbf{B} = B_0 \cos \omega t\, \mathbf{i}_z$ Wb/m^2, find the induced emf around the closed path comprising the straight lines successively connecting the points $(0, 0, 0)$, $(1, 0, 0.01)$, $(1, 1, 0.02)$, $(0, 1, 0.03)$, $(0, 0, 0.04)$, and $(0, 0, 0)$.

2.19. Repeat Problem 2.18 for the closed path comprising the straight lines successively connecting the points $(0, 0, 0)$, $(1, 0, 0.01)$, $(1, 1, 0.02)$, $(0, 1, 0.03)$, $(0, 0, 0.04)$, $(1, 0, 0.05)$, $(1, 1, 0.06)$, $(0, 1, 0.07)$, $(0, 0, 0.08)$, and $(0, 0, 0)$, with a slight kink in the last straight line at the point $(0, 0, 0.04)$ to avoid touching the point.

2.20. A rigid rectangular loop of area A is situated normal to the xy plane and symmetrically about the z axis. It revolves around the z axis at ω_1 rad/s in the sense defined by the curling of the fingers of the right hand when the z axis is grabbed with the thumb pointed in the positive z direction. Find the induced emf around the loop if $\mathbf{B} = B_0 \cos \omega_2 t \, \mathbf{i}_x$ where B_0 is a constant, and show that the induced emf has two frequency components $(\omega_1 + \omega_2)$ and $|\omega_1 - \omega_2|$.

2.21. For the revolving loop of Problem 2.20, find the induced emf around the loop if $\mathbf{B} = B_0(\cos \omega_1 t \, \mathbf{i}_x + \sin \omega_1 t \, \mathbf{i}_y)$.

2.22. For the revolving loop of Problem 2.20, find the induced emf around the loop if $\mathbf{B} = B_0(\cos \omega_1 t \, \mathbf{i}_x - \sin \omega_1 t \, \mathbf{i}_y)$.

2.23. A current I_1 amp flows from infinity to a point charge at the origin through a thin wire along the negative y axis and a current I_2 amp flows from the point charge to infinity through another thin wire along the positive y axis. From considerations of uniqueness of $\oint_C \mathbf{H} \cdot d\mathbf{l}$, find the displacement current emanating from (a) a spherical surface of radius 1 m and having its center at the point (2, 2, 2) and (b) a spherical surface of radius 1 m and having its center at the origin.

2.24. A current density due to flow of charges is given by $\mathbf{J} = y \cos \omega t \, \mathbf{i}_y$ amp/m^2. From consideration of uniqueness of $\oint_C \mathbf{H} \cdot d\mathbf{l}$, find the displacement current emanating from the cubical box bounded by the planes $x = 0$, $x = 1$, $y = 0$, $y = 1$, $z = 0$, and $z = 1$.

2.25. An infinitely long, cylindrical wire of radius a, having the z axis as its axis, carries current in the positive z direction with uniform density J_0 amp/m^2. Find $\mathbf{H}$ both inside and outside the wire.

2.26. An infinitely long, hollow, cylindrical wire of inner radius a and outer radius b, having the z axis as its axis, carries current in the positive z direction with uniform density J_0 amp/m^2. Find $\mathbf{H}$ everywhere.

2.27. An infinitely long, straight wire situated along the z axis carries current I amp in the positive z direction. What are the values of $\int_{(1,0,0)}^{(0,1,0)} \mathbf{H} \cdot d\mathbf{l}$ along (a) the circular path of radius 1 m and centered at the origin and (b) along a straight line path?

2.28. Using the property that $\oint_S \mathbf{B} \cdot d\mathbf{S} = 0$, find $\int \mathbf{B} \cdot d\mathbf{S}$ over that portion of the surface $y = \sin x$ bounded by $x = 0$, $x = \pi$, $z = 0$, and $z = 1$, for $\mathbf{B} = y\mathbf{i}_x - x\mathbf{i}_y$.

2.29. Given $\mathbf{D} = y\mathbf{i}_y$, find the charge contained in the volume of the wedge-shaped box defined by the planes $x = 0$, $x + z = 1$, $y = 0$, $y = 1$, and $z = 0$.

2.30. Given $\rho = xe^{-x^2}$ C/m^3, find the displacement flux emanating from the surface of the cubical box defined by the planes $x = 0$, $x = 1$, $y = 0$, $y = 1$, $z = 0$, and $z = 1$.

2.31. Charge is distributed uniformly along the z axis with density ρ_{L0} C/m. Using Gauss' law for the electric field, find the electric field intensity due to the line charge.

2.32. Charge is distributed uniformly with density ρ_0 C/m³ within a spherical volume of radius a m and having its center at the origin. Using Gauss' law for the electric field, find the electric field intensity both inside and outside the charge distribution.

2.33. A point charge Q C is situated at the origin. What are the values of the displacement flux crossing (a) the spherical surface $x^2 + y^2 + z^2 = 1$, $x > 0$, $y > 0$, and $z > 0$ and (b) the plane surface $x + y + z = 1$, $x > 0$, $y > 0$, and $z > 0$?

2.34. Given $\mathbf{J} = x\mathbf{i}_x$ amp/m², find the time rate of increase of the charge contained in the cubical volume bounded by the planes $x = 0$, $x = 1$, $y = 0$, $y = 1$, $z = 0$, and $z = 1$.

2.35. Given $\mathbf{J} = x\mathbf{i}_x$ amp/m², find the time rate of increase of the charge contained in the volume of the wedge-shaped box that is defined by the planes $x = 0$, $x + z = 1$, $y = 0$, $y = 1$, and $z = 0$.

3. MAXWELL'S EQUATIONS IN DIFFERENTIAL FORM

In Chap. 2 we introduced Maxwell's equations in integral form. We learned that the quantities involved in the formulation of these equations are the scalar quantities, electromotive force, magnetomotive force, magnetic flux, displacement flux, charge, and current, which are related to the field vectors and source densities through line, surface, and volume integrals. Thus the integral forms of Maxwell's equations, while containing all the information pertinent to the interdependence of the field and source quantities over a given region in space, do not permit us to study directly the interaction between the field vectors and their relationships with the source densities at individual points. It is our goal in this chapter to derive the differential forms of Maxwell's equations that apply directly to the field vectors and source densities at a given point.

We shall derive Maxwell's equations in differential form by applying Maxwell's equations in integral form to infinitesimal closed paths, surfaces, and volumes, in the limit that they shrink to points. We will find that the differential equations relate the spatial variations of the field vectors at a given point to their temporal variations and to the charge and current densities at that point. In this process we shall also learn two important operations in vector calculus, known as curl and divergence, and two related theorems, known as Stokes' and divergence theorems.

3.1 FARADAY'S LAW

We recall from the previous chapter that Faraday's law is given in integral form by

$$\oint_C \mathbf{E} \cdot d\mathbf{l} = -\frac{d}{dt} \int_S \mathbf{B} \cdot d\mathbf{S} \tag{3.1}$$

where S is any surface bounded by the closed path C. In the most general case, the electric and magnetic fields have all three components (x, y, and z) and are dependent on all three coordinates (x, y, and z) in addition to time (t). For simplicity, we shall, however, first consider the case in which the electric field has an x component only, which is dependent only on the z coordinate, in addition to time. Thus

$$\mathbf{E} = E_x(z, t)\mathbf{i}_x \tag{3.2}$$

In other words, this simple form of time-varying electric field is everywhere directed in the x direction and it is uniform in planes parallel to the xy plane.

Let us now consider a rectangular path C of infinitesimal size lying in a plane parallel to the xz plane and defined by the points (x, z), $(x, z + \Delta z)$, $(x + \Delta x, z + \Delta z)$, and $(x + \Delta x, z)$ as shown in Fig. 3.1. According to

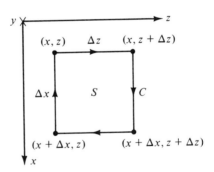

Figure 3.1. Infinitesimal rectangular path lying in a plane parallel to the xz plane.

Faraday's law, the emf around the closed path C is equal to the negative of the time rate of change of the magnetic flux enclosed by C. The emf is given by the line integral of $\mathbf{E}$ around C. Thus evaluating the line integrals of $\mathbf{E}$ along the four sides of the rectangular path, we obtain

$$\int_{(x,z)}^{(x,z+\Delta z)} \mathbf{E} \cdot d\mathbf{l} = 0 \qquad \text{since } E_z = 0 \tag{3.3a}$$

$$\int_{(x,z+\Delta z)}^{(x+\Delta x, z+\Delta z)} \mathbf{E} \cdot d\mathbf{l} = [E_x]_{z+\Delta z} \, \Delta x \qquad (3.3b)$$

$$\int_{(x+\Delta x, z+\Delta z)}^{(x+\Delta x, z)} \mathbf{E} \cdot d\mathbf{l} = 0 \qquad \text{since } E_z = 0 \qquad (3.3c)$$

$$\int_{(x+\Delta x, z)}^{(x, z)} \mathbf{E} \cdot d\mathbf{l} = -[E_x]_z \, \Delta x \qquad (3.3d)$$

Adding up (3.3a)–(3.3d), we obtain

$$\oint_C \mathbf{E} \cdot d\mathbf{l} = [E_x]_{z+\Delta z} \, \Delta x - [E_x]_z \, \Delta x$$
$$= \{[E_x]_{z+\Delta z} - [E_x]_z\} \, \Delta x \qquad (3.4)$$

In (3.3a)–(3.3d) and (3.4), $[E_x]_z$ and $[E_x]_{z+\Delta z}$ denote values of E_x evaluated along the sides of the path for which $z = z$ and $z = z + \Delta z$, respectively.

To find the magnetic flux enclosed by C, let us consider the plane surface S bounded by C. According to the right-hand screw rule, we must use the magnetic flux crossing S toward the positive y direction, that is, into the page, since the path C is traversed in the clockwise sense. The only component of $\mathbf{B}$ normal to the area S is the y component. Also since the area is infinitesimal in size, we can assume B_y to be uniform over the area and equal to its value at (x, z). The required magnetic flux is then given by

$$\int_S \mathbf{B} \cdot d\mathbf{S} = [B_y]_{(x,z)} \, \Delta x \, \Delta z \qquad (3.5)$$

Substituting (3.4) and (3.5) into (3.1) to apply Faraday's law to the rectangular path C under consideration, we get

$$\{[E_x]_{z+\Delta z} - [E_x]_z\} \, \Delta x = -\frac{d}{dt}\{[B_y]_{(x,z)} \, \Delta x \, \Delta z\}$$

or

$$\frac{[E_x]_{z+\Delta z} - [E_x]_z}{\Delta z} = -\frac{\partial [B_y]_{(x,z)}}{\partial t} \qquad (3.6)$$

If we now let the rectangular path shrink to the point (x, z) by letting Δx and Δz tend to zero, we obtain

$$\operatorname*{Lim}_{\substack{\Delta x \to 0 \\ \Delta z \to 0}} \frac{[E_x]_{z+\Delta z} - [E_x]_z}{\Delta z} = -\operatorname*{Lim}_{\substack{\Delta x \to 0 \\ \Delta z \to 0}} \frac{\partial [B_y]_{(x,z)}}{\partial t}$$

or

$$\frac{\partial E_x}{\partial z} = -\frac{\partial B_y}{\partial t} \qquad (3.7)$$

Equation (3.7) is Faraday's law in differential form for the simple case of $\mathbf{E}$ given by (3.2). It relates the variation of E_x with z (space) at a point to the variation of B_y with t (time) at that point. Since the above derivation can be carried out for any arbitrary point (x, y, z), it is valid for all points. It tells us in particular that a time-varying B_y at a point results in an E_x at that point having a differential in the z direction. This is to be expected since if this is not the case, $\oint$ **E** · d**l** around the infinitesimal rectangular path would be zero.

Example 3.1.   Given **B** $= B_0 \cos \omega t$ **i**$_y$ and it is known that **E** has an x component only, let us find E_x.

From (3.6), we have

$$\frac{\partial E_x}{\partial z} = -\frac{\partial B_y}{\partial t} = -\frac{\partial}{\partial t}(B_0 \cos \omega t) = \omega B_0 \sin \omega t$$

$$E_x = \omega B_0 z \sin \omega t$$

We note that the uniform magnetic field gives rise to an electric field varying linearly with z.

Proceeding further, we can verify this result by evaluating $\oint$ **E** · d**l** around the rectangular path of Example 2.8. This rectangular path is reproduced in Fig. 3.2. The required line integral is given by

$$\oint_C \mathbf{E} \cdot d\mathbf{l} = \int_{z=0}^{b} [E_z]_{x=0}\, dz + \int_{x=0}^{a} [E_x]_{z=b}\, dx$$

$$+ \int_{z=b}^{0} [E_z]_{x=a}\, dz + \int_{x=a}^{0} [E_x]_{z=0}\, dx$$

$$= 0 + [\omega B_0\, b \sin \omega t]a + 0 + 0$$

$$= abB_0\omega \sin \omega t$$

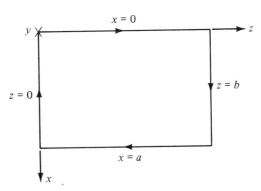

Figure 3.2. Rectangular path of Example 2.8.

which agrees with the result of Example 2.8.   ∎

We shall now proceed to generalize (3.7) for the arbitrary case of the electric field having all three components (x, y, and z), each of them depending on all three coordinates (x, y, and z), in addition to time (t), that is,

$$\mathbf{E} = E_x(x, y, z, t)\mathbf{i}_x + E_y(x, y, z, t)\mathbf{i}_y + E_z(x, y, z, t)\mathbf{i}_z \qquad (3.8)$$

To do this, let us consider the three infinitesimal rectangular paths in planes parallel to the three mutually orthogonal planes of the Cartesian coordinate system, as shown in Fig. 3.3. Evaluating $\oint \mathbf{E} \cdot d\mathbf{l}$ around the closed paths *abcda*, *adefa*, and *afgba*, we get

$$\oint_{abcda} \mathbf{E} \cdot d\mathbf{l} = [E_y]_{(x,z)} \, \Delta y + [E_z]_{(x,y+\Delta y)} \, \Delta z$$
$$- [E_y]_{(x,z+\Delta z)} \, \Delta y - [E_z]_{(x,y)} \, \Delta z \qquad (3.9a)$$

$$\oint_{adefa} \mathbf{E} \cdot d\mathbf{l} = [E_z]_{(x,y)} \, \Delta z + [E_x]_{(y,z+\Delta z)} \, \Delta x$$
$$- [E_z]_{(x+\Delta x,y)} \, \Delta z - [E_x]_{(y,z)} \, \Delta x \qquad (3.9b)$$

$$\oint_{afgba} \mathbf{E} \cdot d\mathbf{l} = [E_x]_{(y,z)} \, \Delta x + [E_y]_{(x+\Delta x,z)} \, \Delta y$$
$$- [E_x]_{(y+\Delta y,z)} \, \Delta x - [E_y]_{(x,z)} \, \Delta y \qquad (3.9c)$$

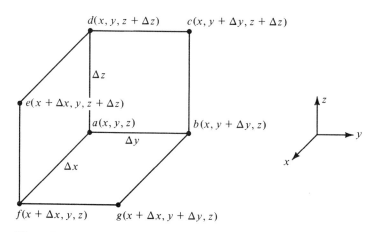

Figure 3.3. Infinitesimal rectangular paths in three mutually orthogonal planes.

In (3.9a)–(3.9c) the subscripts associated with the field components in the various terms on the right sides of the equations denote the value of the coordinates that remain constant along the sides of the closed paths corresponding to the terms. Now, evaluating $\int \mathbf{B} \cdot d\mathbf{S}$ over the surfaces *abcd*, *adef*, and *afgb*, keeping in mind the right-hand screw rule, we have

$$\int_{abcd} \mathbf{B} \cdot d\mathbf{S} = [B_x]_{(x,y,z)} \, \Delta y \, \Delta z \qquad (3.10a)$$

$$\int_{adef} \mathbf{B} \cdot d\mathbf{S} = [B_y]_{(x,y,z)} \, \Delta z \, \Delta x \tag{3.10b}$$

$$\int_{afgb} \mathbf{B} \cdot d\mathbf{S} = [B_z]_{(x,y,z)} \, \Delta x \, \Delta y \tag{3.10c}$$

Applying Faraday's law to each of the three paths by making use of (3.9a)–(3.9c) and (3.10a)–(3.10c) and simplifying, we obtain

$$\frac{[E_z]_{(x,y+\Delta y)} - [E_z]_{(x,y)}}{\Delta y} - \frac{[E_y]_{(x,z+\Delta z)} - [E_y]_{(x,z)}}{\Delta z}$$

$$= -\frac{\partial [B_x]_{(x,y,z)}}{\partial t} \tag{3.11a}$$

$$\frac{[E_x]_{(y,z+\Delta z)} - [E_x]_{(y,z)}}{\Delta z} - \frac{[E_z]_{(x+\Delta x,y)} - [E_z]_{(x,y)}}{\Delta x}$$

$$= -\frac{\partial [B_y]_{(x,y,z)}}{\partial t} \tag{3.11b}$$

$$\frac{[E_y]_{(x+\Delta x,z)} - [E_y]_{(x,z)}}{\Delta x} - \frac{[E_x]_{(y+\Delta y,z)} - [E_x]_{(y,z)}}{\Delta y}$$

$$= -\frac{\partial [B_z]_{(x,y,z)}}{\partial t} \tag{3.11c}$$

If we now let all three paths shrink to the point a by letting Δx, Δy, and Δz tend to zero, (3.11a)–(3.11c) reduce to

$$\frac{\partial E_z}{\partial y} - \frac{\partial E_y}{\partial z} = -\frac{\partial B_x}{\partial t} \tag{3-12a}$$

$$\frac{\partial E_x}{\partial z} - \frac{\partial E_z}{\partial x} = -\frac{\partial B_y}{\partial t} \tag{3.12b}$$

$$\frac{\partial E_y}{\partial x} - \frac{\partial E_x}{\partial y} = -\frac{\partial B_z}{\partial t} \tag{3.12c}$$

Equations (3.12a)–(3.12c) are the differential equations governing the relationships between the space variations of the electric field components and the time variations of the magnetic field components at a point. An examination of one of the three equations is sufficient to reveal the physical meaning of these relationships. For example, (3.12a) tells us that a time-varying B_x at a point results in an electric field at that point having y and z components such that their net right-lateral differential normal to the x direction is nonzero. The right-lateral differential of E_y normal to the x direction is its derivative in the $\mathbf{i}_y \times \mathbf{i}_x$, or $-\mathbf{i}_z$ direction, that is, $\frac{\partial E_y}{\partial(-z)}$ or $-\frac{\partial E_y}{\partial z}$. The right-lateral differential of E_z normal to the x direction is its derivative in the $\mathbf{i}_z \times \mathbf{i}_x$, or $\mathbf{i}_y$

direction, that is, $\frac{\partial E_z}{\partial y}$. Thus the net right-lateral differential of the y and z components of the electric field normal to the x direction is $\left(-\frac{\partial E_y}{\partial z}\right) + \left(\frac{\partial E_z}{\partial y}\right)$, or $\left(\frac{\partial E_z}{\partial y} - \frac{\partial E_y}{\partial z}\right)$. An example in which the net right-lateral differential is zero although the individual derivatives are nonzero is shown in Fig. 3.4(a), whereas Fig. 3.4(b) shows an example in which the net right-lateral differential is nonzero.

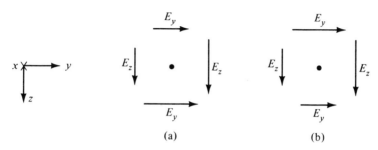

(a)   (b)

Figure 3.4. For illustrating (a) zero, and (b) nonzero net right-lateral differential of E_y and E_z normal to the x direction.

Equations (3.12a)–(3.12c) can be combined into a single vector equation as given by

$$\left(\frac{\partial E_z}{\partial y} - \frac{\partial E_y}{\partial z}\right)\mathbf{i}_x + \left(\frac{\partial E_x}{\partial z} - \frac{\partial E_z}{\partial x}\right)\mathbf{i}_y + \left(\frac{\partial E_y}{\partial x} - \frac{\partial E_x}{\partial y}\right)\mathbf{i}_z$$

$$= -\frac{\partial B_x}{\partial t}\mathbf{i}_x - \frac{\partial B_y}{\partial t}\mathbf{i}_y - \frac{\partial B_z}{\partial t}\mathbf{i}_z \qquad (3.13)$$

This can be expressed in determinant form as

$$\begin{vmatrix} \mathbf{i}_x & \mathbf{i}_y & \mathbf{i}_z \\ \dfrac{\partial}{\partial x} & \dfrac{\partial}{\partial y} & \dfrac{\partial}{\partial z} \\ E_x & E_y & E_z \end{vmatrix} = -\frac{\partial \mathbf{B}}{\partial t} \qquad (3.14)$$

or as

$$\left(\mathbf{i}_x\frac{\partial}{\partial x} + \mathbf{i}_y\frac{\partial}{\partial y} + \mathbf{i}_z\frac{\partial}{\partial z}\right) \times (E_x\mathbf{i}_x + E_y\mathbf{i}_y + E_z\mathbf{i}_z) = -\frac{\partial \mathbf{B}}{\partial t} \qquad (3.15)$$

The left side of (3.14) or (3.15) is known as the "curl of $\mathbf{E}$," denoted as $\nabla \times \mathbf{E}$ (del cross E) where ∇ (del) is the vector operator given by

$$\nabla = \mathbf{i}_x\frac{\partial}{\partial x} + \mathbf{i}_y\frac{\partial}{\partial y} + \mathbf{i}_z\frac{\partial}{\partial z} \qquad (3.16)$$

Thus we have

$$\nabla \times \mathbf{E} = -\frac{\partial \mathbf{B}}{\partial t} \qquad (3.17)$$

Equation (3.17) is Maxwell's equation in differential form corresponding to Faraday's law. We shall discuss curl further in Sec. 3.3.

Example 3.2.   Given $\mathbf{A} = y\mathbf{i}_x - x\mathbf{i}_y$, find $\nabla \times \mathbf{A}$.
From the determinant expansion for the curl of a vector, we have

$$\nabla \times \mathbf{A} = \begin{vmatrix} \mathbf{i}_x & \mathbf{i}_y & \mathbf{i}_z \\ \dfrac{\partial}{\partial x} & \dfrac{\partial}{\partial y} & \dfrac{\partial}{\partial z} \\ y & -x & 0 \end{vmatrix}$$

$$= \mathbf{i}_x \left[-\frac{\partial}{\partial z}(-x) \right] + \mathbf{i}_y \left[\frac{\partial}{\partial z}(y) \right] + \mathbf{i}_z \left[\frac{\partial}{\partial x}(-x) - \frac{\partial}{\partial y}(y) \right]$$

$$= -2\mathbf{i}_z \qquad\qquad \blacksquare$$

3.2   AMPERE'S CIRCUITAL LAW

In the previous section we derived the differential form of Faraday's law from its integral form. In this section we shall derive the differential form of Ampere's circuital law from its integral form in a completely analogous manner. We recall from Sec. 2.4 that Ampere's circuital law in integral form is given by

$$\oint_C \mathbf{H} \cdot d\mathbf{l} = \int_S \mathbf{J} \cdot d\mathbf{S} + \frac{d}{dt} \int_S \mathbf{D} \cdot d\mathbf{S} \qquad (3.18)$$

where S is any surface bounded by the closed path C. For simplicity, we shall first consider the case in which the magnetic field has a y component only, which is dependent only on the z coordinate, in addition to time. Thus

$$\mathbf{H} = H_y(z, t)\mathbf{i}_y \qquad (3.19)$$

In other words, this simple form of the time-varying magnetic field is everywhere directed in the y direction and it is uniform in planes parallel to the xy plane.

Let us now consider a rectangular path C of infinitesimal size lying in a plane parallel to the yz plane and defined by the points (y, z), $(y, z + \Delta z)$, $(y + \Delta y, z + \Delta z)$ and $(y + \Delta y, z)$ as shown in Fig. 3.5. According to Ampere's circuital law, the mmf around the closed path C is equal to the total current enclosed by C. The mmf is given by the line integral of $\mathbf{H}$ around C.

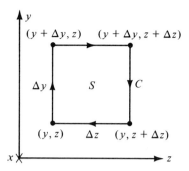

Figure 3.5. Infinitesimal rectangular path lying in a plane parallel to the yz plane.

Thus evaluating the line integrals of **H** along the four sides of the rectangular path, we obtain

$$\oint_C \mathbf{H} \cdot d\mathbf{l} = \int_{(y,z)}^{(y+\Delta y,\,z)} \mathbf{H} \cdot d\mathbf{l} + \int_{(y+\Delta y,\,z)}^{(y+\Delta y,\,z+\Delta z)} \mathbf{H} \cdot d\mathbf{l}$$

$$+ \int_{(y+\Delta y,\,z+\Delta z)}^{(y,\,z+\Delta z)} \mathbf{H} \cdot d\mathbf{l} + \int_{(y,\,z+\Delta z)}^{(y,\,z)} \mathbf{H} \cdot d\mathbf{l}$$

$$= [H_y]_z \,\Delta y + 0 - [H_y]_{z+\Delta z} \,\Delta y + 0$$

$$= -\{[H_y]_{z+\Delta z} - [H_y]_z\}\,\Delta z \qquad (3.20)$$

To find the total current enclosed by C, we consider the plane surface S bounded by C. According to the right-hand screw rule, we must find the current crossing S toward the positive x direction, that is, into the page, since the path is traversed in the clockwise sense. This current consists of two parts:

$$\int_S \mathbf{J} \cdot d\mathbf{S} = [J_x]_{(y,z)}\,\Delta y\,\Delta z \qquad (3.21a)$$

$$\frac{d}{dt}\int_S \mathbf{D} \cdot d\mathbf{S} = \frac{d}{dt}\{[D_x]_{(y,z)}\,\Delta y\,\Delta z\} = \frac{\partial[D_x]_{(y,z)}}{\partial t}\,\Delta y\,\Delta z \qquad (3.21b)$$

where we have assumed that since the area is infinitesimal in size, J_x and D_x are uniform over the area and equal to their values at (y, z).

Substituting (3.20), (3.21a), and (3.21b) into (3.18) to apply Ampere's circuital law to the rectangular path C under consideration, we get

$$-\{[H_y]_{z+\Delta z} - [H_y]_z\}\,\Delta y = \left[J_x + \frac{\partial D_x}{\partial t}\right]_{(y,z)}\Delta y\,\Delta z$$

or

$$\frac{[H_y]_{z+\Delta z} - [H_y]_z}{\Delta z} = -\left[J_x + \frac{\partial D_x}{\partial t}\right]_{(y,z)} \qquad (3.22)$$

If we now let the rectangular path shrink to the point (y, z) by letting Δy and Δz tend to zero, we obtain

$$\underset{\substack{\Delta y \to 0 \\ \Delta z \to 0}}{\mathrm{Lim}} \frac{[H_y]_{z+\Delta z} - [H_y]_z}{\Delta z} = -\underset{\substack{\Delta y \to 0 \\ \Delta z \to 0}}{\mathrm{Lim}} \left[J_x + \frac{\partial D_x}{\partial t} \right]_{(y,z)}$$

or

$$\frac{\partial H_y}{\partial z} = -J_x - \frac{\partial D_x}{\partial t} \qquad (3.23)$$

Equation (3.23) is Ampere's circuital law in differential form for the simple case of **H** given by (3.19). It relates the variation of H_y with z (space) at a point to the current density J_x and to the variation of D_x with t (time) at that point. Since the above derivation can be carried out for any arbitrary point (x, y, z), it is valid at all points. It tells us in particular that a current density J_x or a time-varying D_x or a nonzero combination of the two quantities at a point results in an H_y at that point having a differential in the z direction. This is to be expected since if this is not the case, $\oint \mathbf{H} \cdot d\mathbf{l}$ around the infinitesimal rectangular path would be zero.

Example 3.3. Given $\mathbf{E} = E_0 z \sin \omega t \, \mathbf{i}_x$ and it is known that **J** is zero and **B** has a y component only, let us find B_y.

From (3.23), we have

$$\frac{\partial H_y}{\partial z} = -J_x - \frac{\partial D_x}{\partial t} = 0 - \frac{\partial}{\partial t}(\epsilon_0 E_0 z \sin \omega t) = -\omega \epsilon_0 E_0 z \cos \omega t$$

$$H_y = -\omega \epsilon_0 E_0 \frac{z^2}{2} \cos \omega t$$

$$B_y = \mu_0 H_y = -\omega \mu_0 \epsilon_0 E_0 \frac{z^2}{2} \cos \omega t$$

We note that the electric field varying linearly with z gives rise to a magnetic field proportional to z^2. In Example 3.1, however, an electric field varying linearly with z was found to result from a uniform magnetic field, according to Faraday's law in differential form. The inconsistency of these two results implies that neither the combination of E_x and B_y in Example 3.1 nor the combination of E_x and B_y in this example simultaneously satisfies the two Maxwell's equations in differential form given by (3.7) and (3.23). The pair of E_x and B_y in Example 3.1 satisfies only (3.7), whereas the pair of E_x and B_y in this example satisfies only (3.23). In the following chapter we shall find a pair of solutions for E_x and B_y that simultaneously satisfies the two Maxwell's equations. ∎

Example 3.4. Let us consider the current distribution given by

$$\mathbf{J} = J_0 \mathbf{i}_x \qquad \text{for } -a < z < a$$

as shown in Fig. 3.6(a), where J_0 is a constant, and find the magnetic field everywhere.

Since the current density is independent of x and y, the field is also independent of x and y. Also, since the current density is not a function of time, the field is static. Hence $(\partial D_x/\partial t) = 0$, and we have

$$\frac{\partial H_y}{\partial z} = -J_x$$

Integrating both sides with respect to z, we obtain

$$H_y = -\int_{-\infty}^{z} J_x \, dz + C$$

where C is the constant of integration.

The variation of J_x with z is shown in Fig. 3.6(b). Integrating $-J_x$ with respect to z, that is, finding area under the curve of Fig. 3.6(b) as a function of

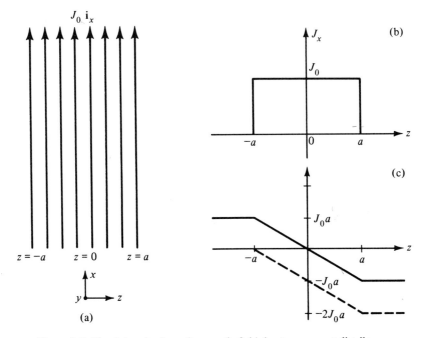

Figure 3.6. The determination of magnetic field due to a current distribution.

z, and taking its negative, we obtain the result shown by the dashed curve in Fig. 3.6(c) for $-\int_{-\infty}^{z} J_x \, dz$. From symmetry considerations, the field must be equal and opposite on either side of the current region $-a < z < a$. Hence we choose the constant of integration C to be equal to $J_0 a$ thereby obtaining the final result for H_y as shown by the solid curve in Fig. 3.6(c). Thus the magnetic field intensity due to the current distribution is given by

$$\mathbf{H} = \begin{cases} J_0 a \mathbf{i}_y & \text{for } z < -a \\ -J_0 z \mathbf{i}_y & \text{for } -a < z < a \\ -J_0 a \mathbf{i}_y & \text{for } z > a \end{cases}$$

The magnetic flux density, $\mathbf{B}$, is equal to $\mu_0 \mathbf{H}$.   ■

We now generalize (3.23) for the arbitrary case of a magnetic field having all three components, each of them depending on all three coordinates, in addition to t, that is,

$$\mathbf{H} = H_x(x, y, z, t)\mathbf{i}_x + H_y(x, y, z, t)\mathbf{i}_y + H_z(x, y, z, t)\mathbf{i}_z \quad (3.24)$$

We do this in exactly the same manner as for the case of Faraday's law by considering the three infinitesimal rectangular paths shown in Fig. 3.3. Applying Ampere's circuital law to each of the three paths and simplifying, we obtain

$$\frac{[H_z]_{(x, y+\Delta y)} - [H_z]_{(x,y)}}{\Delta y} - \frac{[H_y]_{(x, z+\Delta z)} - [H_y]_{(x,z)}}{\Delta z}$$

$$= \left[J_x + \frac{\partial D_x}{\partial t} \right]_{(x,y,z)} \quad (3.25a)$$

$$\frac{[H_x]_{(y, z+\Delta z)} - [H_x]_{(y,z)}}{\Delta z} - \frac{[H_z]_{(x+\Delta x, y)} - [H_z]_{(x,y)}}{\Delta x}$$

$$= \left[J_y + \frac{\partial D_y}{\partial t} \right]_{(x,y,z)} \quad (3.25b)$$

$$\frac{[H_y]_{(x+\Delta x, z)} - [H_y]_{(x,z)}}{\Delta x} - \frac{[H_x]_{(y+\Delta y, z)} - [H_x]_{(y,z)}}{\Delta y}$$

$$= \left[J_z + \frac{\partial D_z}{\partial t} \right]_{(x,y,z)} \quad (3.25c)$$

If we now let all three paths shrink to the point a by letting Δx, Δy, and Δz tend to zero, (3.25a)–(3.25c) reduce to

$$\frac{\partial H_z}{\partial y} - \frac{\partial H_y}{\partial z} = J_x + \frac{\partial D_x}{\partial t} \quad (3.26a)$$

$$\frac{\partial H_x}{\partial z} - \frac{\partial H_z}{\partial x} = J_y + \frac{\partial D_y}{\partial t} \tag{3.26b}$$

$$\frac{\partial H_y}{\partial x} - \frac{\partial H_x}{\partial y} = J_z + \frac{\partial D_z}{\partial t} \tag{3.26c}$$

Equations (3.26a)–(3.26c) are the differential equations governing the relationships between the space variations of the magnetic field components, the components of the current density and the time variations of the electric field components, at a point. They can be interpreted physically in a manner analogous to the interpretation of (3.12a)–(3.12c) in the case of Faraday's law.

Equations (3.26a)–(3.26c) can be combined into a single vector equation in determinant form as given by

$$\begin{vmatrix} \mathbf{i}_x & \mathbf{i}_y & \mathbf{i}_z \\ \frac{\partial}{\partial x} & \frac{\partial}{\partial y} & \frac{\partial}{\partial z} \\ H_x & H_y & H_z \end{vmatrix} = \mathbf{J} + \frac{\partial \mathbf{D}}{\partial t} \tag{3.27}$$

or

$$\nabla \times \mathbf{H} = \mathbf{J} + \frac{\partial \mathbf{D}}{\partial t} \tag{3.28}$$

Equation (3.28) is Maxwell's equation in differential form corresponding to Ampere's circuital law. The quantity $\partial \mathbf{D}/\partial t$ is known as the "displacement current density." We shall discuss curl further in the following section.

3.3   CURL AND STOKES' THEOREM

In Secs. 3.1 and 3.2 we derived the differential forms of Faraday's and Ampere's circuital laws from their integral forms. These differential forms involve a new vector quantity, namely, the "curl" of a vector. In this section we shall introduce the basic definition of curl and then present a physical interpretation of the curl. In order to do this, let us, for simplicity, consider Ampere's circuital law in differential form without the displacement current density term, i.e.,

$$\nabla \times \mathbf{H} = \mathbf{J} \tag{3.29}$$

We wish to express $\nabla \times \mathbf{H}$ at a point in the current region in terms of $\mathbf{H}$ at that point. If we consider an infinitesimal surface $\Delta \mathbf{S}$ at the point and take the dot product of both sides of (3.29) with $\Delta \mathbf{S}$, we get

$$(\nabla \times \mathbf{H}) \cdot \Delta \mathbf{S} = \mathbf{J} \cdot \Delta \mathbf{S} \tag{3.30}$$

But $\mathbf{J} \cdot \Delta \mathbf{S}$ is simply the current crossing the surface $\Delta \mathbf{S}$, and according to Ampere's circuital law in integral form without the displacement current term,

$$\oint_C \mathbf{H} \cdot d\mathbf{l} = \mathbf{J} \cdot \Delta \mathbf{S} \tag{3.31}$$

where C is the closed path bounding $\Delta \mathbf{S}$. Comparing (3.30) and (3.31), we have

$$(\nabla \times \mathbf{H}) \cdot \Delta \mathbf{S} = \oint_C \mathbf{H} \cdot d\mathbf{l}$$

or

$$(\nabla \times \mathbf{H}) \cdot \Delta S \, \mathbf{i}_n = \oint_C \mathbf{H} \cdot d\mathbf{l} \tag{3.32}$$

where $\mathbf{i}_n$ is the unit vector normal to ΔS. Dividing both sides of (3.32) by ΔS, we obtain

$$(\nabla \times \mathbf{H}) \cdot \mathbf{i}_n = \frac{\oint_C \mathbf{H} \cdot d\mathbf{l}}{\Delta S} \tag{3.33}$$

The maximum value of $(\nabla \times \mathbf{H}) \cdot \mathbf{i}_n$, and hence that of the right side of (3.33), occurs when $\mathbf{i}_n$ is oriented parallel to $\nabla \times \mathbf{H}$, that is, when the surface ΔS is oriented normal to the current density vector $\mathbf{J}$. This maximum value is simply $|\nabla \times \mathbf{H}|$. Thus

$$|\nabla \times \mathbf{H}| = \left[\frac{\oint_C \mathbf{H} \cdot d\mathbf{l}}{\Delta S} \right]_{\max} \tag{3.34}$$

Since the direction of $\nabla \times \mathbf{H}$ is the direction of $\mathbf{J}$, or that of the unit vector normal to ΔS, we can then write

$$\nabla \times \mathbf{H} = \left[\frac{\oint_C \mathbf{H} \cdot d\mathbf{l}}{\Delta S} \right]_{\max} \mathbf{i}_n \tag{3.35}$$

Equation (3.35) is only approximate since (3.32) is exact only in the limit that ΔS tends to zero. Thus

$$\nabla \times \mathbf{H} = \lim_{\Delta S \to 0} \left[\frac{\oint_C \mathbf{H} \cdot d\mathbf{l}}{\Delta S} \right]_{\max} \mathbf{i}_n \tag{3.36}$$

Equation (3.36) is the expression for $\nabla \times \mathbf{H}$ at a point in terms of $\mathbf{H}$ at that point. Although we have derived this for the $\mathbf{H}$ vector, it is a general result and, in fact, is often the starting point for the introduction of curl.

Equation (3.36) tells us that in order to find the curl of a vector at a point in that vector field, we first consider an infinitesimal surface at that point and compute the closed line integral or circulation of the vector around the periphery of this surface by orienting the surface such that the circulation is maximum. We then divide the circulation by the area of the surface to obtain the maximum value of the circulation per unit area. Since we need this maximum value of the circulation per unit area in the limit that the area tends to zero, we do this by gradually shrinking the area and making sure that each time we compute the circulation per unit area an orientation for the area that maximizes this quantity is maintained. The limiting value to which the maximum circulation per unit area approaches is the magnitude of the curl. The limiting direction to which the normal vector to the surface approaches is the direction of the curl. The task of computing the curl is simplified if we consider one component of the field at a time and compute the curl corresponding to that component since then it is sufficient if we always maintain the orientation of the surface normal to that component axis. In fact, this is what we did in Secs. 3.1 and 3.2, which led us to the determinant form of curl.

We are now ready to discuss the physical interpretation of the curl. We do this with the aid of a simple device known as the "curl meter." Although the curl meter may take several forms, we shall consider one consisting of a circular disc that floats in water with a paddle wheel attached to the bottom of the disc, as shown in Fig. 3.7. A dot at the periphery on top of the disc serves to indicate any rotational motion of the curl meter about its axis, i.e., the axis of the paddle wheel. Let us now consider a stream of rectangular cross section carrying water in the z direction, as shown in Fig. 3.7(a). Let us assume the velocity $\mathbf{v}$ of the water to be independent of height but increasing uniformly from a value of zero at the banks to a maximum value v_0 at the center, as shown in Fig. 3.7(b), and investigate the behavior of the curl meter when it is placed vertically at different points in the stream. We assume that the size of the curl meter is vanishingly small so that it does not disturb the flow of water as we probe its behavior at different points.

Since exactly in midstream the blades of the paddle wheel lying on either side of the center line are hit by the same velocities, the paddle wheel does not rotate. The curl meter simply slides down the stream without any rotational motion, i.e., with the dot on top of the disc maintaining the same position relative to the center of the disc, as shown in Fig. 3.7(c). At a point to the left of the midstream the blades of the paddle wheel are hit by a greater velocity on the right side than on the left side so that the paddle wheel rotates in the counterclockwise sense. The curl meter rotates in the counterclockwise direction about its axis as it slides down the stream, as indicated by the changing position of the dot on top of the disc relative to the center of the disc, as shown in Fig. 3.7(d). At a point to the right of midstream, the blades of the paddle wheel are hit by a greater velocity on the left side than on the right side

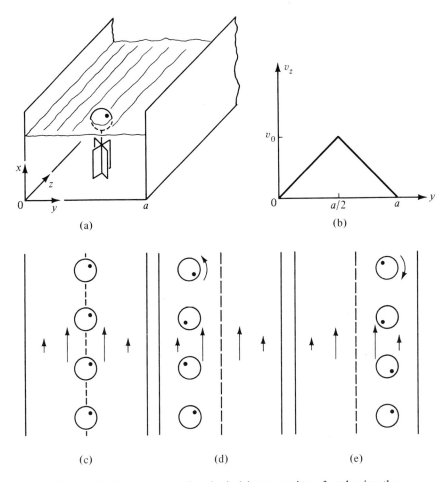

Figure 3.7. For explaining the physical interpretation of curl using the curl meter.

so that the paddle wheel rotates in the clockwise sense. The curl meter rotates in the clockwise direction about its axis as it slides down the stream, as indicated by the changing position of the dot on top of the disc relative to the center of the disc, as shown in Fig. 3.7(e).

To relate the foregoing discussion of the behavior of the curl meter with the curl of the velocity vector field of the water flow, we note that at a point in midstream, the circulation of the velocity vector per unit area in the plane normal to the axis of the paddle wheel, i.e., parallel to the surface of the stream, is zero and hence the component of the curl along that axis, i.e., in the x direction, is zero. At points on either side of midstream, however, the circulation per unit area is not zero in view of the velocity differential along the y direction. Hence the x component of the curl is nonzero at these points. Furthermore, the x component of the curl at points on the right side of midstream is opposite in sign to that on the left side of midstream since the velocity differentials are opposite in sign. These properties are exactly similar to those of the rotational motion of the curl meter.

If we now pick up the curl meter and insert it in the water with its axis parallel to the surface of the stream, the curl meter does not rotate because its blades are hit with the same force on either side of its axis. This behavior of the curl meter is akin to the property that the horizontal component of the curl of the velocity vector is zero since the velocity differential along the x direction is zero.

The foregoing illustration of the physical interpretation of the curl of a vector field can be used to visualize the behavior of electric and magnetic fields. Thus, for example, from

$$\mathbf{\nabla} \times \mathbf{E} = -\frac{\partial \mathbf{B}}{\partial t}$$

we know that at a point in an electromagnetic field at which $\partial \mathbf{B}/\partial t$ is nonzero, there exists an electric field with nonzero circulation per unit area in the plane normal to the vector $\partial \mathbf{B}/\partial t$. Similarly, from

$$\mathbf{\nabla} \times \mathbf{H} = \mathbf{J} + \frac{\partial \mathbf{D}}{\partial t}$$

we know that at a point in an electromagnetic field at which $\mathbf{J} + \partial \mathbf{D}/\partial t$ is nonzero, there exists a magnetic field with nonzero circulation per unit area in the plane normal to the vector $\mathbf{J} + \partial \mathbf{D}/\partial t$.

We shall now derive a useful theorem in vector calculus, the "Stokes' theorem." This relates the closed line integral of a vector field to the surface integral of the curl of that vector field. To derive this theorem, let us consider an arbitrary surface S in a magnetic field region and divide this surface into a number of infinitesimal surfaces $\Delta S_1, \Delta S_2, \Delta S_3, \ldots$, bounded by the contours $C_1, C_2, C_3, \ldots$, respectively. Then, applying (3.32) to each one of these infinitesimal surfaces and adding up, we get

$$\sum_j (\mathbf{\nabla} \times \mathbf{H})_j \cdot \Delta S_j \, \mathbf{i}_{nj} = \oint_{C_1} \mathbf{H} \cdot d\mathbf{l} + \oint_{C_2} \mathbf{H} \cdot d\mathbf{l} + \ldots \qquad (3.37)$$

where $\mathbf{i}_{nj}$ are unit vectors normal to the surfaces ΔS_j chosen in accordance with the right-hand screw rule. In the limit that the number of infinitesimal surfaces tends to infinity, the left side of (3.37) approaches to the surface integral of $\mathbf{\nabla} \times \mathbf{H}$ over the surface S. The right side of (3.37) is simply the closed line integral of $\mathbf{H}$ around the contour C since the contributions to the line integrals from the portions of the contours interior to C cancel, as shown in Fig. 3.8. Thus we get

$$\int_S (\nabla \times \mathbf{H}) \cdot d\mathbf{S} = \oint_C \mathbf{H} \cdot d\mathbf{l} \qquad (3.38)$$

Equation (3.38) is Stokes' theorem. Although we have derived it by considering the $\mathbf{H}$ field, it is general and is applicable for any vector field.

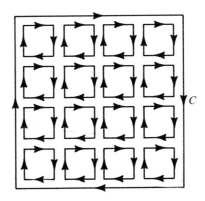

Figure 3.8. For deriving Stokes' theorem.

Example 3.5.   Let us verify Stokes' theorem by considering

$$\mathbf{A} = y\mathbf{i}_x - x\mathbf{i}_y$$

and the closed path C shown in Fig. 3.9.

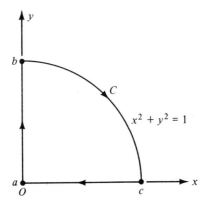

Figure 3.9. A closed path for verifying Stokes' theorem.

We first determine $\oint_C \mathbf{A} \cdot d\mathbf{l}$ by evaluating the line integrals along the three segments of the closed path. To do this, we first note that $\mathbf{A} \cdot d\mathbf{l} = y \, dx - x \, dy$. Then, from a to b, $x = 0$, $dx = 0$, $\mathbf{A} \cdot d\mathbf{l} = 0$

$$\int_a^b \mathbf{A} \cdot d\mathbf{l} = 0$$

From b to c, $x^2 + y^2 = 1$, $y = \sqrt{1 - x^2}$

$$2x \, dx + 2y \, dy = 0, \qquad dy = -\frac{x \, dx}{y} = -\frac{x}{\sqrt{1 - x^2}} \, dx$$

$$\mathbf{A} \cdot d\mathbf{l} = \sqrt{1 - x^2} \, dx + \frac{x^2 \, dx}{\sqrt{1 - x^2}} = \frac{dx}{\sqrt{1 - x^2}}$$

$$\int_b^c \mathbf{A} \cdot d\mathbf{l} = \int_0^1 \frac{dx}{\sqrt{1 - x^2}} = \left[\sin^{-1} x\right]_0^1 = \frac{\pi}{2}$$

From c to a, $y = 0$, $dy = 0$, $\mathbf{A} \cdot d\mathbf{l} = 0$

$$\int_c^a \mathbf{A} \cdot d\mathbf{l} = 0$$

Thus

$$\oint_C \mathbf{A} \cdot d\mathbf{l} = \int_a^b \mathbf{A} \cdot d\mathbf{l} + \int_b^c \mathbf{A} \cdot d\mathbf{l} + \int_c^a \mathbf{A} \cdot d\mathbf{l}$$

$$= 0 + \frac{\pi}{2} + 0 = \frac{\pi}{2}$$

Now, to evaluate $\oint_C \mathbf{A} \cdot d\mathbf{l}$ by using Stokes' theorem, we recall from Example 3.2 that

$$\nabla \times \mathbf{A} = \nabla \times (y\mathbf{i}_x - x\mathbf{i}_y) = -2\mathbf{i}_z$$

For the plane surface S enclosed by C,

$$d\mathbf{S} = -dx \, dy \, \mathbf{i}_z$$

Thus

$$(\nabla \times \mathbf{A}) \cdot d\mathbf{S} = -2\mathbf{i}_z \cdot (-dx \, dy \, \mathbf{i}_z) = 2 \, dx \, dy$$

$$\int_S (\nabla \times \mathbf{A}) \cdot d\mathbf{S} = \int_{x=0}^1 \int_{y=0}^{\sqrt{1-x^2}} 2 \, dx \, dy$$

$$= 2(\text{area enclosed by } C) = 2 \times \frac{\pi}{4} = \frac{\pi}{2}$$

thereby verifying Stokes' theorem.   ■

3.4 GAUSS' LAW FOR THE ELECTRIC FIELD

Thus far we have derived Maxwell's equations in differential form corresponding to the two Maxwell's equations in integral form involving the line integrals of **E** and **H**, that is, Faraday's law and Ampere's circuital law, respectively. The remaining two Maxwell's equations in integral form, namely, Gauss' law for the electric field and Gauss' law for the magnetic field, are concerned with the closed surface integrals of **D** and **B**, respectively. We shall in this and the following sections derive the differential forms of these two equations.

We recall from Sec. 2.6 that Gauss' law for the electric field is given by

$$\oint_S \mathbf{D} \cdot d\mathbf{S} = \int_V \rho \, dv \qquad (3.39)$$

where V is the volume enclosed by the closed surface S. To derive the differential form of this equation, let us consider a rectangular box of infinitesimal sides Δx, Δy, and Δz and defined by the six surfaces $x = x$, $x = x + \Delta x$, $y = y$, $y = y + \Delta y$, $z = z$, and $z = z + \Delta z$, as shown in Fig. 3.10, in a region of electric field

$$\mathbf{D} = \mathbf{D}_x(x, y, z, t)\mathbf{i}_x + D_y(x, y, z, t)\mathbf{i}_y + D_z(x, y, z, t)\mathbf{i}_z \qquad (3.40)$$

and charge of density $\rho(x, y, z, t)$. According to Gauss' law for the electric field, the displacement flux emanating from the box is equal to the charge enclosed by the box. The displacement flux is given by the surface integral of **D** over the surface of the box, which is comprised of six plane surfaces. Thus evaluating the displacement flux emanating out of the box over each of the six plane surfaces of the box, we have

$$\int \mathbf{D} \cdot d\mathbf{S} = -[D_x]_x \, \Delta y \, \Delta z \qquad \text{for the surface } x = x \qquad (3.41a)$$

$$\int \mathbf{D} \cdot d\mathbf{S} = [D_x]_{x+\Delta x} \, \Delta y \, \Delta z \qquad \text{for the surface } x = x + \Delta x \qquad (3.41b)$$

$$\int \mathbf{D} \cdot d\mathbf{S} = -[D_y]_y \, \Delta z \, \Delta x \qquad \text{for the surface } y = y \qquad (3.41c)$$

$$\int \mathbf{D} \cdot d\mathbf{S} = [D_y]_{y+\Delta y} \, \Delta z \, \Delta x \qquad \text{for the surface } y = y + \Delta y \qquad (3.41d)$$

$$\int \mathbf{D} \cdot d\mathbf{S} = -[D_z]_z \, \Delta x \, \Delta y \qquad \text{for the surface } z = z \qquad (3.41e)$$

$$\int \mathbf{D} \cdot d\mathbf{S} = [D_z]_{z+\Delta z} \, \Delta x \, \Delta y \qquad \text{for the surface } z = z + \Delta z \qquad (3.41f)$$

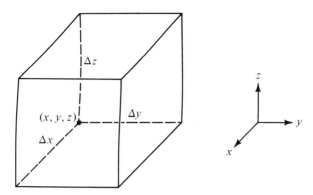

Figure 3.10. An infinitesimal rectangular box.

Adding up (3.41a)–(3.41f), we obtain the total displacement flux emanating from the box to be

$$\oint_S \mathbf{D} \cdot d\mathbf{S} = \{[D_x]_{x+\Delta x} - [D_x]_x\} \, \Delta y \, \Delta z$$
$$+ \{[D_y]_{y+\Delta y} - [D_y]_y\} \, \Delta z \, \Delta x$$
$$+ \{[D_z]_{z+\Delta z} - [D_z]_z\} \, \Delta x \, \Delta y \qquad (3.42)$$

Now the charge enclosed by the rectangular box is given by

$$\int_V \rho \, dv = \rho(x, y, z, t) \cdot \Delta x \, \Delta y \, \Delta z = \rho \, \Delta x \, \Delta y \, \Delta z \qquad (3.43)$$

where we have assumed ρ to be uniform throughout the volume of the box and equal to its value at (x, y, z) since the box is infinitesimal in volume.

Substituting (3.42) and (3.43) into (3.39) to apply Gauss' law for the electric field to the surface of the box under consideration, we get

$$\{[D_x]_{x+\Delta x} - [D_x]_x\} \, \Delta y \, \Delta z + \{[D_y]_{y+\Delta y} - [D_y]_y\} \, \Delta z \, \Delta x$$
$$+ \{[D_z]_{z+\Delta z} - [D_z]_z\} \, \Delta x \, \Delta y = \rho \, \Delta x \, \Delta y \, \Delta z$$

or

$$\frac{[D_x]_{x+\Delta x} - [D_x]_x}{\Delta x} + \frac{[D_y]_{y+\Delta y} - [D_y]_y}{\Delta y} + \frac{[D_z]_{z+\Delta z} - [D_z]_z}{\Delta z} = \rho \qquad (3.44)$$

If we now let the box shrink to the point (x, y, z) by letting Δx, Δy, and Δz tend to zero, we obtain

$$\underset{\Delta x \to 0}{\text{Lim}} \frac{[D_x]_{x+\Delta x} - [D_x]_x}{\Delta x} + \underset{\Delta y \to 0}{\text{Lim}} \frac{[D_y]_{y+\Delta y} - [D_y]_y}{\Delta y}$$
$$+ \underset{\Delta z \to 0}{\text{Lim}} \frac{[D_z]_{z+\Delta z} - [D_z]_z}{\Delta z} = \underset{\substack{\Delta x \to 0 \\ \Delta y \to 0 \\ \Delta z \to 0}}{\text{Lim}} \rho$$

or

$$\frac{\partial D_x}{\partial x} + \frac{\partial D_y}{\partial y} + \frac{\partial D_z}{\partial z} = \rho \qquad (3.45)$$

Equation (3.45) tells us that the net longitudinal differential of the components of **D**, that is, the algebraic sum of the derivatives of the components of **D** along their respective directions is equal to the charge density at that point. Conversely, a charge density at a point results in an electric field, having components of **D** such that their net longitudinal differential is nonzero. An example in which the net longitudinal differential is zero although some of the individual derivatives are nonzero is shown in Fig. 3.11(a). Fig. 3.11(b) shows an example in which the net longitudinal differential is nonzero. Equation (3.45) can be written in vector notation as

$$\left(\mathbf{i}_x \frac{\partial}{\partial x} + \mathbf{i}_y \frac{\partial}{\partial y} + \mathbf{i}_z \frac{\partial}{\partial z}\right) \cdot (D_x \mathbf{i}_x + D_y \mathbf{i}_y + D_z \mathbf{i}_z) = \rho \qquad (3.46)$$

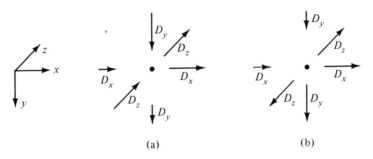

(a) (b)

Figure 3.11. For illustrating (a) zero, and (b) nonzero net longitudinal differential of the components of **D**.

The left side of (3.46) is known as the "divergence of **D**," denoted as **∇ · D** (del dot **D**). Thus we have

$$\mathbf{\nabla} \cdot \mathbf{D} = \rho \qquad (3.47)$$

Equation (3.47) is Maxwell's equation in differential form corresponding to Gauss' law for the electric field. We shall discuss divergence further in Sec. 3.6.

Example 3.6. Given $\mathbf{A} = 3x\mathbf{i}_x + (y - 3)\mathbf{i}_y + (2 - z)\mathbf{i}_z$, find $\mathbf{\nabla} \cdot \mathbf{A}$.
From the expansion for the divergence of a vector, we have

$$\mathbf{\nabla} \cdot \mathbf{A} = \left(\mathbf{i}_x \frac{\partial}{\partial x} + \mathbf{i}_y \frac{\partial}{\partial y} + \mathbf{i}_z \frac{\partial}{\partial z}\right) \cdot [3x\mathbf{i}_x + (y - 3)\mathbf{i}_y + (2 - z)\mathbf{i}_z]$$

$$= \frac{\partial}{\partial x}(3x) + \frac{\partial}{\partial y}(y - 3) + \frac{\partial}{\partial z}(2 - z)$$

$$= 3 + 1 - 1 = 3 \qquad \blacksquare$$

Example 3.7.   Let us consider the charge distribution given by

$$\rho = \begin{cases} -\rho_0 & \text{for } -a < x < 0 \\ \;\;\rho_0 & \text{for } 0 < x < a \end{cases}$$

as shown in Fig. 3.12(a), where ρ_0 is a constant, and find the electric field everywhere.

Since the charge density is independent of y and z, the field is also independent of y and z, thereby giving us $\partial D_y/\partial y = \partial D_z/\partial z = 0$ and reducing Gauss' law for the electric field to

$$\frac{\partial D_x}{\partial x} = \rho$$

Integrating both sides with respect to x, we obtain

$$D_x = \int_{-\infty}^{x} \rho \, dx + C$$

where C is the constant of integration.

The variation of ρ with x is shown in Fig. 3.12(b). Integrating ρ with respect to x, that is, finding the area under the curve of Fig. 3.12(b) as a

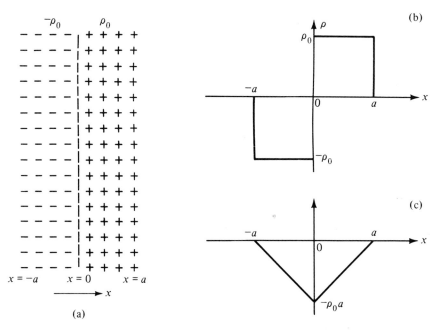

Figure 3.12. The determination of electric field due to a charge distribution.

function of x, we obtain the result shown in Fig. 3.12(c) for $\int_{-\infty}^{x} \rho \, dx$. The constant of integration C is zero since the symmetry of the field required by the symmetry of the charge distribution is already satisfied by the curve of Fig. 3.12(c). Thus the displacement flux density due to the charge distribution is given by

$$
\mathbf{D} = \begin{cases}
0 & \text{for } x < -a \\
-\rho_0(x + a)\mathbf{i}_x & \text{for } -a < x < 0 \\
\rho_0(x - a)\mathbf{i}_x & \text{for } 0 < x < a \\
0 & \text{for } x > a
\end{cases}
$$

The electric field intensity, $\mathbf{E}$, is equal to $\mathbf{D}/\epsilon_0$.   ■

3.5   GAUSS' LAW FOR THE MAGNETIC FIELD

In the previous section we derived the differential form of Gauss' law for the electric field from its integral form. In this section we shall derive the differential form of Gauss' law for the magnetic field from its integral form. We recall from Sec. 2.5 that Gauss' law for the magnetic field in integral form is given by

$$
\oint_S \mathbf{B} \cdot d\mathbf{S} = 0 \tag{3.48}
$$

where S is any closed surface. This equation states that the magnetic flux emanating from a closed surface is zero. Thus considering an infinitesimal rectangular box as shown in Fig. 3.10 in a region of magnetic field

$$
\mathbf{B} = B_x(x, y, z, t)\mathbf{i}_x + B_y(x, y, z, t)\mathbf{i}_y + B_z(x, y, z, t)\mathbf{i}_z \tag{3.49}
$$

and evaluating the magnetic flux emanating out of the box in a manner similar to that of the evaluation of the displacement flux in the previous section, and substituting in (3.48), we obtain

$$
\{[B_x]_{x+\Delta x} - [B_x]_x\} \, \Delta y \, \Delta z + \{[B_y]_{y+\Delta y} - [B_y]_y\} \, \Delta z \, \Delta x
$$
$$
+ \{[B_z]_{z+\Delta z} - [B_z]_z\} \, \Delta x \, \Delta y = 0 \tag{3.50}
$$

Dividing (3.50) on both sides by $\Delta x \, \Delta y \, \Delta z$ and letting Δx, Δy, and Δz tend to zero, thereby shrinking the box to the point (x, y, z), we obtain

$$
\lim_{\Delta x \to 0} \frac{[B_x]_{x+\Delta x} - [B_x]_x}{\Delta x} + \lim_{\Delta y \to 0} \frac{[B_y]_{y+\Delta y} - [B_y]_y}{\Delta y}
$$
$$
+ \lim_{\Delta z \to 0} \frac{[B_z]_{z+\Delta z} - [B_z]_z}{\Delta z} = 0
$$

or

$$\frac{\partial B_x}{\partial x} + \frac{\partial B_y}{\partial y} + \frac{\partial B_z}{\partial z} = 0 \qquad (3.51)$$

Equation (3.51) tells us that the net longitudinal differential of the components of **B** is zero. In vector form it is given by

$$\mathbf{V} \cdot \mathbf{B} = 0 \qquad (3.52)$$

Equation (3.52) is Maxwell's equation in differential form corresponding to Gauss' law for the magnetic field. We shall discuss divergence further in the following section.

Example 3.8. Determine if the vector $\mathbf{A} = y\mathbf{i}_x - x\mathbf{i}_y$ can represent a magnetic field **B**.

From (3.52), we note that a given vector can be realized as a magnetic field **B** if its divergence is zero. For $\mathbf{A} = y\mathbf{i}_x - x\mathbf{i}_y$,

$$\mathbf{V} \cdot \mathbf{A} = \frac{\partial}{\partial x}(y) + \frac{\partial}{\partial y}(-x) + \frac{\partial}{\partial z}(0) = 0$$

Hence the given vector can represent a magnetic field **B**. ∎

3.6 DIVERGENCE AND THE DIVERGENCE THEOREM

In Secs. 3.4 and 3.5 we derived the differential forms of Gauss' laws for the electric and magnetic fields from their integral forms. These differential forms involve a new quantity, namely, the "divergence" of a vector. The divergence of a vector is a scalar as compared to the vector nature of the curl of a vector. In this section we shall introduce the basic definition of divergence and then present a physical interpretation for the divergence. In order to do this, let us consider Gauss' law for the electric field in differential form, that is,

$$\mathbf{V} \cdot \mathbf{D} = \rho \qquad (3.53)$$

We wish to express $\mathbf{V} \cdot \mathbf{D}$ at a point in the charge region in terms of **D** at that point. If we consider an infinitesimal volume Δv at the point and multiply both sides of (3.53) by Δv, we get

$$(\mathbf{V} \cdot \mathbf{D}) \, \Delta v = \rho \, \Delta v \qquad (3.54)$$

But $\rho \, \Delta v$ is simply the charge contained in the volume Δv, and according to Gauss' law for the electric field in integral form,

$$\oint_s \mathbf{D} \cdot d\mathbf{S} = \rho \, \Delta v \qquad (3.55)$$

where S is the closed surface bounding Δv. Comparing (3.54) and (3.55), we have

$$(\nabla \cdot \mathbf{D}) \Delta v = \oint_S \mathbf{D} \cdot d\mathbf{S} \tag{3.56}$$

Dividing both sides of (3.56) by Δv, we obtain

$$\nabla \cdot \mathbf{D} = \frac{\oint_S \mathbf{D} \cdot d\mathbf{S}}{\Delta v} \tag{3.57}$$

Equation (3.57) is only approximate since (3.56) is exact only in the limit that Δv tends to zero. Thus

$$\nabla \cdot \mathbf{D} = \lim_{\Delta v \to 0} \frac{\oint_S \mathbf{D} \cdot d\mathbf{S}}{\Delta v} \tag{3.58}$$

Equation (3.58) is the expression for $\nabla \cdot \mathbf{D}$ at a point in terms of $\mathbf{D}$ at that point. Although we have derived this for the $\mathbf{D}$ vector, it is a general result and, in fact, is often the starting point for the introduction of divergence.

Equation (3.58) tells us that in order to find the divergence of a vector at a point in that vector field, we first consider an infinitesimal volume at that point and compute the surface integral of the vector over the surface bounding that volume, that is, the outward flux of the vector field emanating from that volume. We then divide the flux by the volume to obtain the flux per unit volume. Since we need this flux per unit volume in the limit that the volume tends to zero, we do this by gradually shrinking the volume. The limiting value to which the flux per unit volume approaches is the value of the divergence of the vector field at the point to which the volume is shrunk.

We are now ready to discuss the physical interpretation of the divergence. To simplify this task, we shall consider the differential form of the law of conservation of charge given in integral form by (2.39), or

$$\oint_S \mathbf{J} \cdot d\mathbf{S} = -\frac{d}{dt} \int_V \rho \, dv \tag{3.59}$$

where S is the surface bounding the volume V. Applying (3.59) to an infinitesimal volume Δv, we have

$$\oint_S \mathbf{J} \cdot d\mathbf{S} = -\frac{d}{dt}(\rho \, \Delta v) = -\frac{\partial \rho}{\partial t} \Delta v$$

or

$$\frac{\oint_S \mathbf{J} \cdot d\mathbf{S}}{\Delta v} = -\frac{\partial \rho}{\partial t} \tag{3.60}$$

Now taking the limit on both sides of (3.60) as Δv tends to zero, we obtain

$$\underset{\Delta v \to 0}{\text{Lim}} \frac{\oint \mathbf{J} \cdot d\mathbf{S}}{\Delta v} = \underset{\Delta v \to 0}{\text{Lim}} - \frac{\partial \rho}{\partial t} \tag{3.61}$$

or

$$\mathbf{\nabla} \cdot \mathbf{J} = -\frac{\partial \rho}{\partial t} \tag{3.62}$$

or

$$\mathbf{\nabla} \cdot \mathbf{J} + \frac{\partial \rho}{\partial t} = 0 \tag{3.63}$$

Equation (3.63), which is the differential form of the law of conservation of charge, is familiarly known as the "continuity equation." It tells us that the divergence of the current density vector at a point is equal to the time rate of decrease of the charge density at that point.

Let us now investigate three different cases: (a) positive value, (b) negative value, and (c) zero value of the time rate of decrease of the charge density at a point, that is, the divergence of the current density vector at that point. We shall do this with the aid of a simple device which we shall call the "divergence meter." The divergence meter can be imagined to be a tiny, elastic balloon enclosing the point and that expands when hit by charges streaming outward from the point and contracts when acted upon by charges streaming inward toward the point. For case (a), that is, when the time rate of decrease of the charge density at the point is positive, there is a net amount of charge streaming out of the point in a given time, resulting in a net current flow outward from the point that will make the imaginary balloon expand. For case (b), that is, when the time rate of decrease of the charge density at the point is negative or the time rate of increase of the charge density is positive, there is a net amount of charge streaming toward the point in a given time, resulting in a net current flow toward the point and the imaginary balloon will contract. For case (c), that is, when the time rate of decrease of the charge density at the point is zero, the balloon will remain unaffected since the charge is streaming out of the point at exactly the same rate as it is streaming into the point. These three cases are illustrated in Figs. 3.13(a), (b), and (c), respectively.

Generalizing the foregoing discussion to the physical interpretation of the divergence of any vector field at a point, we can imagine the vector field to be a velocity field of streaming charges acting upon the divergence meter and obtain in most cases a qualitative picture of the divergence of the vector field. If the divergence meter expands, the divergence is positive and a source of the flux of the vector field exists at that point. If the divergence meter contracts, the divergence is negative and a sink of the flux of the vector field exists at that point. If the divergence meter remains unaffected, the divergence is zero and neither a source nor a sink of the flux of the vector field exists at that point.

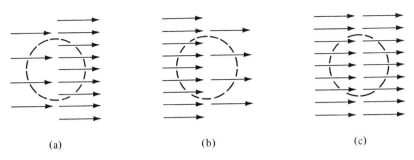

(a)   (b)   (c)

Figure 3.13. For explaining the physical interpretation of divergence using the divergence meter.

Alternatively, there can exist at the point pairs of sources and sinks of equal strengths.

We shall now derive a useful theorem in vector calculus, "the divergence theorem." This relates the closed surface integral of the vector field to the volume integral of the divergence of that vector field. To derive this theorem, let us consider an arbitrary volume V in an electric field region and divide this volume into a number of infinitesimal volumes $\Delta v_1, \Delta v_2, \Delta v_3, \ldots$, bounded by the surfaces $S_1, S_2, S_3, \ldots$, respectively. Then, applying (3.56) to each one of these infinitesimal volumes and adding up, we get

$$\sum_j (\nabla \cdot \mathbf{D})_j \, \Delta v_j = \oint_{S_1} \mathbf{D} \cdot d\mathbf{S} + \oint_{S_2} \mathbf{D} \cdot d\mathbf{S} + \ldots \qquad (3.64)$$

In the limit that the number of the infinitesimal volumes tends to infinity, the left side of (3.64) approaches to the volume integral of $\nabla \cdot \mathbf{D}$ over the volume V. The right side of (3.64) is simply the closed surface integral of $\mathbf{D}$ over S since the contribution to the surface integrals from the portions of the surfaces interior to S cancel, as shown in Fig. 3.14. Thus we get

$$\int_V (\nabla \cdot \mathbf{D}) \, dv = \oint_S \mathbf{D} \cdot d\mathbf{S} \qquad (3.65)$$

Equation (3.65) is the divergence theorem. Although we have derived it by considering the $\mathbf{D}$ field, it is general and is applicable for any vector field.

Example 3.9.   Let us verify the divergence theorem by considering

$$\mathbf{A} = 3x\mathbf{i}_x + (y - 3)\mathbf{i}_y + (2 - z)\mathbf{i}_z$$

and the closed surface of the box bounded by the planes $x = 0, x = 1, y = 0$, $y = 2, z = 0$, and $z = 3$.

We first determine $\oint_S \mathbf{A} \cdot d\mathbf{S}$ by evaluating the surface integrals over the

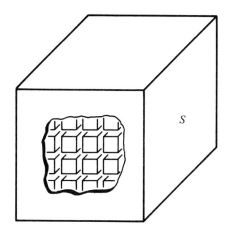

Figure 3.14. For deriving the divergence theorem.

six surfaces of the rectangular box. Thus for the surface $x = 0$,

$$\mathbf{A} = (y - 3)\mathbf{i}_y + (2 - z)\mathbf{i}_z, \qquad d\mathbf{S} = -dy\, dz\, \mathbf{i}_x$$

$$\mathbf{A} \cdot d\mathbf{S} = 0$$

$$\int \mathbf{A} \cdot d\mathbf{S} = 0$$

For the surface $x = 1$,

$$\mathbf{A} = 3\mathbf{i}_x + (y - 3)\mathbf{i}_y + (2 - z)\mathbf{i}_z, \qquad d\mathbf{S} = dy\, dz\, \mathbf{i}_x$$

$$\mathbf{A} \cdot d\mathbf{S} = 3\, dy\, dz$$

$$\int \mathbf{A} \cdot d\mathbf{S} = \int_{z=0}^{3} \int_{y=0}^{2} 3\, dy\, dz = 18$$

For the surface $y = 0$,

$$\mathbf{A} = 3x\mathbf{i}_x - 3\mathbf{i}_y + (2 - z)\mathbf{i}_z, \qquad d\mathbf{S} = -dz\, dx\, \mathbf{i}_y$$

$$\mathbf{A} \cdot d\mathbf{S} = 3\, dz\, dx$$

$$\int \mathbf{A} \cdot d\mathbf{S} = \int_{x=0}^{1} \int_{z=0}^{3} 3\, dz\, dx = 9$$

For the surface $y = 2$,

$$\mathbf{A} = 3x\mathbf{i}_x - \mathbf{i}_y + (2 - z)\mathbf{i}_z, \qquad d\mathbf{S} = dz\, dx\, \mathbf{i}_y$$

$$\mathbf{A} \cdot d\mathbf{S} = -dz\, dx$$

$$\int \mathbf{A} \cdot d\mathbf{S} = \int_{x=0}^{1} \int_{z=0}^{3} - dz\, dx = -3$$

For the surface $z = 0$,

$$\mathbf{A} = 3x\mathbf{i}_x + (y - 3)\mathbf{i}_y + 2\mathbf{i}_z, \qquad d\mathbf{S} = -dx\,dy\,\mathbf{i}_z$$

$$\mathbf{A} \cdot d\mathbf{S} = -2\,dx\,dy$$

$$\int \mathbf{A} \cdot d\mathbf{S} = \int_{y=0}^{2} \int_{x=0}^{1} -2\,dx\,dy = -4$$

For the surface $z = 3$,

$$\mathbf{A} = 3x\mathbf{i}_x + (y - 3)\mathbf{i}_y - \mathbf{i}_z, \qquad d\mathbf{S} = dx\,dy\,\mathbf{i}_z$$

$$\mathbf{A} \cdot d\mathbf{S} = -dx\,dy$$

$$\int \mathbf{A} \cdot d\mathbf{S} = \int_{y=0}^{2} \int_{x=0}^{1} -dx\,dy = -2$$

Thus

$$\oint_S \mathbf{A} \cdot d\mathbf{S} = 0 + 18 + 9 - 3 - 4 - 2 = 18$$

Now, to evaluate $\oint_S \mathbf{A} \cdot d\mathbf{S}$ by using the divergence theorem, we recall from Example 3.6 that

$$\mathbf{\nabla} \cdot \mathbf{A} = \mathbf{\nabla} \cdot [3x\mathbf{i}_x + (y - 3)\mathbf{i}_y + (2 - z)\mathbf{i}_z] = 3$$

For the volume enclosed by the rectangular box,

$$\int (\mathbf{\nabla} \cdot \mathbf{A})\,dv = \int_{z=0}^{3} \int_{y=0}^{2} \int_{x=0}^{1} 3\,dx\,dy\,dz = 18$$

thereby verifying the divergence theorem.   ∎

3.7 SUMMARY

We have in this chapter derived the differential forms of Maxwell's equations from their integral forms, which we introduced in the previous chapter. For the general case of electric and magnetic fields having all three components (x, y, z), each of them dependent on all coordinates (x, y, z), and time (t), Maxwell's equations in differential form are given as follows in words and in mathematical form.

Faraday's Law: The curl of the electric field intensity is equal to the negative of the time derivative of the magnetic flux density, that is,

$$\mathbf{\nabla} \times \mathbf{E} = -\frac{\partial \mathbf{B}}{\partial t} \tag{3.66}$$

AMPERE'S CIRCUITAL LAW: The curl of the magnetic field intensity is equal to the sum of the current density due to flow of charges and the displacement current density, which is the time derivative of the displacement flux density, that is,

$$\nabla \times \mathbf{H} = \mathbf{J} + \frac{\partial \mathbf{D}}{\partial t} \tag{3.67}$$

GAUSS' LAW FOR THE ELECTRIC FIELD: The divergence of the displacement flux density is equal to the charge density, that is,

$$\nabla \cdot \mathbf{D} = \rho \tag{3.68}$$

GAUSS' LAW FOR THE MAGNETIC FIELD: The divergence of the magnetic flux density is equal to zero, that is,

$$\nabla \cdot \mathbf{B} = 0 \tag{3.69}$$

Auxiliary to (3.66)–(3.69), the continuity equation is given by

$$\nabla \cdot \mathbf{J} + \frac{\partial \rho}{\partial t} = 0 \tag{3.70}$$

This equation, which is the differential form of the law of conservation of charge, states that the sum of the divergence of the current density due to flow of charges and the time derivative of the charge density is equal to zero. Also, we recall that

$$\mathbf{D} = \epsilon_0 \mathbf{E} \tag{3.71}$$

$$\mathbf{H} = \frac{\mathbf{B}}{\mu_0} \tag{3.72}$$

which relate $\mathbf{D}$ and $\mathbf{H}$ to $\mathbf{E}$ and $\mathbf{B}$, respectively, for free space.

We have learned that the basic definitions of curl and divergence, which have enabled us to discuss their physical interpretations with the aid of the curl and divergence meters, are

$$\nabla \times \mathbf{A} = \operatorname*{Lim}_{\Delta S \to 0} \left[\frac{\oint_C \mathbf{A} \cdot d\mathbf{l}}{\Delta S} \right]_{max} \mathbf{i}_n$$

$$\nabla \cdot \mathbf{A} = \operatorname*{Lim}_{\Delta v \to 0} \frac{\oint_S \mathbf{A} \cdot d\mathbf{S}}{\Delta v}$$

Thus the curl of a vector field at a point is a vector whose magnitude is the circulation of that vector field per unit area with the area oriented so as to maximize this quantity and in the limit that the area shrinks to the point. The direction of the vector is normal to the area in the aforementioned limit and in the right-hand sense. The divergence of a vector field at a point is a scalar quantity equal to the net outward flux of that vector field per unit volume in the limit that the volume shrinks to the point. In Cartesian coordinates the expansions for curl and divergence are

$$\mathbf{\nabla} \times \mathbf{A} = \begin{vmatrix} \mathbf{i}_x & \mathbf{i}_y & \mathbf{i}_z \\ \dfrac{\partial}{\partial x} & \dfrac{\partial}{\partial y} & \dfrac{\partial}{\partial z} \\ A_x & A_y & A_z \end{vmatrix}$$

$$= \left(\frac{\partial A_z}{\partial y} - \frac{\partial A_y}{\partial z}\right)\mathbf{i}_x + \left(\frac{\partial A_x}{\partial z} - \frac{\partial A_z}{\partial x}\right)\mathbf{i}_y + \left(\frac{\partial A_y}{\partial x} - \frac{\partial A_x}{\partial y}\right)\mathbf{i}_z$$

$$\mathbf{\nabla} \cdot \mathbf{A} = \frac{\partial A_x}{\partial x} + \frac{\partial A_y}{\partial y} + \frac{\partial A_z}{\partial z}$$

Thus Maxwell's equations in differential form relate the spatial variations of the field vectors at a point to their temporal variations and to the charge and current densities at that point.

We have also learned two theorems associated with curl and divergence. These are the Stokes' theorem and the divergence theorem given, respectively, by

$$\oint_C \mathbf{A} \cdot d\mathbf{l} = \int_S (\mathbf{\nabla} \times \mathbf{A}) \cdot d\mathbf{S}$$

$$\oint_S \mathbf{A} \cdot d\mathbf{S} = \int_V (\mathbf{\nabla} \cdot \mathbf{A}) \, dv$$

Stokes' theorem enables us to replace the line integral of a vector around a closed path by the surface integral of the curl of that vector over any surface bounded by that closed path, and vice versa. The divergence theorem enables us to replace the surface integral of a vector over a closed surface by the volume integral of the divergence of that vector over the volume bounded by the closed surface, and vice versa.

In Chap. 2 we learned that all Maxwell's equations in integral form are not independent. Since Maxwell's equations in differential form are derived from their integral forms, it follows that the same is true for these equations. In fact, by noting that (see Problem 3.32),

$$\mathbf{\nabla} \cdot \mathbf{\nabla} \times \mathbf{A} \equiv 0 \tag{3.73}$$

and applying it to (3.66), we obtain

$$\mathbf{\nabla} \cdot \left(-\frac{\partial \mathbf{B}}{\partial t}\right) = \mathbf{\nabla} \cdot \mathbf{\nabla} \times \mathbf{E} = 0$$

$$\frac{\partial}{\partial t}(\mathbf{\nabla} \cdot \mathbf{B}) = 0$$

$$\mathbf{\nabla} \cdot \mathbf{B} = \text{constant with time} \tag{3.74}$$

Similarly, applying (3.73) to (3.67), we obtain

$$\mathbf{V} \cdot \left(\mathbf{J} + \frac{\partial \mathbf{D}}{\partial t} \right) = \mathbf{V} \cdot \mathbf{V} \times \mathbf{H} = 0$$

$$\mathbf{V} \cdot \mathbf{J} + \frac{\partial}{\partial t}(\mathbf{V} \cdot \mathbf{D}) = 0$$

Using (3.70), we then have

$$-\frac{\partial \rho}{\partial t} + \frac{\partial}{\partial t}(\mathbf{V} \cdot \mathbf{D}) = 0$$

$$\frac{\partial}{\partial t}(\mathbf{V} \cdot \mathbf{D} - \rho) = 0$$

$$\mathbf{V} \cdot \mathbf{D} - \rho = \text{constant with time} \tag{3.75}$$

Since for any given point in space, the constants on the right sides of (3.74) and (3.75) can be made equal to zero at some instant of time, it follows that they are zero forever, giving us (3.69) and (3.68), respectively. Thus (3.69) follows from (3.66), whereas (3.68) follows from (3.67) with the aid of (3.70). Finally, for the simple, special case in which

$$\mathbf{E} = E_x(z, t)\mathbf{i}_x$$

$$\mathbf{H} = H_y(z, t)\mathbf{i}_y$$

the two Maxwell's curl equations reduce to

$$\frac{\partial E_x}{\partial z} = -\frac{\partial B_y}{\partial t} \tag{3.76}$$

$$\frac{\partial H_y}{\partial z} = -J_x - \frac{\partial D_x}{\partial t} \tag{3.77}$$

In fact, we derived these equations first and then the general equations (3.66) and (3.67). We will be using (3.76) and (3.77) in the following chapters to study the phenomenon of electromagnetic wave propagation resulting from the interdependence between the space-variations and time-variations of the electric and magnetic fields.

REVIEW QUESTIONS

3.1. State Faraday's law in differential form for the simple case of $\mathbf{E} = E_x(z, t)\mathbf{i}_x$. How is it derived from Faraday's law in integral form?

3.2. Discuss the physical interpretation of Faraday's law in differential form for the simple case of $\mathbf{E} = E_x(z, t)\mathbf{i}_x$.

3.3. State Faraday's law in differential form for the general case of an arbitrary electric field. How is it derived from its integral form?

3.4. What is meant by the net right-lateral differential of the x and y components of a vector normal to the z direction?

3.5. Give an example in which the net right-lateral differential of E_y and E_z normal to the x direction is zero although the individual derivatives are nonzero.

3.6. If at a point in space B_y varies with time but B_x and B_z do not, what can we say about the components of **E** at that point?

3.7. What is the determinant expansion for the curl of a vector?

3.8. What is the significance of the curl of a vector being equal to zero?

3.9. State Ampere's circuital law in differential form for the simple case of $\mathbf{H} = H_y(z, t)\mathbf{i}_y$. How is it derived from Ampere's circuital law in integral form?

3.10. Discuss the physical interpretation of Ampere's circuital law in differential form for the simple case of $\mathbf{H} = H_y(z, t)\mathbf{i}_y$.

3.11. State Ampere's circuital law in differential form for the general case of an arbitrary magnetic field. How is it derived from its integral form?

3.12. What is the significance of a nonzero net right-lateral differential of H_x and H_y normal to the z direction at a point in space?

3.13. If a pair of **E** and **B** at a point satisfies Faraday's law in differential form, does it necessarily follow that it also satisfies Ampere's circuital law in differential form and vice versa?

3.14. State and briefly discuss the basic definition of the curl of a vector.

3.15. What is a curl meter? How does it help visualize the behavior of the curl of a vector field?

3.16. Provide two examples of physical phenomena in which the curl of a vector field is nonzero.

3.17. State Stokes' theorem and discuss its application.

3.18. State Gauss' law for the electric field in differential form. How is it derived from its integral form?

3.19. What is meant by the net longitudinal differential of the components of a vector field?

3.20. Give an example in which the net longitudinal differential of the components of a vector is zero, although the individual derivatives are nonzero.

3.21. What is the expansion for the divergence of a vector?

3.22. State Gauss' law for the magnetic field in differential form. How is it derived from its integral form?

3.23. How can you determine if a given vector can represent a magnetic field?

3.24. State and briefly discuss the basic definition of the divergence of a vector.

3.25. What is a divergence meter? How does it help visualize the behavior of the divergence of a vector field?

3.26. Provide two examples of physical phenomena in which the divergence of a vector field is nonzero.

3.27. State the continuity equation and discuss its physical interpretation.

3.28. Distinguish between the physical interpretations of the divergence and the curl of a vector field by means of examples.

3.29. State the divergence theorem and discuss its application.

3.30. What is the divergence of the curl of a vector?

3.31. Summarize Maxwell's equations in differential form.

3.32. Are all Maxwell's equations in differential form independent? If not, which of them are independent?

PROBLEMS

3.1. Given $\mathbf{B} = B_0 z \cos \omega t \, \mathbf{i}_y$ and it is known that $\mathbf{E}$ has only an x component, find $\mathbf{E}$ by using Faraday's law in differential form. Then verify your result by applying Faraday's law in integral form to the rectangular closed path, in the xz plane, defined by $x = 0$, $x = a$, $z = 0$, and $z = b$.

3.2. Assuming $\mathbf{E} = E_y(z, t)\mathbf{i}_y$ and considering a rectangular closed path in the yz plane, carry out the derivation of Faraday's law in differential form similar to that in the text.

3.3. Find the curls of the following vector fields:
(a) $zx\mathbf{i}_x + xy\mathbf{i}_y + yz\mathbf{i}_z$; (b) $ye^{-x}\mathbf{i}_x - e^{-x}\mathbf{i}_y$.

3.4. For $\mathbf{A} = xy^2\mathbf{i}_x + x^2\mathbf{i}_y$, (a) find the net right-lateral differential of A_x and A_y normal to the z direction at the point $(2, 1, 0)$, and (b) find the locus of the points at which the net right-lateral differential of A_x and A_y normal to the z direction is zero.

3.5. Given $\mathbf{E} = 10 \cos (6\pi \times 10^8 t - 2\pi z) \, \mathbf{i}_x$ V/m, find $\mathbf{B}$ by using Faraday's law in differential form.

3.6. Show that the curl of $\left(\mathbf{i}_x \dfrac{\partial}{\partial x} + \mathbf{i}_y \dfrac{\partial}{\partial y} + \mathbf{i}_z \dfrac{\partial}{\partial z} \right) f$, that is, ∇f, where f is any scalar function of x, y, and z, is zero. Then find the scalar function for which $\nabla f = y\mathbf{i}_x + x\mathbf{i}_y$.

3.7. Given $\mathbf{E} = E_0 z^2 \sin \omega t \, \mathbf{i}_x$ and it is known that $\mathbf{J}$ is zero and $\mathbf{B}$ has only a y component, find $\mathbf{B}$ by using Ampere's circuital law in differential form. Then find $\mathbf{E}$ from $\mathbf{B}$ by using Faraday's law in differential form. Comment on your result.

3.8. Assuming $\mathbf{H} = H_x(z, t) \, \mathbf{i}_x$ and considering a rectangular closed path in the xz plane, carry out the derivation of Ampere's circuital law in differential form similar to that in the text.

3.9. Given $\mathbf{B} = \dfrac{10^{-7}}{3} \cos (6\pi \times 10^8 t - 2\pi z) \, \mathbf{i}_y$ Wb/m^2 and it is known that $\mathbf{J} = 0$, find $\mathbf{E}$ by using Ampere's circuital law in differential form. Then find $\mathbf{B}$ from $\mathbf{E}$ by using Faraday's law in differential form. Comment on your result.

3.10. Assuming $\mathbf{J} = 0$, determine which of the following pairs of E_x and H_y simultaneously satisfy the two Maxwell's equations in differential form given by (3.7) and (3.23):

(a) $E_x = 10 \cos 2\pi z \cos 6\pi \times 10^8 t$, $\quad H_y = \dfrac{1}{12\pi} \sin 2\pi z \sin 6\pi \times 10^8 t$

(b) $E_x = (t - z\sqrt{\mu_0 \epsilon_0})$ $\quad H_y = \sqrt{\dfrac{\epsilon_0}{\mu_0}}(t - z\sqrt{\mu_0 \epsilon_0})$

(c) $E_x = z^2 \sin \omega t$, $\quad H_y = -\dfrac{\omega \epsilon_0}{3} z^3 \cos \omega t$

3.11. A current distribution is given by

$$\mathbf{J} = \begin{cases} -J_0 \mathbf{i}_x & \text{for } -a < z < 0 \\ J_0 \mathbf{i}_x & \text{for } 0 < z < a \end{cases}$$

where J_0 is a constant. Using Ampere's circuital law in differential form and symmetry considerations, find the magnetic field everywhere.

3.12. A current distribution is given by

$$\mathbf{J} = J_0 \left(1 - \frac{|z|}{a}\right)\mathbf{i}_x \qquad \text{for } -a < z < a$$

where J_0 is a constant. Using Ampere's circuital law in differential form and symmetry considerations, find the magnetic field everywhere.

3.13. Assume that the velocity of water in the stream of Fig. 3.7(a) decreases linearly from a maximum at the top surface to zero at the bottom surface, with the velocity at the top surface given by Fig. 3.7(b). Discuss the curl of the velocity vector field with the aid of the curl meter.

3.14. For the vector field $\mathbf{r} = x\mathbf{i}_x + y\mathbf{i}_y + z\mathbf{i}_z$, discuss the behavior of the curl meter and verify your reasoning by evaluating the curl of $\mathbf{r}$.

3.15. Discuss the curl of the vector field $y\mathbf{i}_x - x\mathbf{i}_y$ with the aid of the curl meter.

3.16. Verify Stokes' theorem for the vector field $\mathbf{A} = y\mathbf{i}_x + z\mathbf{i}_y + x\mathbf{i}_z$ and the closed path comprising the straight lines from $(1, 0, 0)$ to $(0, 1, 0)$, from $(0, 1, 0)$ to $(0, 0, 1)$, and from $(0, 0, 1)$ to $(1, 0, 0)$.

3.17. Verify Stokes' theorem for the vector field $\mathbf{A} = e^{-y}\mathbf{i}_x - xe^{-y}\mathbf{i}_y$ and any closed path of your choice.

3.18. For the vector $\mathbf{A} = yz\mathbf{i}_x + zx\mathbf{i}_y + xy\mathbf{i}_z$, use Stokes' theorem to show that $\oint_C \mathbf{A} \cdot d\mathbf{l}$ is zero for any closed path C. Then evaluate $\int \mathbf{A} \cdot d\mathbf{l}$ from the origin to the point $(1, 1, 2)$ along the curve $x = \sqrt{2} \sin t$, $y = \sqrt{2} \sin t$, $z = (8/\pi)t$.

3.19. Find the divergences of the following vector fields:
(a) $3xy^2\mathbf{i}_x + 3x^2y\mathbf{i}_y + z^3\mathbf{i}_z$; (b) $2xy\mathbf{i}_x - y^2\mathbf{i}_y$.

3.20. For $\mathbf{A} = xy\mathbf{i}_x + yz\mathbf{i}_y + zx\mathbf{i}_z$, (a) find the net longitudinal differential of the components of $\mathbf{A}$ at the point $(1, 1, 1)$, and (b) find the locus of the points at which the net longitudinal differential of the components of $\mathbf{A}$ is zero.

3.21. For each of the following vectors, find the curl and the divergence and discuss your results: (a) $xy\mathbf{i}_x$; (b) $y\mathbf{i}_x$; (c) $x\mathbf{i}_x$; (d) $y\mathbf{i}_x + x\mathbf{i}_y$.

3.22. A charge distribution is given by

$$\rho = \rho_0\left(1 - \frac{|x|}{a}\right) \qquad \text{for } -a < x < a$$

where ρ_0 is a constant. Using Gauss' law for the electric field in differential form and symmetry considerations, find the electric field everywhere.

3.23. A charge distribution is given by

$$\rho = \rho_0\frac{x}{a} \qquad \text{for } -a < x < a$$

where ρ_0 is a constant. Using Gauss' law for the electric field in differential form and symmetry considerations, find the electric field everywhere.

3.24. Given $\mathbf{D} = x^2y\mathbf{i}_x - y^3\mathbf{i}_y$, find the charge density at (a) the point (2, 1, 0) and (b) the point (3, 2, 0).

3.25. Determine which of the following vectors can represent a magnetic flux density vector $\mathbf{B}$: (a) $y\mathbf{i}_x - x\mathbf{i}_y$; (b) $x\mathbf{i}_x + y\mathbf{i}_y$; (c) $z^3 \cos \omega t\, \mathbf{i}_y$.

3.26. Given $\mathbf{J} = e^{-x^2}\mathbf{i}_x$, find the time rate of decrease of the charge density at (a) the point (0, 0, 0) and (b) the point (1, 0, 0).

3.27. For the vector field $\mathbf{r} = x\mathbf{i}_x + y\mathbf{i}_y + z\mathbf{i}_z$, discuss the behavior of the divergence meter, and verify your reasoning by evaluating the divergence of $\mathbf{r}$.

3.28. Discuss the divergence of the vector field $y\mathbf{i}_x - x\mathbf{i}_y$ with the aid of the divergence meter.

3.29. Verify the divergence theorem for the vector field $\mathbf{A} = x\mathbf{i}_x + y\mathbf{i}_y + z\mathbf{i}_z$ and the closed surface bounding the volume within the hemisphere of radius unity above the xy plane and centered at the origin.

3.30. Verify the divergence theorem for the vector field $\mathbf{A} = xy\mathbf{i}_x + yz\mathbf{i}_y + zx\mathbf{i}_z$ and the closed surface of the volume bounded by the planes $x = 0$, $x = 1$, $y = 0$, $y = 1$, $z = 0$, and $z = 1$

3.31. For the vector $\mathbf{A} = y^2\mathbf{i}_y - 2yz\mathbf{i}_z$, use the divergence theorem to show that $\oint_s \mathbf{A} \cdot d\mathbf{S}$ is zero for any closed surface S. Then evaluate $\int \mathbf{A} \cdot d\mathbf{S}$ over the surface $x + y + z = 1$, $x > 0$, $y > 0$, $z > 0$.

3.32. Show that $\nabla \cdot \nabla \times \mathbf{A} = 0$ for any $\mathbf{A}$ in two ways: (a) by evaluating $\nabla \cdot \nabla \times \mathbf{A}$ in Cartesian coordinates, and (b) by using Stokes' and divergence theorems.

4. WAVE PROPAGATION IN FREE SPACE

In Chaps. 2 and 3 we learned Maxwell's equations in integral form and in differential form. We now have the knowledge of the fundamental laws of electromagnetics that enable us to embark upon the study of the elements of their applications. Many of these applications are based on electromagnetic wave phenomena, and hence it is necessary to gain an understanding of the basic principles of wave propagation, which is our goal in this chapter. In particular, we shall consider wave propagation in free space. We shall then in the next chapter consider the interaction of the wave fields with materials to extend the application of Maxwell's equations to material media and discuss wave propagation in material media.

We shall employ an approach in this chapter that will enable us not only to learn how the coupling between space-variations and time-variations of the electric and magnetic fields, as indicated by Maxwell's equations, results in wave motion but also to illustrate the basic principle of radiation of waves from an antenna, which will be treated in detail in Chap. 8. In this process, we will also learn several techniques of analysis pertinent to field problems. We shall augment our discussion of radiation and propagation of waves by considering such examples as the principle of an antenna array and the Doppler effect. Finally, we shall discuss power flow and energy storage associated with the wave motion and introduce the Poynting vector.

4.1 THE INFINITE PLANE CURRENT SHEET

In Chap. 3 we learned that the space-variations of the electric and magnetic field components are related to the time-variations of the magnetic and electric field components, respectively, through Maxwell's equations. This interdependence gives rise to the phenomenon of electromagnetic wave propagation. In the general case, electromagnetic wave propagation involves electric and magnetic fields having more than one component, each dependent on all three coordinates, in addition to time. However, a simple and very useful type of wave that serves as a building block in the study of electromagnetic waves consists of electric and magnetic fields that are perpendicular to each other and to the direction of propagation and are uniform in planes perpendicular to the direction of propagation. These waves are known as "uniform plane waves." By orienting the coordinate axes such that the electric field is in the x direction, the magnetic field is in the y direction, and the direction of propagation is in the z direction, as shown in Fig. 4.1, we have

$$\mathbf{E} = E_x(z, t)\mathbf{i}_x \tag{4.1}$$

$$\mathbf{H} = H_y(z, t)\mathbf{i}_y \tag{4.2}$$

Uniform plane waves do not exist in practice because they cannot be produced by finite-sized antennas. At large distances from physical antennas and ground, however, the waves can be approximated as uniform plane waves. Furthermore, the principles of guiding of electromagnetic waves along transmission lines and waveguides and the principles of many other wave phenomena can be studied basically in terms of uniform plane waves. Hence it is very important that we understand the principles of uniform plane wave propagation.

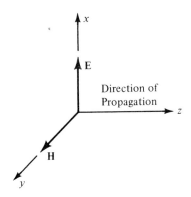

Figure 4.1. Directions of electric and magnetic fields and direction of propagation for a simple case of uniform plane wave.

In order to illustrate the phenomenon of interaction of electric and magnetic fields giving rise to uniform plane electromagnetic wave propagation, and the principle of radiation of electromagnetic waves from an antenna, we shall consider a simple, idealized, hypothetical source. This source consists of an infinite sheet lying in the xy plane, as shown in Fig. 4.2. On this infinite plane

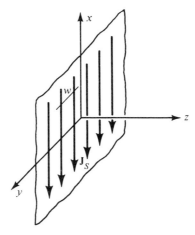

Figure 4.2. Infinite plane sheet in the xy plane carrying surface current of uniform density.

sheet a uniformly distributed current varying sinusoidally with time flows in the negative x direction. Since the current is distributed on a surface, we talk of surface current density in order to express the current distribution mathematically. The surface current density, denoted by the symbol $\mathbf{J}_S$, is a vector quantity having the magnitude equal to the current per unit width (amp/m) crossing an infinitesimally long line, on the surface, oriented so as to maximize the current. The direction of $\mathbf{J}_S$ is then normal to the line and toward the side of the current flow. In the present case, the surface current density is given by

$$\mathbf{J}_S = -J_{S0} \cos \omega t\, \mathbf{i}_x \qquad \text{for } z = 0 \qquad (4.3)$$

where J_{S0} is a constant and ω is the radian frequency of the sinusoidal time-variation of the current density.

Because of the uniformity of the surface current density on the infinite sheet, if we consider any line of width w parallel to the y axis, as shown in Fig. 4.2, the current crossing that line is simply given by w times the current density, that is, $wJ_{S0} \cos \omega t$. If the current density is non-uniform, we have to perform an integration along the width of the line in order to find the current crossing the line. In view of the sinusoidal time-variation of the current density, the current crossing the width w actually alternates between negative

x and positive x directions, that is, downward and upward. The time history of the current flow for one period of the sinusoidal variation is illustrated in Fig. 4.3, with the lengths of the lines indicating the magnitudes of the current.

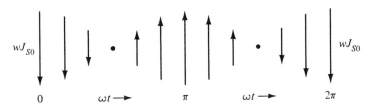

Figure 4.3. Time history of current flow across a line of width w parallel to the y axis for the current sheet of Fig. 4.2.

4.2   MAGNETIC FIELD ADJACENT TO THE CURRENT SHEET

In the previous section we introduced the infinite current sheet lying in the xy plane and upon which a surface current flows with density given by

$$\mathbf{J}_S = -J_{S0} \cos \omega t \, \mathbf{i}_x \qquad (4.4)$$

Our goal is to find the electromagnetic field due to this time-varying current distribution. In order to do this, we have to solve Faraday's and Ampere's circuital laws simultaneously. Since we have here only an x component of the current density independent of x and y, the equations of interest are

$$\frac{\partial E_x}{\partial z} = -\frac{\partial B_y}{\partial t} \qquad (4.5)$$

$$\frac{\partial H_y}{\partial z} = -\left(J_x + \frac{\partial D_x}{\partial t}\right) \qquad (4.6)$$

The quantity J_x on the right side of (4.6) represents volume current density whereas we now have a surface current density. Furthermore, in the free space on either side of the current sheet the current density is zero and the differential equations reduce to

$$\frac{\partial E_x}{\partial z} = -\frac{\partial B_y}{\partial t} \qquad (4.7)$$

$$\frac{\partial H_y}{\partial z} = -\frac{\partial D_x}{\partial t} \qquad (4.8)$$

To obtain the solutions for E_x and H_y on either side of the current sheet, we therefore have to solve these two differential equations simultaneoulsy.

To obtain a start on the solution, however, we need to consider the surface current distribution and find the magnetic field immediately adjacent to the current sheet. This is done by making use of Ampere's circuital law in integral form given by

$$\oint_{c} \mathbf{H} \cdot d\mathbf{l} = \int_{S} \mathbf{J} \cdot d\mathbf{S} + \frac{d}{dt} \int_{S} \mathbf{D} \cdot d\mathbf{S} \qquad (4.9)$$

and applying it to a rectangular closed path *abcda*, as shown in Fig. 4.4, with the sides *ab* and *cd* lying immediately adjacent to the current sheet, that is, touching the current sheet, and on either side of it. This choice of the rectangular path is not arbitrary but is intentionally chosen to achieve the task of finding the required magnetic field. First, we note from (4.6) that an *x*-directed current density gives rise to a magnetic field in the *y* direction. At the source of the current, this magnetic field must also have a differential in the third direction, namely, the *z* direction. In fact, from symmetry considerations, we can say that H_y on *ab* and *cd* must be equal in magnitude and opposite in direction.

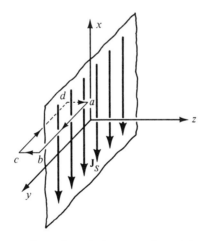

Figure 4.4. Rectangular path enclosing a portion of the current on the infinite plane current sheet.

If we now consider the line integral of **H** around the rectangular path *abcda*, we have

$$\int_{abcda} \mathbf{H} \cdot d\mathbf{l} = \int_{a}^{b} \mathbf{H} \cdot d\mathbf{l} + \int_{b}^{c} \mathbf{H} \cdot d\mathbf{l} + \int_{c}^{d} \mathbf{H} \cdot d\mathbf{l} + \int_{d}^{a} \mathbf{H} \cdot d\mathbf{l} \quad (4.10)$$

The second and the fourth integrals on the right side of (4.10) are, however, equal to zero since **H** is normal to the sides *bc* and *da* and furthermore *bc* and *da* are infinitesimally small. The first and third integrals on the right side of (4.10) are given by

$$\int_a^b \mathbf{H} \cdot d\mathbf{l} = [H_y]_{ab}(ab)$$

$$\int_c^d \mathbf{H} \cdot d\mathbf{l} = -[H_y]_{cd}(cd)$$

Thus

$$\oint_{abcda} \mathbf{H} \cdot d\mathbf{l} = [H_y]_{ab}(ab) - [H_y]_{cd}(cd) = 2[H_y]_{ab}(ab) \qquad (4.11)$$

since $[H_y]_{cd} = -[H_y]_{ab}$.

We have just evaluated the left side of (4.9) for the particular problem under consideration here. To complete the task of finding the magnetic field adjacent to the current sheet, we now evaluate the right side of (4.9), which consists of two terms. The second term is, however, zero since the area enclosed by the rectangular path is zero in view of the infinitesimally small thickness of the current sheet. The first term is not zero since there is a current flowing on the sheet. Thus the first term is simply equal to the current enclosed by the path *abcda* in the right-hand sense, that is, the current crossing the width *ab* toward the negative *x* direction. This is equal to the surface current density multiplied by the width *ab*, that is, $J_{S0} \cos \omega t$ (*ab*). Thus substituting for the quantities on either side of (4.9), we have

$$2[H_y]_{ab}(ab) = J_{S0} \cos \omega t \, (ab)$$

or

$$[H_y]_{ab} = \frac{J_{S0}}{2} \cos \omega t \qquad (4.12)$$

It then follows that

$$[H_y]_{cd} = -\frac{J_{S0}}{2} \cos \omega t \qquad (4.13)$$

Thus immediately adjacent to the current sheet the magnetic field intensity has a magnitude $\frac{J_{S0}}{2} \cos \omega t$ and is directed in the positive *y* direction on the side $z > 0$ and in the negative *y* direction on the side $z < 0$. This is illustrated in Fig. 4.5. It is cautioned that this result is true only for points right next to the current sheet since if we consider points at some distance from the current sheet, the second term on the right side of (4.9) will no longer be zero.

The technique we have used here for finding the magnetic field adjacent to the time-varying current sheet by using Ampere's circuital law in integral form is a standard procedure for finding the static electric and magnetic fields due to static charge and current distributions, possessing certain sym-

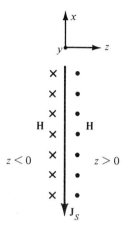

Figure 4.5. Magnetic field adjacent to and on either side of the infinite plane current sheet.

metries, by using Gauss' law for the electric field and Ampere's circuital law in integral forms, respectively, as we have already demonstrated in Chap. 2. Since for the static field case the terms involving time derivatives are zero, Ampere's circuital law simplifies to

$$\oint_C \mathbf{H} \cdot d\mathbf{l} = \int_S \mathbf{J} \cdot d\mathbf{S}$$

Hence, if the current distribution were not varying with time, then in order to compute the magnetic field we can choose a rectangular path of any width bc and it would still enclose the same current, namely, the current on the sheet. Thus the magnetic field would be independent of the distance away from the sheet on either side of it. There are several problems in static fields that can be solved in this manner. We shall not discuss these here; instead, we shall include a few cases in the problems for the interested reader and we shall continue with the derivation of the electromagnetic field due to our time-varying current sheet in the following section.

4.3 SUCCESSIVE SOLUTION OF MAXWELL'S EQUATIONS*

In the preceding section we found the magnetic field ajdacent to the infinite plane sheet of current introduced in Sec. 4.1. Now, to find the solutions for the fields everywhere on either side of the current sheet, let us first

---

*This section may be omitted without loss of continuity.

consider the region $z > 0$. In this region the fields simultaneously satisfy the two differential equations (4.7) and (4.8) and with the constraint that the magnetic field at $z = 0$ is given by (4.12). To find the solutions for these differential equations, we have a choice of starting with the solution for H_y given by (4.12) and solving them successively and repeatedly in a step-by-step manner until the solutions satisfy both differential equations or of combining the two differential equations into one and then solving the single equation subject to the constraint at $z = 0$. Although it is somewhat longer and tedious, we shall use the first approach in this section in order to obtain a feeling for the mechanism of interaction between the electric and magnetic fields. We shall consider the second and more conventional approach in the following section.

To simplify the task of the repetitive solution of the two differential equations (4.7) and (4.8) we shall employ the phasor technique. Thus by letting

$$E_x(z, t) = \text{Re} \, [\bar{E}_x(z) e^{j\omega t}] \tag{4.14}$$

$$H_y(z, t) = \text{Re} \, [\bar{H}_y(z) e^{j\omega t}] \tag{4.15}$$

where Re stands for "real part of" and $\bar{E}_x(z)$ and $\bar{H}_y(z)$ are the phasors corresponding to the time functions $E_x(z, t)$ and $H_y(z, t)$, respectively, and replacing the time functions in (4.7) and (4.8) by the corresponding phasor functions and $\partial/\partial t$ by $j\omega$, we obtain the differential equations for the phasor functions as

$$\frac{\partial \bar{E}_x}{\partial z} = -j\omega \bar{B}_y = -j\omega \mu_0 \bar{H}_y \tag{4.16}$$

$$\frac{\partial \bar{H}_y}{\partial z} = -j\omega \bar{D}_x = -j\omega \epsilon_0 \bar{E}_x \tag{4.17}$$

We also note that since (4.12) can be written as

$$[H_y]_{ab} = \text{Re} \left(\frac{J_{S0}}{2} e^{j\omega t} \right)$$

the solution for the phasor $\bar{H}_y$ at $z = 0$ is given by

$$[\bar{H}_y]_{z=0} = \frac{J_{S0}}{2} \tag{4.18}$$

We start with (4.18) and solve (4.16) and (4.17) successively and repeatedly, and after obtaining the final solutions for $\bar{E}_x$ and $\bar{H}_y$, we put them in (4.14) and (4.15), respectively, to obtain the solutions for the real fields.

Thus starting with (4.18) and substituting it in (4.16), we get

$$\frac{\partial \bar{E}_x}{\partial z} = -j\omega\mu_0\frac{J_{S0}}{2}$$

Integrating both sides of this equation with respect to z, we have

$$\bar{E}_x = -j\omega\mu_0\frac{J_{S0}z}{2} + \bar{C}$$

where $\bar{C}$ is the constant of integration. This constant of integration must, however, be equal to $[\bar{E}_x]_{z=0}$ since the first term on the right side tends to zero as $z \rightarrow 0$. Thus

$$\bar{E}_x = -j\omega\mu_0\frac{J_{S0}z}{2} + [\bar{E}_x]_{z=0} \qquad (4.19)$$

Now, substituting (4.19) into (4.17), we obtain

$$\frac{\partial \bar{H}_y}{\partial z} = -j\omega\epsilon_0\left\{-j\omega\mu_0\frac{J_{S0}z}{2} + [\bar{E}_x]_{z=0}\right\}$$

$$= -j\omega\epsilon_0[\bar{E}_x]_{z=0} - \omega^2\mu_0\epsilon_0\frac{J_{S0}z}{2}$$

$$\bar{H}_y = -j\omega\epsilon_0 z[\bar{E}_x]_{z=0} - \omega^2\mu_0\epsilon_0\frac{J_{S0}z^2}{4} + [\bar{H}_y]_{z=0}$$

$$= -j\omega\epsilon_0 z[\bar{E}_x]_{z=0} - \omega^2\mu_0\epsilon_0\frac{J_{S0}z^2}{4} + \frac{J_{S0}}{2}$$

$$= -j\omega\epsilon_0 z[\bar{E}_x]_{z=0} + \frac{J_{S0}}{2}\left(1 - \frac{\omega^2\mu_0\epsilon_0 z^2}{2}\right) \qquad (4.20)$$

We have thus obtained a second-order solution for $\bar{H}_y$, which, however, does not satisfy (4.16) together with the solution for $\bar{E}_x$ given by (4.19). Hence we must continue the step-by-step solution by substituting (4.20) into (4.16) and finding a higher-order solution for $\bar{E}_x$ and so on. Thus by substituting (4.20) into (4.16), we get

$$\frac{\partial \bar{E}_x}{\partial z} = -j\omega\mu_0\left\{-j\omega\epsilon_0 z[\bar{E}_x]_{z=0} + \frac{J_{S0}}{2}\left(1 - \frac{\omega^2\mu_0\epsilon_0 z^2}{2}\right)\right\}$$

$$= -\omega^2\mu_0\epsilon_0 z[\bar{E}_x]_{z=0} - j\omega\mu_0\frac{J_{S0}}{2}\left(1 - \frac{\omega^2\mu_0\epsilon_0 z^2}{2}\right)$$

$$\bar{E}_x = -\omega^2\mu_0\epsilon_0\frac{z^2}{2}[\bar{E}_x]_{z=0} - j\omega\mu_0\frac{J_{S0}}{2}\left(z - \frac{\omega^2\mu_0\epsilon_0 z^3}{6}\right) + [\bar{E}_x]_{z=0}$$

$$= [\bar{E}_x]_{z=0}\left(1 - \frac{\omega^2\mu_0\epsilon_0 z^2}{2}\right) - \frac{j\omega\mu_0 J_{S0}}{2}\left(z - \frac{\omega^2\mu_0\epsilon_0 z^3}{6}\right) \qquad (4.21)$$

From (4.17), we then have

$$\frac{\partial \bar{H}_y}{\partial z} = -j\omega\epsilon_0[\bar{E}_x]_{z=0}\left(1 - \frac{\omega^2\mu_0\epsilon_0 z^2}{2}\right) - \frac{\omega^2\mu_0\epsilon_0 J_{S0}}{2}\left(z - \frac{\omega^2\mu_0\epsilon_0 z^3}{6}\right)$$

$$\bar{H}_y = -j\omega\epsilon_0[\bar{E}_x]_{z=0}\left(z - \frac{\omega^2\mu_0\epsilon_0 z^3}{6}\right)$$

$$- \frac{\omega^2\mu_0\epsilon_0 J_{S0}}{2}\left(\frac{z^2}{2} - \frac{\omega^2\mu_0\epsilon_0 z^4}{24}\right) + [\bar{H}_y]_{z=0}$$

$$= -j\omega\epsilon_0[\bar{E}_x]_{z=0}\left(z - \frac{\omega^2\mu_0\epsilon_0 z^3}{6}\right)$$

$$+ \frac{J_{S0}}{2}\left(1 - \frac{\omega^2\mu_0\epsilon_0 z^2}{2} + \frac{\omega^4\mu_0^2\epsilon_0^2 z^4}{24}\right) \tag{4.22}$$

Continuing in this manner, we will get infinite series expressions for $\bar{E}_x$ and $\bar{H}_y$ as follows:

$$\bar{E}_x = [\bar{E}_x]_{z=0}\left[1 - \frac{(\beta z)^2}{2!} + \frac{(\beta z)^4}{4!} - \cdots\right]$$

$$- j\frac{\eta_0 J_{S0}}{2}\left[\beta z - \frac{(\beta z)^3}{3!} + \frac{(\beta z)^5}{5!} - \cdots\right] \tag{4.23}$$

$$\bar{H}_y = -j\frac{1}{\eta_0}[\bar{E}_x]_{z=0}\left[\beta z - \frac{(\beta z)^3}{3!} + \frac{(\beta z)^5}{5!} - \cdots\right]$$

$$+ \frac{J_{S0}}{2}\left[1 - \frac{(\beta z)^2}{2!} + \frac{(\beta z)^4}{4!} - \cdots\right] \tag{4.24}$$

where we have introduced the notations

$$\beta = \omega\sqrt{\mu_0\epsilon_0} \tag{4.25}$$

$$\eta_0 = \sqrt{\frac{\mu_0}{\epsilon_0}} \tag{4.26}$$

It is left to the student to verify that the two expressions (4.23) and (4.24) simultaneously satisfy the two differential equations (4.16) and (4.17). Now, noting that

$$\cos\beta z = 1 - \frac{(\beta z)^2}{2!} + \frac{(\beta z)^4}{4!} - \cdots$$

$$\sin\beta z = \beta z - \frac{(\beta z)^3}{3!} + \frac{(\beta z)^5}{5!} + \cdots$$

and substituting into (4.23) and (4.24), we have

$$\bar{E}_x = [\bar{E}_x]_{z=0} \cos \beta z - j\frac{\eta_0 J_{S0}}{2} \sin \beta z \tag{4.27}$$

$$\bar{H}_y = -j\frac{1}{\eta_0}[\bar{E}_x]_{z=0} \sin \beta z + \frac{J_{S0}}{2} \cos \beta z \tag{4.28}$$

We now obtain the expressions for the real fields by putting (4.27) and (4.28) into (4.14) and (4.15), respectively. Thus

$$E_x(z, t) = \text{Re}\left\{[\bar{E}_x]_{z=0} \cos \beta z \, e^{j\omega t} - j\frac{\eta_0 J_{S0}}{2} \sin \beta z \, e^{j\omega t}\right\}$$

$$= \cos \beta z \, \text{Re}\,\{[\bar{E}_x]_{z=0}e^{j\omega t}\} + \frac{\eta_0 J_{S0}}{2} \sin \beta z \, \text{Re}\,[e^{j(\omega t - \pi/2)}]$$

$$= \cos \beta z \,(C \cos \omega t + D \sin \omega t) + \frac{\eta_0 J_{S0}}{2} \sin \beta z \sin \omega t \tag{4.29}$$

$$H_y(z, t) = \text{Re}\left\{-j\frac{1}{\eta_0}[\bar{E}_x]_{z=0} \sin \beta z \, e^{j\omega t} + \frac{J_{S0}}{2} \cos \beta z \, e^{j\omega t}\right\}$$

$$= \frac{1}{\eta_0} \sin \beta z \, \text{Re}\,\{[\bar{E}_x]_{z=0} \, e^{j(\omega t - \pi/2)}\} + \frac{J_{S0}}{2} \cos \beta z \, \text{Re}\,[e^{j\omega t}]$$

$$= \frac{1}{\eta_0} \sin \beta z \,(C \sin \omega t - D \cos \omega t) + \frac{J_{S0}}{2} \cos \beta z \cos \omega t \tag{4.30}$$

where we have replaced the quantity $\text{Re}\,\{[\bar{E}_x]_{z=0}e^{j\omega t}\}$ by $(C \cos \omega t + D \sin \omega t)$ in which C and D are arbitrary constants to be determined. Making use of trigonometric identities and proceeding further, we write (4.29) and (4.30) as

$$E_x(z, t) = \frac{2C + \eta_0 J_{S0}}{4} \cos (\omega t - \beta z) + \frac{2C - \eta_0 J_{S0}}{4} \cos (\omega t + \beta z)$$

$$+ \frac{D}{2} \sin (\omega t - \beta z) + \frac{D}{2} \sin (\omega t + \beta z) \tag{4.31}$$

$$H_y(z, t) = \frac{2C + \eta J_{S0}}{4\eta_0} \cos (\omega t - \beta z) - \frac{2C - \eta_0 J_{S0}}{4\eta_0} \cos (\omega t + \beta z)$$

$$+ \frac{D}{2\eta_0} \sin (\omega t - \beta z) - \frac{D}{2\eta_0} \sin (\omega t + \beta z) \tag{4.32}$$

Equation (4.32) is the solution for H_y which together with the solution for E_x given by (4.31) satisfies the two differential equations (4.7) and (4.8) and which reduces to (4.12) for $z = 0$. Likewise, we can obtain the solutions for H_y and E_x for the region $z < 0$ by starting with $[H_y]_{z=0-}$ given by (4.13) and proceeding in a similar manner. We shall however proceed with the evaluation of the constants C and D in (4.31) and (4.32). In order to do this, we first have to understand the meanings of the functions $\cos(\omega t \mp \beta z)$ and $\sin(\omega t \mp \beta z)$. We shall do this in Sec. 4.5.

4.4   SOLUTION BY WAVE EQUATION

In Sec. 4.3 we found the solutions to the two simultaneous differential equations (4.7) and (4.8) by solving them successively and repeatedly in a step-by-step manner. In this section we shall consider an alternative and more conventional method by combining the two equations into a single equation and then solving it. We recall that the two simultaneous differential equations to be satisfied in the free space on either side of the current sheet are

$$\frac{\partial E_x}{\partial z} = -\frac{\partial B_y}{\partial t} = -\mu_0 \frac{\partial H_y}{\partial t} \tag{4.33}$$

$$\frac{\partial H_y}{\partial z} = -\frac{\partial D_x}{\partial t} = -\epsilon_0 \frac{\partial E_x}{\partial t} \tag{4.34}$$

Differentiating (4.33) with respect to z and then substituting for $\partial H_y/\partial z$ from (4.34), we obtain

$$\frac{\partial^2 E_x}{\partial z^2} = -\mu_0 \frac{\partial}{\partial z}\left(\frac{\partial H_y}{\partial t}\right) = -\mu_0 \frac{\partial}{\partial t}\left(\frac{\partial H_y}{\partial z}\right) = -\mu_0 \frac{\partial}{\partial t}\left(-\epsilon_0 \frac{\partial E_x}{\partial t}\right)$$

or

$$\frac{\partial^2 E_x}{\partial z^2} = \mu_0 \epsilon_0 \frac{\partial^2 E_x}{\partial t^2} \tag{4.35}$$

We have thus eliminated H_y from (4.33) and (4.34) and obtained a single second-order partial differential equation involving E_x only.

Equation (4.35) is known as the "wave equation." A technique of solving this equation is the "separation of variables" technique. Since it is a differential equation involving two variables z and t, the technique consists of assuming that the required solution is the product of two functions, one of which is a function of z only and the second is a function of t only. Denoting these functions to be Z and T, respectively, we have

$$E_x(z, t) = Z(z)\,T(t) \tag{4.36}$$

Substituting (4.36) into (4.35) and dividing throughout by $\mu_0 \epsilon_0 Z(z)\,T(t)$, we obtain

$$\frac{1}{\mu_0 \epsilon_0 Z}\frac{d^2 Z}{dz^2} = \frac{1}{T}\frac{d^2 T}{dt^2} \tag{4.37}$$

In (4.37) the left side is a function of z only and the right side is a function of t only. In order for this to be satisfied, they both must be equal to a constant. Hence setting them equal to a constant, say α^2, we have

$$\frac{d^2Z}{dz^2} = \alpha^2 \mu_0 \epsilon_0 Z \tag{4.38a}$$

$$\frac{d^2T}{dt^2} = \alpha^2 T \tag{4.38b}$$

We have thus obtained two ordinary differential equations involving separately the two variables z and t; hence the technique is known as the "separation of variables" technique.

The constant α^2 in (4.38a) and (4.38b) is not arbitrary since for the case of the sinusoidally time-varying current source the fields must also be sinusoidally time-varying with the same frequency although not necessarily in phase with the source. Thus the solution for $T(t)$ must be of the form

$$T(t) = A \cos \omega t + B \sin \omega t \tag{4.39}$$

where A and B are arbitrary constants to be determined. Substitution of (4.39) into (4.38b) gives us $\alpha^2 = -\omega^2$. The solution for (4.38a) is then given by

$$\begin{aligned} Z(z) &= A' \cos \omega\sqrt{\mu_0\epsilon_0}z + B' \sin \omega\sqrt{\mu_0\epsilon_0}z \\ &= A' \cos \beta z + B' \sin \beta z \end{aligned} \tag{4.40}$$

where A' and B' are arbitrary constants to be determined and we have defined

$$\beta = \omega\sqrt{\mu_0\epsilon_0} \tag{4.41}$$

The solution for E_x is then given by

$$\begin{aligned} E_x &= (A' \cos \beta z + B' \sin \beta z)(A \cos \omega t + B \sin \omega t) \\ &= C \cos \beta z \cos \omega t + D \cos \beta z \sin \omega t \\ &\quad + C' \sin \beta z \cos \omega t + D' \sin \beta z \sin \omega t \end{aligned} \tag{4.42}$$

The corresponding solution for H_y can be obtained by substituting (4.42) into one of the two equations (4.33) and (4.34). Thus using (4.34), we get

$$\frac{\partial H_y}{\partial z} = -\epsilon_0[-\omega C \cos \beta z \sin \omega t + \omega D \cos \beta z \cos \omega t$$

$$-\omega C' \sin \beta z \sin \omega t + \omega D' \sin \beta z \cos \omega t]$$

$$H_y = \frac{\omega\epsilon_0}{\beta}[C \sin \beta z \sin \omega t - D \sin \beta z \cos \omega t$$

$$-C' \cos \beta z \sin \omega t + D' \cos \beta z \cos \omega t]$$

Defining

$$\eta_0 = \frac{\beta}{\omega\epsilon_0} = \frac{\omega\sqrt{\mu_0\epsilon_0}}{\omega\epsilon_0} = \sqrt{\frac{\mu_0}{\epsilon_0}} \qquad (4.43)$$

we have

$$H_y = \frac{1}{\eta_0}[C \sin \beta z \sin \omega t - D \sin \beta z \cos \omega t$$
$$- C' \cos \beta z \sin \omega t + D' \cos \beta z \cos \omega t] \qquad (4.44)$$

Equation (4.44) is the general solution for H_y valid on both sides of the current sheet. In order to deduce the arbitrary constants, we first recall that the magnetic field adjacent to the current sheet is given by

$$H_y = \begin{cases} \dfrac{J_{s0}}{2} \cos \omega t & \text{for } z = 0+ \\[3mm] -\dfrac{J_{s0}}{2} \cos \omega t & \text{for } z = 0- \end{cases} \qquad (4.45)$$

Thus for $z > 0$,

$$\frac{1}{\eta_0}[-C' \sin \omega t + D' \cos \omega t] = \frac{J_{s0}}{2} \cos \omega t$$

or

$$C' = 0 \quad \text{and} \quad D' = \frac{\eta_0 J_{s0}}{2}$$

giving us

$$H_y = \frac{J_{s0}}{2} \cos \beta z \cos \omega t + \frac{1}{\eta_0} \sin \beta z (C \sin \omega t - D \cos \omega t) \qquad (4.46)$$

$$E_x = \frac{\eta_0 J_{s0}}{2} \sin \beta z \sin \omega t + \cos \beta z (C \cos \omega t + D \sin \omega t) \qquad (4.47)$$

Making use of trigonometric identities and proceeding further, we write (4.47) and (4.46) as

$$E_x(z, t) = \frac{2C + \eta_0 J_{s0}}{4} \cos(\omega t - \beta z) + \frac{2C - \eta_0 J_{s0}}{4} \cos(\omega t + \beta z)$$
$$+ \frac{D}{2} \sin(\omega t - \beta z) + \frac{D}{2} \sin(\omega t + \beta z) \qquad (4.48)$$

$$H_y(z, t) = \frac{2C + \eta_0 J_{s0}}{4\eta_0} \cos(\omega t - \beta z) - \frac{2C - \eta_0 J_{s0}}{4\eta_0} \cos(\omega t + \beta z)$$
$$+ \frac{D}{2\eta_0} \sin(\omega t - \beta z) - \frac{D}{2\eta_0} \sin(\omega t + \beta z) \qquad (4.49)$$

Equation (4.49) is the solution for H_y which together with the solution for E_x given by (4.48) satisfies the two differential equations (4.7) and (4.8) and which reduces to (4.12) for $z = 0$. Similarly, we can obtain the solutions for H_y and E_x for the region $z < 0$ by using the value of $[H_y]_{z=0^-}$ to evaluate C' and D' in (4.44). We shall, however, proceed with the evaluation of the constants C and D in (4.48) and (4.49). In order to do this, we first have to understand the meanings of the functions $\cos(\omega t \mp \beta z)$ and $\sin(\omega t \mp \beta z)$. We shall do this in the following section.

4.5   UNIFORM PLANE WAVES

In the previous two sections we derived the solutions for E_x and H_y, due to the infinite plane sheet of sinusoidally time-varying uniform current density, for the region $z > 0$. These solutions consist of the functions $\cos(\omega t \mp \beta z)$ and $\sin(\omega t \mp \beta z)$, which are dependent on both time and distance. Let us first consider the function $\cos(\omega t - \beta z)$. To understand the behavior of this function, we note that for a fixed value of time it varies in a cosinusoidal manner with the distance z. Let us therefore consider three values of time $t = 0$, $t = \pi/4\omega$, and $t = \pi/2\omega$ and examine the sketches of this function versus z for these three times. By noting that

$$\text{for } t = 0, \quad \cos(\omega t - \beta z) = \cos(-\beta z) = \cos \beta z$$

$$\text{for } t = \frac{\pi}{4\omega}, \quad \cos(\omega t - \beta z) = \cos\left(\frac{\pi}{4} - \beta z\right)$$

$$\text{for } t = \frac{\pi}{2\omega}, \quad \cos(\omega t - \beta z) = \cos\left(\frac{\pi}{2} - \beta z\right) = \sin \beta z$$

we draw the sketches of the three functions as shown in Fig. 4.6.

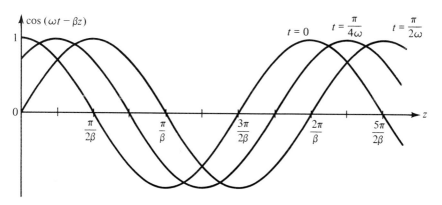

Figure 4.6. Sketches of the function $\cos(\omega t - \beta z)$ versus z for three values of t.

It is evident from Fig. 4.6 that the sketch of the function for $t = \pi/4\omega$ is a replica of the function for $t = 0$ except that it is shifted by a distance of $\pi/4\beta$ toward the positive z direction. Similarly, the sketch of the function for $t = \pi/2\omega$ is a replica of the function for $t = 0$ except that it is shifted by a distance of $\pi/2\beta$ toward the positive z direction. Thus as time progresses, the function shifts bodily to the right, that is, toward increasing values of z. In fact, we can even find the velocity with which the function is traveling by dividing the distance moved by the time elapsed. This gives

$$\text{velocity} = \frac{\pi/\beta - \pi/2\beta}{\pi/2\omega - 0} = \frac{\omega}{\beta} = \frac{\omega}{\omega\sqrt{\mu_0\epsilon_0}}$$

$$= \frac{1}{\sqrt{\mu_0\epsilon_0}} = \frac{1}{\sqrt{4\pi \times 10^{-7} \times 10^{-9}/36\pi}}$$

$$= 3 \times 10^8 \text{ m/s}$$

which is the velocity of light in free space. Thus the function $\cos(\omega t - \beta z)$ represents a "traveling wave" moving with a velocity ω/β toward the direction of increasing z. The wave is also known as the "positive going" or "(+) wave."

Similarly, by considering three values of time $t = 0$, $t = \pi/4\omega$, and $t = \pi/2\omega$ for the function $\cos(\omega t + \beta z)$, we obtain the sketches shown in Fig. 4.7. An examination of these sketches reveals that $\cos(\omega t + \beta z)$ represents a "traveling wave" moving with a velocity ω/β toward the direction of decreasing values of z. The wave is also known as the "negative going" or "(−) wave." Since the sine functions are cosine functions shifted in phase by $\pi/2$, it follows that $\sin(\omega t - \beta z)$ and $\sin(\omega t + \beta z)$ represent traveling waves moving in the positive and negative z directions, respectively.

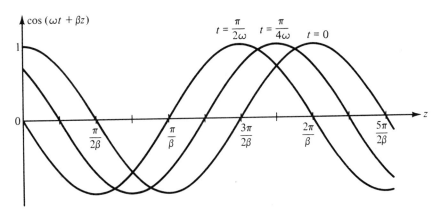

Figure 4.7. Sketches of the function $\cos(\omega t + \beta z)$ versus z for three values of t.

Returning to the solutions for E_x and H_y given by (4.31) and (4.32) or (4.48) and (4.49), we now know that these solutions consist of superpositions of traveling waves propagating away from and toward the current sheet. In the region $z > 0$ we, however, have to rule out traveling waves propagating toward the current sheet because such a situation requires a source of waves to the right of the sheet or an object that reflects the wave back toward the sheet. Thus we have

$$D = 0$$

$$2C - \eta_0 J_{S0} = 0 \quad \text{or} \quad C = \frac{\eta_0 J_{S0}}{2}$$

which give us finally

$$\left. \begin{array}{l} E_x = \dfrac{\eta_0 J_{S0}}{2} \cos{(\omega t - \beta z)} \\[2mm] H_y = \dfrac{J_{S0}}{2} \cos{(\omega t - \beta z)} \end{array} \right\} \text{ for } z > 0 \qquad (4.50)$$

Having found the solutions for the fields in the region $z > 0$, we can now consider the solutions for the fields in the region $z < 0$. From our discussion of the functions $\cos{(\omega t \mp \beta z)}$, we know that these solutions must be of the form $\cos{(\omega t + \beta z)}$ since this function represents a traveling wave progressing in the negative z direction, that is, away from the sheet in the region $z < 0$. Recalling that the magnetic field adjacent to the current sheet and to the left of it is given by

$$[H_y]_{z=0^-} = -\frac{J_{S0}}{2} \cos{\omega t}$$

we get

$$H_y = -\frac{J_{S0}}{2} \cos{(\omega t + \beta z)} \qquad \text{for } z < 0 \qquad (4.51\text{a})$$

The corresponding E_x can be obtained by simply substituting the result just obtained for H_y into one of the two differential equations (4.7) and (4.8). Thus using (4.7), we obtain

$$\frac{\partial E_x}{\partial z} = -\frac{\partial B_y}{\partial t} = -\frac{\mu_0 J_{S0}}{2} \omega \sin{(\omega t + \beta z)}$$

$$E_x = \frac{\mu_0 J_{S0}}{2} \frac{\omega}{\beta} \cos{(\omega t + \beta z)}$$

$$= \frac{\eta_0 J_{S0}}{2} \cos{(\omega t + \beta z)} \qquad \text{for } z < 0 \qquad (4.51\text{b})$$

Combining (4.50) and (4.51), we find that the solution for the electromagnetic field due to the infinite plane current sheet in the xy plane characterized by

$$\mathbf{J}_S = -J_{S0} \cos \omega t \, \mathbf{i}_x$$

is given by

$$\mathbf{E} = \frac{\eta_0 J_{S0}}{2} \cos (\omega t \mp \beta z) \, \mathbf{i}_x \qquad \text{for } z \gtrless 0 \qquad (4.52\text{a})$$

$$\mathbf{H} = \pm \frac{J_{S0}}{2} \cos (\omega t \mp \beta z) \, \mathbf{i}_y \qquad \text{for } z \gtrless 0 \qquad (4.52\text{b})$$

These results are illustrated in Fig. 4.8, which shows sketches of the current density on the sheet and the distance-variation of the electric and magnetic fields on either side of the current sheet for a few values of t. It can be seen from these sketches that the phenomenon is one of electromagnetic waves "radiating" away from the current sheet to either side of it, in step with the time-variation of the current density on the sheet.

The solutions that we have just obtained for the fields due to the time-varying infinite plane current sheet are said to correspond to "uniform plane electromagnetic waves" propagating away from the current sheet to either side of it. The terminology arises from the fact that the fields are *uniform* (that is, they do not vary with position) over the *planes* $z =$ constant. Thus the phase of the fields, that is, the quantity $(\omega t \pm \beta z)$, as well as the amplitudes of the fields, is uniform over the planes $z =$ constant. The magnitude of the rate of change of phase with distance z for any fixed time is β. The quantity β is therefore known as the "phase constant." Since the velocity of propagation of the wave, that is, ω/β, is the velocity with which a given constant phase progresses along the z direction, that is, along the direction of propagation, it is known as the "phase velocity" and is denoted by the symbol v_p. Thus

$$v_p = \frac{\omega}{\beta} \qquad (4.53)$$

The distance in which the phase changes by 2π radians for a fixed time is $2\pi/\beta$. This quantity is known as the "wavelength" and is denoted by the symbol λ. Thus

$$\lambda = \frac{2\pi}{\beta} \qquad (4.54)$$

Substituting (4.53) into (4.54), we obtain

$$\lambda = \frac{2\pi}{\omega/v_p} = \frac{v_p}{f}$$

or

$$\lambda f = v_p \qquad (4.55)$$

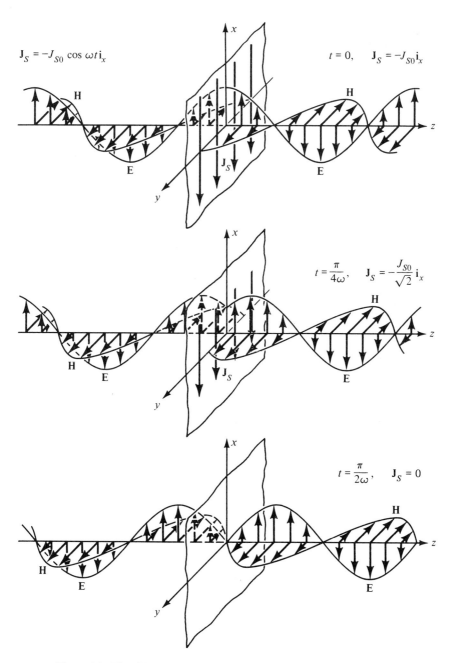

Figure 4.8. Time history of uniform plane electromagnetic wave radiating away from an infinite plane current sheet in free space.

Equation (4.55) is a simple relationship between the wavelength λ which is a parameter governing the variation of the field with distance for a fixed time and the frequency f which is a parameter governing the variation of the field with time for a fixed value of z. Since for free space $v_p = 3 \times 10^8$ m/s, we have

$$\lambda \text{ in meters} \times f \text{ in Hz} = 3 \times 10^8$$

$$\lambda \text{ in meters} \times f \text{ in MHz} = 300 \tag{4.56}$$

Other properties of uniform plane waves evident from (4.52) are that the electric and magnetic fields have components lying in the planes of constant phase and perpendicular to each other and to the direction of propagation. In fact, the cross product of $\mathbf{E}$ and $\mathbf{H}$ results in a vector that is directed along the direction of propagation, as can be seen by noting that

$$\mathbf{E} \times \mathbf{H} = E_x \mathbf{i}_x \times H_y \mathbf{i}_y$$

$$= \pm \frac{\eta_0 J_{s0}^2}{4} \cos^2 (\omega t \mp \beta z) \mathbf{i}_z \quad \text{for } z \gtrless 0 \tag{4.57}$$

Finally, we note that the ratio of E_x to H_y is given by

$$\frac{E_x}{H_y} = \begin{cases} \eta_0 \text{ for } z > 0, \text{ i.e., for the } (+) \text{ wave} \\ -\eta_0 \text{ for } z < 0, \text{ i.e., for the } (-) \text{ wave} \end{cases} \tag{4.58}$$

The quantity η_0 which is equal to $\sqrt{\mu_0/\epsilon_0}$ is known as the "intrinsic imped-ance" of free space. Its value is given by

$$\eta_0 = \sqrt{\frac{(4\pi \times 10^{-7}) \text{ H/m}}{(10^{-9}/36\pi) \text{ F/m}}} = \sqrt{(144\pi^2 \times 10^2) \text{ H/F}}$$

$$= 120\pi \text{ ohms} = 377 \text{ ohms} \tag{4.59}$$

Example 4.1. The electric field of a uniform plane wave is given by $\mathbf{E} = 10 \cos (3\pi \times 10^8 t - \pi z) \mathbf{i}_x$ V/m. Let us identify the various parameters asso-ciated with the uniform plane wave.

We recognize that

$$\omega = 3\pi \times 10^8 \text{ rad/s}$$

$$f = \frac{\omega}{2\pi} = 1.5 \times 10^8 \text{ Hz} = 150 \text{ MHz}$$

$$\beta = \pi \text{ rad/m}$$

$$\lambda = \frac{2\pi}{\beta} = 2 \text{ m}$$

$$v_p = \frac{\omega}{\beta} = \frac{3\pi \times 10^8}{\pi} = 3 \times 10^8 \text{ m/s}$$

Also, $\lambda f = v_p = 2 \times 1.5 \times 10^8 = 3 \times 10^8$ m/s. From (4.58) and since the given field represents a $(+)$ wave,

$$\mathbf{H} = \frac{E_x}{\eta_0}\mathbf{i}_y = \frac{10}{377}\cos(3\pi \times 10^8 t - \pi z)\,\mathbf{i}_y \text{ amp/m} \quad \blacksquare$$

Example 4.2. An antenna array consists of two or more antenna elements spaced appropriately and excited with currents having the appropriate amplitudes and phases in order to obtain a desired radiation characteristic. To illustrate the principle of an antenna array, let us consider two infinite plane parallel current sheets, spaced $\lambda/4$ apart and carrying currents of equal amplitudes but out of phase by $\pi/2$ as given by the densities

$$\mathbf{J}_{S1} = -J_{S0}\cos\omega t\,\mathbf{i}_x \qquad z = 0$$

$$\mathbf{J}_{S2} = -J_{S0}\sin\omega t\,\mathbf{i}_x \qquad z = \frac{\lambda}{4}$$

and find the electric field due to the array of the two current sheets.

We apply the result given by (4.52) to each current sheet separately and then use superposition to find the required total electric field due to the array of the two current sheets. Thus for the current sheet in the $z = 0$ plane, we have

$$\mathbf{E}_1 = \begin{cases} \dfrac{\eta_0 J_{S0}}{2}\cos(\omega t - \beta z)\,\mathbf{i}_x & \text{for } z > 0 \\[2mm] \dfrac{\eta_0 J_{S0}}{2}\cos(\omega t + \beta z)\,\mathbf{i}_x & \text{for } z < 0 \end{cases}$$

For the current sheet in the $z = \lambda/4$ plane, we have

$$\mathbf{E}_2 = \begin{cases} \dfrac{\eta_0 J_{S0}}{2}\sin\left[\omega t - \beta\left(z - \dfrac{\lambda}{4}\right)\right]\mathbf{i}_x & \text{for } z > \dfrac{\lambda}{4} \\[3mm] \dfrac{\eta_0 J_{S0}}{2}\sin\left[\omega t + \beta\left(z - \dfrac{\lambda}{4}\right)\right]\mathbf{i}_x & \text{for } z < \dfrac{\lambda}{4} \end{cases}$$

$$= \begin{cases} \dfrac{\eta_0 J_{S0}}{2}\sin\left(\omega t - \beta z + \dfrac{\pi}{2}\right)\mathbf{i}_x & \text{for } z > \dfrac{\lambda}{4} \\[3mm] \dfrac{\eta_0 J_{S0}}{2}\sin\left(\omega t + \beta z - \dfrac{\pi}{2}\right)\mathbf{i}_x & \text{for } z < \dfrac{\lambda}{4} \end{cases}$$

$$= \begin{cases} \dfrac{\eta_0 J_{S0}}{2}\cos(\omega t - \beta z)\,\mathbf{i}_x & \text{for } z > \dfrac{\lambda}{4} \\[3mm] -\dfrac{\eta_0 J_{S0}}{2}\cos(\omega t + \beta z)\,\mathbf{i}_x & \text{for } z < \dfrac{\lambda}{4} \end{cases}$$

Now, using superposition, we find the total electric field due to the two current sheets to be

$$\mathbf{E} = \mathbf{E}_1 + \mathbf{E}_2$$

$$= \begin{cases} \eta_0 J_{s0} \cos{(\omega t - \beta z)}\,\mathbf{i}_x & \text{for } z > \dfrac{\lambda}{4} \\[2mm] \eta_0 J_{s0} \sin{\omega t} \sin{\beta z}\,\mathbf{i}_x & \text{for } 0 < z < \dfrac{\lambda}{4} \\[2mm] 0 & \text{for } z < 0 \end{cases}$$

Thus the total field is zero in the region $z < 0$ and hence there is no radiation toward that side of the array. In the region $z > \lambda/4$ the total field is twice that of the field due to a single sheet. The phenomenon is illustrated in Fig. 4.9, which shows sketches of the individual fields E_{x1} and E_{x2} and the total field $E_x = E_{x1} + E_{x2}$ for a few values of t. The result that we have obtained here for the total field due to the array of two current sheets, spaced $\lambda/4$ apart and fed with currents of equal amplitudes but out of phase by $\pi/2$, is said to correspond to an "endfire" radiation pattern.   ■

Returning now to the solution for the electromagnetic field given by (4.52), let us ask ourselves the question, "How does the phase associated with the wave change with time as viewed by a moving observer?" To answer this question, let us consider the $(+)$ wave and an observer moving along the positive z direction with a velocity v_0 m/s, starting at $z = z_0$ at $t = 0$. Then the position of the observer as a function of time is given by $z = z_0 + v_0 t$ and the phase of the wave at that position is given by

$$\begin{aligned} \phi_{\text{obs}} &= \omega t - \beta(z_0 + v_0 t) \\ &= (\omega - \beta v_0)t - \beta z_0 \end{aligned} \tag{4.60}$$

Ignoring relativistic effects, the rate of change of phase with time or the radian frequency of the wave viewed by the moving observer is

$$\begin{aligned} \omega_{\text{obs}} &= \frac{d}{dt}[(\omega - \beta v_0)t - \beta z_0] \\ &= \omega - \beta v_0 = \omega - \frac{\omega}{v_p} v_0 \\ &= \omega\left(1 - \frac{v_0}{v_p}\right) \end{aligned}$$

or

$$\begin{aligned} f_{\text{obs}} &= \frac{\omega_{\text{obs}}}{2\pi} = f\left(1 - \frac{v_0}{v_p}\right) \\ &= f - \frac{v_0}{\lambda} \end{aligned} \tag{4.61}$$

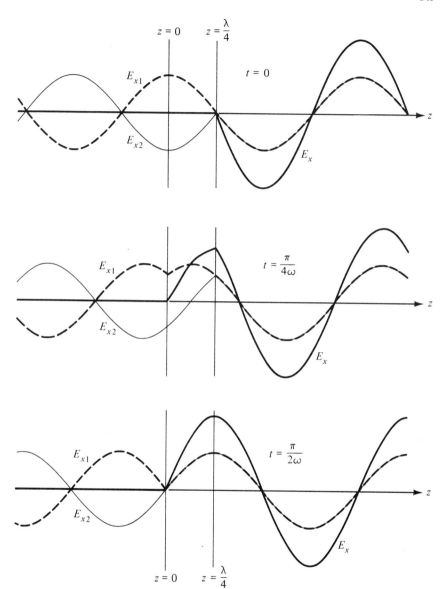

Figure 4.9. Time history of individual fields and the total field due to an array of two infinite plane parallel current sheets.

Thus the moving observer views a frequency that is different from that of the source of the wave. This phenomenon of a shift in the frequency of the wave is known as the "Doppler shift." For an observer moving along the direction of propagation, the Doppler-shifted frequency is less than the

actual frequency by the amount $f v_0/v_p$ or v_0/λ. For an observer moving opposite to the direction of propagation of the wave, the Doppler-shifted frequency
is higher than the actual frequency by the same amount. The situation is
illustrated in Fig. 4.10 which depicts the wave motion as viewed by a sta-

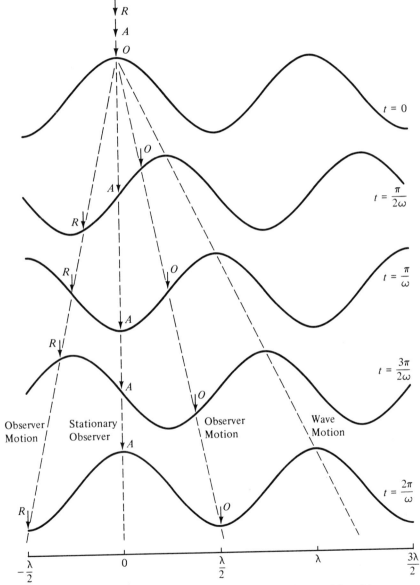

Figure 4.10. Wave motion as viewed by a stationary observer (A) and two
moving observers (O and R).

tionary observer (A) and two moving observers (O and R), one moving along, and the other moving opposite to the direction of propagation of the wave with a velocity which, for simplicity, is assumed to be one-half the phase velocity of the wave. From the series of sketches for one period of the wave, it can be seen that observer A views a complete cycle of the wave whereas observer O views only one-half cycle of the wave and observer R views one and one-half cycles of the wave during that period. Thus the stationary observer A views the same frequency as that of the wave, but moving observer O views a frequency that is one-half that of the wave and moving observer R views a frequency that is one and one-half that of the wave. The Doppler shifts are greater when relativistic effects are included.

Example 4.3.   Let us consider an automotive radar operating at a frequency $f = 9\text{ GHz} = 9000\text{ MHz}$ and determine the Doppler shift due to an automobile directly approaching the radar at a speed of 100 km/hr.

For the given frequency, $\lambda = \dfrac{300}{9000}\text{ m} = \dfrac{1}{30}\text{ m}$. Since $v_0 = 100\text{ km/hr} = \dfrac{10^5}{3600}\text{ m/s}$, the Doppler shift in frequency as given by (4.61) is

$$\Delta f_D = \frac{v_0}{\lambda} = \frac{10^5}{3600 \times (1/30)} = 833.3\text{ Hz}$$

Since the automotive radar operates on the signal reflected from the moving automobile, the actual Doppler shift is $2 \times 833.3\text{ Hz}$ or 1666.6 Hz. ∎

To discuss the phenomenon of Doppler shift further, let us consider the case of a satellite that transmits electromagnetic waves at a radian frequency ω and a receiver on the earth's surface. For simplicity, we shall consider the earth to be plane and the satellite orbit to be horizontal at a height h above the earth, as shown in Fig. 4.11. Let the satellite be overhead at $t = 0$ and its

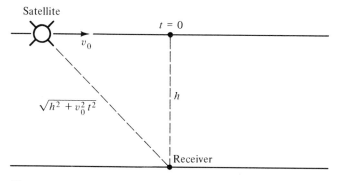

Figure 4.11. For the discussion of Doppler shift of a satellite signal.

velocity be v_0 so that its horizontal distance from the overhead point is $v_0 t$. Although the satellite is a point source, the waves at a large distance from it are approximately uniform plane waves with the constant phase surfaces normal to the line joining the point of observation to the satellite. In the present case, the distance between the satellite and the receiver is $\sqrt{h^2 + v_0^2 t^2}$. The phase of the wave as observed at the receiver is therefore given by

$$\phi_{\text{obs}} = \omega t - \beta \sqrt{h^2 + v_0^2 t^2} + \phi_0 \tag{4.62}$$

where ϕ_0 is the phase of the field at the satellite when it is at the overhead point. Thus the Doppler-shifted frequency observed at the receiver is given by

$$\omega_D = \frac{d\phi_{\text{obs}}}{dt} = \frac{d}{dt}(\omega t - \beta \sqrt{h^2 + v_0^2 t^2} + \phi_0)$$

$$= \omega - \frac{\beta v_0^2 t}{\sqrt{h^2 + v_0^2 t^2}}$$

$$= \omega \left(1 - \frac{v_0}{v_p} \frac{v_0 t}{\sqrt{h^2 + v_0^2 t^2}} \right) \tag{4.63}$$

A sketch of the variation of ω_D with t is shown in Fig. 4.12. Note that when the satellite is overhead, there is no Doppler shift.

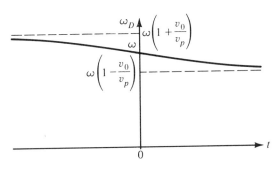

Figure 4.12. Doppler-shifted frequency versus time for the satellite signal of Fig. 4.11.

Example 4.4.   Let us consider a satellite at a height of 1000 km, operating at a frequency $f = 40$ MHz, and with an orbital velocity $v_0 = 7$ km/s and find the maximum Doppler shift.

From (4.63), the maximum Doppler shift is given by

$$[\Delta f_D]_{\text{max}} = f \frac{v_0}{v_p} = 40 \times 10^6 \times \frac{7 \times 10^3}{3 \times 10^8} = 933.3 \text{ Hz}$$

Proceeding further, the Doppler shifts are $933.3/\sqrt{5}$ and $933.3/\sqrt{2}$ or 417.4 Hz and 660.0 Hz when the horizontal distances of the satellite from the overhead point are 500 km and 1000 km, respectively.   ■

4.6 POYNTING VECTOR AND ENERGY STORAGE

In the preceding section we found the solution for the electromagnetic field due to an infinite plane current sheet situated in the $z = 0$ plane. For a surface current flowing in the negative x direction, we found the electric field on the sheet to be directed in the positive x direction. Since the current is flowing against the force due to the electric field, a certain amount of work must be done by the source of the current in order to maintain the current flow on the sheet. Let us consider a rectangular area of length Δx and width Δy on the current sheet as shown in Fig. 4.13. Since the cur-

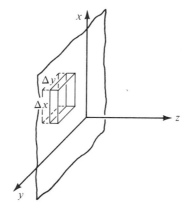

Figure 4.13. For the determination of power flow density associated with the electromagnetic field.

rent density is $J_{S0} \cos \omega t$, the charge crossing the width Δy in time dt is $dq = J_{S0} \, \Delta y \cos \omega t \, dt$ coulombs. The force exerted on this charge by the electric field is given by

$$\mathbf{F} = dq \, \mathbf{E} = J_{S0} \, \Delta y \cos \omega t \, dt \, E_x \mathbf{i}_x \qquad (4.64)$$

The amount of work required to be done against the electric field in displacing this charge by the distance Δx is

$$dw = F_x \, \Delta x = J_{S0} E_x \cos \omega t \, dt \, \Delta x \, \Delta y \qquad (4.65)$$

Thus the power supplied by the source of the current in maintaining the surface current over the area $\Delta x \, \Delta y$ is

$$\frac{dw}{dt} = J_{S0} E_x \cos \omega t \, \Delta x \, \Delta y \tag{4.66}$$

Recalling that E_x on the sheet is $\eta_0 \dfrac{J_{S0}}{2} \cos \omega t$, we obtain

$$\frac{dw}{dt} = \eta_0 \frac{J_{S0}^2}{2} \cos^2 \omega t \, \Delta x \, \Delta y \tag{4.67}$$

We would expect the power given by (4.67) to be carried by the electromagnetic wave, half of it to either side of the current sheet. To investigate this, we note that the quantity $\mathbf{E} \times \mathbf{H}$ has the units of

$$\frac{\text{newtons}}{\text{coulomb}} \times \frac{\text{amperes}}{\text{meter}} = \frac{\text{newtons}}{\text{coulomb}} \times \frac{\text{coulomb}}{\text{second-meter}} \times \frac{\text{meter}}{\text{meter}}$$

$$= \frac{\text{newton-meters}}{\text{second}} \times \frac{1}{(\text{meter})^2} = \frac{\text{watts}}{(\text{meter})^2}$$

which represents power density. Let us then consider the rectangular box enclosing the area $\Delta x \, \Delta y$ on the current sheet and with its sides almost touching the current sheet on either side of it, as shown in Fig. 4.13. Recalling that $\mathbf{E} \times \mathbf{H}$ is given by (4.57) and evaluating the surface integral of $\mathbf{E} \times \mathbf{H}$ over the surface of the rectangular box, we obtain the power flow out of the box as

$$\oint \mathbf{E} \times \mathbf{H} \cdot d\mathbf{S} = \eta_0 \frac{J_{S0}^2}{4} \cos^2 \omega t \, \mathbf{i}_z \cdot \Delta x \, \Delta y \, \mathbf{i}_z$$

$$+ \left(-\eta_0 \frac{J_{S0}^2}{4} \cos^2 \omega t \, \mathbf{i}_z \right) \cdot (-\Delta x \, \Delta y \, \mathbf{i}_z)$$

$$= \eta_0 \frac{J_{S0}^2}{2} \cos^2 \omega t \, \Delta x \, \Delta y \tag{4.68}$$

This result is exactly equal to the power supplied by the current source as given by (4.67).

We now interpret the quantity $\mathbf{E} \times \mathbf{H}$ as the power flow density vector associated with the electromagnetic field. It is known as the "Poynting vector" after J. H. Poynting and is denoted by the symbol $\mathbf{P}$. Although we have here introduced the Poynting vector by considering the specific case of the electromagnetic field due to the infinite plane current sheet, the interpretation that $\oint_s \mathbf{E} \times \mathbf{H} \cdot d\mathbf{S}$ is equal to the power flow out of the closed surface S is applicable in the general case.

Example 4.5.   Far from a physical antenna, that is, at a distance of several wavelengths from the antenna, the radiated electromagnetic waves are approximately uniform plane waves with their constant phase surfaces lying normal to the radial directions away from the antenna, as shown for two directions in Fig. 4.14. We wish to show from the Poynting vector and physical considerations that the electric and magnetic fields due to the antenna vary inversely proportional to the radial distance away from the antenna.

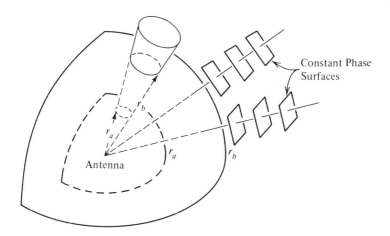

Figure 4.14. Radiation of electromagnetic waves far from a physical antenna.

From considerations of electric and magnetic fields of a uniform plane wave, the Poynting vector is directed everywhere in the radial direction indicating power flow radially away from the antenna and is proportional to the square of the magnitude of the electric field intensity. Let us now consider two spherical surfaces of radii r_a and r_b and centered at the antenna and insert a cone through these two surfaces such that the vertex is at the antenna, as shown in Fig. 4.14. Then the power crossing the portion of the spherical surface of radius r_b inside the cone must be the same as the power crossing the portion of the spherical surface of radius r_a inside the cone. Since these surface areas are proportional to the square of the radius and since the surface integral of the Poynting vector gives the power, the Poynting vector must be inversely proportional to the square of the radius. This in turn means that the electric field intensity and hence the magnetic field intensity must be inversely proportional to the radius.

Thus from these simple considerations we have established that far from a radiating antenna the electromagnetic field is inversely proportional to the radial distance away from the antenna. This reduction of the field intensity inversely proportional to the distance is known as the "free space reduction."

For example, let us consider communication from earth to the moon. The distance from the earth to the moon is approximately 38×10^4 km or 38×10^7 m. Hence the free space reduction factor for the field intensity is $10^{-7}/38$ or, in terms of decibels, the reduction is $20 \log_{10} 38 \times 10^7$, or 171.6 db.   ∎

Returning to the electromagnetic field due to the infinite plane current sheet, let us consider the region $z > 0$. The magnitude of the Poynting vector in this region is given by

$$P_z = E_x H_y = \eta_0 \frac{J_{S0}^2}{4} \cos^2 (\omega t - \beta z) \qquad (4.69)$$

The variation of P_z with z for $t = 0$ is shown in Fig. 4.15. If we now consider a rectangluar box lying between $z = z$ and $z = z + \Delta z$ planes and having dimensions Δx and Δy in the x and y directions, respectively, we would in general obtain a nonzero result for the power flowing out of the box, since $\partial P_z/\partial z$ is not everywhere zero. Thus there is some energy stored in the volume of the box. We then ask ourselves the question, "Where does this energy reside?" A convenient way of interpretation is to attribute the energy storage to the electric and magnetic fields.

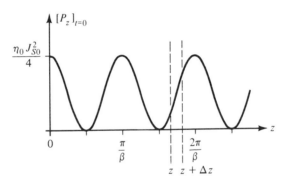

Figure 4.15. For the discussion of energy storage in electric and magnetic fields.

To discuss the energy storage in the electric and magnetic fields further, we evaluate the power flow out of the rectangular box. Thus

$$\oint_S \mathbf{P} \cdot d\mathbf{S} = [P_z]_{z+\Delta z} \, \Delta x \, \Delta y - [P_z]_z \, \Delta x \, \Delta y$$

$$= \frac{[P_z]_{z+\Delta z} - [P_z]_z}{\Delta z} \, \Delta x \, \Delta y \, \Delta z$$

$$= \frac{\partial P_z}{\partial z} \, \Delta v \qquad (4.70)$$

where Δv is the volume of the box. Letting P_z equal $E_x H_y$ and using (4.7) and (4.8), we obtain

$$\oint_s \mathbf{P} \cdot d\mathbf{S} = \frac{\partial}{\partial z} [E_x H_y] \, \Delta v$$

$$= \left(H_y \frac{\partial E_x}{\partial z} + E_x \frac{\partial H_y}{\partial z} \right) \Delta v$$

$$= \left(-H_y \frac{\partial B_y}{\partial t} - E_x \frac{\partial D_x}{\partial t} \right) \Delta v$$

$$= -\mu_0 H_y \frac{\partial H_y}{\partial t} \, \Delta v - \epsilon_0 E_x \frac{\partial E_x}{\partial t} \, \Delta v$$

$$= -\frac{\partial}{\partial t} \left(\frac{1}{2} \mu_0 H_y^2 \, \Delta v \right) - \frac{\partial}{\partial t} \left(\frac{1}{2} \epsilon_0 E_x^2 \, \Delta v \right) \tag{4.71}$$

Equation (4.71), which is known as Poynting's theorem, tells us that the power flow out of the box is equal to the sum of the time rates of decrease of the quantities $\frac{1}{2} \epsilon_0 E_x^2 \, \Delta v$ and $\frac{1}{2} \mu_0 H_y^2 \, \Delta v$. These quantities are obviously the energies stored in the electric and magnetic fields, respectively, in the volume of the box. It then follows that the energy densities associated with the electric and magnetic fields are $\frac{1}{2} \epsilon_0 E_x^2$ and $\frac{1}{2} \mu_0 H_y^2$, respectively. It is left to the student to verify that the quantities $\frac{1}{2} \epsilon_0 E^2$ and $\frac{1}{2} \mu_0 H^2$ do indeed have the units J/m³. Once again, although we have obtained these results by considering the particular case of the uniform plane wave, they hold in general.

Summarizing our discussion in this section, we have introduced the Poynting vector $\mathbf{P} = \mathbf{E} \times \mathbf{H}$ as the power flow density associated with the electromagnetic field characterized by the electric and magnetic fields, $\mathbf{E}$ and $\mathbf{H}$, respectively. The surface integral of $\mathbf{P}$ over a closed surface always gives the correct result for the power flow out of that surface. There is energy storage associated with the electric and magnetic fields with the energy densities given by

$$w_e = \frac{1}{2} \epsilon_0 E^2 \tag{4.72}$$

and

$$w_m = \frac{1}{2} \mu_0 H^2 \tag{4.73}$$

respectively.

4.7 SUMMARY

In this chapter we studied the principles of uniform plane wave propagation in free space. Uniform plane waves are a building block in the study of electromagnetic wave propagation. They are the simplest type of solutions resulting from the coupling of the electric and magnetic fields in Maxwell's curl equations. We learned that uniform plane waves have their electric and magnetic fields perpendicular to each other and to the direction of propagation. The fields are *uniform* in the *planes* perpendicular to the direction of propagation.

We obtained the uniform plane wave solution to Maxwell's equations by considering an infinite plane current sheet in the xy plane with uniform surface current density given by

$$\mathbf{J}_S = -J_{S0} \cos \omega t \, \mathbf{i}_x \, \text{amp/m} \qquad (4.74)$$

and deriving the electromagnetic field due to the current sheet to be given by

$$\mathbf{E} = \frac{\eta_0 J_{S0}}{2} \cos (\omega t \mp \beta z) \, \mathbf{i}_x \qquad \text{for } z \gtrless 0 \qquad (4.75a)$$

$$\mathbf{H} = \pm \frac{J_{S0}}{2} \cos (\omega t \mp \beta z) \, \mathbf{i}_y \qquad \text{for } z \gtrless 0 \qquad (4.75b)$$

In (4.75a) and (4.75b), $\cos (\omega t - \beta z)$ represents wave motion in the positive z direction, whereas $\cos (\omega t + \beta z)$ represents wave motion in the negative z direction. Thus (4.75a) and (4.75b) correspond to waves propagating away from the current sheet to either side of it. Since the fields are independent of x and y, they represent uniform plane waves.

The quantity $\beta \, (= \omega \sqrt{\mu_0 \epsilon_0})$ is the phase constant, that is, the magnitude of the rate of change of phase with distance along the direction of propagation, for a fixed time. The phase velocity v_p, that is, the velocity with which a particular constant phase progresses along the direction of propagation, is given by

$$v_p = \frac{\omega}{\beta} \qquad (4.76)$$

The wavelength λ, that is, the distance along the direction of propagation in which the phase changes by 2π radians, for a fixed time, is given by

$$\lambda = \frac{2\pi}{\beta} \qquad (4.77)$$

The wavelength is related to the frequency f in a simple manner as given by

$$v_p = \lambda f \qquad (4.78)$$

which follows from (4.76) and (4.77). The quantity $\eta_0 \, (= \sqrt{\mu_0/\epsilon_0})$ is the intrinsic impedance of free space. It is the ratio of the magnitude of $\mathbf{E}$ to the magnitude of $\mathbf{H}$ and has a value of 120π ohms.

In the process of deriving the electromagnetic field due to the infinite plane current sheet, we used two approaches and learned several useful techniques. These are discussed in the following:

1. The determination of the magnetic field adjacent to the current sheet by employing Ampere's circuital law in integral form: This is a common procedure used in the computation of static fields due to charge and current distributions possessing certain symmetries. In Chap. 6 we shall derive the "boundary conditions," that is, the relationships between the fields on either side of an interface between two different media, by applying Maxwell's equations in integral form to closed paths and surfaces straddling the boundary as we have done here in the case of the current sheet.

2. The successive, step-by-step solution of the two Maxwell's curl equations, to obtain the final solution consistent with the two equations, starting with the solution obtained for the field adjacent to the current sheet: This technique provided us a feel for the phenomenon of "radiation" of electromagnetic waves resulting from the time-varying current distribution and the interaction between the electric and magnetic fields. We shall use this kind of approach and the knowledge gained on wave propagation to obtain in Chap. 8 the complete electromagnetic field due to an elemental antenna, which forms the basis for the study of physical antennas.

3. The solution of wave equation by the separation of variables technique: This is the standard technique employed in the solution of partial differential equations involving multiple variables. We shall use it in Chap. 9 to solve Laplace's equation in two dimensions.

4. The application of phasor technique for the solution of the differential equations: The phasor technique is a convenient tool for analyzing sinusoidal steady-state problems as we learned in Chap. 1. We shall continue to use it in the following chapters.

We also learned that there is power flow and energy storage associated with the wave propagation that accounts for the work done in maintaining the current flow on the sheet. The power flow density is given by the Poynting vector

$$\mathbf{P} = \mathbf{E} \times \mathbf{H}$$

and the energy densities associated with the electric and magnetic fields are given, respectively, by

$$w_e = \frac{1}{2} \epsilon_0 E^2$$

$$w_m = \frac{1}{2} \mu_0 H^2$$

The surface integral of the Poynting vector over a given closed surface gives the total power flow out of the volume bounded by that surface.

Finally, we have augmented our study of uniform plane wave propagation in free space by illustrating (a) the principle of an antenna array, (b) the Doppler effect, and (c) the inverse distance dependence of the fields far from a physical antenna.

REVIEW QUESTIONS

4.1. What is a uniform plane wave?

4.2. Why is the study of uniform plane waves important?

4.3. How is the surface current density vector defined? Distinguish it from the volume current density vector.

4.4. How do you find the current crossing a given line on a sheet of surface current?

4.5. Why is it that Ampere's circuital law in integral form is used to find the magnetic field adjacent to the current sheet of Fig. 4.2?

4.6. Why is the path chosen to evaluate the magnetic field in Fig. 4.4 rectangular?

4.7. Outline the application of Ampere's circuital law in integral form to find the magnetic field adjacent to the current sheet of Fig. 4.2.

4.8. Why is the displacement current enclosed by the rectangular path *abcda* in Fig. 4.4 equal to zero?

4.9. How would you use Ampere's circuital law in differential form to find the magnetic field adjacent to the current sheet?

4.10. If the current density on the infinite plane current sheet of Fig. 4.2 were directed in the positive y direction, what would be the directions of the magnetic field adjacent to the current sheet and on either side of it?

4.11. Why are the results given by (4.12) and (4.13) for the magnetic field not valid for points at some distance from the current sheet?

4.12. Under what conditions would a result obtained for the magnetic field adjacent to the infinite plane current sheet of Fig. 4.2 be valid at points distant from the current sheet?

4.13. Briefly outline the procedure involved in the successive solution of Maxwell's equations.

4.14. How does the technique of successive solution of Maxwell's equations reveal the interaction between the electric and magnetic fields giving rise to wave propagation?

4.15. State the wave equation for the case of $\mathbf{E} = E_x(z, t)\mathbf{i}_x$. How is it derived?

4.16. Briefly outline the separation of variables technique of solving the wave equation.

4.17. Discuss how the function $\cos(\omega t - \beta z)$ represents a traveling wave propagating in the positive z direction.

4.18. Discuss how the function $\cos(\omega t + \beta z)$ represents a traveling wave propagating in the negative z direction.

4.19. Give some examples of nonsinusoidally time-varying functions representing traveling waves propagating in the positive z direction.

4.20. Discuss how the solution for the electromagnetic field given by (4.52) corresponds to that of a uniform plane wave.

4.21. Why is the quantity β in $\cos(\omega t - \beta z)$ known as the phase constant?

4.22. What is phase velocity? How is it related to the radian frequency and the phase constant of the wave?

4.23. Define wavelength. How is it related to the phase constant?

4.24. What is the relationship between frequency, wavelength, and phase velocity? What is the wavelength in free space for a frequency of 15 MHz?

4.25. What is the direction of propagation for a uniform plane wave having its electric field in the negative y direction and its magnetic field in the positive z direction?

4.26. What is the direction of the magnetic field for a uniform plane wave having its electric field in the positive z direction and propagating in the positive x direction?

4.27. What is intrinsic impedance? What is its value for free space?

4.28. Discuss the principle of an antenna array.

4.29. What should be the spacing and the relative phase angle of the current densities for an array of two infinite, plane, parallel current sheets of uniform densities, equal in magnitude, to confine their radiation to the region between the two sheets?

4.30. What is the Doppler effect? Illustrate with some examples.

4.31. When is the Doppler shift of a satellite signal frequency zero? Why?

4.32. How can a Doppler shift be observed for the case of a stationary transmitter and a stationary receiver?

4.33. Why is a certain amount of work involved in maintaining current flow on the sheet of Fig. 4.2? How is this work accounted for?

4.34. What is a Poynting vector? What is its physical significance?

4.35. What is the physical interpretation of the surface integral of the Poynting vector over a closed surface?

4.36. Discuss how the fields far from a physical antenna vary inversely proportional to the distance from the antenna.

4.37. Discuss the interpretation of energy storage in the electric and magnetic fields of a uniform plane wave.

4.38. What are the energy densities associated with the electric and magnetic fields?

PROBLEMS

4.1. An infinite plane sheet lying in the $z = 0$ plane carries a current of uniform density $\mathbf{J}_S = -0.1\mathbf{i}_x$ amp/m. Find the currents crossing the following straight lines: (a) from $(0, 0, 0)$ to $(0, 2, 0)$; (b) from $(0, 0, 0)$ to $(2, 0, 0)$; (c) from $(0, 0, 0)$ to $(2, 2, 0)$.

4.2. An infinite plane sheet lying in the $z = 0$ plane carries a current of non-uniform density $\mathbf{J}_S = -0.1e^{-|y|}\mathbf{i}_x$ amp/m. Find the currents crossing the following straight lines: (a) from $(0, 0, 0)$ to $(0, 1, 0)$; (b) from $(0, 0, 0)$ to $(0, \infty, 0)$; (c) from $(0, 0, 0)$ to $(1, 1, 0)$.

4.3. An infinite plane sheet lying in the $z = 0$ plane carries a current of uniform density

$$\mathbf{J}_S = (-0.1 \cos \omega t \, \mathbf{i}_x + 0.1 \sin \omega t \, \mathbf{i}_y) \text{ amp/m}$$

Find the currents crossing the following straight lines: (a) from $(0, 0, 0)$ to $(0, 2, 0)$; (b) from $(0, 0, 0)$ to $(2, 0, 0)$; (c) from $(0, 0, 0)$ to $(2, 2, 0)$.

4.4. An infinite plane sheet lying in the $z = 0$ plane carries a current of uniform density

$$\mathbf{J}_S = (-0.2 \cos \omega t \, \mathbf{i}_x + 0.2 \sin \omega t \, \mathbf{i}_y) \text{ amp/m}$$

Find the magnetic field intensities adjacent to the sheet and on either side of it. What is the polarization of the field?

4.5. An infinite plane sheet lying in the $z = 0$ plane carries a current of non-uniform density $\mathbf{J}_S = -0.2e^{-|y|} \cos \omega t \, \mathbf{i}_x$ amp/m. Find the magnetic field intensities adjacent to the current sheet and on either side of it at (a) the point $(0, 1, 0)$ and (b) the point $(2, 2, 0)$.

4.6. Current flows with uniform density $\mathbf{J} = J_0\mathbf{i}_x$ amp/m² in the region $|z| < a$. Using Ampere's circuital law in integral form and symmetry considerations, find $\mathbf{H}$ everywhere.

4.7. Current flows with nonuniform density $\mathbf{J} = J_0(1 - |z|/a)\mathbf{i}_x$ amp/m² in the region $|z| < a$, where J_0 is a constant. Using Ampere's circuital law in integral form and symmetry considerations, find $\mathbf{H}$ everywhere.

4.8. For an infinite plane sheet of charge lying in the xy plane with uniform surface charge density ρ_{s0} C/m², find the electric field intensity on both sides of the sheet by using Gauss' law for the electric field in integral form and symmetry considerations.

4.9. Charge is distributed with uniform density $\rho = \rho_0$ C/m³ in the region $|x| < a$. Using Gauss' law for the electric field in integral form and symmetry considerations, find **E** everywhere.

4.10. Charge is distributed with nonuniform density $\rho = \rho_0(1 - |x|/a)$ C/m³ in the region $|x| < a$, where ρ_0 is a constant. Using Gauss' law for the electric field in integral form and symmetry considerations, find **E** everywhere.

4.11. Verify that expressions (4.23) and (4.24) simultaneously satisfy the differential equations (4.16) and (4.17).

4.12. For the infinite plane current sheet in the $z = 0$ plane carrying surface current of density $\mathbf{J}_S = -J_{S0}t\mathbf{i}_x$ amp/m, where J_{S0} is a constant, find the magnetic field adjacent to the current sheet. Then use the method of successive solution of Maxwell's equations to show that for $z > 0$,

$$E_x = \left(\frac{2C + \eta_0 J_{S0}}{4}\right)(t - z\sqrt{\mu_0\epsilon_0}) + \left(\frac{2C - \eta_0 J_{S0}}{4}\right)(t + z\sqrt{\mu_0\epsilon_0})$$

$$H_y = \left(\frac{2C + \eta_0 J_{S0}}{4\eta_0}\right)(t - z\sqrt{\mu_0\epsilon_0}) - \left(\frac{2C - \eta_0 J_{S0}}{4\eta_0}\right)(t + z\sqrt{\mu_0\epsilon_0})$$

where C is a constant.

4.13. For the infinite plane current sheet in the $z = 0$ plane carrying surface current of density $\mathbf{J}_S = -J_{S0}t^2\mathbf{i}_x$ amp/m, where J_{S0} is a constant, find the magnetic field adjacent to the current sheet. Then use the method of successive solution of Maxwell's equations to show that for $z > 0$,

$$E_x = \left(\frac{2C + \eta_0 J_{S0}}{4}\right)(t - z\sqrt{\mu_0\epsilon_0})^2 + \left(\frac{2C - \eta_0 J_{S0}}{4}\right)(t + z\sqrt{\mu_0\epsilon_0})^2$$

$$H_y = \left(\frac{2C + \eta_0 J_{S0}}{4\eta_0}\right)(t - z\sqrt{\mu_0\epsilon_0})^2 - \left(\frac{2C - \eta_0 J_{S0}}{4\eta_0}\right)(t + z\sqrt{\mu_0\epsilon_0})^2$$

where C is a constant.

4.14. Verify that expressions (4.48) and (4.49) simultaneously satisfy the differential equations (4.7) and (4.8), and that (4.49) reduces to (4.12) for $z = 0+$.

4.15. Show that $(t - z\sqrt{\mu_0\epsilon_0})^2$ and $(t + z\sqrt{\mu_0\epsilon_0})^2$ are solutions of the wave equation. With the aid of sketches, discuss the nature of these functions.

4.16. For arbitrary time-variation of the fields, show that the solutions for the differential equations (4.33) and (4.34) are

$$E_x = Af(t - z\sqrt{\mu_0\epsilon_0}) + Bg(t + z\sqrt{\mu_0\epsilon_0})$$

$$H_y = \frac{1}{\eta_0}[Af(t - z\sqrt{\mu_0\epsilon_0}) - Bg(t + z\sqrt{\mu_0\epsilon_0})]$$

where A and B are arbitrary constants. Discuss the nature of the functions $f(t - z\sqrt{\mu_0\epsilon_0})$ and $g(t + z\sqrt{\mu_0\epsilon_0})$.

4.17. In Problems 4.12 and 4.13, evaluate the constant C and obtain the solutions for E_x and H_y in the region $z > 0$. Then write the solutions for E_x and H_y in the region $z < 0$.

4.18. The electric field intensity of a uniform plane wave is given by

$$\mathbf{E} = 37.7 \cos (6\pi \times 10^8 t + 2\pi z) \, \mathbf{i}_y \, \text{V/m}.$$

Find (a) the frequency, (b) the wavelength, (c) the phase velocity, (d) the direction of propagation of the wave, and (e) the associated magnetic field intensity vector **H**.

4.19. An infinite plane sheet lying in the $z = 0$ plane carries a surface current of density

$$\mathbf{J}_S = (-0.2 \cos 6\pi \times 10^8 t \, \mathbf{i}_x - 0.1 \cos 12\pi \times 10^8 t \, \mathbf{i}_x) \, \text{amp/m}$$

Find the expressions for the electric and magnetic fields on either side of the sheet.

4.20. An infinite plane sheet lying in the $z = 0$ plane carries a surface current of density $\mathbf{J}_S = -J_S(t)\mathbf{i}_x$, where $J_S(t)$ is the periodic function shown in Fig. 4.16. Find and sketch (a) H_y versus t for $z = 0+$, (b) E_x versus t for $z = 150$ m, and (c) E_x versus z for $t = 1$ μs.

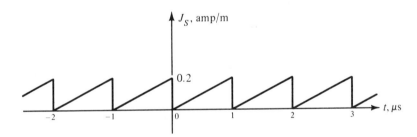

Figure 4.16. For Problem 4.20.

4.21. The time-variation of the electric field intensity E_x in the $z = 600$ m plane of a uniform plane wave propagating away from an infinite plane current sheet lying in the $z = 0$ plane is given by the periodic function shown in Fig. 4.17.

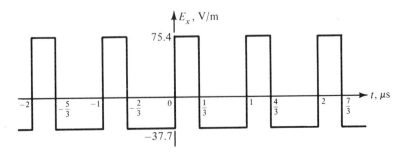

Figure 4.17. For Problem 4.21.

Find and sketch (a) E_x versus t for $z = 200$ m, (b) E_x versus z for $t = 0$, and (c) H_y versus z for $t = \frac{1}{3}$ μs.

4.22. The time-variation of the electric field intensity E_x in the $z = 300$ m plane of a uniform plane wave propagating away from an infinite plane current sheet lying in the $z = 0$ plane is given by the aperiodic function shown in Fig. 4.18. Find and sketch (a) E_x versus t for $z = 600$ m, (b) E_x versus z for $t = 1$ μsec, and (c) H_y versus z for $t = 2$ μsec.

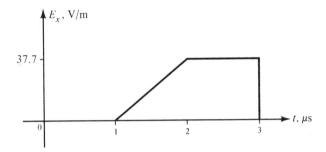

Figure 4.18. For Problem 4.22.

4.23. An array is formed by two infinite plane parallel current sheets with the current densities given by

$$\mathbf{J}_{S1} = -J_{S0} \cos \omega t \, \mathbf{i}_x \qquad z = 0$$

$$\mathbf{J}_{S2} = -J_{S0} \cos \omega t \, \mathbf{i}_x \qquad z = \frac{\lambda}{2}$$

where J_{S0} is a constant. Find the electric field intensity in all three regions: (a) $z < 0$; (b) $0 < z < \lambda/2$; (c) $z > \lambda/2$.

4.24. Determine the spacing, relative amplitudes, and phase angles of current densities for an array of two infinite plane parallel current sheets required to obtain a radiation characteristic such that the field radiated to one side of the array is twice that of the field radiated to the other side of the array.

4.25. For an array of two infinite plane parallel current sheets with the current densities given by

$$\mathbf{J}_{S1} = -J_{S0} \cos \omega t \, \mathbf{i}_x \qquad z = 0$$

$$\mathbf{J}_{S2} = -J_{S0} \cos \omega t \, \mathbf{i}_y \qquad z = \frac{\lambda}{2}$$

where J_{S0} is a constant, find the electric field in all three regions: (a) $z < 0$; (b) $0 < z < \lambda/2$; (c) $z > \lambda/2$. Discuss the polarization of the field in all three regions.

4.26. For an array of two infinite plane parallel current sheets with the current densities given by

$$\mathbf{J}_{S1} = -J_{S0} \cos \omega t \, \mathbf{i}_x \qquad z = 0$$

$$\mathbf{J}_{S2} = -J_{S0} \cos \omega t \, \mathbf{i}_y \qquad z = \frac{\lambda}{4}$$

where J_{s0} is a constant, find the electric field in all three regions: (a) $z < 0$; (b) $0 < z < \lambda/4$; (c) $z > \lambda/4$. Discuss the polarization of the field in all three regions.

4.27. The electric field intensity of a uniform plane wave is given by

$$\mathbf{E} = 37.7 \cos (6\pi \times 10^8 t - 2\pi z) \, \mathbf{i}_x \text{ V/m}$$

(a) What is the Doppler shift for an observer moving in the positive x direction? (b) Find the Doppler shift for an observer moving in the positive z direction with a velocity 3 km/s. (c) Find the magnitude of the Doppler shift for an observer moving along the straight line path $x = y = z$ with a velocity 3 km/s.

4.28. Consider an observer moving on the circumference of a circle of radius a in the xz plane with an angular velocity ω_0 rad/s, in the field of a uniform plane wave of frequency f propagating in the positive z direction. Find and sketch the Doppler shift observed by the moving observer as a function of position on the circle.

4.29. An experimental rocket is fired with an initial velocity v_0 m/s and making an angle of 45° with the horizontal. Communication is maintained between the rocket and the launching site. Show that the received frequency when the rocket is at its apogee is Doppler shifted by the amount $0.6324 \, v_0 f/c$ where f is the transmitted frequency and c is the velocity of light in free space. Assume plane earth.

4.30. Show that the time-average value of the magnitude of the Poynting vector given by (4.69) is one-half its peak value. For an antenna radiating a time-average power of 150 kW, find the peak value of the electric field intensity at a distance of 100 km from the antenna. Assume the antenna to be radiating equally in all directions.

4.31. The electric field of a uniform plane wave propagating in the positive z direction is given by

$$\mathbf{E} = E_0 \cos (\omega t - \beta z) \, \mathbf{i}_x + E_0 \sin (\omega t - \beta z) \, \mathbf{i}_y$$

where E_0 is a constant. (a) Find the corresponding magnetic field $\mathbf{H}$. (b) Find the Poynting vector.

4.32. Show that the quantities $\frac{1}{2} \epsilon_0 E^2$ and $\frac{1}{2} \mu_0 H^2$ have the units joules/m³.

4.33. Show that the energy is stored equally in the electric and magnetic fields of a traveling wave.

5. WAVE PROPAGATION IN MATERIAL MEDIA

In Chapter 4 we introduced wave propagation in free space by considering the infinite plane current sheet of uniform, sinusoidally time-varying current density. We learned that the solution for the electromagnetic field due to the infinite plane current sheet represents uniform plane electromagnetic waves propagating away from the sheet to either side of it. With the knowledge of the principles of uniform plane wave propagation in free space, we are now ready to consider wave propagation in material media, which is our goal in this chapter. Materials contain charged particles that respond to applied electric and magnetic fields and give rise to currents, which modify the properties of wave propagation from those associated with free space.

We shall learn that there are three basic phenomena resulting from the interaction of the charged particles with the electric and magnetic fields. These are conduction, polarization, and magnetization. Although a given material may exhibit all three properties, it is classified as a conductor, a dielectric, or a magnetic material depending on whether conduction, polarization, or magnetization is the predominant phenomenon. Thus we shall introduce these three kinds of materials one at a time and develop a set of relations known as the constitutive relations which enable us to avoid the necessity of explicitly taking into account the interaction of the charged particles with the fields. We shall then use these constitutive relations together with Maxwell's equations to first discuss uniform plane wave propagation in a general material medium and then consider several special cases.

5.1 CONDUCTORS

We recall that the classical model of an atom postulates a tightly bound, positively charged nucleus surrounded by a diffuse cloud of electrons spinning and orbiting around the nucleus. In the absence of an applied electromagnetic field, the force of attraction between the positively charged nucleus and the negatively charged electrons is balanced by the outward centrifugal force to maintain stable electronic orbits. The electrons can be divided into "bound" electrons and "free" or "conduction" electrons. The bound electrons can be displaced but not removed from the influence of the nucleus. The conduction electrons are constantly under thermal agitation, being released from the parent atom at one point and recaptured by another atom at a different point.

In the absence of an applied field, the motion of the conduction electrons is completely random; the average thermal velocity on a "macroscopic" scale, that is, over volumes large compared with atomic dimensions, is zero so that there is no net current and the electron cloud maintains a fixed position. With the application of an electromagnetic field, an additional velocity is superimposed on the random velocities, predominatly due to the electric force. This causes drift of the average position of the electrons in a direction opposite to that of the applied electric field. Due to the frictional mechanism provided by collisions of the electrons with the atomic lattice, the electrons, instead of accelerating under the influence of the electric field, drift with an average drift velocity proportional in magnitude to the applied electric field. This phenomenon is known as "conduction," and the resulting current due to the electron drift is known as the "conduction current."

In certain materials a large number of electrons may take part in the conduction process, but in certain other materials only a very few or negligible number of electrons may participate in conduction. The former class of materials is known as "conductors," and the latter class is known as "dielectrics" or "insulators." If the number of free electrons participating in conduction is N_e per cubic meter of the material, then the conduction current density is given by

$$\mathbf{J}_c = N_e e \mathbf{v}_d \qquad (5.1)$$

where e is the charge of an electron, and $\mathbf{v}_d$ is the drift velocity of the electrons. The drift velocity varies from one conductor to another, depending on the average time between successive collisions of the electrons with the atomic lattice. It is related to the applied electric field in the manner

$$\mathbf{v}_d = -\mu_e \mathbf{E} \qquad (5.2)$$

where μ_e is known as the "mobility" of the electron. Substituting (5.2) into (5.1), we obtain

$$\mathbf{J}_c = -\mu_e N_e e\mathbf{E} = \mu_e N_e |e| \mathbf{E} \qquad (5.3)$$

Semiconductors are characterized by drift of "holes," that is, vacancies created by detachment of electrons from covalent bonds, in addition to the drift of electrons. If N_e and N_h are the number of electrons and holes, respectively, per cubic meter of the material and if μ_e and μ_h are the electron and hole mobilities, respectively, then the conduction current density in the semiconductor is given by

$$\mathbf{J}_c = (\mu_e N_e |e| + \mu_h N_h |e|)\mathbf{E} \qquad (5.4)$$

Defining a quantity σ, known as the "conductivity" of the material, as given by

$$\sigma = \begin{cases} \mu_e N_e |e| & \text{for conductors} \\ \mu_e N_e |e| + \mu_h N_h |e| & \text{for semiconductors} \end{cases} \qquad (5.5)$$

we obtain the simple and important relationship

$$\mathbf{J}_c = \sigma \mathbf{E} \qquad (5.6)$$

for the conduction current density in a material. Equation (5.6) is known as Ohm's law applicable at a point from which follows the familiar form of Ohm's law used in circuit theory. The units of σ are mhos/meter where a mho ("ohm" spelled in reverse and having the symbol $\mho$) is an ampere per volt. Values of σ for a few materials are listed in Table 5.1. In considering electromagnetic wave propagation in conducting media, the conduction current density given by (5.6) must be employed for the current density term on the right side of Ampere's circuital law. Thus Maxwell's curl equation for **H** for

TABLE 5.1. Conductivities of Some Materials

Material	Conductivity mhos/m	Material	Conductivity mhos/m
Silver	6.1×10^7	Sea water	4
Copper	5.8×10^7	Intrinsic germanium	2.2
Gold	4.1×10^7	Intrinsic silicon	1.6×10^{-3}
Aluminum	3.5×10^7	Fresh water	10^{-3}
Tungsten	1.8×10^7	Distilled water	2×10^{-4}
Brass	1.5×10^7	Dry earth	10^{-5}
Solder	7.0×10^6	Bakelite	10^{-9}
Lead	4.8×10^6	Glass	10^{-10}–10^{-14}
Constantin	2.0×10^6	Mica	10^{-11}–10^{-15}
Mercury	1.0×10^6	Fused quartz	0.4×10^{-17}

a conducting medium is given by

$$\nabla \times \mathbf{H} = \mathbf{J}_c + \frac{\partial \mathbf{D}}{\partial t} = \sigma \mathbf{E} + \frac{\partial \mathbf{D}}{\partial t} \tag{5.7}$$

5.2 DIELECTRICS

In the previous section we learned that conductors are characterized by abundance of "conduction" or "free" electrons that give rise to conduction current under the influence of an applied electric field. In this section we turn our attention to dielectric materials in which the "bound" electrons are predominant. Under the application of an external electric field, the bound electrons of an atom are displaced such that the centroid of the electron cloud is separated from the centroid of the nucleus. The atom is then said to be "polarized," thereby creating an "electric dipole," as shown in Fig. 5.1(a). This kind of polarization is called "electronic polarization." The schematic representation of an electric dipole is shown in Fig. 5.1(b). The strength of the dipole is defined by the electric dipole moment $\mathbf{p}$ given by

$$\mathbf{p} = Q\mathbf{d} \tag{5.8}$$

where $\mathbf{d}$ is the vector displacement between the centroids of the positive and negative charges, each of magnitude Q coulombs.

In certain dielectric materials, polarization may exist in the molecular structure of the material even under the application of no external electric field. The polarization of individual atoms and molecules, however, is randomly oriented, and hence the net polarization on a "macroscopic" scale is zero. The application of an external field results in torques acting on the "microscopic" dipoles, as shown in Fig. 5.2, to convert the initially random polarization into a partially coherent one along the field, on a macroscopic scale. This kind of polarization is known as "orientational polarization." A third kind of polarization known as "ionic polarization" results from the separation of positive and negative ions in molecules formed by the transfer of electrons from one atom to another in the molecule. Certain materials

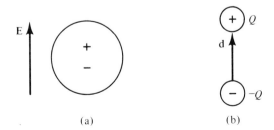

(a)   (b)

Figure 5.1. (a) An electric dipole. (b) Schematic representation of an electric dipole.

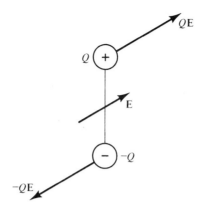

Figure 5.2. Torque acting on an electric dipole in an external electric field.

exhibit permanent polarization, that is, polarization even in the absence of an applied electric field. Electrets, when allowed to solidify in the applied electric field, become permanently polarized and ferroelectric materials exhibit spontaneous, permanent polarization.

On a macroscopic scale, we define a vector $\mathbf{P}$, called the "polarization vector," as the "electric dipole moment per unit volume." Thus if N denotes the number of molecules per unit volume of the material, then there are $N\,\Delta v$ molecules in a volume Δv and

$$\mathbf{P} = \frac{1}{\Delta v}\sum_{j=1}^{N\Delta v}\mathbf{p}_j = N\mathbf{p} \tag{5.9}$$

where $\mathbf{p}$ is the average dipole moment per molecule. The units of $\mathbf{P}$ are coulomb-meter/meter^3 or coulombs per square meter. It is found that for many dielectric materials the polarization vector is related to the electric field $\mathbf{E}$ in the dielectric in the simple manner given by

$$\mathbf{P} = \epsilon_0 \chi_e \mathbf{E} \tag{5.10}$$

where χ_e, a dimensionless parameter, is known as the "electric susceptibility." The quantity χ_e is a measure of the ability of the material to become polarized and differs from one dielectric to another.

To discuss the influence of polarization in the dielectric upon electromagnetic wave propagation in the dielectric medium, let us consider the case of the infinite plane current sheet of Fig. 4.8, radiating uniform plane waves, except that now the space on either side of the current sheet is a dielectric medium instead of being free space. The electric field in the medium induces polarization. The polarization in turn acts together with other factors to govern the behavior of the electromagnetic field. For the case under consideration, the electric field is entirely in the x direction and uniform in x and y.

Thus the induced electric dipoles are all oriented in the x direction, on a macroscopic scale, with the dipole moment per unit volume given by

$$\mathbf{P} = P_x \mathbf{i}_x = \epsilon_0 \chi_e E_x \mathbf{i}_x \qquad (5.11)$$

where E_x is understood to be a function of z and t.

If we now consider an infinitesimal surface of area $\Delta y\, \Delta z$ parallel to the yz plane, we can write E_x associated with that infinitesimal area to be equal to $E_0 \cos \omega t$ where E_0 is a constant. The time history of the induced dipoles associated with that area can be sketched for one complete period of the current source, as shown in Fig. 5.3. In view of the cosinusoidal variation of

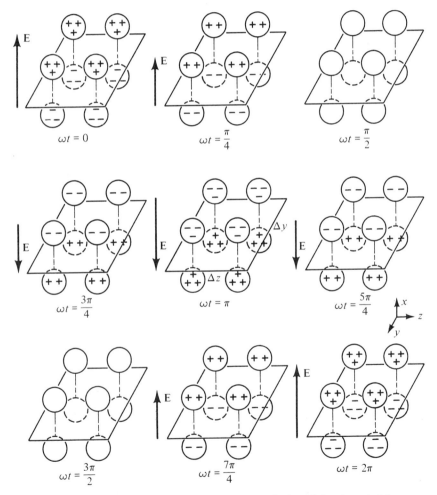

Figure 5.3. Time history of induced electric dipoles in a dielectric material under the influence of a sinusoidally time-varying electric field.

the electric field with time, the dipole moment of the individual dipoles varies in a cosinusoidal manner with maximum strength in the positive x direction at $t = 0$, decreasing sinusoidally to zero strength at $t = \pi/2\omega$ and then reversing to the negative x direction, increasing to maximum strength in that direction at $t = \pi/\omega$, and so on.

The arrangement can be considered as two plane sheets of equal and opposite time-varying charges displaced by the amount δ in the x direction, as shown in Fig. 5.4. To find the magnitude of either charge, we note that the dipole moment per unit volume is

$$P_x = \epsilon_0 \chi_e E_0 \cos \omega t \tag{5.12}$$

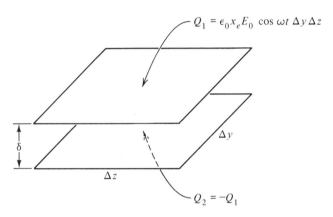

$$Q_1 = \epsilon_0 \chi_e E_0 \cos \omega t\, \Delta y\, \Delta z$$

$$Q_2 = -Q_1$$

Figure 5.4. Two plane sheets of equal and opposite time-varying charges equivalent to the phenomenon depicted in Fig. 5.3.

Since the total volume occupied by the dipoles is $\delta\, \Delta y\, \Delta z$, the total dipole moment associated with the dipoles is $\epsilon_0 \chi_e E_0 \cos \omega t\, (\delta\, \Delta y\, \Delta z)$. The dipole moment associated with two equal and opposite sheet charges is equal to the magnitude of either sheet charge multiplied by the displacement between the two sheets. Hence we obtain the magnitude of either sheet charge to be $\epsilon_0 \chi_e E_0 \cos \omega t\, \Delta y\, \Delta z$. Thus we have a situation in which a sheet charge $Q_1 = \epsilon_0 \chi_e E_0 \cos \omega t\, \Delta y\, \Delta z$ is above the surface and a sheet charge $Q_2 = -Q_1 = -\epsilon_0 \chi_e E_0 \cos \omega t\, \Delta y\, \Delta z$ is below the surface. This is equivalent to a current flowing across the surface, since the charges are varying with time.

We call this current the "polarization current" since it results from the time variation of the electric dipole moments induced in the dielectric due to polarization. The polarization current crossing the surface in the positive x direction, that is, from below to above, is

$$I_{px} = \frac{dQ_1}{dt} = -\epsilon_0 \chi_e E_0 \omega \sin \omega t\, \Delta y\, \Delta z \tag{5.13}$$

where the subscript p denotes polarization. By dividing I_{px} by $\Delta y\, \Delta z$ and letting the area tend to zero, we obtain the polarization current density associated with the points on the surface as

$$J_{px} = \lim_{\substack{\Delta y \to 0 \\ \Delta z \to 0}} \frac{I_{px}}{\Delta y\, \Delta z} = -\epsilon_0 \chi_e E_0 \omega \sin \omega t$$

$$= \frac{\partial}{\partial t}(\epsilon_0 \chi_e E_0 \cos \omega t) = \frac{\partial P_x}{\partial t} \tag{5.14}$$

or

$$\mathbf{J}_p = \frac{\partial \mathbf{P}}{\partial t} \tag{5.15}$$

Although we have deduced this result by considering the special case of the infinite plane current sheet, it is valid in general.

In considering electromagnetic wave propagation in a dielectric medium, the polarization current density given by (5.15) must be included with the current density term on the right side of Ampere's circuital law. Thus considering Ampere's circuital law in differential form for the general case given by (3.28), we have

$$\nabla \times \mathbf{H} = \mathbf{J} + \mathbf{J}_p + \frac{\partial}{\partial t}(\epsilon_0 \mathbf{E}) \tag{5.16}$$

Substituting (5.15) into (5.16), we get

$$\nabla \times \mathbf{H} = \mathbf{J} + \frac{\partial \mathbf{P}}{\partial t} + \frac{\partial}{\partial t}(\epsilon_0 \mathbf{E})$$

$$= \mathbf{J} + \frac{\partial}{\partial t}(\epsilon_0 \mathbf{E} + \mathbf{P}) \tag{5.17}$$

In order to make (5.17) consistent with the corresponding equation for free space given by (3.28), we now revise the definition of the displacement vector $\mathbf{D}$ to read as

$$\mathbf{D} = \epsilon_0 \mathbf{E} + \mathbf{P} \tag{5.18}$$

Substituting for $\mathbf{P}$ by using (5.10), we obtain

$$\mathbf{D} = \epsilon_0 \mathbf{E} + \epsilon_0 \chi_e \mathbf{E}$$

$$= \epsilon_0 (1 + \chi_e)\mathbf{E}$$

$$= \epsilon_0 \epsilon_r \mathbf{E}$$

$$= \epsilon \mathbf{E} \tag{5.19}$$

where we define

$$\epsilon_r = 1 + \chi_e \qquad (5.20)$$

and

$$\epsilon = \epsilon_0 \epsilon_r \qquad (5.21)$$

The quantity ϵ_r is known as the "relative permittivity" or "dielectric constant" of the dielectric, and ϵ is the "permittivity" of the dielectric. The new definition for **D** permits the use of the same Maxwell's equations as for free space with ϵ_0 replaced by ϵ and without the need for explicitly considering the polarization current density. The permittivity ϵ takes into account the effects of polarization, and there is no need to consider them when we use ϵ for ϵ_0! The relative permittivity is an experimentally measurable parameter and its values for several dielectric materials are listed in Table 5.2.

TABLE 5.2. Relative Permittivities of Some Materials

Material	Relative Permittivity	Material	Relative Permittivity
Air	1.0006	Dry earth	5
Paper	2.0–3.0	Mica	6
Teflon	2.1	Neoprene	6.7
Polystyrene	2.56	Wet earth	10
Plexiglass	2.6–3.5	Ethyl alcohol	24.3
Nylon	3.5	Glycerol	42.5
Fused quartz	3.8	Distilled water	81
Bakelite	4.9	Titanium dioxide	100

Equation (5.19) governs the relationship between **D** and **E** for dielectric materials. Dielectrics for which ϵ is independent of the magnitude as well as the direction of **E** as indicated by (5.19) are known as "linear isotropic dielectrics." For certain dielectric materials, each component of the polarization vector can be dependent on all components of the electric field intensity. For such materials, known as "anisotropic dielectric materials," **D** is not in general parallel to **E** and the relationship between these two quantities is expressed in the form of a matrix equation as

$$\begin{bmatrix} D_x \\ D_y \\ D_z \end{bmatrix} = \begin{bmatrix} \epsilon_{xx} & \epsilon_{xy} & \epsilon_{xz} \\ \epsilon_{yx} & \epsilon_{yy} & \epsilon_{yz} \\ \epsilon_{zx} & \epsilon_{zy} & \epsilon_{zz} \end{bmatrix} \begin{bmatrix} E_x \\ E_y \\ E_z \end{bmatrix} \qquad (5.22)$$

The square matrix in (5.22) is known as the "permittivity tensor" of the anisotropic dielectric.

Example 5.1. An anisotropic dielectric material is characterized by the permittivity tensor

$$[\epsilon] = \begin{bmatrix} 7\epsilon_0 & 2\epsilon_0 & 0 \\ 2\epsilon_0 & 4\epsilon_0 & 0 \\ 0 & 0 & 3\epsilon_0 \end{bmatrix}$$

Let us find **D** for several cases of **E**.
Substituting the given permittivity matrix in (5.22), we obtain

$$D_x = 7\epsilon_0 E_x + 2\epsilon_0 E_y$$

$$D_y = 2\epsilon_0 E_x + 4\epsilon_0 E_y$$

$$D_z = 3\epsilon_0 E_z$$

For $\mathbf{E} = E_0 \cos \omega t\, \mathbf{i}_z$, $\mathbf{D} = 3\epsilon_0 E_0 \cos \omega t\, \mathbf{i}_z$; **D** is parallel to **E**.
For $\mathbf{E} = E_0 \cos \omega t\, \mathbf{i}_x$, $\mathbf{D} = 7\epsilon_0 E_0 \cos \omega t\, \mathbf{i}_x + 2\epsilon_0 E_0 \cos \omega t\, \mathbf{i}_y$; **D** is not parallel to **E**.
For $\mathbf{E} = E_0 \cos \omega t\, \mathbf{i}_y$, $\mathbf{D} = 2\epsilon_0 E_0 \cos \omega t\, \mathbf{i}_x + 4\epsilon_0 E_0 \cos \omega t\, \mathbf{i}_y$; **D** is not parallel to **E**.
For $\mathbf{E} = E_0 \cos \omega t\, (\mathbf{i}_x + 2\mathbf{i}_y)$, $\mathbf{D} = 11\epsilon_0 E_0 \cos \omega t\, \mathbf{i}_x + 10\epsilon_0 E_0 \cos \omega t\, \mathbf{i}_y$; **D** is not parallel to **E**.
For $\mathbf{E} = E_0 \cos \omega t\, (2\mathbf{i}_x + \mathbf{i}_y)$, $\mathbf{D} = 16\epsilon_0 E_0 \cos \omega t\, \mathbf{i}_x + 8\epsilon_0 E_0 \cos \omega t\, \mathbf{i}_y = 8\epsilon_0 \mathbf{E}$; **D** is parallel to **E** and the dielectric behaves "effectively" in the same manner as an isotropic dielectric having the permittivity $8\epsilon_0$, that is, the "effective permittivity" of the anisotropic dielectric for this case is $8\epsilon_0$.
Thus we find that in general **D** is not parallel to **E** but for certain polarizations of **E**, **D** is parallel to **E**. These polarizations are known as the characteristic polarizations. ∎

5.3   MAGNETIC MATERIALS

The important characteristic of magnetic materials is "magnetization." Magnetization is the phenomenon by means of which the orbital and spin motions of electrons are influenced by an external magnetic field. An electronic orbit is equivalent to a current loop, which is the magnetic analog of an electric dipole. The schematic representation of a magnetic dipole as seen from along its axis and from a point in its plane are shown in Figs. 5.5(a) and 5.5(b), respectively. The strength of the dipole is defined by the magnetic dipole moment **m** given by

$$\mathbf{m} = IA\mathbf{i}_n \tag{5.23}$$

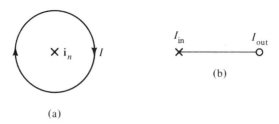

Figure 5.5. Schematic representation of a magnetic dipole as seen from (a) along its axis and (b) a point in its plane.

where A is the area enclosed by the current loop and i_n is the unit vector normal to the plane of the loop and directed in the right-hand sense.

In many materials the net magnetic moment of each atom is zero, that is, on the average, the magnetic dipole moments corresponding to the various electronic orbital and spin motions add up to zero. An external magnetic field has the effect of inducing a net dipole moment by changing the angular velocities of the electronic orbits, thereby magnetizing the material. This kind of magnetization, known as "diamagnetism," is in fact prevalent in all materials. In certain materials known as "paramagnetic materials," the individual atoms possess net nonzero magnetic moments even in the absence of an external magnetic field. These "permanent" magnetic moments of the individual atoms are, however, randomly oriented so that the net magnetization on a macroscopic scale is zero. An applied magnetic field has the effect of exerting torques on the individual permanent dipoles as shown in Fig. 5.6 to convert, on a macroscopic scale, the initially random alignment into a partially coherent one along the magnetic field, that is, with the normal to the current loop directed along the magnetic field. This kind of magnetization is known as "paramagnetism." Certain materials known as "ferromagnetic," "antiferromagnetic," and "ferrimagnetic" materials exhibit permanent magnetization, that is, magnetization even in the absence of an applied magnetic field.

On a macroscopic scale we define a vector **M**, called the "magnetization

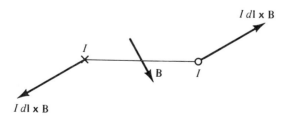

Figure 5.6. Torque acting on a magnetic dipole in an external magnetic field.

vector," as the "magnetic dipole moment per unit volume." Thus if N denotes the number of molecules per unit volume of the material, then there are $N \, \Delta v$ molecules in a volume Δv and

$$\mathbf{M} = \frac{1}{\Delta v} \sum_{j=1}^{N \, \Delta v} \mathbf{m}_j = N\mathbf{m} \qquad (5.24)$$

where $\mathbf{m}$ is the average dipole moment per molecule. The units of $\mathbf{M}$ are ampere-meter^2/meter^3 or amperes per meter. It is found that for many magnetic materials, the magnetization vector is related to the magnetic field $\mathbf{B}$ in the material in the simple manner given by

$$\mathbf{M} = \frac{\chi_m}{1 + \chi_m} \frac{\mathbf{B}}{\mu_0} \qquad (5.25)$$

where χ_m, a dimensionless parameter, is known as the "magnetic susceptibility." The quantity χ_m is a measure of the ability of the material to become magnetized and differs from one magnetic material to another.

To discuss the influence of magnetization in the material on electromagnetic wave propagation in the magnetic material medium, let us consider the case of the infinite plane current sheet of Fig. 4.8, radiating uniform plane waves, except that now the space on either side of the current sheet possesses magnetic material properties in addition to dielectric properties. The magnetic field in the medium induces magnetization. The magnetization in turn acts together with other factors to govern the behavior of the electromagnetic field. For the case under consideration, the magnetic field is entirely in the y direction and uniform in x and y. Thus the induced dipoles are all oriented with their axes in the y direction, on a macroscopic scale, with the dipole moment per unit volume given by

$$\mathbf{M} = M_y \mathbf{i}_y = \frac{\chi_m}{1 + \chi_m} \frac{B_y}{\mu_0} \mathbf{i}_y \qquad (5.26)$$

where B_y is understood to be a function of z and t.

Let us now consider an infinitesimal surface of area $\Delta y \, \Delta z$ parallel to the yz plane and the magnetic dipoles associated with the two areas $\Delta y \, \Delta z$ to the left and to the right of the center of this area as shown in Fig. 5.7(a). Since B_y is a function of z, we can assume the dipoles in the left area to have a different moment than the dipoles in the right area for any given time. If the dimension of an individual dipole is δ in the x direction, then the total dipole moment associated with the dipoles in the left area is $[M_y]_{z - \Delta z/2} \, \delta \, \Delta y \, \Delta z$ and the total dipole moment associated with the dipoles in the right area is $[M_y]_{z + \Delta z/2} \, \delta \, \Delta y \, \Delta z$.

The arrangement of dipoles can be considered to be equivalent to two rectangular surface current loops as shown in Fig. 5.7(b) with the left side

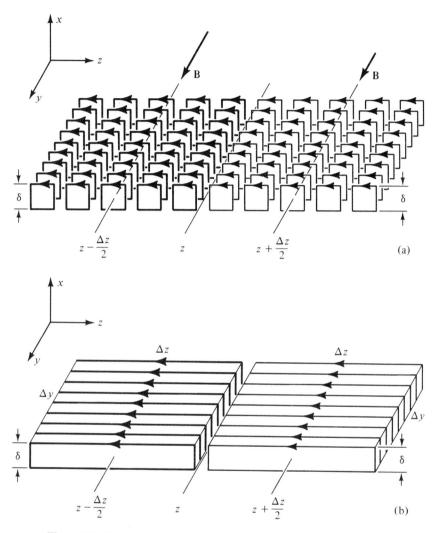

Figure 5.7. (a) Induced magnetic dipoles in a magnetic material. (b) Equivalent surface current loops.

current loop having a dipole moment $[M_y]_{z-\Delta z/2}\,\delta\,\Delta y\,\Delta z$ and the right side current loop having a dipole moment $[M_y]_{z+\Delta z/2}\,\delta\,\Delta y\,\Delta z$. Since the magnetic dipole moment of a rectangular surface current loop is simply equal to the product of the surface current and the cross-sectional area of the loop, the surface current associated with the left loop is $[M_y]_{z-\Delta z/2}\,\Delta y$ and the surface current associated with the right loop is $[M_y]_{z+\Delta z/2}\,\Delta y$. Thus we have a situation in which a current equal to $[M_y]_{z-\Delta z/2}\,\Delta y$ is crossing the area $\Delta y\,\Delta z$ in the positive x direction, and a current equal to $[M_y]_{z+\Delta z/2}\,\Delta y$ is crossing the same area in the negative x direction. This is equivalent to a net current flowing across the surface.

We call this current the "magnetization current" since it results from the space variation of the magnetic dipole moments induced in the magnetic material due to magnetization. The net magnetization current crossing the surface in the positive x direction is

$$I_{mx} = [M_y]_{z-\Delta z/2} \, \Delta y - [M_y]_{z+\Delta z/2} \, \Delta y \qquad (5.27)$$

where the subscript m denotes magnetization. By dividing I_{mx} by $\Delta y \, \Delta z$ and letting the area tend to zero, we obtain the magnetization current density associated with the points on the surface as

$$J_{mx} = \operatorname*{Lim}_{\substack{\Delta y \to 0 \\ \Delta z \to 0}} \frac{I_{mx}}{\Delta y \, \Delta z} = \operatorname*{Lim}_{\Delta z \to 0} \frac{[M_y]_{z-\Delta z/2} - [M_y]_{z+\Delta z/2}}{\Delta z}$$

$$= -\frac{\partial M_y}{\partial z} \qquad (5.28)$$

or

$$J_{mx} \mathbf{i}_x = \begin{vmatrix} \mathbf{i}_x & \mathbf{i}_y & \mathbf{i}_z \\ \dfrac{\partial}{\partial x} & \dfrac{\partial}{\partial y} & \dfrac{\partial}{\partial z} \\ 0 & M_y & 0 \end{vmatrix}$$

or

$$\mathbf{J}_m = \nabla \times \mathbf{M} \qquad (5.29)$$

Although we have deduced this result by considering the special case of the infinite plane current sheet, it is valid in general.

In considering electromagnetic wave propagation in a magnetic material medium, the magnetization current density given by (5.29) must be included with the current density term on the right side of Ampere's circuital law. Thus considering Ampere's circuital law in differential form for the general case given by (3.28), we have

$$\nabla \times \frac{\mathbf{B}}{\mu_0} = \mathbf{J} + \mathbf{J}_m + \frac{\partial \mathbf{D}}{\partial t} \qquad (5.30)$$

Substituting (5.29) into (5.30), we get

$$\nabla \times \frac{\mathbf{B}}{\mu_0} = \mathbf{J} + \nabla \times \mathbf{M} + \frac{\partial \mathbf{D}}{\partial t}$$

or

$$\nabla \times \left(\frac{\mathbf{B}}{\mu_0} - \mathbf{M} \right) = \mathbf{J} + \frac{\partial \mathbf{D}}{\partial t} \qquad (5.31)$$

In order to make (5.31) consistent with the corresponding equation for free space given by (3.28), we now revise the definition of the magnetic field intensity vector $\mathbf{H}$ to read as

$$\mathbf{H} = \frac{\mathbf{B}}{\mu_0} - \mathbf{M} \tag{5.32}$$

Substituting for $\mathbf{M}$ by using (5.25), we obtain

$$\begin{aligned}
\mathbf{H} &= \frac{\mathbf{B}}{\mu_0} - \frac{\chi_m}{1 + \chi_m} \frac{\mathbf{B}}{\mu_0} \\
&= \frac{\mathbf{B}}{\mu_0(1 + \chi_m)} \\
&= \frac{\mathbf{B}}{\mu_0 \mu_r} \\
&= \frac{\mathbf{B}}{\mu} \tag{5.33}
\end{aligned}$$

where we define

$$\mu_r = 1 + \chi_m \tag{5.34}$$

and

$$\mu = \mu_0 \mu_r \tag{5.35}$$

The quantity μ_r is known as the "relative permeability" of the magnetic material and μ is the "permeability" of the magnetic material. The new definition for $\mathbf{H}$ permits the use of the same Maxwell's equations as for free space with μ_0 replaced by μ and without the need for explicitly considering the magnetization current density. The permeability μ takes into account the effects of magnetization, and there is no need to consider them when we use μ for μ_0! For anisotropic magnetic materials, $\mathbf{H}$ is not in general parallel to $\mathbf{B}$ and the relationship between the two quantities is expressed in the form of a matrix equation as given by

$$\begin{bmatrix} B_x \\ B_y \\ B_z \end{bmatrix} = \begin{bmatrix} \mu_{xx} & \mu_{xy} & \mu_{xz} \\ \mu_{yx} & \mu_{yy} & \mu_{yz} \\ \mu_{zx} & \mu_{zy} & \mu_{zz} \end{bmatrix} \begin{bmatrix} H_x \\ H_y \\ H_z \end{bmatrix} \tag{5.36}$$

just as in the case of the relationship between $\mathbf{D}$ and $\mathbf{E}$ for anisotropic dielectric materials.

For many materials for which the relationship between $\mathbf{H}$ and $\mathbf{B}$ is linear, the relative permeability does not differ appreciably from unity, unlike the case of linear dielectric materials, for which the relative permittivity can be very large, as shown in Table 5.2. In fact, for diamagnetic materials, the magnetic susceptibility χ_m is a small negative number of the order -10^{-4} to -10^{-8} whereas for paramagnetic materials, χ_m is a small positive number of the order 10^{-3} to 10^{-7}. Ferromagnetic materials, however, possess large values of relative permeability on the order of several hundreds, thousands, or more. The relationship between **B** and **H** for these materials is nonlinear, resulting in a nonunique value of μ_r for a given material. In fact, these materials are characterized by hysteresis, that is, the relationship between **B** and **H** dependent on the past history of the material.

A typical curve of B versus H, known as the "B–H curve" or the "hysteresis curve" for a ferromagnetic material, is shown in Fig. 5.8. If we start with an unmagnetized sample of the material in which both B and H are initially zero, corresponding to point a in Fig. 5.8, and then magnetize the material,

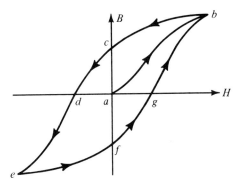

Figure 5.8. Hysteresis curve for a ferromagnetic material.

the manner in which magnetization is built up initially to saturation is given by the portion ab of the curve. If the magnetization is now decreased gradually and then reversed in polarity, the curve does not retrace ab backward but instead follows along bcd until saturation is reached in the opposite direction at point e. A decrease in the magnetization back to zero followed by a reversal back to the original polarity brings the point back to b along the curve through the points f and g, thereby completing the loop. A continuous repetition of the process thereafter would simply make the point trace the hysteresis loop $bcdefgb$ repeatedly.

5.4 WAVE EQUATION AND SOLUTION

In the previous three sections we introduced conductors, dielectrics, and magnetic materials. We found that conductors are characterized by conduction current, dielectrics are characterized by polarization current, and magnetic materials are characterized by magnetization current. The conduction current density is related to the electric field intensity through the conductivity σ of the conductor. To take into account the effects of polarization, we modified the relationship between **D** and **E** by introducing the permittivity ϵ of the dielectric. Similarly, to take into account the effects of magnetization, we modified the relationship between **H** and **B** by introducing the permeability μ of the magnetic material. The three pertinent relations, known as the "constitutive relations," are

$$\mathbf{J}_c = \sigma \mathbf{E} \qquad (5.37a)$$

$$\mathbf{D} = \epsilon \mathbf{E} \qquad (5.37b)$$

$$\mathbf{H} = \frac{\mathbf{B}}{\mu} \qquad (5.37c)$$

A given material may possess all three properties although usually one of them is predominant. Hence in this section we shall consider a material medium characterized by σ, ϵ, and μ. The Maxwell's curl equations for such a medium are

$$\nabla \times \mathbf{E} = -\frac{\partial \mathbf{B}}{\partial t} = -\mu \frac{\partial \mathbf{H}}{\partial t} \qquad (5.38)$$

$$\nabla \times \mathbf{H} = \mathbf{J} + \frac{\partial \mathbf{D}}{\partial t} = \mathbf{J}_c + \frac{\partial \mathbf{D}}{\partial t} = \sigma \mathbf{E} + \epsilon \frac{\partial \mathbf{E}}{\partial t} \qquad (5.39)$$

To discuss electromagnetic wave propagation in the material medium, let us consider the infinite plane current sheet of Fig. 4.8, except that now the medium on either side of the sheet is a material instead of free space, as shown in Fig. 5.9.

The electric and magnetic fields for the simple case of the infinite plane

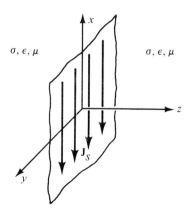

Figure 5.9. Infinite plane current sheet imbedded in a material medium.

current sheet in the $z = 0$ plane and carrying uniformly distributed current in the negative x direction as given by

$$\mathbf{J}_S = -J_{S0} \cos \omega t \, \mathbf{i}_x \tag{5.40}$$

are of the form

$$\mathbf{E} = E_x(z, t)\mathbf{i}_x \tag{5.41a}$$

$$\mathbf{H} = H_y(z, t)\mathbf{i}_y \tag{5.41b}$$

The corresponding simplified forms of the Maxwell's curl equations are

$$\frac{\partial E_x}{\partial z} = -\mu \frac{\partial H_y}{\partial t} \tag{5.42}$$

$$\frac{\partial H_y}{\partial z} = -\sigma E_x - \epsilon \frac{\partial E_x}{\partial t} \tag{5.43}$$

We shall make use of the phasor technique to solve these equations. Thus letting

$$E_x(z, t) = \text{Re} \left[\bar{E}_x(z) \, e^{j\omega t} \right] \tag{5.44a}$$

$$H_y(z, t) = \text{Re} \left[\bar{H}_y(z) \, e^{j\omega t} \right] \tag{5.44b}$$

and replacing E_x and H_y in (5.42) and (5.43) by their phasors $\bar{E}_x$ and $\bar{H}_y$, respectively, and $\partial/\partial t$ by $j\omega$, we obtain the corresponding differential equations for the phasors $\bar{E}_x$ and $\bar{H}_y$ as

$$\frac{\partial \bar{E}_x}{\partial z} = -j\omega\mu\bar{H}_y \tag{5.45}$$

$$\frac{\partial \bar{H}_y}{\partial z} = -\sigma\bar{E}_x - j\omega\epsilon\bar{E}_x = -(\sigma + j\omega\epsilon)\bar{E}_x \tag{5.46}$$

Differentiating (5.45) with respect to z and using (5.46), we obtain

$$\frac{\partial^2 \bar{E}_x}{\partial z^2} = -j\omega\mu \frac{\partial \bar{H}_y}{\partial z} = j\omega\mu(\sigma + j\omega\epsilon)\bar{E}_x \tag{5.47}$$

Defining

$$\bar{\gamma} = \sqrt{j\omega\mu(\sigma + j\omega\epsilon)} \tag{5.48}$$

and substituting in (5.47), we have

$$\frac{\partial^2 \bar{E}_x}{\partial z^2} = \bar{\gamma}^2 \bar{E}_x \tag{5.49}$$

Equation (5.49) is the wave equation for $\bar{E}_x$ in the material medium and its solution is given by

$$\bar{E}_x(z) = \bar{A}e^{-\bar{\gamma}z} + \bar{B}e^{\bar{\gamma}z} \tag{5.50}$$

where $\bar{A}$ and $\bar{B}$ are arbitrary constants. Noting that $\bar{\gamma}$ is a complex number and hence can be written as

$$\bar{\gamma} = \alpha + j\beta \tag{5.51}$$

and also writing $\bar{A}$ and $\bar{B}$ in exponential form as $Ae^{j\theta}$ and $Be^{j\phi}$, respectively, we have

$$\bar{E}_x(z) = Ae^{j\theta}e^{-\alpha z}e^{-j\beta z} + Be^{j\phi}e^{\alpha z}e^{j\beta z}$$

or

$$\begin{aligned}
E_x(z, t) &= \text{Re}\,[\bar{E}_x(z)\,e^{j\omega t}] \\
&= \text{Re}\,[Ae^{j\theta}e^{-\alpha z}e^{-j\beta z}e^{j\omega t} + Be^{j\phi}e^{\alpha z}e^{j\beta z}e^{j\omega t}] \\
&= Ae^{-\alpha z}\cos(\omega t - \beta z + \theta) + Be^{\alpha z}\cos(\omega t + \beta z + \phi)
\end{aligned} \tag{5.52}$$

We now recognize the two terms on the right side of (5.52) as representing uniform plane waves propagating in the positive z and negative z directions, respectively, with phase constant β, in view of the factors $\cos(\omega t - \beta z + \theta)$ and $\cos(\omega t + \beta z + \phi)$, respectively. They are, however, multiplied by the factors $e^{-\alpha z}$ and $e^{\alpha z}$, respectively. Hence the peak amplitude of the field differs from one constant phase surface to another. Since there cannot be a positive going wave in the region $z < 0$, that is, to the left of the current sheet, and since there cannot be a negative going wave in the region $z > 0$, that is, to the right of the current sheet, the solution for the electric field is given by

$$E_x(z, t) = \begin{cases} Ae^{-\alpha z}\cos(\omega t - \beta z + \theta) & \text{for } z > 0 \\ Be^{\alpha z}\cos(\omega t + \beta z + \phi) & \text{for } z < 0 \end{cases} \tag{5.53}$$

To discuss how the peak amplitude of E_x varies with z on either side of the current sheet, we note that since σ, ϵ, and μ are all positive, the phase angle of $j\omega\mu(\sigma + j\omega\epsilon)$ lies between 90° and 180° and hence the phase angle of $\bar{\gamma}$ lies between 45° and 90°, making α and β positive quantities. This means that $e^{-\alpha z}$ decreases with increasing value of z, that is, in the positive z direction, and $e^{\alpha z}$ decreases with decreasing value of z, that is, in the negative z direction. Thus the exponential factors $e^{-\alpha z}$ and $e^{\alpha z}$ associated with the solutions for E_x in (5.53) have the effect of reducing the amplitude of the field, that is, attenuating it as it propagates away from the sheet to either side of it. For this reason, the quantity α is known as the "attenuation constant." The attenuation per unit length is equal to e^{α}. In terms of decibels, this is equal to $20\log_{10} e^{\alpha}$ or 8.686α db. The units of α are nepers per meter. The quantity $\bar{\gamma}$ is known as the "propagation constant" since its real and imaginary parts, α and β, together determine the propagation characteristics, that is, attenuation and phase shift of the wave.

Returning now to the expression for $\bar{\gamma}$ given by (5.48), we can obtain the expressions for α and β by squaring it on both sides and equating the real and imaginary parts on both sides. Thus

$$\bar{\gamma}^2 = (\alpha + j\beta)^2 = j\omega\mu(\sigma + j\omega\epsilon)$$

or

$$\alpha^2 - \beta^2 = -\omega^2\mu\epsilon \tag{5.54a}$$

$$2\alpha\beta = \omega\mu\sigma \tag{5.54b}$$

Now, squaring (5.54a) and (5.54b) and adding and then taking the square root, we obtain

$$\alpha^2 + \beta^2 = \omega^2\mu\epsilon\sqrt{1 + \left(\frac{\sigma}{\omega\epsilon}\right)^2} \tag{5.55}$$

From (5.54a) and (5.55), we then have

$$\alpha^2 = \frac{1}{2}\left[-\omega^2\mu\epsilon + \omega^2\mu\epsilon\sqrt{1 + \left(\frac{\sigma}{\omega\epsilon}\right)^2}\right]$$

$$\beta^2 = \frac{1}{2}\left[\omega^2\mu\epsilon + \omega^2\mu\epsilon\sqrt{1 + \left(\frac{\sigma}{\omega\epsilon}\right)^2}\right]$$

Since α and β are both positive, we finally get

$$\alpha = \frac{\omega\sqrt{\mu\epsilon}}{\sqrt{2}}\left[\sqrt{1 + \left(\frac{\sigma}{\omega\epsilon}\right)^2} - 1\right]^{1/2} \tag{5.56}$$

$$\beta = \frac{\omega\sqrt{\mu\epsilon}}{\sqrt{2}}\left[\sqrt{1 + \left(\frac{\sigma}{\omega\epsilon}\right)^2} + 1\right]^{1/2} \tag{5.57}$$

We note from (5.56) and (5.57) that α and β are both dependent on σ through the factor $\sigma/\omega\epsilon$. This factor, known as the "loss tangent," is the ratio of the magnitude of the conduction current density $\sigma\bar{E}_x$ to the magnitude of the displacement current density $j\omega\epsilon\bar{E}_x$ in the material medium. In practice, the loss tangent is, however, not simply inversely proportional to ω since both σ and ϵ are generally functions of frequency.

The phase velocity of the wave along the direction of propagation is given by

$$v_p = \frac{\omega}{\beta} = \frac{\sqrt{2}}{\sqrt{\mu\epsilon}}\left[\sqrt{1 + \left(\frac{\sigma}{\omega\epsilon}\right)^2} + 1\right]^{-1/2} \tag{5.58}$$

We note that the phase velocity is dependent on the frequency of the wave. Thus waves of different frequencies travel with different phase velocities, that is, they undergo different rates of change of phase with z at any fixed time.

This characteristic of the material medium gives rise to a phenomenon known as "dispersion." We shall discuss dispersion in Chap. 7. The wavelength in the medium is given by

$$\lambda = \frac{2\pi}{\beta} = \frac{\sqrt{2}}{f\sqrt{\mu\epsilon}}\left[\sqrt{1 + \left(\frac{\sigma}{\omega\epsilon}\right)^2} + 1\right]^{-1/2} \tag{5.59}$$

Having found the solution for the electric field of the wave and discussed its general properties, we now turn to the solution for the corresponding magnetic field by substituting for E_x in (5.45). Thus

$$\bar{H}_y = -\frac{1}{j\omega\mu}\frac{\partial \bar{E}_x}{\partial z} = \frac{\bar{\gamma}}{j\omega\mu}(\bar{A}e^{-\bar{\gamma}z} - \bar{B}e^{\bar{\gamma}z})$$

$$= \sqrt{\frac{\sigma + j\omega\epsilon}{j\omega\mu}}(\bar{A}e^{-\bar{\gamma}z} - \bar{B}e^{\bar{\gamma}z})$$

$$= \frac{1}{\bar{\eta}}(\bar{A}e^{-\bar{\gamma}z} - \bar{B}e^{\bar{\gamma}z}) \tag{5.60}$$

where

$$\bar{\eta} = \sqrt{\frac{j\omega\mu}{\sigma + j\omega\epsilon}} \tag{5.61}$$

is the intrinsic impedance of the medium. Writing

$$\bar{\eta} = |\bar{\eta}|e^{j\tau} \tag{5.62}$$

we obtain the solution for $H_y(z, t)$ as

$$H_y(z, t) = \text{Re}\,[\bar{H}_y(z)\,e^{j\omega t}]$$

$$= \text{Re}\left[\frac{1}{|\bar{\eta}|e^{j\tau}}Ae^{j\theta}e^{-\alpha z}e^{-j\beta z}e^{j\omega t} - \frac{1}{|\bar{\eta}|e^{j\tau}}Be^{j\phi}e^{\alpha z}e^{j\beta z}e^{j\omega t}\right]$$

$$= \frac{A}{|\bar{\eta}|}e^{-\alpha z}\cos{(\omega t - \beta z + \theta - \tau)} - \frac{B}{|\bar{\eta}|}e^{\alpha z}\cos{(\omega t + \beta z + \phi - \tau)} \tag{5.63}$$

Remembering that the first and second terms on the right side of (5.63) correspond to (+) and (−) waves, respectively, and hence represent the solutions for the magnetic field in the regions $z > 0$ and $z < 0$, respectively, and recalling that the solution for H_y adjacent to the current sheet is given by

$$H_y = \begin{cases} \dfrac{J_{S0}}{2}\cos\omega t & \text{for } z = 0+ \\[2ex] -\dfrac{J_{S0}}{2}\cos\omega t & \text{for } z = 0- \end{cases} \tag{5.64}$$

we obtain

$$A = \frac{|\bar{\eta}| J_{S0}}{2}, \quad \theta = \tau \qquad (5.65a)$$

$$B = \frac{|\bar{\eta}| J_{S0}}{2}, \quad \phi = \tau \qquad (5.65b)$$

Thus the electromagnetic field due to the infinite plane current sheet in the xy plane having

$$\mathbf{J}_S = -J_{S0} \cos \omega t \, \mathbf{i}_x$$

and with a material medium characterized by σ, ϵ, and μ on either side of it is given by

$$\mathbf{E}(z, t) = \frac{|\bar{\eta}| J_{S0}}{2} e^{\mp \alpha z} \cos (\omega t \mp \beta z + \tau) \, \mathbf{i}_x \qquad \text{for } z \gtrless 0 \qquad (5.66a)$$

$$\mathbf{H}(z, t) = \pm \frac{J_{S0}}{2} e^{\mp \alpha z} \cos (\omega t \mp \beta z) \, \mathbf{i}_y \qquad \text{for } z \gtrless 0 \qquad (5.66b)$$

We note from (5.66a) and (5.66b) that wave propagation in the material medium is characterized by phase difference between **E** and **H** in addition to attenuation. These properties are illustrated in Fig. 5.10, which shows sketches of the current density on the sheet and the distance–variation of the electric and magnetic fields on either side of the current sheet for a few values of t.

Since the fields are attenuated as they progress in their respective directions of propagation, the medium is characterized by power dissipation. In fact, by evaluating the power flow out of a rectangular box lying between z and $z + \Delta z$ and having dimensions Δx and Δy in the x and y directions, respectively, as was done in Sect. 4.6, we obtain

$$\oint_S \mathbf{P} \cdot d\mathbf{S} = \frac{\partial P_z}{\partial z} \Delta x \, \Delta y \, \Delta z = \frac{\partial}{\partial z} (E_x H_y) \, \Delta v$$

$$= \left(E_x \frac{\partial H_y}{\partial z} + H_y \frac{\partial E_x}{\partial z} \right) \Delta v$$

$$= \left[E_x \left(-\sigma E_x - \epsilon \frac{\partial E_x}{\partial t} \right) + H_y \left(-\mu \frac{\partial H_y}{\partial t} \right) \right] \Delta v$$

$$= -\sigma E_x^2 \, \Delta v - \frac{\partial}{\partial t} \left(\frac{1}{2} \epsilon E_x^2 \, \Delta v \right) - \frac{\partial}{\partial t} \left(\frac{1}{2} \mu H_y^2 \, \Delta v \right) \qquad (5.67)$$

The quantity $\sigma E_x^2 \, \Delta v$ is obviously the power dissipated in the volume Δv due to attenuation and the quantities $\frac{1}{2} \epsilon E_x^2 \, \Delta v$ and $\frac{1}{2} \mu H_y^2 \, \Delta v$ are the energies

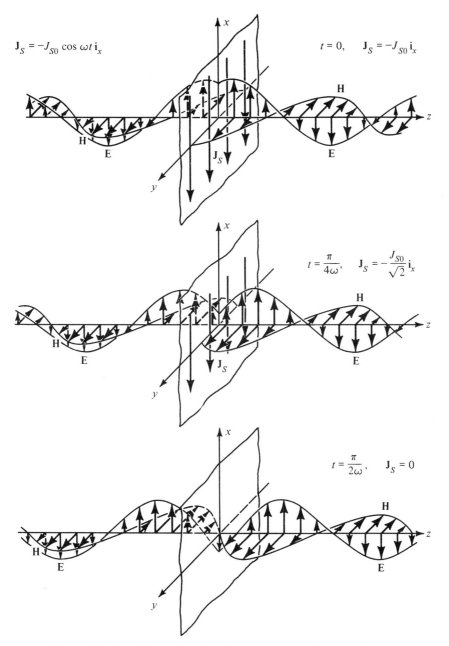

Figure 5.10. Time history of uniform plane electromagnetic wave radiating away from an infinite plane current sheet imbedded in a material medium.

183

stored in the electric and magnetic fields, respectively, in the volume Δv. It then follows that the power dissipation density, the stored energy density associated with the electric field and the stored energy density associated with the magnetic field are given by

$$P_d = \sigma E_x^2 \tag{5.68}$$

$$w_e = \frac{1}{2}\epsilon E_x^2 \tag{5.69}$$

and

$$w_m = \frac{1}{2}\mu H_y^2 \tag{5.70}$$

respectively. Equation (5.67) is the generalization, to the material medium, of the Poynting's theorem given by (4.71) for free space.

5.5   UNIFORM PLANE WAVES IN DIELECTRICS

In the previous section we discussed electromagnetic wave propagation for the general case of a material medium characterized by conductivity σ, permittivity ϵ, and permeability μ. We found general expressions for the attenuation constant α, the phase constant β, the phase velocity v_p, the wavelength λ, and the intrinsic impedance $\bar{\eta}$. These are given by (5.56), (5.57), (5.58), (5.59), and (5.61), respectively. For $\sigma = 0$, the medium is a "perfect dielectric," having the propagation characteristics

$$\alpha = 0 \tag{5.71a}$$

$$\beta = \omega\sqrt{\mu\epsilon} \tag{5.71b}$$

$$v_p = \frac{1}{\sqrt{\mu\epsilon}} \tag{5.71c}$$

$$\lambda = \frac{1}{f\sqrt{\mu\epsilon}} \tag{5.71d}$$

$$\bar{\eta} = \sqrt{\frac{\mu}{\epsilon}} \tag{5.71e}$$

Thus the waves propagate without attenuation as in free space but with ϵ_0 and μ_0 replaced by ϵ and μ, respectively. For nonzero σ, there are two special cases: (a) imperfect dielectrics or poor conductors and (b) good conductors. The first case is characterized by conduction current small in magnitude compared to the displacement current; the second case is characterized by just the opposite. We shall consider the first case in this section and the second case in the following section.

Thus considering the case of "imperfect dielectrics," we have $|\sigma \bar{E}_x| \ll |j\omega\epsilon \bar{E}_x|$, or $\sigma/\omega\epsilon \ll 1$. We can then obtain approximate expressions for α, β, v_p, λ, and $\bar{\eta}$ as follows:

$$\alpha = \frac{\omega\sqrt{\mu\epsilon}}{\sqrt{2}}\left[\sqrt{1 + \left(\frac{\sigma}{\omega\epsilon}\right)^2} - 1\right]^{1/2}$$

$$= \frac{\omega\sqrt{\mu\epsilon}}{\sqrt{2}}\left[1 + \frac{\sigma^2}{2\omega^2\epsilon^2} - \frac{\sigma^4}{8\omega^4\epsilon^4} + \cdots - 1\right]^{1/2}$$

$$\approx \frac{\omega\sqrt{\mu\epsilon}}{\sqrt{2}}\frac{\sigma}{\sqrt{2}\,\omega\epsilon}\left[1 - \frac{\sigma^2}{4\omega^2\epsilon^2}\right]^{1/2}$$

$$\approx \frac{\sigma}{2}\sqrt{\frac{\mu}{\epsilon}}\left(1 - \frac{\sigma^2}{8\omega^2\epsilon^2}\right) \tag{5.72a}$$

$$\beta = \frac{\omega\sqrt{\mu\epsilon}}{\sqrt{2}}\left[\sqrt{1 + \left(\frac{\sigma}{\omega\epsilon}\right)^2} + 1\right]^{1/2}$$

$$\approx \frac{\omega\sqrt{\mu\epsilon}}{\sqrt{2}}\left[2 + \frac{\sigma^2}{2\omega^2\epsilon^2}\right]^{1/2}$$

$$\approx \omega\sqrt{\mu\epsilon}\left(1 + \frac{\sigma^2}{8\omega^2\epsilon^2}\right) \tag{5.72b}$$

$$v_p = \frac{\sqrt{2}}{\sqrt{\mu\epsilon}}\left[\sqrt{1 + \left(\frac{\sigma}{\omega\epsilon}\right)^2} + 1\right]^{-1/2}$$

$$\approx \frac{\sqrt{2}}{\sqrt{\mu\epsilon}}\left[2 + \frac{\sigma^2}{2\omega^2\epsilon^2}\right]^{-1/2}$$

$$\approx \frac{1}{\sqrt{\mu\epsilon}}\left(1 - \frac{\sigma^2}{8\omega^2\epsilon^2}\right) \tag{5.72c}$$

$$\lambda = \frac{\sqrt{2}}{f\sqrt{\mu\epsilon}}\left[\sqrt{1 + \left(\frac{\sigma}{\omega\epsilon}\right)^2} + 1\right]^{-1/2}$$

$$\approx \frac{1}{f\sqrt{\mu\epsilon}}\left(1 - \frac{\sigma^2}{8\omega^2\epsilon^2}\right) \tag{5.72d}$$

$$\bar{\eta} = \sqrt{\frac{j\omega\mu}{\sigma + j\omega\epsilon}} = \sqrt{\frac{j\omega\mu}{j\omega\epsilon}}\left(1 - j\frac{\sigma}{\omega\epsilon}\right)^{-1/2}$$

$$= \sqrt{\frac{\mu}{\epsilon}}\left[1 + j\frac{\sigma}{2\omega\epsilon} - \frac{3}{8}\frac{\sigma^2}{\omega^2\epsilon^2} - \cdots\right]$$

$$\approx \sqrt{\frac{\mu}{\epsilon}}\left[\left(1 - \frac{3}{8}\frac{\sigma^2}{\omega^2\epsilon^2}\right) + j\frac{\sigma}{2\omega\epsilon}\right] \tag{5.72e}$$

In (5.72a)–(5.72e) we have retained all terms up to and including the second power in $\sigma/\omega\epsilon$ and have neglected all higher-order terms. For a value of $\sigma/\omega\epsilon$ equal to 0.1, the quantities β, v_p, and λ are different from those for the corresponding perfect dielectric case by a factor of only 0.01/8 or $\frac{1}{800}$ whereas the intrinsic impedance has a real part differing from the intrinsic impedance of the perfect dielectric medium by a factor of $\frac{3}{800}$ and an imaginary part which is $\frac{1}{20}$ of the intrinsic impedance of the perfect dielectric medium. Thus the only significant feature different from the perfect dielectric case is the attenuation.

Example 5.2.   Let us consider that a material can be classified as a dielectric for $\sigma/\omega\epsilon < 0.1$ and compute the values of the several propagation parameters for three materials: mica, dry earth, and sea water.

Denoting the frequency for which $\sigma/\omega\epsilon = 1$ as f_q, we have $f_q = \sigma/2\pi\epsilon$, assuming that σ and ϵ are independent of frequency. Values of σ, ϵ, and f_q and approximate values of the several propagation parameters for $f > 10f_q$ are listed in Table 5.3, in which c is the velocity of light in free space and β_0

TABLE 5.3.   Values of Several Propagation Parameters for Three Materials for the Dielectric Range of Frequencies

Material	σ $\mho/m$	ϵ_r	f_q Hz	α Np/m	β/β_0	v_p/c	λ/λ_0	$\bar{\eta}$ ohms
Mica	10^{-11}	6	3×10^{-2}	77×10^{-11}	2.45	0.408	0.408	153.9
Dry earth	10^{-5}	5	3.6×10^4	84×10^{-5}	2.24	0.447	0.447	168.6
Sea water	4	80	0.9×10^9	84.3	8.94	0.112	0.112	42.15

and λ_0 are the phase constant and wavelength in free space for the frequency of operation. It can be seen from Table 5.3 that mica behaves as a dielectric for almost any frequency, but sea water can be classified as a dielectric only for frequencies above approximately 10 GHz. We also note that because of the low value of α, mica is a good dielectric, but the high value of α for sea water makes it a poor dielectric.   ∎

5.6   UNIFORM PLANE WAVES IN CONDUCTORS

In the previous section we considered the special case of imperfect dielectrics. Turning now to the case of "good conductors," we have $|\sigma\bar{E}_x| \gg |j\omega\epsilon\bar{E}_x|$, or $\sigma/\omega\epsilon \gg 1$. We can then obtain approximate expressions for α, β, v_p, λ, and η as follows:

$$\alpha = \frac{\omega\sqrt{\mu\epsilon}}{\sqrt{2}}\left[\sqrt{1 + \left(\frac{\sigma}{\omega\epsilon}\right)^2} - 1\right]^{1/2}$$

$$\approx \frac{\omega\sqrt{\mu\epsilon}}{\sqrt{2}}\sqrt{\frac{\sigma}{\omega\epsilon}} = \sqrt{\frac{\omega\mu\sigma}{2}}$$

$$= \sqrt{\pi f\mu\sigma} \tag{5.73a}$$

$$\beta = \frac{\omega\sqrt{\mu\epsilon}}{\sqrt{2}}\left[\sqrt{1 + \left(\frac{\sigma}{\omega\epsilon}\right)^2} + 1\right]^{1/2}$$

$$\approx \frac{\omega\sqrt{\mu\epsilon}}{\sqrt{2}}\sqrt{\frac{\sigma}{\omega\epsilon}}$$

$$= \sqrt{\pi f \mu \sigma} \tag{5.73b}$$

$$v_p = \frac{\sqrt{2}}{\sqrt{\mu\epsilon}}\left[\sqrt{1 + \left(\frac{\sigma}{\omega\epsilon}\right)^2} + 1\right]^{-1/2}$$

$$\approx \frac{\sqrt{2}}{\sqrt{\mu\epsilon}}\sqrt{\frac{\omega\epsilon}{\sigma}} = \sqrt{\frac{2\omega}{\mu\sigma}}$$

$$= \sqrt{\frac{4\pi f}{\mu\sigma}} \tag{5.73c}$$

$$\lambda = \frac{\sqrt{2}}{f\sqrt{\mu\epsilon}}\left[\sqrt{1 + \left(\frac{\sigma}{\omega\epsilon}\right)^2} + 1\right]^{-1/2}$$

$$\approx \sqrt{\frac{4\pi}{f\mu\sigma}} \tag{5.73d}$$

$$\bar{\eta} = \sqrt{\frac{j\omega\mu}{\sigma + j\omega\epsilon}} \approx \sqrt{\frac{j\omega\mu}{\sigma}}$$

$$= (1 + j)\sqrt{\frac{\pi f \mu}{\sigma}} \tag{5.73e}$$

We note that α, β, v_p, and $\bar{\eta}$ are proportional to $\sqrt{f}$, provided that σ and μ are constants.

To discuss the propagation characteristics of a wave inside a good conductor, let us consider the case of copper. The constants for copper are $\sigma = 5.80 \times 10^7$ mho/m, $\epsilon = \epsilon_0$, and $\mu = \mu_0$. Hence the frequency at which σ is equal to $\omega\epsilon$ for copper is equal to $5.8 \times 10^7/2\pi\epsilon_0$ or 1.04×10^{18} Hz. Thus at frequencies of even several gigahertz, copper behaves like an excellent conductor. To obtain an idea of the attenuation of the wave inside the conductor, we note that the attenuation undergone in a distance of one wavelength is equal to $e^{-\alpha\lambda}$ or $e^{-2\pi}$. In terms of decibels, this is equal to $20 \log_{10} e^{2\pi} = 54.58$ db. In fact, the field is attenuated by a factor e^{-1} or 0.368 in a distance equal to $1/\alpha$. This distance is known as the "skin depth" and is denoted by the symbol δ. From (5.73a), we obtain

$$\delta = \frac{1}{\sqrt{\pi f \mu \sigma}} \tag{5.74}$$

The skin depth for copper is equal to

$$\frac{1}{\sqrt{\pi f \times 4\pi \times 10^{-7} \times 5.8 \times 10^7}} = \frac{0.066}{\sqrt{f}}\text{ m.}$$

Thus in copper the fields are attenuated by a factor e^{-1} in a distance of 0.066 mm even at the low frequency of 1 MHz, thereby resulting in the concentration of the fields near to the skin of the conductor. This phenomenon is known as the "skin effect." It also explains "shielding" by conductors.

To discuss further the characteristics of wave propagation in a good conductor, we note that the ratio of the wavelength in the conducting medium to the wavelength in a dielectric medium having the same ϵ and μ as those of the conductor is given by

$$\frac{\lambda_{\text{conductor}}}{\lambda_{\text{dielectric}}} \approx \frac{\sqrt{4\pi/f\mu\sigma}}{1/f\sqrt{\mu\epsilon}} = \sqrt{\frac{4\pi f\epsilon}{\sigma}} = \sqrt{\frac{2\omega\epsilon}{\sigma}} \qquad (5.75)$$

Since $\sigma/\omega\epsilon \gg 1$, $\lambda_{\text{conductor}} \ll \lambda_{\text{dielectric}}$. For example, for sea water, $\sigma = 4$ mhos/m, $\epsilon = 80\epsilon_0$, and $\mu = \mu_0$ so that the ratio of the two wavelengths for $f = 25$ kHz is equal to 0.00745. Thus for $f = 25$ kHz, the wavelength in sea water is $\frac{1}{134}$ of the wavelength in a dielectric having the same ϵ and μ as those of sea water and a still smaller fraction of the wavelength in free space. Furthermore, the lower the frequency, the smaller is this fraction. Since it is the electrical length, that is, the length in terms of the wavelength, instead of the physical length that determines the radiation efficiency of an antenna, this means that antennas of much shorter length can be used in sea water than in free space. Together with the property that $\alpha \propto \sqrt{f}$, this illustrates that low frequencies are more suitable than high frequencies for communication under water, and with underwater objects.

Equation (5.73e) tells us that the intrinsic impedance of a good conductor has a phase angle of 45°. Hence the electric and magnetic fields in the medium are out of phase by 45°. The magnitude of the intrinsic impedance is given by

$$|\bar{\eta}| = \left|(1 + j)\sqrt{\frac{\pi f\mu}{\sigma}}\right| = \sqrt{\frac{2\pi f\mu}{\sigma}} \qquad (5.76)$$

As a numerical example, for copper, this quantity is equal to

$$\sqrt{\frac{2\pi f \times 4\pi \times 10^{-7}}{5.8 \times 10^7}} = 3.69 \times 10^{-7}\sqrt{f} \text{ ohms}$$

Thus the intrinsic impedance of copper has as low a magnitude as 0.369 ohms even at a frequency of 10^{12} Hz. In fact, by recognizing that

$$|\bar{\eta}| = \sqrt{\frac{2\pi f\mu}{\sigma}} = \sqrt{\frac{\omega\epsilon}{\sigma}}\sqrt{\frac{\mu}{\epsilon}} \qquad (5.77)$$

we note that the magnitude of the intrinsic impedance of a good conductor medium is a small fraction of the intrinsic impedance of a dielectric medium having the same ϵ and μ. It follows that for the same electric field, the magnetic field inside a good conductor is much larger than the magnetic field inside a dielectric having the same ϵ and μ as those of the conductor.

Finally, for $\sigma = \infty$, the medium is a "perfect conductor," an idealization of the good conductor. From (5.74), we note that the skin depth is then equal to zero and that there is no penetration of the fields. Thus no fields can exist inside a perfect conductor.

5.7 SUMMARY

In this chapter we studied the principles of uniform plane wave propagation in a material medium. Material media can be classified as (a) conductors, (b) dielectrics, and (c) magnetic materials, depending on the nature of the response of the charged particles in the materials to applied fields. Conductors are characterized by conduction which is the phenomenon of steady drift of free electrons under the influence of an applied electric field. Dielectrics are characterized by polarization which is the phenomenon of the creation and net alignment of electric dipoles, formed by the displacement of the centroids of the electron clouds from the centroids of the nucleii of the atoms, along the direction of an applied electric field. Magnetic materials are characterized by magnetization which is the phenomenon of net alignment of the axes of the magnetic dipoles, formed by the electron orbital and spin motion around the nucleii of the atoms, along the direction of an applied magnetic field.

Under the influence of applied electromagnetic wave fields, all three phenomena described above give rise to currents in the material which in turn influence the wave propagation. These currents are known as the conduction, polarization, and magnetization currents, respectively, for conductors, dielectrics, and magnetic materials. They must be taken into account in the first term on the right side of Ampere's circuital law, that is, $\int_S \mathbf{J} \cdot d\mathbf{S}$ in the case of the integral form and $\mathbf{J}$ in the case of the differential form. The conduction current density is given by

$$\mathbf{J}_c = \sigma \mathbf{E} \tag{5.78}$$

where σ is the conductivity of the material. The conduction current is taken into account explicitly by replacing $\mathbf{J}$ by $\mathbf{J}_c$. The polarization and magnetization currents are taken into account implicitly by revising the definitions of the displacement flux density vector and the magnetic field intensity vector to read as

$$\mathbf{D} = \epsilon_0 \mathbf{E} + \mathbf{P} \tag{5.79}$$

$$\mathbf{H} = \frac{\mathbf{B}}{\mu_0} - \mathbf{M} \tag{5.80}$$

where **P** and **M** are the polarization and magnetization vectors, respectively. For linear isotropic materials, (5.79) and (5.80) simplify to

$$\mathbf{D} = \epsilon \mathbf{E} \tag{5.81}$$

$$\mathbf{H} = \frac{\mathbf{B}}{\mu} \tag{5.82}$$

where

$$\epsilon = \epsilon_0 \epsilon_r$$

$$\mu = \mu_0 \mu_r$$

are the permittivity and the permeability, respectively, of the material. The quantities ϵ_r and μ_r are the relative permittivity and the relative permeability, respectively, of the material. The parameters σ, ϵ, and μ vary from one material to another and are in general dependent on the frequency of the wave. Equations (5.78), (5.81), and (5.82) are known as the constitutive relations. For anisotropic materials, these relations are expressed in the form of matrix equations with the material parameters represented by tensors.

Together with Maxwell's equations, the constitutive relations govern the behavior of the electromagnetic field in a material medium. Thus Maxwell's curl equations for a material medium are given by

$$\mathbf{V} \times \mathbf{E} = -\frac{\partial \mathbf{B}}{\partial t} = -\mu \frac{\partial \mathbf{H}}{\partial t}$$

$$\mathbf{V} \times \mathbf{H} = \mathbf{J}_c + \frac{\partial \mathbf{D}}{\partial t} = \sigma \mathbf{E} + \epsilon \frac{\partial \mathbf{E}}{\partial t}$$

We made use of these equations for the simple case of $\mathbf{E} = E_x(z, t)\mathbf{i}_x$ and $\mathbf{H} = H_y(z, t)\mathbf{i}_y$ to obtain the uniform plane wave solution by considering the infinite plane current sheet in the xy plane with uniform surface current density

$$\mathbf{J}_S = -\mathbf{J}_{S0} \cos \omega t \, \mathbf{i}_x$$

and with a material medium on either side of it and finding the electromagnetic field due to the current sheet to be given by

$$\mathbf{E} = \frac{|\bar{\eta}|J_{S0}}{2} e^{\mp \alpha z} \cos(\omega t \mp \beta z + \tau)\, \mathbf{i}_x \qquad \text{for } z \gtrless 0 \tag{5.83a}$$

$$\mathbf{H} = \pm \frac{J_{S0}}{2} e^{\mp \alpha z} \cos(\omega t \mp \beta z)\, \mathbf{i}_y \qquad \text{for } z \gtrless 0 \tag{5.83b}$$

In (5.83a–b), α and β are the attenuation and phase constants given, respectively, by the real and imaginary parts of the propagation constant, $\bar{\gamma}$. Thus

$$\bar{\gamma} = \alpha + j\beta = \sqrt{j\omega\mu(\sigma + j\omega\epsilon)}$$

The quantities $|\bar{\eta}|$ and τ are the magnitude and phase angle, respectively, of the intrinsic impedance, $\bar{\eta}$, of the medium. Thus

$$\bar{\eta} = |\bar{\eta}|\, e^{j\tau} = \sqrt{\frac{j\omega\mu}{\sigma + j\omega\epsilon}}$$

The uniform plane wave solution given by (5.83a–b) tells us that the wave propagation in the material medium is characterized by attenuation as indicated by $e^{\mp\alpha z}$ and phase difference between $\mathbf{E}$ and $\mathbf{H}$ by the amount τ. We learned that the attenuation of the wave results from power dissipation due to conduction current flow in the medium. The power dissipation density is given by

$$p_d = \sigma E_x^2$$

The stored energy densities associated with the electric and magnetic fields in the medium are given by

$$w_e = \frac{1}{2}\epsilon E^2$$

$$w_m = \frac{1}{2}\mu H^2$$

Having discussed uniform plane wave propagation for the general case of a medium characterized by σ, ϵ, and μ, we then considered several special cases. These are discussed in the following:

PERFECT DIELECTRICS: For these materials, $\sigma = 0$. Wave propagation occurs without attenuation as in free space but with the propagation parameters governed by ϵ and μ instead of ϵ_0 and μ_0, respectively.

IMPERFECT DIELECTRICS: A material is classified as an imperfect dielectric for $\sigma \ll \omega\epsilon$, that is, conduction current density is small in magnitude compared to the displacement current density. The only significant feature of wave propagation in an imperfect dielectric as compared to that in a perfect dielectric is the attenuation undergone by the wave.

GOOD CONDUCTORS: A material is classified as a good conductor for $\sigma \gg \omega\epsilon$, that is, conduction current density is large in magnitude compared to the displacement current density. Wave propagation in a good conductor medium is characterized by attenuation and phase constants both equal to $\sqrt{\pi f \mu \sigma}$. Thus for large values of f and/or σ, the fields do not penetrate very deeply into the conductor. This phenomenon is known as the skin effect. From considerations of the frequency dependence of the attenuation and wavelength for a fixed σ, we learned that low frequencies are more suitable for communication with underwater objects. We also learned that the intrinsic impedance of a good conductor medium is very low in magnitude compared to that of a dielectric medium having the same ϵ and μ.

PERFECT CONDUCTORS: These are idealizations of good conductors in the limit $\sigma \rightarrow \infty$. For $\sigma = \infty$, the skin depth, that is, the distance in which the fields inside a conductor are attenuated by a factor e^{-1}, is zero and hence there can be no penetration of fields into a perfect conductor.

REVIEW QUESTIONS

5.1. Distinguish between bound electrons and free electrons in an atom.

5.2. Briefly describe the phenomenon of conduction.

5.3. State Ohms' law applicable at a point. How is it taken into account in Maxwell's equations?

5.4. Briefly describe the phenomenon of polarization in a dielectric material.

5.5. What is an electric dipole? How is its strength defined?

5.6. What are the different kinds of polarization in a dielectric?

5.7. What is the polarization vector? How is it related to the electric field intensity?

5.8. Discuss how polarization current arises in a dielectric material.

5.9. State the relationship between polarization current density and electric field intensity. How is it taken into account in Maxwell's equations?

5.10. What is the revised definition of $\mathbf{D}$?

5.11. State the relationship between $\mathbf{D}$ and $\mathbf{E}$ in a dielectric material. How does it simplify the solution of field problems involving dielectrics?

5.12. What is an anisotropic dielectric material?

5.13. When can an effective permittivity be defined for an anisotropic dielectric material?

5.14. Briefly describe the phenomenon of magnetization.

5.15. What is a magnetic dipole? How is its strength defined?

5.16. What are the different kinds of magnetic materials?

5.17. What is the magnetization vector? How is it related to the magnetic flux density?

5.18. Discuss how magnetization current arises in a magnetic material.

5.19. State the relationship between magnetization current density and magnetic flux density. How is it taken into account in Maxwell's equations?

5.20. What is the revised definition of $\mathbf{H}$?

5.21. State the relationship between $\mathbf{H}$ and $\mathbf{B}$ for a magnetic material. How does it simplify the solution of field problems involving magnetic materials?

5.22. What is an anisotropic magnetic material?

5.23. Discuss the relationship between B and H for a ferromagnetic material.

5.24. Summarize the constitutive relations for a material medium.

5.25. What is the propagation constant for a material medium? Discuss the significance of its real and imaginary parts.

5.26. Discuss the consequence of the frequency dependence of the phase velocity of a wave in a material medium.

5.27. What is loss tangent? Discuss its significance.

5.28. What is the intrinsic impedance of a material medium? What is the consequence of its complex nature?

5.29. How do you account for the attenuation undergone by the wave in a material medium?

5.30. What is the power dissipation density in a medium characterized by nonzero conductivity?

5.31. What are the stored energy densities associated with electric and magnetic fields in a material medium?

5.32. What is the condition for a medium to be a perfect dielectric? How do the characteristics of wave propagation in a perfect dielectric medium differ from those of wave propagation in free space?

5.33. What is the criterion for a material to be an imperfect dielectric? What is the significant feature of wave propagation in an imperfect dielectric as compared to that in a perfect dielectric?

5.34. Give two examples of materials that behave as good dielectrics for frequencies down to almost zero.

5.35. What is the criterion for a material to be a good conductor?

5.36. Give two examples of materials that behave as good conductors for frequencies of up to several gigahertz.

5.37. What is skin effect? Discuss skin depth, giving some numerical values.

5.38. Why are low-frequency waves more suitable than high-frequency waves for communication with underwater objects?

5.39. Discuss the consequence of the low intrinsic impedance of a good conductor as compared to that of a dielectric medium having the same ϵ and μ.

5.40. Why can there be no fields inside a perfect conductor?

PROBLEMS

5.1. Find the electric field intensity required to produce a current of 0.1 amp crossing an area of 1 cm² normal to the field for the following materials: (a) copper, (b) aluminum, and (c) sea water. Then find the voltage drop along a length of 1 cm parallel to the field, and find the ratio of the voltage drop to the current (resistance) for each material.

5.2. The free electron density in silver is 5.80×10^{28} m^{-3}. (a) Find the mobility of the electron for silver. (b) Find the drift velocity of the electrons for an applied electric field of intensity 0.1 V/m.

5.3. Use the continuity equation, Ohm's law, and Gauss' law for the electric field to show that the time variation of the charge density at a point inside a conductor is governed by the differential equation

$$\frac{\partial \rho}{\partial t} + \frac{\sigma}{\epsilon_0}\rho = 0$$

Then show that the charge density inside the conductor decays exponentially with a time constant ϵ_0/σ. Compute the value of the time constant for copper.

5.4. Show that the torque acting on an electric dipole of moment **p** due to an applied electric field **E** is **p** × **E**.

5.5. For an applied electric field $\mathbf{E} = 0.1 \cos 2\pi \times 10^9 t\, \mathbf{i}_x$ V/m, find the polarization current crossing an area of 1 cm^2 normal to the field for the following materials: (a) polystyrene, (b) mica, and (c) distilled water.

5.6. For the anisotropic dielectric material having the permittivity tensor given in Example 5.1, find **D** for $\mathbf{E} = E_0 (\cos \omega t\, \mathbf{i}_x + \sin \omega t\, \mathbf{i}_y)$. Comment on your result.

5.7. An anisotropic dielectric material is characterized by the permittivity tensor

$$[\epsilon] = \epsilon_0 \begin{bmatrix} 4 & 2 & 2 \\ 2 & 4 & 2 \\ 2 & 2 & 4 \end{bmatrix}$$

(a) Find **D** for $\mathbf{E} = E_0\mathbf{i}_x$. (b) Find **D** for $\mathbf{E} = E_0(\mathbf{i}_x + \mathbf{i}_y + \mathbf{i}_z)$. (c) Find **E** which produces $\mathbf{D} = 4\epsilon_0 E_0\mathbf{i}_x$.

5.8. An anisotropic dielectric material is characterized by the permittivity tensor

$$[\epsilon] = \begin{bmatrix} \epsilon_{xx} & \epsilon_{xy} & 0 \\ \epsilon_{yx} & \epsilon_{yy} & 0 \\ 0 & 0 & \epsilon_{zz} \end{bmatrix}$$

For $\mathbf{E} = (E_x\mathbf{i}_x + E_y\mathbf{i}_y)\cos \omega t$, find the value(s) of E_y/E_x for which **D** is parallel to **E**. Find the effective permittivity for each case.

5.9. Find the magnetic dipole moment of an electron in circular orbit of radius a normal to a uniform magnetic field of flux density B_0. Compute its value for $a = 10^{-3}$ m and $B_0 = 5 \times 10^{-5}$ Wb/m^2.

5.10. Show that the torque acting on a magnetic dipole of moment **m** due to an applied magnetic field **B** is **m** × **B**. For simplicity, consider a rectangular loop in the xy plane and $\mathbf{B} = B_x\mathbf{i}_x + B_y\mathbf{i}_y + B_z\mathbf{i}_z$.

5.11. For an applied magnetic field $\mathbf{B} = 10^{-6} \cos 2\pi z\, \mathbf{i}_y$ Wb/m², find the magnetization current crossing an area 1 cm² normal to the x direction for a magnetic material having $\chi_m = 10^{-3}$.

5.12. An anisotropic magnetic material is characterized by the permeability tensor

$$[\mu] = \mu_0 \begin{bmatrix} 7 & 6 & 0 \\ 6 & 12 & 0 \\ 0 & 0 & 3 \end{bmatrix}$$

Find the effective permeability for $\mathbf{H} = H_0(3\mathbf{i}_x - 2\mathbf{i}_y) \cos \omega t$.

5.13. Obtain the wave equation for $\bar{H}_y$ similar to that for $\bar{E}_x$ given by (5.49).

5.14. Obtain the expression for the attenuation per wavelength undergone by a uniform plane wave in a material medium characterized by σ, ϵ, and μ. Using the logarithmic scale for $\sigma/\omega\epsilon$, plot the attenuation per wavelength in decibels versus $\sigma/\omega\epsilon$.

5.15. For dry earth, $\sigma = 10^{-5}$ mho/m, $\epsilon = 5\epsilon_0$, and $\mu = \mu_0$. Compute α, β, v_p, λ, and $\bar{\eta}$ for $f = 100$ kHz.

5.16. Obtain the expressions for the real and imaginary parts of the intrinsic impedance of a material medium given by (5.61).

5.17. An infinite plane sheet lying in the xy plane carries current of uniform density

$$\mathbf{J}_S = -0.1 \cos 2\pi \times 10^6 t\, \mathbf{i}_x \text{ amp/m}$$

The medium on either side of the sheet is characterized by $\sigma = 10^{-3}$ mho/m, $\epsilon = 18\epsilon_0$, and $\mu = \mu_0$. Find $\mathbf{E}$ and $\mathbf{H}$ on either side of the current sheet.

5.18. Repeat Problem 5.17 for

$$\mathbf{J}_S = -0.1(\cos 2\pi \times 10^6 t\, \mathbf{i}_x + \cos 4\pi \times 10^6 t\, \mathbf{i}_x) \text{ amp/m}$$

5.19. For an array of two infinite plane parallel current sheets of uniform densities situated in a medium characterized by $\sigma = 10^{-3}$ mho/m, $\epsilon = 18\epsilon_0$, and $\mu = \mu_0$, find the spacing and the relative amplitudes and phase angles of the current densities to obtain an endfire radiation characteristic for $f = 10^6$ Hz.

5.20. Show that energy is not stored equally in the electric and magnetic fields in a material medium for $\sigma \neq 0$.

5.21. The electric field of a uniform plane wave propagating in a perfect dielectric medium having $\mu = \mu_0$ is given by

$$\mathbf{E} = 10 \cos (6\pi \times 10^7 t - 0.4\pi z)\, \mathbf{i}_x \text{ V/m}$$

Find (a) the frequency, (b) the wavelength, (c) the phase velocity, (d) the permittivity of the medium, and (e) the associated magnetic field vector $\mathbf{H}$.

5.22. The electric and magnetic fields of a uniform plane wave propagating in a perfect dielectric medium are given by

$$E = 10 \cos (6\pi \times 10^7 t - 0.8\pi z) \, i_x \text{ V/m}$$

$$H = \frac{1}{6\pi} \cos (6\pi \times 10^7 t - 0.8\pi z) \, i_y \text{ amp/m}$$

Find the permittivity and the permeability of the medium.

5.23. An infinite plane sheet situated in the xy plane carries a current of uniform density

$$J_S = -0.2 \cos 3\pi \times 10^7 t \, i_x \text{ amp/m}$$

The medium on either side of the current sheet is a perfect dielectric having $\epsilon = 8\epsilon_0$ and $\mu = 2\mu_0$. (a) Find H, B, M, and J_m for $z > 0$. (b) Find E, D, P, and J_p for $z > 0$.

5.24. Compute f_q for each of the following materials: (a) fused quartz, (b) bakelite, and (c) distilled water. Then compute for the imperfect dielectric range of frequencies the values of α, β, v_p, λ, and $\bar{\eta}$ for each material.

5.25. For uniform plane wave propagation in fresh water ($\sigma = 10^{-3}$ mho/m, $\epsilon = 80\epsilon_0$, $\mu = \mu_0$), find α, β, v_p, λ, and $\bar{\eta}$ for two frequencies: (a) 100 MHz, and (b) 10 kHz.

5.26. Show that for a given material, the ratio of the attenuation constant for the good conductor range of frequencies to the attenuation constant for the imperfect dielectric range of frequencies is equal to $\sqrt{2\omega\epsilon/\sigma}$ where ω is in the good conductor range of frequencies.

5.27. For a 25-kHz wave propagating in sea water, find the Doppler shift observed by an observer, moving with a velocity 10 m/s along the direction of propagation of the wave.

6. TRANSMISSION LINES

In Chap. 4 we studied the principles of uniform plane wave propagation in free space. In Chap. 5 we extended the study of wave propagation to material media. In both chapters we were concerned with propagation in unbounded media. In this and the next chapters we shall consider guided wave propagation, that is, propagation of waves between boundaries. The boundaries are generally provided by conductors, whereas the media between the boundaries are generally dielectrics. There are two kinds of waveguiding systems. These are transmission lines and waveguides. A transmission line consists of two or more parallel conductors, whereas a waveguide is generally made up of one conductor. Our goal in particular in this chapter is to learn the principles of transmission lines.

We shall introduce the transmission line by considering a uniform plane wave and placing two parallel plane, perfect conductors such that the fields remain unaltered by satisfying the "boundary conditions" on the perfect conductor surfaces, which we will derive at the outset. The wave is then guided between and parallel to the conductors, thus leading to the parallel-plate line. We shall learn to represent a line by the "distributed" parameter equivalent circuit and discuss wave propagation on the line in terms of voltage and current. We shall learn to compute the circuit parameters for the parallel-plate line and then extend the computation to the general case of a line of arbitrary cross section. We shall discuss the "standing wave" phenomenon by considering the short-circuited line and reflection and transmission of waves at the junction between two lines in cascade.

6.1 BOUNDARY CONDITIONS ON A PERFECT CONDUCTOR SURFACE

In Sec. 5.6 we learned that the fields inside a perfect conductor are zero, as illustrated in Fig. 6.1. In this section we shall use this property to derive the "boundary conditions" for the fields on the surface of a perfect conductor.

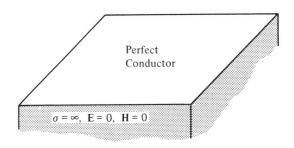

Figure 6.1. Showing that the fields inside a perfect conductor are zero.

Boundary conditions are simply a set of relationships relating the field components at a point adjacent to and on one side of the boundary between two different media to the field components at a corresponding point adjacent to and on the other side of the boundary. These relationships arise from the fact that Maxwell's equations in integral form involve closed paths and surfaces and they must be satisfied for all possible closed paths and surfaces whether they lie entirely in one medium or encompass a portion of the boundary between two different media. In the latter case, Maxwell's equations in integral form must be satisfied collectively by the fields on either side of the boundary, thereby resulting in the boundary conditions. To derive these boundary conditions, we recall that Maxwell's equations in integral form are given by

$$\oint_C \mathbf{E} \cdot d\mathbf{l} = -\frac{d}{dt} \int_S \mathbf{B} \cdot d\mathbf{S} \tag{6.1a}$$

$$\oint_C \mathbf{H} \cdot d\mathbf{l} = \int_S \mathbf{J} \cdot d\mathbf{S} + \frac{d}{dt} \int \mathbf{D} \cdot d\mathbf{S} \tag{6.1b}$$

$$\oint_S \mathbf{D} \cdot d\mathbf{S} = \int_V \rho \, dv \tag{6.1c}$$

$$\oint_S \mathbf{B} \cdot d\mathbf{S} = 0 \tag{6.1d}$$

We shall apply these equations, one at a time, to a closed path or a closed surface encompassing the surface of a perfect conductor and derive the corresponding boundary conditions.

Considering Faraday's law in integral form, that is, (6.1a) first and applying it to an infinitesimal rectangular closed path *abcda* chosen such that *ab* and *cd* are very close to and on either side of the perfect conductor surface as shown in Fig. 6.2, we have

$$\oint_{abcda} \mathbf{E} \cdot d\mathbf{l} = -\frac{d}{dt} \int_{abcd} \mathbf{B} \cdot d\mathbf{S}$$

or

$$\int_a^b \mathbf{E} \cdot d\mathbf{l} + \int_b^c \mathbf{E} \cdot d\mathbf{l} + \int_c^d \mathbf{E} \cdot d\mathbf{l} + \int_d^a \mathbf{E} \cdot d\mathbf{l} = -\frac{d}{dt} \int_{abcd} \mathbf{B} \cdot d\mathbf{S} \quad (6.2)$$

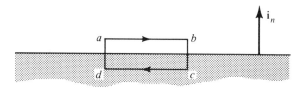

Figure 6.2. For deriving the boundary condition for the tangential component of **E** on a perfect conductor surface.

But $\int_c^d \mathbf{E} \cdot d\mathbf{l} = 0$ since **E** is zero inside the perfect conductor. If we now let *ad* and *bc* → 0 by making *ab* and *cd* almost touch each other but remaining on either side of the boundary, the quantities $\int_b^c \mathbf{E} \cdot d\mathbf{l}$, $\int_d^a \mathbf{E} \cdot d\mathbf{l}$, and $\int_{abcd} \mathbf{B} \cdot d\mathbf{S}$ all tend to zero, leaving us

$$\int_a^b \mathbf{E} \cdot d\mathbf{l} = 0 \qquad (6.3)$$

Since *ab* is infinitesimal in size, we can write (6.3) as

$$E_{ab}(ab) = 0 \qquad (6.4)$$

where E_{ab} is the component of **E** on the perfect conductor surface along the line *ab*. Thus we obtain

$$E_{ab} = 0 \qquad (6.5)$$

Since we can choose the rectangle *abcda* with any orientation, it follows that E_{ab} is zero for any orientation of *ab*. Hence we obtain the first boundary condition that "the tangential component of **E** at a point on a perfect conductor surface is equal to zero." We can express this statement concisely in vector form as

$$\mathbf{i}_n \times \mathbf{E} = 0 \tag{6.6}$$

on the perfect conductor surface where $\mathbf{i}_n$ is the unit normal vector to the conductor surface, as shown in Fig. 6.2.

Considering next Ampere's circuital law in integral form, that is, (6.1b), and applying it to the rectangular path $abcda$ of Fig. 6.2, we have

$$\oint_{abcda} \mathbf{H} \cdot d\mathbf{l} = \int_{abcd} \mathbf{J} \cdot d\mathbf{S} + \frac{d}{dt} \int_{abcd} \mathbf{D} \cdot d\mathbf{S}$$

or

$$\int_a^b \mathbf{H} \cdot d\mathbf{l} + \int_b^c \mathbf{H} \cdot d\mathbf{l} + \int_c^d \mathbf{H} \cdot d\mathbf{l} + \int_d^a \mathbf{H} \cdot d\mathbf{l}$$

$$= \int_{abcd} \mathbf{J} \cdot d\mathbf{S} + \frac{d}{dt} \int_{abcd} \mathbf{D} \cdot d\mathbf{S} \tag{6.7}$$

But $\int_c^d \mathbf{H} \cdot d\mathbf{l} = 0$ since $\mathbf{H}$ is zero inside the perfect conductor. If we now let ad and $bc \longrightarrow 0$ as before, the quantities $\int_b^c \mathbf{H} \cdot d\mathbf{l}$, $\int_d^a \mathbf{H} \cdot d\mathbf{l}$, and $\int_{abcd} \mathbf{D} \cdot d\mathbf{S}$ all tend to zero, but $\int_{abcd} \mathbf{J} \cdot d\mathbf{S}$ does not necessarily tend to zero since there can be a surface current enclosed by the area $abcd$ although the area $abcd$ tends to zero, as shown in Fig. 6.3(a). If α is the angle between the surface

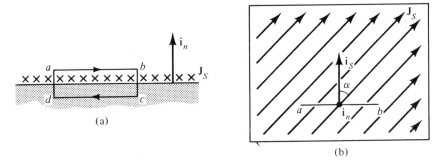

Figure 6.3. For deriving the boundary condition for the tangential component of $\mathbf{H}$ on a perfect conductor surface.

current density vector $\mathbf{J}_S$ and the unit normal vector $\mathbf{i}_S$ to the area $abcd$, directed in the right-hand sense, as shown in Fig. 6.3(b), then

$$\int_{abcd} \mathbf{J} \cdot d\mathbf{S} = J_S(ab \cos \alpha) \tag{6.8}$$

Thus we obtain

$$\int_a^b \mathbf{H} \cdot d\mathbf{l} = J_S(ab \cos \alpha)$$

or

$$H_{ab}(ab) = J_S(ab \cos \alpha)$$

$$H_{ab} = J_S \cos \alpha \qquad (6.9)$$

The maximum value of H_{ab}, that is, the tangential component H_t of $\mathbf{H}$ on the conductor surface is obtained for α equal to zero, that is, when ab is oriented perpendicular to $\mathbf{J}_S$ and then

$$H_t = J_S \qquad (6.10)$$

Hence we obtain the second boundary condition that "the tangential component of $\mathbf{H}$ at a point on a perfect conductor surface is perpendicular (in the right-hand sense) to the surface current density at that point and is equal in magnitude to the surface current density." We can express this statement concisely in vector form as

$$\mathbf{i}_n \times \mathbf{H} = \mathbf{J}_S \qquad (6.11)$$

on the perfect conductor surface where $\mathbf{i}_n$ is again the unit normal vector to the conductor surface pointing out of the conductor, as shown in Fig. 6.3(a).

Considering now Gauss' law for the electric field in integral form, that is, (6.1c), and applying it to an infinitesimal rectangular box $abcdefgh$ chosen such that the surfaces $abcd$ and $efgh$ are very close to and on either side of the perfect conductor surface, as shown in Fig. 6.4, we have

$$\oint_{\substack{\text{surface} \\ \text{of the} \\ \text{box}}} \mathbf{D} \cdot d\mathbf{S} = \int_{\substack{\text{volume} \\ \text{of the} \\ \text{box}}} \rho \, dv$$

or

$$\int_{abcd} \mathbf{D} \cdot d\mathbf{S} + \int_{\substack{\text{side} \\ \text{surfaces}}} \mathbf{D} \cdot d\mathbf{S} + \int_{efgh} \mathbf{D} \cdot d\mathbf{S} = \int_{\substack{\text{volume} \\ \text{of the} \\ \text{box}}} \rho \, dv \qquad (6.12)$$

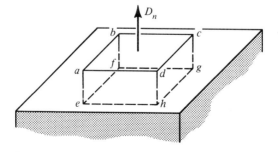

Figure 6.4. For deriving the boundary conditions for the normal components of $\mathbf{D}$ and $\mathbf{B}$ on a perfect conductor surface.

But $\int_{efgh} \mathbf{D} \cdot d\mathbf{S} = 0$ since $\mathbf{D}$ is zero inside the perfect conductor. If we now let the side surfaces $\rightarrow 0$ by making *abcd* and *efgh* almost touch each other but remaining on either side of the boundary, $\int_{\substack{\text{side}\\\text{surfaces}}} \mathbf{D} \cdot d\mathbf{S}$ tends to zero and $\int_{\substack{\text{volume}\\\text{of the box}}} \rho\, dv$ tends to the surface charge enclosed by the box. If the surface charge density is ρ_S, then the surface charge enclosed by the box is $\rho_S(abcd)$. Thus we obtain

$$\int_{abcd} \mathbf{D} \cdot d\mathbf{S} = \rho_S(abcd)$$

or

$$D_n(abcd) = \rho_S$$
$$D_n = \rho_S \tag{6.13}$$

where D_n is the normal component of $\mathbf{D}$. Hence we obtain the third boundary condition that "the normal component of $\mathbf{D}$ at a point on a perfect conductor surface is equal to the surface charge density at that point." We can express this statement concisely in vector form as

$$\mathbf{i}_n \cdot \mathbf{D} = \rho_S \tag{6.14}$$

on the perfect conductor surface.

Considering finally Gauss' law for the magnetic field in integral form, that is, (6.1d), and applying it to the rectangular box *abcdefgh* of Fig. 6.4, we have

$$\oint_{\substack{\text{surface}\\\text{of the}\\\text{box}}} \mathbf{B} \cdot d\mathbf{S} = 0$$

or

$$\int_{abcd} \mathbf{B} \cdot d\mathbf{S} + \int_{\substack{\text{side}\\\text{surfaces}}} \mathbf{B} \cdot d\mathbf{S} + \int_{efgh} \mathbf{B} \cdot d\mathbf{S} = 0 \tag{6.15}$$

But $\int_{efgh} \mathbf{B} \cdot d\mathbf{S} = 0$ since $\mathbf{B}$ is zero inside the perfect conductor surface. If we now let the side surfaces $\rightarrow 0$ as before, $\int_{\substack{\text{side}\\\text{surfaces}}} \mathbf{B} \cdot d\mathbf{S}$ tends to zero. Thus we obtain

$$\int_{abcd} \mathbf{B} \cdot d\mathbf{S} = 0$$

or

$$B_n(abcd) = 0$$
$$B_n = 0 \tag{6.16}$$

where B_n is the normal component of $\mathbf{B}$. Hence we obtain the fourth boundary condition that "the normal component of **B** at a point on a perfect conductor surface is equal to zero." We can express this statement concisely in vector form as

$$\mathbf{i}_n \cdot \mathbf{B} = 0 \qquad (6.17)$$

on the perfect conductor surface.

Summarizing the four boundary conditions for the field components on a perfect conductor surface, we have

$$\mathbf{i}_n \times \mathbf{E} = 0$$
$$\mathbf{i}_n \times \mathbf{H} = \mathbf{J}_S$$
$$\mathbf{i}_n \cdot \mathbf{D} = \rho_S$$
$$\mathbf{i}_n \cdot \mathbf{B} = 0$$

where $\mathbf{i}_n$ is the unit normal vector pointing out of the conductor, $\mathbf{J}_S$ is the surface current density, and ρ_S is the surface charge density on the conductor surface.

Example 6.1. Let us consider a perfect dielectric medium $z < 0$ bounded by a perfect conductor $z > 0$, as shown in Fig. 6.5. Let the fields in the dielectric medium be given by the superposition of $(+)$ and $(-)$ uniform plane waves propagating normal to the conductor surface, that is,

$$\mathbf{E} = E_1 \cos(\omega t - \beta z)\,\mathbf{i}_x + E_2 \cos(\omega t + \beta z)\,\mathbf{i}_x$$

$$\mathbf{H} = \frac{E_1}{\eta} \cos(\omega t - \beta z)\,\mathbf{i}_y - \frac{E_2}{\eta} \cos(\omega t + \beta z)\,\mathbf{i}_y$$

where $\beta = \omega\sqrt{\mu\epsilon}$ and $\eta = \sqrt{\mu/\epsilon}$. We wish to investigate the relationship between E_2 and E_1.

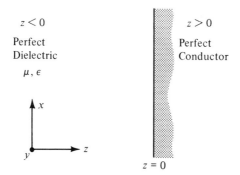

Figure 6.5. A perfect dielectric medium bounded by a perfect conductor.

Since E_x is tangential to the perfect conductor surface, the boundary condition for the tangential component of $\mathbf{E}$ given by (6.6) requires that

$$[E_x]_{z=0} = 0$$

or

$$[E_1 \cos (\omega t - \beta z) + E_2 \cos (\omega t + \beta z)]_{z=0} = 0$$

$$E_1 \cos \omega t + E_2 \cos \omega t = 0 \quad \text{for all } t$$

Thus we obtain the required relationship to be

$$E_2 = -E_1$$

Proceeding further, we obtain the total electric field in the dielectric as given by

$$\mathbf{E} = E_1 \cos (\omega t - \beta z)\, \mathbf{i}_x - E_1 \cos (\omega t + \beta z)\, \mathbf{i}_x$$
$$= 2E_1 \sin \omega t \sin \beta z\, \mathbf{i}_x$$

and the total magnetic field in the dielectric as given by

$$\mathbf{H} = \frac{E_1}{\eta} \cos (\omega t - \beta z)\, \mathbf{i}_y + \frac{E_1}{\eta} \cos (\omega t + \beta z)\, \mathbf{i}_y$$
$$= \frac{2E_1}{\eta} \cos \omega t \cos \beta z\, \mathbf{i}_y$$

These expressions for $\mathbf{E}$ and $\mathbf{H}$ correspond to standing waves. We shall discuss the standing wave phenomenon in Sec. 6.4.

Now, from the boundary condition for the tangential component of $\mathbf{H}$ given by (6.11), we obtain

$$[\mathbf{J}_S]_{z=0} = \mathbf{i}_n \times [\mathbf{H}]_{z=0} = -\mathbf{i}_z \times [\mathbf{H}]_{z=0}$$
$$= -\mathbf{i}_z \times \frac{2E_1}{\eta} \cos \omega t\, \mathbf{i}_y$$
$$= \frac{2E_1}{\eta} \cos \omega t\, \mathbf{i}_x \qquad \blacksquare$$

6.2 PARALLEL-PLATE TRANSMISSION LINE

In the previous section we introduced the boundary conditions for the field components on the surface of a perfect conductor. We learned that the tangential component of the electric field intensity and the normal component of the magnetic field intensity are zero on the perfect conductor surface. Let us now consider the uniform plane electromagnetic wave propagating in the z direction and having an x component only of the electric field and a y component only of the magnetic field, that is,

$$\mathbf{E} = E_x(z, t)\,\mathbf{i}_x$$

$$\mathbf{H} = H_y(z, t)\,\mathbf{i}_y$$

and place perfectly conducting sheets in two planes $x = 0$ and $x = d$, as shown in Fig. 6.6. Since the electric field is completely normal and the magnetic field is completely tangential to the sheets, the two boundary conditions referred to above are satisfied, and hence the wave will simply propagate, as though the sheets were not present, being guided by the sheets. We then have a simple case of transmission line, namely, the parallel-plate transmission line.

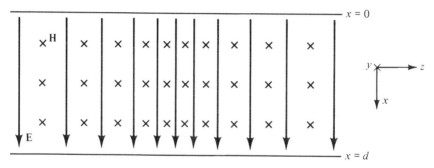

Figure 6.6. Uniform plane electromagnetic wave propagating between two perfectly conducting sheets.

According to the remaining two boundary conditions, there must be charges and currents on the conductors. The charge densities on the two plates are

$$[\rho_S]_{x=0} = [\mathbf{i}_n \cdot \mathbf{D}]_{x=0} = \mathbf{i}_x \cdot \epsilon E_x \mathbf{i}_x = \epsilon E_x \qquad (6.18a)$$

$$[\rho_S]_{x=d} = [\mathbf{i}_n \cdot \mathbf{D}]_{x=d} = -\mathbf{i}_x \cdot \epsilon E_x \mathbf{i}_x = -\epsilon E_x \qquad (6.18b)$$

where ϵ is the permittivity of the medium between the two plates. The current densities on the two plates are

$$[\mathbf{J}_S]_{x=0} = [\mathbf{i}_n \times \mathbf{H}]_{x=0} = \mathbf{i}_x \times H_y \mathbf{i}_y = H_y \mathbf{i}_z \qquad (6.19a)$$

$$[\mathbf{J}_S]_{x=d} = [\mathbf{i}_n \times \mathbf{H}]_{x=d} = -\mathbf{i}_x \times H_y \mathbf{i}_y = -H_y \mathbf{i}_z \qquad (6.19b)$$

In addition, there is conduction current in the medium between the plates flowing from one plate to the other with density given by

$$\mathbf{J}_c = \sigma \mathbf{E} = \sigma E_x \mathbf{i}_x \qquad (6.20)$$

where σ is the conductivity of the medium. In (6.18)–(6.20) it is understood that the charge and current densities are functions of z and t as E_x and H_y are. Thus the wave propagation along the transmission line is supported by charges and currents on the plates, varying with time and distance along the line, as shown in Fig. 6.7.

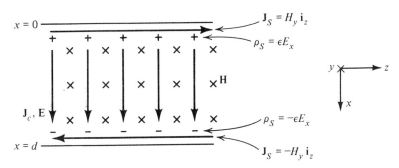

Figure 6.7. Charges and currents on the plates of a parallel-plate transmission line.

Let us now consider finitely sized plates having width w in the y direction, as shown in Fig. 6.8(a), and neglect fringing of the fields at the edges or assume that the structure is part of a much larger-sized configuration. By considering a constant z plane, that is, a plane "transverse" to the direction of propagation of the wave, as shown in Fig. 6.8(b), we can find the voltage between the two conductors in terms of the line integral of the electric field intensity evaluated along any path in that plane between the two conductors. Since the electric field is directed in the x direction and since it is uniform in that plane, this voltage is given by

$$V(z, t) = \int_{x=0}^{d} E_x(z, t)\, dx = E_x(z, t) \int_{x=0}^{d} dx = dE_x(z, t) \qquad (6.21a)$$

Thus each transverse plane is characterized by a voltage between the two conductors which is related simply to the electric field as given by (6.21a). Each transverse plane is also characterized by a current I flowing in the positive z direction on the upper conductor and in the negative z direction on the lower conductor. From Fig. 6.8(b), we can see that this current is given by

$$I(z, t) = \int_{y=0}^{w} J_S(z, t)\, dy = \int_{y=0}^{w} H_y(z, t)\, dy = H_y(z, t) \int_{y=0}^{w} dy$$

$$= wH_y(z, t) \qquad\qquad (6.21b)$$

since H_y is uniform in the cross-sectional plane. Thus the current crossing

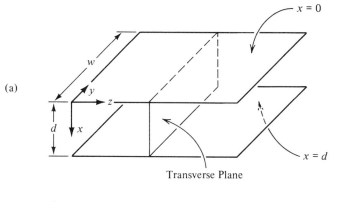

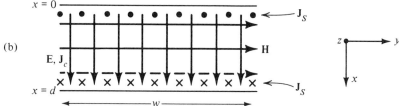

Figure 6.8. (a) Parallel-plate transmission line. (b) A transverse plane of the parallel-plate transmission line.

a given transverse plane is related simply to the magnetic field in that plane as given by (6.21b).

Proceeding further, we can find the power flow down the line by evaluating the surface integral of the Poynting vector over a given transverse plane. Thus

$$P(z, t) = \int_{\substack{\text{transverse} \\ \text{plane}}} (\mathbf{E} \times \mathbf{H}) \cdot d\mathbf{S}$$

$$= \int_{x=0}^{d} \int_{y=0}^{w} E_x(z, t)H_y(z, t) \mathbf{i}_z \cdot dx \, dy \, \mathbf{i}_z$$

$$= \int_{x=0}^{d} \int_{y=0}^{w} \frac{V(z, t)}{d} \frac{I(z, t)}{w} \, dx \, dy$$

$$= V(z, t)I(z, t) \tag{6.22}$$

which is the familiar relationship employed in circuit theory.

We now recall from Sec. 5.4 that E_x and H_y satisfy the two differential equations

$$\frac{\partial E_x}{\partial z} = -\frac{\partial B_y}{\partial t} = -\mu \frac{\partial H_y}{\partial t} \tag{6.23a}$$

$$\frac{\partial H_y}{\partial z} = -J_{cx} - \frac{\partial D_x}{\partial t} = -\sigma E_x - \epsilon \frac{\partial E_x}{\partial t} \tag{6.23b}$$

From (6.21a) and (6.21b), however, we have

$$E_x = \frac{V}{d} \qquad (6.24a)$$

$$H_y = \frac{I}{w} \qquad (6.24b)$$

Substituting for E_x and H_y in (6.23a) and (6.23b) from (6.24a) and (6.24b), respectively, we now obtain two differential equations for voltage and current along the line as

$$\frac{\partial}{\partial z}\left(\frac{V}{d}\right) = -\mu \frac{\partial}{\partial t}\left(\frac{I}{w}\right) \qquad (6.25a)$$

$$\frac{\partial}{\partial z}\left(\frac{I}{w}\right) = -\sigma\left(\frac{V}{d}\right) - \epsilon \frac{\partial}{\partial t}\left(\frac{V}{d}\right) \qquad (6.25b)$$

or

$$\frac{\partial V}{\partial z} = -\left(\frac{\mu d}{w}\right)\frac{\partial I}{\partial t} \qquad (6.26a)$$

$$\frac{\partial I}{\partial z} = -\left(\frac{\sigma w}{d}\right)V - \left(\frac{\epsilon w}{d}\right)\frac{\partial V}{\partial t} \qquad (6.26b)$$

These equations are known as the "transmission-line equations." They characterize the wave propagation along the line in terms of line voltage and line current instead of in terms of the fields.

We now define three quantities familiarly known as the "circuit parameters." These are the inductance, the capacitance, and the conductance (reciprocal of resistance) per unit length of the transmission line in the z direction and are denoted by the symbols $\mathcal{L}$, $\mathcal{C}$, and $\mathcal{G}$, respectively. The inductance per unit length, having the units henries per meter (H/m), is defined as the ratio of the magnetic flux per unit length at any value of z to the line current at that value of z. Noting from Fig. 6.8 that the cross-sectional area normal to the magnetic field lines and per unit length in the z direction is $(d)(1)$ or d, we find the magnetic flux per unit length to be $B_y d$ or $\mu H_y d$. Since the line current is $H_y w$, we then have

$$\mathcal{L} = \frac{\mu H_y d}{H_y w} = \frac{\mu d}{w} \qquad (6.27a)$$

The capacitance per unit length, having the units farads per meter (F/m), is defined as the ratio of the charge per unit length on either plate at any value of z to the line voltage at that value of z. Noting from Fig. 6.8 that the cross-sectional area normal to the electric field lines and per unit length in the z direction is $(w)(1)$ or w, we find the charge per unit length to be $\rho_s w$ or

$\epsilon E_x w$. Since the line voltage is $E_x d$, we then have

$$\mathcal{C} = \frac{\epsilon E_x w}{E_x d} = \frac{\epsilon w}{d} \tag{6.27b}$$

The conductance per unit length, having the units mhos per meter ($\mho/m$), is defined as the ratio of the conduction current per unit length flowing from one plate to the other at any value of z to the line voltage at that value of z. Noting from Fig. 6.8 that the cross-sectional area normal to the conduction current flow and per unit length in the z direction is $(w)(1)$ or w, we find the conduction current per unit length to be $J_{cx} w$ or $\sigma E_x w$. We then have

$$\mathcal{G} = \frac{\sigma E_x w}{E_x d} = \frac{\sigma w}{d} \tag{6.27c}$$

We note that $\mathcal{L}$, $\mathcal{C}$, and $\mathcal{G}$ are purely dependent on the dimensions of the line and are independent of E_x and H_y. We further note that

$$\mathcal{L}\mathcal{C} = \mu\epsilon \tag{6.28a}$$

$$\frac{\mathcal{G}}{\mathcal{C}} = \frac{\sigma}{\epsilon} \tag{6.28b}$$

We now recognize the quantities in parentheses in (6.26a) and (6.26b) to be $\mathcal{L}$, $\mathcal{G}$, and $\mathcal{C}$, respectively, of the line. Thus we obtain the transmission-line equations in terms of these parameters as

$$\frac{\partial V}{\partial z} = -\mathcal{L}\frac{\partial I}{\partial t} \tag{6.29a}$$

$$\frac{\partial I}{\partial z} = -\mathcal{G}V - \mathcal{C}\frac{\partial V}{\partial t} \tag{6.29b}$$

These equations permit us to discuss wave propagation along the line in terms of circuit quantities instead of in terms of field quantities. It should, however, not be forgotten that the actual phenomenon is one of electromagnetic waves guided by the conductors of the line.

It is customary to represent a transmission line by means of its circuit equivalent, derived from the transmission-line equations (6.29a) and (6.29b). To do this, let us consider a section of infinitesimal length Δz along the line between z and $z + \Delta z$. From (6.29a), we then have

$$\lim_{\Delta z \to 0} \frac{V(z + \Delta z, t) - V(z, t)}{\Delta z} = -\mathcal{L}\frac{\partial I(z, t)}{\partial t}$$

or, for $\Delta z \to 0$,

$$V(z + \Delta z, t) - V(z, t) = -\mathcal{L} \, \Delta z \, \frac{\partial I(z, t)}{\partial t} \qquad (6.30)$$

This equation can be represented by the circuit equivalent shown in Fig. 6.9(a) since it satisfies Kirchhoff's voltage law written around the loop $abcda$. Similarly, from (6.29b), we have

$$\lim_{\Delta z \to 0} \frac{I(z + \Delta z, t) - I(z, t)}{\Delta z} = \lim_{\Delta z \to 0} \left[-\mathcal{G}V(z + \Delta z, t) - \mathcal{C} \frac{\partial V(z + \Delta z, t)}{\partial t} \right]$$

or, for $\Delta z \to 0$,

$$I(z + \Delta z, t) - I(z, t) = -\mathcal{G} \, \Delta z \, V(z + \Delta z, t) - \mathcal{C} \, \Delta z \, \frac{\partial V(z + \Delta z, t)}{\partial t} \qquad (6.31)$$

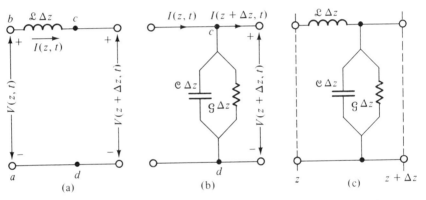

Figure 6.9. Development of circuit equivalent for an infinitesimal length Δz of a transmission line.

This equation can be represented by the circuit equivalent shown in Fig. 6.9(b) since it satisfies Kirchhoff's current law written for node c. Combining the two equations, we then obtain the equivalent circuit shown in Fig. 6.9(c) for a section Δz of the line. It then follows that the circuit representation for a portion of length l of the line consists of an infinite number of such sections in cascade, as shown in Fig. 6.10. Such a circuit is known as a "distributed circuit" as opposed to the "lumped circuits" that are familiar in circuit theory. The distributed circuit notion arises from the fact that the inductance, capacitance, and conductance are distributed uniformly and overlappingly along the line.

A more physical interpretation of the distributed circuit concept follows from energy considerations. We know that the uniform plane wave propagation between the conductors of the line is characterized by energy storage

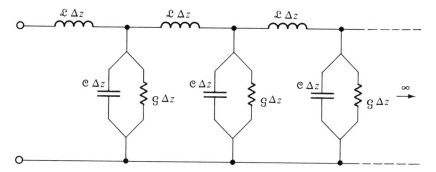

Figure 6.10. Distributed circuit representation of a transmission line.

in the electric and magnetic fields and power dissipation due to the conduction current flow. If we consider a section Δz of the line, the energy stored in the electric field in this section is given by

$$W_e = \frac{1}{2}\epsilon E_x^2 \text{ (volume)} = \frac{1}{2}\epsilon E_x^2 (dw \, \Delta z)$$

$$= \frac{1}{2}\frac{\epsilon w}{d}(E_x d)^2 \, \Delta z = \frac{1}{2}\mathfrak{C} \, \Delta z \, V^2 \qquad (6.32)$$

The energy stored in the magnetic field in that section is given by

$$W_m = \frac{1}{2}\mu H_y^2 \text{ (volume)} = \frac{1}{2}\mu H_y^2 (dw \, \Delta z)$$

$$= \frac{1}{2}\frac{\mu d}{w}(H_y w)^2 \, \Delta z = \frac{1}{2}\mathfrak{L} \, \Delta z \, I^2 \qquad (6.33)$$

The power dissipated due to conduction current flow in that section is given by

$$P_d = \sigma E_x^2 \text{ (volume)} = \sigma E_x^2 (dw \, \Delta z)$$

$$= \frac{\sigma w}{d}(E_x d)^2 \, \Delta z = \mathfrak{G} \, \Delta z \, V^2 \qquad (6.34)$$

Thus we note that $\mathfrak{L}$, $\mathfrak{C}$, and $\mathfrak{G}$ are elements associated with energy storage in the magnetic field, energy storage in the electric field, and power dissipation due to the conduction current flow in the dielectric, respectively, for a given infinitesimal section of the line. Since these phenomena occur continuously and since they overlap, the inductance, capacitance, and conductance must be distributed uniformly and overlappingly along the line. In actual practice, the conductors of the transmission line are imperfect, resulting in slight penetration of the fields into the conductors, in accordance with the skin effect phenomenon. This gives rise to power dissipation and magnetic field energy storage in the conductors, which are taken into account by including a resistance and additional inductance in the series branch of the transmission-line equivalent circuit (see Problem 6.9).

6.3 TRANSMISSION LINE WITH AN ARBITRARY CROSS SECTION

In the previous section we considered the parallel-plate transmission line made up of perfectly conducting sheets lying in the planes $x = 0$ and $x =$ d so that the boundary conditions of zero tangential component of the electric field and zero normal component of the magnetic field are satisfied by the uniform plane wave characterized by the fields

$$\mathbf{E} = E_x(z, t)\,\mathbf{i}_x$$
$$\mathbf{H} = H_y(z, t)\,\mathbf{i}_y$$

thereby leading to the situation in which the uniform plane wave is guided by the conductors of the transmission line. In the general case, however, the conductors of the transmission line have arbitrary cross sections and the fields consist of both x and y components and are dependent on x and y coordinates in addition to the z coordinate. Thus the fields between the conductors are given by

$$\mathbf{E} = E_x(x, y, z, t)\,\mathbf{i}_x + E_y(x, y, z, t)\,\mathbf{i}_y$$
$$\mathbf{H} = H_x(x, y, z, t)\,\mathbf{i}_x + H_y(x, y, z, t)\,\mathbf{i}_y$$

These fields are no longer uniform in x and y but are directed entirely transverse to the direction of propagation, that is, the z axis, which is the axis of the transmission line. Hence they are known as "transverse electromagnetic waves," or "TEM waves." The uniform plane waves are simply a special case of the transverse electromagnetic waves.

To extend the computation of the transmission line parameters $\mathcal{L}$, $\mathcal{C}$, and $\mathcal{G}$ to the general case, let us consider a transmission line made up of parallel, perfect conductors of arbitrary cross sections, as shown by the cross-sectional view in Fig. 6.11(a). Let us assume that the inner conductor is positive with respect to the outer conductor and that the current flows along the positive z direction (into the page) on the inner conductor and along the negative z direction (out of the page) on the outer conductor. We can then draw a "field map," that is, a graphical sketch of the direction lines of the fields between the conductors, from the following considerations: (a) The electric field lines must originate on the inner conductor and be normal to it and must terminate on the outer conductor and be normal to it since the tangential component of the electric field on a perfect conductor surface must be zero. (b) The magnetic field lines must be everywhere perpendicular to the electric field lines; although this can be shown by a rigorous mathematical proof, it is intuitively obvious since, first, the magnetic field lines must be tangential near the conductor surfaces and, second, at any arbitrary point the fields

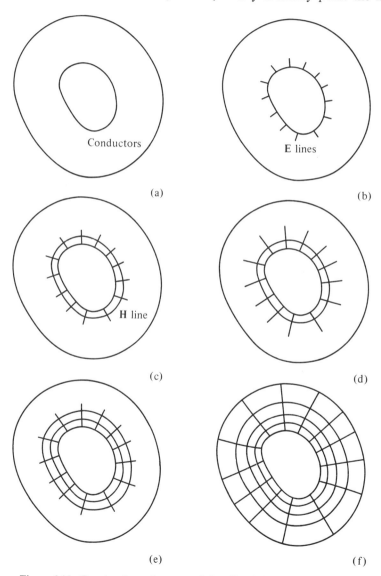

Figure 6.11. Construction of a transmission line field map consisting of curvilinear rectangles.

correspond to those of a locally uniform plane wave. Thus suppose that we start with the inner conductor and draw several lines normal to it at several points on the surface as shown in Fig. 6.11(b). We can then draw a curved line displaced from the conductor surface and such that it is perpendicular everywhere to the electric field lines of Fig. 6.11(b), as shown in Fig. 6.11(c). This contour represents a magnetic field line and forms the basis for further extension of the electric field lines, as shown in Fig. 6.11(d). A second magnetic field line can then be drawn so that it is everywhere perpendicular to the extended electric field lines, as shown in Fig. 6.11(e). This procedure is continued until the entire cross section between the conductors is filled with two sets of orthogonal contours, as shown in Fig. 6.11(f), thereby resulting in a field map made up of curvilinear rectangles.

By drawing the field lines with very small spacings, we can make the rectangles so small that each of them can be considered to be the cross section of a parallel-plate line. In fact, by choosing the spacings appropriately, we can even make them a set of squares. If we now replace the magnetic field lines by perfect conductors, since it does not violate any boundary condition, it can be seen that the arrangement can be viewed as the parallel combination, in the angular direction, of m number of series combinations of n number of parallel-plate lines in the radial direction, where m is the number of squares in the angular direction, that is, along a magnetic field line, and n is the number of squares in the radial direction, that is, along an electric field line. We can then find simple expressions for $\mathcal{L}$, $\mathcal{C}$, and $\mathcal{G}$ of the line in the following manner.

Let us for simplicity consider the field map of Fig. 6.12, consisting of eight segments $1, 2, \ldots, 8$ in the angular direction and two segments a and b in the radial direction. The arrangement is then a parallel combination, in the angular direction, of eight series combinations of two lines in the radial direction, each having a curvilinear rectangular cross section. Let $I_1, I_2, \ldots,$ I_8 be the currents associated with the segments $1, 2, \ldots, 8$, respectively, and let ψ_a and ψ_b be the magnetic fluxes per unit length in the z direction associated with the segments a and b, respectively. Then the inductance per unit length of the transmission line is given by

$$
\mathcal{L} = \frac{\psi}{I} = \frac{\psi_a + \psi_b}{I_1 + I_2 + \ldots + I_8}
$$

$$
= \frac{1}{\dfrac{I_1}{\psi_a} + \dfrac{I_2}{\psi_a} + \ldots + \dfrac{I_8}{\psi_a}} + \frac{1}{\dfrac{I_1}{\psi_b} + \dfrac{I_2}{\psi_b} \ldots + \dfrac{I_8}{\psi_b}}
$$

$$
= \frac{1}{\dfrac{1}{\mathcal{L}_{1a}} + \dfrac{1}{\mathcal{L}_{2a}} + \ldots + \dfrac{1}{\mathcal{L}_{8a}}} + \frac{1}{\dfrac{1}{\mathcal{L}_{1b}} + \dfrac{1}{\mathcal{L}_{2b}} + \ldots + \dfrac{1}{\mathcal{L}_{8b}}} \qquad (6.35a)
$$

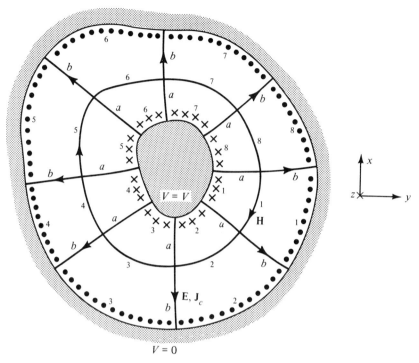

Figure 6.12. For deriving the expressions for the transmission-line parameters from the field map.

Let $Q_1, Q_2, \ldots, Q_8$ be the charges per unit length in the z direction associated with the segments $1, 2, \ldots, 8$, respectively, and let V_a and V_b be the voltages associated with the segments a and b, respectively. Then the capacitance per unit length of the transmission line is given by

$$\mathfrak{C} = \frac{Q}{V} = \frac{Q_1 + Q_2 + \ldots + Q_8}{V_a + V_b}$$

$$= \frac{1}{\dfrac{V_a}{Q_1} + \dfrac{V_b}{Q_1}} + \frac{1}{\dfrac{V_a}{Q_2} + \dfrac{V_b}{Q_2}} + \ldots + \frac{1}{\dfrac{V_a}{Q_8} + \dfrac{V_b}{Q_8}}$$

$$= \frac{1}{\dfrac{1}{\mathfrak{C}_{1a}} + \dfrac{1}{\mathfrak{C}_{1b}}} + \frac{1}{\dfrac{1}{\mathfrak{C}_{2a}} + \dfrac{1}{\mathfrak{C}_{2b}}} + \ldots + \frac{1}{\dfrac{1}{\mathfrak{C}_{8a}} + \dfrac{1}{\mathfrak{C}_{8b}}} \qquad (6.35b)$$

Let $I_{c1}, I_{c2}, \ldots, I_{c8}$ be the conduction currents per unit length in the z direction associated with the segments $1, 2, \ldots, 8$, respectively. Then the conductance per unit length of the transmission line is given by

$$\mathcal{G} = \frac{I_c}{V} = \frac{I_{c1} + I_{c2} + \dots + I_{c8}}{V_a + V_b}$$

$$= \frac{1}{\dfrac{V_a}{I_{c1}} + \dfrac{V_b}{I_{c1}}} + \frac{1}{\dfrac{V_a}{I_{c2}} + \dfrac{V_b}{I_{c2}}} + \dots + \frac{1}{\dfrac{V_a}{I_{c8}} + \dfrac{V_b}{I_{c8}}}$$

$$= \frac{1}{\dfrac{1}{\mathcal{G}_{1a}} + \dfrac{1}{\mathcal{G}_{1b}}} + \frac{1}{\dfrac{1}{\mathcal{G}_{2a}} + \dfrac{1}{\mathcal{G}_{2b}}} + \dots + \frac{1}{\dfrac{1}{\mathcal{G}_{8a}} + \dfrac{1}{\mathcal{G}_{8b}}} \qquad (6.35c)$$

Generalizing the expressions (6.35a), (6.35b), and (6.35c) to m segments in the angular direction and n segments in the radial direction, we obtain

$$\mathcal{L} = \sum_{j=1}^{n} \frac{1}{\sum_{i=1}^{m} \dfrac{1}{\mathcal{L}_{ij}}} \qquad (6.36a)$$

$$\mathcal{C} = \sum_{i=1}^{m} \frac{1}{\sum_{j=1}^{n} \dfrac{1}{\mathcal{C}_{ij}}} \qquad (6.36b)$$

$$\mathcal{G} = \sum_{i=1}^{m} \frac{1}{\sum_{j=1}^{n} \dfrac{1}{\mathcal{G}_{ij}}} \qquad (6.36c)$$

where $\mathcal{L}_{ij}$, $\mathcal{C}_{ij}$, and $\mathcal{G}_{ij}$ are the inductance, capacitance, and conductance per unit length corresponding to the rectangle ij. If the map consists of curvilinear squares, then $\mathcal{L}_{ij}$, $\mathcal{C}_{ij}$, and $\mathcal{G}_{ij}$ are equal to μ, ϵ, and σ, respectively, according to (6.27a), (6.27b), and (6.27c), respectively, since the width w of the plates is equal to the spacing d of the plates for each square. Thus we obtain simple expressions for $\mathcal{L}$, $\mathcal{C}$, and $\mathcal{G}$ as given by

$$\mathcal{L} = \mu \frac{n}{m} \qquad (6.37a)$$

$$\mathcal{C} = \epsilon \frac{m}{n} \qquad (6.37b)$$

$$\mathcal{G} = \sigma \frac{m}{n} \qquad (6.37c)$$

The computation of $\mathcal{L}$, $\mathcal{C}$, and $\mathcal{G}$ then consists of sketching a field map consisting of curvilinear squares, counting the number of squares in each direction, and substituting these values in (6.37a), (6.37b), and (6.37c). Note that once again

$$\mathcal{LC} = \mu\epsilon \qquad (6.38a)$$

$$\frac{\mathcal{G}}{\mathcal{C}} = \frac{\sigma}{\epsilon} \qquad (6.38b)$$

We shall now consider an example of the application of the curvilinear squares technique.

Example 6.2. The coaxial cable is a transmission line made up of parallel, coaxial, cylindrical conductors. Let the radius of the inner conductor be a and that of the outer conductor be b. We wish to find expressions for $\mathcal{L}$, $\mathcal{C}$, and $\mathcal{G}$ of the coaxial cable by using the curvilinear squares technique.

Figure 6.13 shows the cross-sectional view of the coaxial cable and the field map. In view of the symmetry associated with the conductor configuration, the construction of the field map is simplified in this case. The electric field lines are radial lines from one conductor to the other, and the magnetic field lines are circles concentric with the conductors, as shown in the figure.

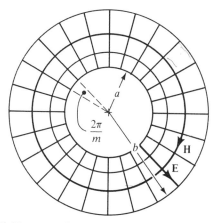

Figure 6.13. Field map consisting of curvilinear squares for a coaxial cable.

Let the number of curvilinear squares in the angular direction be m. Then to find the number of curvilinear squares in the radial direction, we note that the angle subtended at the center of the conductors by adjacent pairs of electric field lines is equal to $2\pi/m$. Hence at any arbitrary radius r between the two conductors, the side of the curvilinear square is equal to $r(2\pi/m)$. The number of squares in an infinitesimal distance dr in the radial direction is then equal to $\dfrac{dr}{r(2\pi/m)}$ or $\dfrac{m}{2\pi}\dfrac{dr}{r}$. The total number of squares in the radial direction from the inner to the outer conductor is given by

$$n = \int_{r=a}^{b} \frac{m}{2\pi}\frac{dr}{r} = \frac{m}{2\pi}\ln\frac{b}{a}$$

The required expressions for $\mathcal{L}$, $\mathcal{C}$, and $\mathcal{G}$ are then given by

$$\mathcal{L} = \mu\frac{n}{m} = \frac{\mu}{2\pi}\ln\frac{b}{a} \tag{6.39a}$$

$$\mathbb{C} = \epsilon \frac{m}{n} = \frac{2\pi\epsilon}{\ln(b/a)} \qquad (6.39\text{b})$$

$$\mathcal{G} = \sigma \frac{m}{n} = \frac{2\pi\sigma}{\ln(b/a)} \qquad (6.39\text{c})$$

These expressions are exact. We have been able to obtain exact expressions in this case because of the geometry involved. When the geometry is not so simple, we can only obtain approximate values for $\mathcal{L}$, $\mathbb{C}$, and $\mathcal{G}$. ∎

We have just discussed an example of the determination of the transmission-line parameters $\mathcal{L}$, $\mathbb{C}$, and $\mathcal{G}$ for a coaxial cable. There are other configurations having different cross sections for which one can obtain the parameters either by the curvilinear squares technique or by other analytical or experimental techniques. We shall, however, not pursue the discussion of these techniques any further. With the understanding that different transmission lines are characterized by different values of $\mathcal{L}$, $\mathbb{C}$, and $\mathcal{G}$, which can be computed from the formulas, we now recall that the voltage and current on the line are governed by the transmission-line equations

$$\frac{\partial V}{\partial z} = -\mathcal{L}\frac{\partial I}{\partial t} \qquad (6.40\text{a})$$

$$\frac{\partial I}{\partial z} = -\mathcal{G}V - \mathbb{C}\frac{\partial V}{\partial t} \qquad (6.40\text{b})$$

For the sinusoidally time-varying case, the corresponding differential equations for the phasor voltage $\bar{V}$ and phasor current $\bar{I}$ are given by

$$\frac{\partial \bar{V}}{\partial z} = -j\omega\mathcal{L}\bar{I} \qquad (6.41\text{a})$$

$$\frac{\partial \bar{I}}{\partial z} = -\mathcal{G}\bar{V} - j\omega\mathbb{C}\bar{V} = -(\mathcal{G} + j\omega\mathbb{C})\bar{V} \qquad (6.41\text{b})$$

Combining (6.41a) and (6.41b) by eliminating $\bar{I}$, we obtain the wave equation for $\bar{V}$ as

$$\frac{\partial^2 \bar{V}}{\partial z^2} = -j\omega\mathcal{L}\frac{\partial \bar{I}}{\partial z} = j\omega\mathcal{L}(\mathcal{G} + j\omega\mathbb{C})\bar{V}$$

$$= \bar{\gamma}^2\bar{V} \qquad (6.42)$$

where

$$\bar{\gamma} = \sqrt{j\omega\mathcal{L}(\mathcal{G} + j\omega\mathbb{C})} \qquad (6.43)$$

is the propagation constant associated with the wave propagation on the line. The solution for $\bar{V}$ is given by

$$\bar{V}(z) = \bar{A}e^{-\bar{\gamma}z} + \bar{B}e^{\bar{\gamma}z} \qquad (6.44)$$

where $\bar{A}$ and $\bar{B}$ are arbitrary constants to be determined by the boundary conditions. The corresponding solution for $\bar{I}$ is then given by

$$\bar{I}(z) = -\frac{1}{j\omega\mathcal{L}}\frac{\partial \bar{V}}{\partial z} = -\frac{1}{j\omega\mathcal{L}}(-\bar{\gamma}\bar{A}e^{-\bar{\gamma}z} + \bar{\gamma}\bar{B}e^{\bar{\gamma}z})$$

$$= \sqrt{\frac{\mathcal{G}+j\omega\mathcal{C}}{j\omega\mathcal{L}}}(\bar{A}e^{-\bar{\gamma}z} - \bar{B}e^{\bar{\gamma}z})$$

$$= \frac{1}{\bar{Z}_0}(\bar{A}e^{-\bar{\gamma}z} - \bar{B}e^{\bar{\gamma}z}) \qquad (6.45)$$

where

$$\bar{Z}_0 = \sqrt{\frac{j\omega\mathcal{L}}{\mathcal{G}+j\omega\mathcal{C}}} \qquad (6.46)$$

is known as the "characteristic impedance" of the transmission line.

The solutions for the line voltage and line current given by (6.44) and (6.45), respectively, represent the superposition of $(+)$ and $(-)$ waves, that is, waves propagating in the positive z and negative z directions, respectively. They are completely analogous to the solutions for the electric and magnetic fields in the medium between the conductors of the line. In fact, the propagation constant given by (6.43) is the same as the propagation constant $\sqrt{j\omega\mu(\sigma + j\omega\epsilon)}$, as it should be. The characteristic impedance of the line is analogous to (but not equal to) the intrinsic impedance of the material medium between the conductors of the line. We note that for a perfect dielectric medium between the conductors, that is, for $\sigma = 0$, $\mathcal{G} = 0$ and

$$\bar{Z}_0 = Z_0 = \sqrt{\frac{\mathcal{L}}{\mathcal{C}}} \qquad (6.47)$$

is purely real. For example, for the coaxial cable of Example 6.2, with a perfect dielectric between the conductors,

$$Z_0 = \sqrt{\frac{\mathcal{L}}{\mathcal{C}}} = \sqrt{\frac{\mu}{2\pi}\ln\frac{b}{a}\bigg/\frac{2\pi\epsilon}{\ln(b/a)}}$$

$$= \frac{1}{2\pi}\sqrt{\frac{\mu}{\epsilon}}\ln\frac{b}{a} \qquad (6.48)$$

For $\mu = \mu_0$, $\epsilon = 2.25\epsilon_0$, and $b/a = 3.67$, the characteristic impedance of the coaxial cable is approximately 52 ohms.

6.4 SHORT-CIRCUITED TRANSMISSION LINE

In the previous section we found the general solutions for the complex voltage and complex current $\bar{V}$ and $\bar{I}$, respectively, on a transmission line. For a "lossless line," that is, for a line consisting of a perfect dielectric medium between the conductors, $\mathcal{G} = 0$, and

$$\bar{\gamma} = \alpha + j\beta = \sqrt{j\omega\mathcal{L}\cdot j\omega\mathcal{C}} = j\omega\sqrt{\mathcal{L}\mathcal{C}} \qquad (6.49)$$

Thus the attenuation constant α is equal to zero, which is to be expected, and the phase constant β is equal to $\omega\sqrt{\mathcal{L}\mathcal{C}}$. We can then write the solutions for $\bar{V}$ and $\bar{I}$ as

$$\bar{V}(z) = \bar{A}e^{-j\beta z} + \bar{B}e^{j\beta z} \qquad (6.50a)$$

$$\bar{I}(z) = \frac{1}{Z_0}(\bar{A}e^{-j\beta z} - \bar{B}e^{j\beta z}) \qquad (6.50b)$$

where $Z_0 = \sqrt{\mathcal{L}/\mathcal{C}}$ as given by (6.47).

Let us now consider a lossless line short circuited at the far end $z = 0$, as shown in Fig. 6.14(a), in which the double-ruled lines represent the conductors of the transmission line. In actuality, the arrangement may consist, for example, of a perfectly conducting rectangular sheet joining the two conductors of a parallel-plate line as in Fig. 6.14(b) or a perfectly conducting ring-shaped sheet joining the two conductors of a coaxial cable as in Fig. 6.14(c). We shall assume that the line is driven by a voltage generator of frequency ω at the left end $z = -l$ so that waves are set up on the line. The short circuit at $z = 0$ requires that the tangential electric field on the surface of the conductor comprising the short circuit be zero. Since the voltage between the conductors of the line is proportional to this electric field which is transverse to them, it follows that the voltage across the short circuit has to be zero. Thus we have

$$\bar{V}(0) = 0 \qquad (6.51)$$

Applying the boundary condition given by (6.51) to the general solution for $\bar{V}$ given by (6.50a), we have

$$\bar{V}(0) = \bar{A}e^{-j\beta(0)} + \bar{B}e^{j\beta(0)} = 0$$

or

$$\bar{B} = -\bar{A} \qquad (6.52)$$

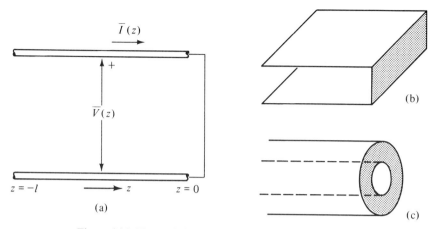

Figure 6.14. Transmission line short-circuited at the far end.

Thus we find that the short circuit gives rise to a $(-)$ or reflected wave whose voltage is exactly the negative of the $(+)$ or incident wave voltage, at the short circuit. Substituting this result in (6.50a) and (6.50b), we get the particular solutions for the complex voltage and current on the short-circuited line to be

$$\bar{V}(z) = \bar{A}e^{-j\beta z} - \bar{A}e^{j\beta z} = -2j\bar{A}\sin\beta z \tag{6.53a}$$

$$\bar{I}(z) = \frac{1}{Z_0}(\bar{A}e^{-j\beta z} + \bar{A}e^{j\beta z}) = \frac{2\bar{A}}{Z_0}\cos\beta z \tag{6.53b}$$

The real voltage and current are then given by

$$V(z, t) = \text{Re}[\bar{V}(z)e^{j\omega t}] = \text{Re}(2e^{-j\pi/2}Ae^{j\theta}\sin\beta z\, e^{j\omega t})$$
$$= 2A\sin\beta z\sin(\omega t + \theta) \tag{6.54a}$$

$$I(z, t) = \text{Re}[\bar{I}(z)e^{j\omega t}] = \text{Re}\left[\frac{2}{Z_0}Ae^{j\theta}\cos\beta z\, e^{j\omega t}\right]$$
$$= \frac{2A}{Z_0}\cos\beta z\cos(\omega t + \theta) \tag{6.54b}$$

where we have replaced $\bar{A}$ by $Ae^{j\theta}$ and $-j$ by $e^{-j\pi/2}$. The instantaneous power flow down the line is given by

$$P(z, t) = V(z, t)I(z, t)$$
$$= \frac{4A^2}{Z_0}\sin\beta z\cos\beta z\sin(\omega t + \theta)\cos(\omega t + \theta)$$
$$= \frac{A^2}{Z_0}\sin 2\beta z\sin 2(\omega t + \theta) \tag{6.54c}$$

These results for the voltage, current, and power flow on the short-circuited line given by (6.54a), (6.54b), and (6.54c), respectively, are illustrated in Fig. 6.15, which shows the variation of each of these quantities with distance from the short circuit for several values of time. The numbers 1, 2, 3, . . . , 9 beside the curves in Fig. 6.15 represent the order of the curves

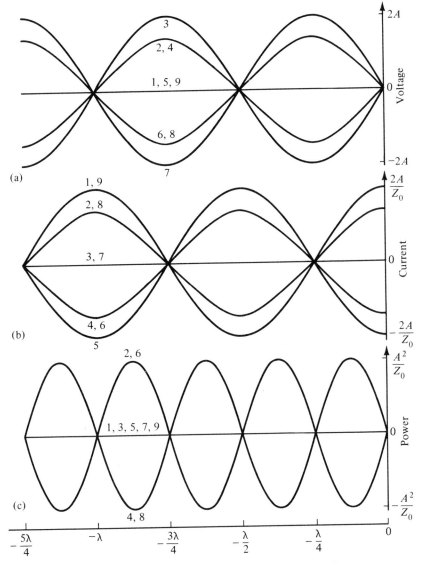

Figure 6.15. Time variations of voltage, current, and power flow associated with standing waves on a short-circuited transmission line.

corresponding to values of $(\omega t + \theta)$ equal to $0, \pi/4, \pi/2, \ldots, 2\pi$, It can be seen that the phenomenon is one in which the voltage, current, and power flow oscillate sinusoidally with time with different amplitudes at different locations on the line, unlike in the case of traveling waves in which a given point on the waveform progresses in distance with time. These waves are therefore known as "standing waves." In particular, they represent "complete standing waves" in view of the zero amplitudes of the voltage, current, and power flow at certain locations on the line, as shown by Fig. 6.15.

The line voltage amplitude is zero for values of z given by $\sin \beta z = 0$ or $\beta z = -m\pi$, $m = 1, 2, 3, \ldots$, or $z = -m\lambda/2$, $m = 1, 2, 3, \ldots$, that is at multiples of $\lambda/2$ from the short circuit. The line current amplitude is zero for values of z given by $\cos \beta z = 0$ or $\beta z = -(2m + 1)\pi/2$, $m = 0, 1, 2, 3, \ldots$, or $z = -(2m + 1)\lambda/4$, $m = 0, 1, 2, 3, \ldots$, that is, at odd multiples of $\lambda/4$ from the short circuit. The power flow amplitude is zero for values of z given by $\sin 2\beta z = 0$ or $\beta z = -m\pi/2$, $m = 1, 2, 3, \ldots$, or $z = -m\lambda/4$, $m = 1, 2, 3, \ldots$, that is, at multiples of $\lambda/4$ from the short circuit. Proceeding further, we find that the time-average power flow down the line, that is, power flow averaged over one period of the source voltage, is

$$\langle P \rangle = \frac{1}{T} \int_{t=0}^{T} P(z, t)\, dt = \frac{\omega}{2\pi} \int_{t=0}^{2\pi/\omega} P(z, t)\, dt$$

$$= \frac{\omega}{2\pi} \frac{A^2}{Z_0} \sin 2\beta z \int_{t=0}^{2\pi/\omega} \sin 2(\omega t + \theta)\, dt = 0 \qquad (6.55)$$

Thus the time average power flow down the line is zero at all points on the line. This is characteristic of complete standing waves.

From (6.53a) and (6.53b) or (6.54a) and (6.54b), or from Figs. 6.15(a) and 6.15(b), we find that the amplitudes of the sinusoidal time-variations of the line voltage and line current as functions of distance along the line are

$$|\bar{V}(z)| = 2A |\sin \beta z| = 2A \left| \sin \frac{2\pi}{\lambda} z \right| \qquad (6.56a)$$

$$|\bar{I}(z)| = \frac{2A}{Z_0} |\cos \beta z| = \frac{2A}{Z_0} \left| \cos \frac{2\pi}{\lambda} z \right| \qquad (6.56b)$$

Sketches of these quantities versus z are shown in Fig. 6.16. These are known as the "standing wave patterns." They are the patterns of line voltage and line current one would obtain by connecting an a.c. voltmeter between the conductors of the line and an a.c. ammeter in series with one of the conductors of the line and observing their readings at various points along the line. Alternatively, one can sample the electric and magnetic fields by means of probes.

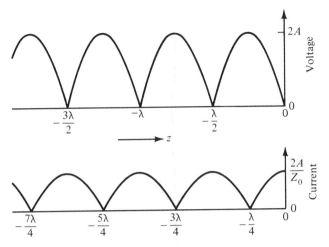

Figure 6.16. Standing wave patterns for voltage and current on a short-circuited line.

Returning now to the solutions for $\bar{V}(z)$ and $\bar{I}(z)$ given by (6.53a) and (6.53b), respectively, we can find the input impedance of the short-circuited line of length l by taking the ratio of the complex line voltage to the complex line current at the input $z = -l$. Thus

$$\bar{Z}_{\text{in}} = \frac{\bar{V}(-l)}{\bar{I}(-l)} = \frac{-2j\bar{A}\sin\beta(-l)}{\dfrac{2\bar{A}}{Z_0}\cos\beta(-l)}$$

$$= jZ_0\tan\beta l = jZ_0\tan\frac{2\pi}{\lambda}l$$

$$= jZ_0\tan\frac{2\pi f}{v_p}l \tag{6.57}$$

We note from (6.57) that the input impedance of the short-circuited line is purely reactive. As the frequency is varied from a low value upward, the input reactance changes from inductive to capacitive and back to inductive, and so on, as illustrated in Fig. 6.17. The input reactance is zero for values of frequency equal to multiples of $v_p/2l$. These are the frequencies for which l is equal to multiples of $\lambda/2$ so that the line voltage is zero at the input and hence the input sees a short circuit. The input reactance is infinity for values of frequency equal to odd multiples of $v_p/4l$. These are the frequencies for which l is equal to odd multiples of $\lambda/4$ so that the line current is zero at the input and hence the input sees an open circuit.

Example 6.3. From the foregoing discussion of the input reactance of the short-circuited line, we note that as the frequency of the generator is varied

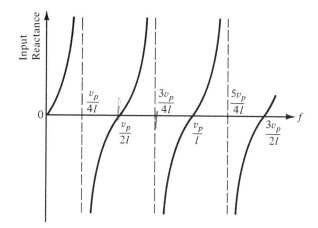

Figure 6.17. Variation of the input reactance of a short-circuited transmission line with frequency.

continuously upward, the current drawn from it undergoes alternatively maxima and minima corresponding to zero input reactance and infinite input reactance conditions, respectively. This behavior can be utilized for determining the location of a short circuit in the line.

Since the difference between a pair of consecutive frequencies for which the input reactance values are zero and infinity is $v_p/4l$, as can be seen from Fig. 6.17, it follows that the difference between successive frequencies for which the currents drawn from the generator are maxima and minima is $v_p/4l$. As a numerical example, if for an air dielectric line, it is found that as the frequency is varied from 50 MHz upward, the current reaches a minimum for 50.01 MHz and then a maximum for 50.04 MHz, then the distance l of the short circuit from the generator is given by

$$\frac{v_p}{4l} = (50.04 - 50.01) \times 10^6 = 0.03 \times 10^6 = 3 \times 10^4$$

Since $v_p = 3 \times 10^8$ m/s, it follows that

$$l = \frac{3 \times 10^8}{4 \times 3 \times 10^4} = 2500 \text{ m} = 2.5 \text{ km} \qquad \blacksquare$$

Example 6.4. We found that the input impedance of a short-circuited line of length l is given by

$$\bar{Z}_{in} = jZ_0 \tan \beta l$$

Let us investigate the low-frequency behavior of this input impedance.

First, we note that for any arbitrary value of βl,

$$\tan \beta l = \beta l + \frac{1}{3}(\beta l)^3 + \frac{2}{15}(\beta l)^5 + \cdots$$

For $\beta l \ll 1$, i.e., $\frac{2\pi}{\lambda}l \ll 1$ or $l \ll \frac{\lambda}{2\pi}$ or $f \ll \frac{v_p}{2\pi l}$,

$$\tan \beta l \approx \beta l$$

$$\bar{Z}_{in} \approx jZ_0\beta l = j\sqrt{\frac{\mathcal{L}}{\mathcal{C}}}\,\omega\sqrt{\mathcal{L}\mathcal{C}}\,l = j\omega\mathcal{L}l$$

Thus for frequencies $f \ll v_p/2\pi l$, the short-circuited line as seen from its input behaves essentially like a single inductor of value $\mathcal{L}l$, the total inductance of the line, as shown in Fig. 6.18(a).

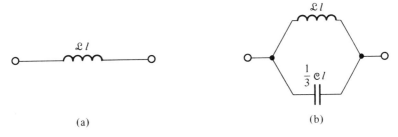

(a)   (b)

Figure 6.18. Equivalent circuits for the input behavior of a short-circuited transmission line.

Proceeding further, we observe that if the frequency is slightly beyond the range for which the above approximation is valid, then

$$\tan \beta l \approx \beta l + \frac{1}{3}(\beta l)^3$$

$$\bar{Z}_{in} \approx jZ_0\left(\beta l + \frac{1}{3}\beta^3 l^3\right)$$

$$= j\sqrt{\frac{\mathcal{L}}{\mathcal{C}}}\left(\omega\sqrt{\mathcal{L}\mathcal{C}}\,l + \frac{1}{3}\omega^3\mathcal{L}^{3/2}\mathcal{C}^{3/2}l^3\right)$$

$$= j\omega\mathcal{L}l\left(1 + \frac{1}{3}\omega^2\mathcal{L}\mathcal{C}l^2\right)$$

$$\bar{Y}_{in} = \frac{1}{\bar{Z}_{in}} = \frac{1}{j\omega\mathcal{L}l}\left(1 + \frac{1}{3}\omega^2\mathcal{L}\mathcal{C}l^2\right)^{-1}$$

$$\approx \frac{1}{j\omega\mathcal{L}l}\left(1 - \frac{1}{3}\omega^2\mathcal{L}\mathcal{C}l^2\right)$$

$$= \frac{1}{j\omega\mathcal{L}l} + j\frac{1}{3}\omega\mathcal{C}l$$

Thus for frequencies somewhat above those for which the approximation $f \ll v_p/2\pi l$ is valid, the short-circuited line as seen from its input behaves like an inductor of value $\mathcal{L}l$ in parallel with a capacitance of value $\frac{1}{3}\mathcal{C}l$, as shown in Fig. 6.18(b).

These findings illustrate that a physical structure that can be considered as an inductor at low frequencies $f \ll v_p/2\pi l$ no longer behaves like an inductor if the frequency is increased beyond that range. In fact, it has a "stray" capacitance associated with it. As the frequency is still increased, the equivalent circuit becomes further complicated. Thus conventional circuit theory considerations of physical structures are strictly valid only for $f \ll v_p/2\pi l$, or $l \ll \lambda/2\pi$. ∎

6.5 BOUNDARY CONDITIONS AT A DIELECTRIC DISCONTINUITY

In Sec. 6.1 we derived the boundary conditions for the field components at a perfect conductor surface by applying Maxwell's equations in integral form to infinitesimal closed paths and closed surfaces encompassing the boundary and by using the fact that the fields inside the perfect conductor are zero. In this section we shall derive the boundary conditions at an interface between two different perfect dielectric media by similarly considering the Maxwell's equations in integral form one at a time. We shall note, however, that fields exist on either side of the boundary and that there cannot be any surface charge or surface current on the boundary in view of the perfect dielectric nature of the two media. We shall then use these boundary conditions in the following section to study reflection and transmission at the junction of two transmission lines having different dielectrics.

Thus let us consider a plane boundary between two different dielectric media 1 and 2 characterized by ϵ_1, μ_1 and ϵ_2, μ_2, respectively, as shown in Fig. 6.19. Then, applying Faraday's law in integral form (6.1a) to the infinitesimal rectangular path $abcda$ as shown in Fig. 6.19, we have

$$\int_a^b \mathbf{E} \cdot d\mathbf{l} + \int_b^c \mathbf{E} \cdot d\mathbf{l} + \int_c^d \mathbf{E} \cdot d\mathbf{l} + \int_d^a \mathbf{E} \cdot d\mathbf{l} = -\frac{d}{dt} \int_{abcd} \mathbf{B} \cdot d\mathbf{S}$$

(6.58)

In the limit that ad and $bc \rightarrow 0$, we obtain

$$E_{ab}(ab) + E_{cd}(cd) = 0$$

$$E_{ab}(ab) - E_{dc}(cd) = 0 \quad \text{or} \quad E_{ab} = E_{dc}$$

(6.59)

Since this is true for any orientation of the rectangle, it follows that "the tangential component of **E** is continuous at the dielectric interface." Thus

$$E_{t1} = E_{t2} \quad \text{or} \quad \mathbf{i}_n \times (\mathbf{E}_1 - \mathbf{E}_2) = 0 \tag{6.60}$$

where the subscript t denotes "tangential" and $\mathbf{i}_n$ is the unit normal vector to the boundary, as shown in Fig. 6.19.

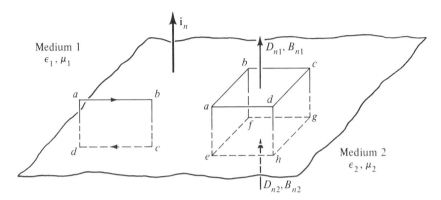

Figure 6.19. For deriving the boundary conditions at the interface between two perfect dielectric media.

Similarly, applying Ampere's circuital law in integral form (6.1b) to the rectangular path *abcda*, we have

$$\int_a^b \mathbf{H} \cdot d\mathbf{l} + \int_b^c \mathbf{H} \cdot d\mathbf{l} + \int_c^d \mathbf{H} \cdot d\mathbf{l} + \int_d^a \mathbf{H} \cdot d\mathbf{l}$$

$$= \int_{abcd} \mathbf{J} \cdot d\mathbf{S} + \frac{d}{dt} \int_{abcd} \mathbf{D} \cdot d\mathbf{S} \tag{6.61}$$

In the limit that *ad* and *bc* → 0 and noting that there is no current enclosed by *abcd*, we obtain

$$H_{ab}(ab) + H_{cd}(cd) = 0$$

$$H_{ab}(ab) - H_{dc}(cd) = 0 \quad \text{or} \quad H_{ab} = H_{dc} \tag{6.62}$$

Since this is true for any orientation of the rectangle, it follows that "the tangential component of **H** is continuous at the dielectric interface." Thus

$$H_{t1} = H_{t2} \quad \text{or} \quad \mathbf{i}_n \times (\mathbf{H}_1 - \mathbf{H}_2) = 0 \tag{6.63}$$

Considering next Gauss' law for the electric field in integral form (6.1c) and applying it to the infinitesimal rectangular box *abcdefgh*, as shown in Fig. 6.19, we have

$$\int_{abcd} \mathbf{D} \cdot d\mathbf{S} + \int_{\substack{\text{side} \\ \text{surfaces}}} \mathbf{D} \cdot d\mathbf{S} + \int_{efgh} \mathbf{D} \cdot d\mathbf{S} = \int_{\substack{\text{volume} \\ \text{of the} \\ \text{box}}} \rho \, dv \qquad (6.64)$$

In the limit that the side surfaces $\rightarrow 0$ and noting that there is no charge enclosed by the box, we obtain

$$D_{n1}(abcd) - D_{n2}(efgh) = 0 \quad \text{or} \quad D_{n1} = D_{n2} \qquad (6.65)$$

where the subscript n denotes "normal," and D_{n1} and D_{n2} are both directed into medium 1. Thus "the normal component of $\mathbf{D}$ is continuous at the dielectric interface." In vector form, we have

$$\mathbf{i}_n \cdot (\mathbf{D}_1 - \mathbf{D}_2) = 0 \qquad (6.66)$$

Similarly, applying Gauss' law for the magnetic field in integral form (6.1d) to the rectangular box *abcdefgh*, we have

$$\int_{abcd} \mathbf{B} \cdot d\mathbf{S} + \int_{\substack{\text{side} \\ \text{surfaces}}} \mathbf{B} \cdot d\mathbf{S} + \int_{efgh} \mathbf{B} \cdot d\mathbf{S} = 0 \qquad (6.67)$$

In the limit that the side surfaces $\rightarrow 0$, we obtain

$$B_{n1}(abcd) - B_{n2}(efgh) = 0 \quad \text{or} \quad B_{n1} = B_{n2} \qquad (6.68)$$

Thus "the normal component of $\mathbf{B}$ is continuous at the dielectric interface." In vector form, we have

$$\mathbf{i}_n \cdot (\mathbf{B}_1 - \mathbf{B}_2) = 0 \qquad (6.69)$$

Summarizing the boundary conditions for the field components at a dielectric interface, we have

$$\mathbf{i}_n \times (\mathbf{E}_1 - \mathbf{E}_2) = 0$$
$$\mathbf{i}_n \times (\mathbf{H}_1 - \mathbf{H}_2) = 0$$
$$\mathbf{i}_n \cdot (\mathbf{D}_1 - \mathbf{D}_2) = 0$$
$$\mathbf{i}_n \cdot (\mathbf{B}_1 - \mathbf{B}_2) = 0$$

Example 6.5. At a particular instant of time the fields at point 1 in Fig. 6.20 are given by

$$\mathbf{E}_1 = E_0(3\mathbf{i}_x + \mathbf{i}_z)$$
$$\mathbf{H}_1 = H_0(2\mathbf{i}_y)$$

where E_0 and H_0 are constants. Let us find the fields at point 2, lying adjacent to point 1 and on the other side of the interface between media 1 and 2.

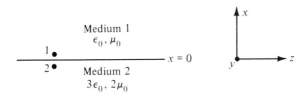

Figure 6.20. For illustrating the application of boundary conditions at the interface between two perfect dielectric media.

From (6.66), we have

$$D_{2x} = D_{1x} = \epsilon_0(3E_0) = 3\epsilon_0 E_0$$

$$E_{2x} = \frac{D_{2x}}{3\epsilon_0} = \frac{3\epsilon_0 E_0}{3\epsilon_0} = E_0$$

From (6.60), we get

$$E_{2y} = E_{1y} = 0$$

$$E_{2z} = E_{1z} = E_0$$

From (6.69), we obtain

$$B_{2x} = B_{1x} = \mu_0(0) = 0$$

$$H_{2x} = \frac{B_{2x}}{2\mu_0} = 0$$

From (6.63), we find

$$H_{2y} = H_{1y} = 2H_0$$

$$H_{2z} = H_{1z} = 0$$

Thus we obtain the rqeuired fields at point 2 to be

$$\mathbf{E}_2 = E_0(\mathbf{i}_x + \mathbf{i}_z)$$

$$\mathbf{H}_2 = H_0(2\mathbf{i}_y) \qquad\qquad \blacksquare$$

6.6 TRANSMISSION-LINE DISCONTINUITY

Let us now consider the case of two transmission lines 1 and 2 having different characteristic impedances Z_{01} and Z_{02}, respectively, and phase constants β_1 and β_2, respectively, connected in cascade and driven by a generator at the left end of line 1, as shown in Fig. 6.21(a). Physically, the arrangement may, for example, consist of two parallel-plate lines or two coaxial cables of

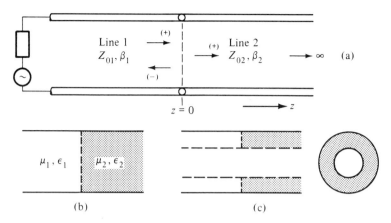

Figure 6.21. Two transmission lines connected in cascade.

different dielectrics in cascade, as shown in Figs. 6.21(b) and 6.21(c), respectively. In view of the discontinuity at the junction $z = 0$ between the two lines, the incident (+) wave on the junction sets up a reflected (−) wave in line 1 and a transmitted (+) wave in line 2. We shall assume that line 2 is infinitely long so that there is no (−) wave in that line.

We can now write the solutions for the complex voltage and complex current in line 1 as

$$\bar{V}_1(z) = \bar{V}_1^+ e^{-j\beta_1 z} + \bar{V}_1^- e^{j\beta_1 z} \tag{6.70a}$$

$$\bar{I}_1(z) = \bar{I}_1^+ e^{-j\beta_1 z} + \bar{I}_1^- e^{j\beta_1 z}$$

$$= \frac{1}{Z_{01}}(\bar{V}_1^+ e^{-j\beta_1 z} - \bar{V}_1^- e^{j\beta_1 z}) \tag{6.70b}$$

where $\bar{V}_1^+$, $\bar{V}_1^-$, $\bar{I}_1^+$, and $\bar{I}_1^-$ are the (+) and (−) wave voltages and currents at $z = 0-$ in line 1, that is, just to the left of the junction. The solutions for the complex voltage and current in line 2 are

$$\bar{V}_2(z) = \bar{V}_2^+ e^{-j\beta_2 z} \tag{6.71a}$$

$$\bar{I}_2(z) = \bar{I}_2^+ e^{-j\beta_2 z} = \frac{1}{Z_{02}}\bar{V}_2^+ e^{-j\beta_2 z} \tag{6.71b}$$

where $\bar{V}_2^+$ and $\bar{I}_2^+$ are the (+) wave voltage and current at $z = 0+$ in line 2, that is, just to the right of the junction.

At the junction the boundary conditions (6.60) and (6.63) require that the components of **E** and **H** tangential to the dielectric interface be continuous, as shown, for example, for the parallel-plate arrangement in Fig.

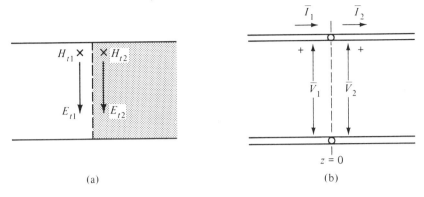

Figure 6.22. Application of boundary conditions at the junction between two transmission lines.

(a)   (b)

6.22(a). These are, in fact, the only components present since the transmission line fields are entirely transverse to the direction of propagation. Now, since the line voltage and current are related to these electric and magnetic fields, respectively, it then follows that the line voltage and line current be continuous at the junction, as shown in Fig. 6.22(b). Thus we obtain the boundary conditions at the junction in terms of line voltage and line current as

$$[\bar{V}_1]_{z=0-} = [\bar{V}_2]_{z=0+} \tag{6.72a}$$

$$[\bar{I}_1]_{z=0-} = [\bar{I}_2]_{z=0+} \tag{6.72b}$$

Applying these boundary conditions to the solutions given by (6.70a) and (6.70b), we obtain

$$\bar{V}_1^+ + \bar{V}_1^- = \bar{V}_2^+ \tag{6.73a}$$

$$\frac{1}{Z_{01}}(\bar{V}_1^+ - \bar{V}_1^-) = \frac{1}{Z_{02}}\bar{V}_2^+ \tag{6.73b}$$

Eliminating $\bar{V}_2^+$ from (6.73a) and (6.73b), we get

$$\bar{V}_1^+\left(\frac{1}{Z_{02}} - \frac{1}{Z_{01}}\right) + \bar{V}_1^-\left(\frac{1}{Z_{02}} + \frac{1}{Z_{01}}\right) = 0$$

or

$$\bar{V}_1^- = \bar{V}_1^+\frac{Z_{02} - Z_{01}}{Z_{02} + Z_{01}} \tag{6.74}$$

We now define the voltage reflection coefficient at the junction, Γ_V, as the ratio of the reflected wave voltage ($\bar{V}_1^-$) at the junction to the incident wave voltage ($\bar{V}_1^+$) at the junction. Thus

$$\Gamma_V = \frac{\bar{V}_1^-}{\bar{V}_1^+} = \frac{Z_{02} - Z_{01}}{Z_{02} + Z_{01}} \tag{6.75}$$

The current reflection coefficient at the junction, Γ_I, which is the ratio of the reflected wave current ($\bar{I}_1^-$) at the junction to the incident wave current ($\bar{I}_1^+$) at the junction is then given by

$$\Gamma_I = \frac{\bar{I}_1^-}{\bar{I}_1^+} = \frac{-\bar{V}_1^-/Z_{01}}{\bar{V}_1^+/Z_{01}} = -\frac{\bar{V}_1^-}{\bar{V}_1^+} = -\Gamma_V \tag{6.76}$$

We also define the voltage transmission coefficient at the junction, τ_V, as the ratio of the transmitted wave voltage ($\bar{V}_2^+$) at the junction to the incident wave voltage ($\bar{V}_1^+$) at the junction. Thus

$$\tau_V = \frac{\bar{V}_2^+}{\bar{V}_1^+} = \frac{\bar{V}_1^+ + \bar{V}_1^-}{\bar{V}_1^+} = 1 + \frac{\bar{V}_1^-}{\bar{V}_1^+} = 1 + \Gamma_V \tag{6.77}$$

The current transmission coefficient at the junction, τ_I, which is the ratio of the transmitted wave current ($\bar{I}_2^+$) at the junction to the incident wave current ($\bar{I}_1^+$) at the junction is given by

$$\tau_I = \frac{\bar{I}_2^+}{\bar{I}_1^+} = \frac{\bar{I}_1^+ + \bar{I}_1^-}{\bar{I}_1^+} = 1 + \frac{\bar{I}_1^-}{\bar{I}_1^+} = 1 - \Gamma_V \tag{6.78}$$

We note that for $Z_{02} = Z_{01}$, $\Gamma_V = 0$, $\Gamma_I = 0$, $\tau_V = 1$, and $\tau_I = 1$. Thus the incident wave is entirely transmitted as we may expect since there is no discontinuity at the junction.

Example 6.6. Let us consider the junction of two lines having characteristic impedances $Z_{01} = 50$ ohms and $Z_{02} = 75$ ohms, as shown in Fig. 6.23, and compute the various quantities.

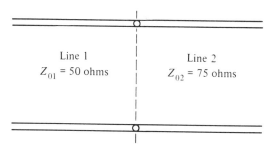

Figure 6.23. For the computation of several quantities pertinent to reflection and transmission at the junction between two transmission lines.

From (6.75)–(6.78), we have

$$\Gamma_V = \frac{75-50}{75+50} = \frac{25}{125} = \frac{1}{5}; \qquad \bar{V}_1^- = \frac{1}{5}\bar{V}_1^+$$

$$\Gamma_I = -\Gamma_V = -\frac{1}{5}; \qquad \bar{I}_1^- = -\frac{1}{5}\bar{I}_1^+$$

$$\tau_V = 1 + \Gamma_V = 1 + \frac{1}{5} = \frac{6}{5}; \qquad \bar{V}_2^+ = \frac{6}{5}\bar{V}_1^+$$

$$\tau_I = 1 - \Gamma_V = 1 - \frac{1}{5} = \frac{4}{5}; \qquad \bar{I}_2^+ = \frac{4}{5}\bar{I}_1^+$$

The fact that the transmitted wave voltage is greater than the incident wave voltage should not be of concern since it is the power balance that must be satisfied at the junction. We can verify this by noting that if the incident power on the junction is P_i, then

$$\text{reflected power, } P_r = \Gamma_V\Gamma_I P_i = -\frac{1}{25}P_i$$

$$\text{transmitted power, } P_t = \tau_V\tau_I P_i = \frac{24}{25}P_i$$

Recognizing that the minus sign for P_r signifies power flow in the negative z direction, we find that power balance is indeed satisfied at the junction. ∎

Returning now to the solutions for the voltage and current in line 1 given by (6.70a) and (6.70b), respectively, we obtain by replacing $\bar{V}_1^-$ by $\Gamma_V\bar{V}_1^+$,

$$\bar{V}_1(z) = \bar{V}_1^+ e^{-j\beta_1 z} + \Gamma_V\bar{V}_1^+ e^{j\beta_1 z}$$
$$= \bar{V}_1^+ e^{-j\beta_1 z}(1 + \Gamma_V e^{j2\beta_1 z}) \tag{6.79a}$$

$$\bar{I}_1(z) = \frac{1}{Z_{01}}(\bar{V}_1^+ e^{-j\beta_1 z} - \Gamma_V\bar{V}_1^+ e^{j\beta_1 z})$$
$$= \frac{\bar{V}_1^+}{Z_{01}}e^{-j\beta_1 z}(1 - \Gamma_V e^{j2\beta_1 z}) \tag{6.79b}$$

The amplitudes of the sinusoidal time-variations of the line voltage and line current as functions of distance along the line are then given by

$$|\bar{V}_1(z)| = |\bar{V}_1^+||e^{-j\beta_1 z}||1 + \Gamma_V e^{j2\beta_1 z}|$$
$$= |\bar{V}_1^+||1 + \Gamma_V \cos 2\beta_1 z + j\Gamma_V \sin 2\beta_1 z|$$
$$= |\bar{V}_1^+|\sqrt{1 + \Gamma_V^2 + 2\Gamma_V \cos 2\beta_1 z} \tag{6.80a}$$

$$|\bar{I}_1(z)| = \frac{|\bar{V}_1^+|}{Z_{01}}|e^{-j\beta_1 z}||1 - \Gamma_V e^{j2\beta_1 z}|$$

$$= \frac{|\bar{V}_1^+|}{Z_{01}}|1 - \Gamma_V \cos 2\beta_1 z - j\Gamma_V \sin 2\beta_1 z|$$

$$= \frac{|\bar{V}_1^+|}{Z_{01}}\sqrt{1 + \Gamma_V^2 - 2\Gamma_V \cos 2\beta_1 z} \qquad (6.80\text{b})$$

From (6.80a) and (6.80b), we note the following:

1. The line voltage amplitude undergoes alternate maxima and minima equal to $|\bar{V}_1^+|(1 + |\Gamma_V|)$ and $|\bar{V}_1^+|(1 - |\Gamma_V|)$, respectively. The line voltage amplitude at $z = 0$ is a maximum or minimum depending on whether Γ_V is positive or negative. The distance between a voltage maximum and the adjacent voltage minimum is $\pi/2\beta_1$ or $\lambda_1/4$.

2. The line current amplitude undergoes alternate maxima and minima equal to $\frac{|\bar{V}_1^+|}{Z_{01}}(1 + |\Gamma_V|)$ and $\frac{|\bar{V}_1^+|}{Z_{01}}(1 - |\Gamma_V|)$, respectively. The line current amplitude at $z = 0$ is a minimum or maximum depending on whether Γ_V is positive or negative. The distance between a current maximum and the adjacent current minimum is $\pi/2\beta_1$ or $\lambda_1/4$.

Knowing these properties of the line voltage and current amplitudes, we now sketch the voltage and current standing wave patterns, as shown in Fig. 6.24, assuming $\Gamma_V > 0$. Since these standing wave patterns do not contain perfect nulls, as in the case of the short-circuited line of Sec. 6.4, these are said to correspond to "partial standing waves."

We now define a quantity known as the "standing wave ratio" (SWR) as the ratio of the maximum voltage, $V_{\max}$, to the minimum voltage, $V_{\min}$, of the standing wave pattern. Thus we find that

$$\text{SWR} = \frac{V_{\max}}{V_{\min}} = \frac{|\bar{V}_1^+|(1 + |\Gamma_V|)}{|\bar{V}_1^+|(1 - |\Gamma_V|)} = \frac{1 + |\Gamma_V|}{1 - |\Gamma_V|} \qquad (6.81)$$

The SWR is an important parameter in transmission-line matching. It is an indicator of the degree of the existence of standing waves on the line. We shall, however, not pursue the topic here any further. Finally, we note that for the case of Example 6.6, the SWR in line 1 is $\left(1 + \frac{1}{5}\right)\big/\left(1 - \frac{1}{5}\right)$ or 1.5. The SWR in line 2 is, of course, equal to 1 since there is no reflected wave in that line.

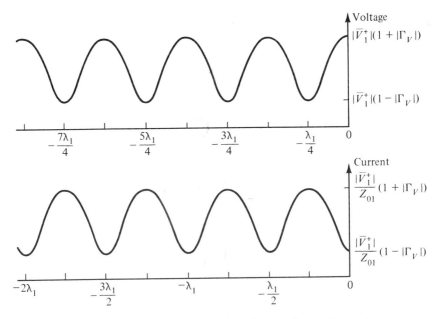

Figure 6.24. Standing wave patterns for voltage and current on a transmission line terminated by another transmission line.

6.7 SUMMARY

In this chapter we studied the principles of transmission lines by extending our knowledge of uniform plane wave propagation gained in the previous two chapters. To introduce the transmission line, we first derived the boundary conditions required to be satisfied by the field components at a perfect conductor surface. These boundary conditions, which follow from the application of Maxwell's equations in integral form to infinitesimal closed paths and surfaces straddling the boundary and from the property that the fields inside a perfect conductor are zero, are given in vector form by

$$\mathbf{i}_n \times \mathbf{E} = 0 \qquad (6.82a)$$

$$\mathbf{i}_n \times \mathbf{H} = \mathbf{J}_S \qquad (6.82b)$$

$$\mathbf{i}_n \cdot \mathbf{D} = \rho_S \qquad (6.82c)$$

$$\mathbf{i}_n \cdot \mathbf{B} = 0 \qquad (6.82d)$$

where $\mathbf{i}_n$ is the unit normal vector to the conductor surface and directed into the field region. Equations (6.82a) and (6.82d) state that the electric field be completely normal and that the magnetic field be completely tangential at a point on the conductor surface. The normal displacement flux density and the tangential magnetic field intensity are then related to the surface charge density and the surface current density as given by (6.82c) and (6.82b), respectively.

We used the boundary conditions (6.82a)–(6.82d) to illustrate that the placing of perfect conductors in planes normal to the electric field and hence tangential to the magnetic field of a uniform plane wave does not alter the field distribution and the wave is simply guided between and parallel to the conductors supported by the charges and currents on the conductors, as though they were not present, thereby constituting a parallel-plate transmission line. We then showed that wave propagation on a transmission line can be discussed in terms of voltage and current, which are related to the electric and magnetic fields, respectively, by deriving the "transmission-line equations"

$$\frac{\partial V}{\partial z} = -\mathcal{L}\frac{\partial I}{\partial t} \tag{6.83a}$$

$$\frac{\partial I}{\partial z} = -\mathcal{G}V - \mathcal{C}\frac{\partial V}{\partial t} \tag{6.83b}$$

which then led us to the concept of the distributed circuit.

The parameters $\mathcal{L}$, $\mathcal{C}$, and $\mathcal{G}$ in (6.83a) and (6.83b) are the inductance, capacitance, and conductance per unit length of line, which differ from one line to another. For the parallel-plate line having width w of the plates and spacing d between the plates, they are given by

$$\mathcal{L} = \frac{\mu d}{w}$$

$$\mathcal{C} = \frac{\epsilon w}{d}$$

$$\mathcal{G} = \frac{\sigma w}{d}$$

where μ, ϵ, and σ are the material parameters of the medium between the plates, and fringing of the fields is neglected. We learned how to compute $\mathcal{L}$, $\mathcal{C}$, and $\mathcal{G}$ for a line of arbitrary cross section by constructing a field map of the transverse electromagnetic wave fields, consisting of curvilinear squares in the cross-sectional plane of the line. If m is the number of squares tangential to the conductors and n is the number of squares normal to the conductors, then

$$\mathcal{L} = \mu\frac{n}{m}$$

$$\mathcal{C} = \epsilon \frac{m}{n}$$

$$\mathcal{G} = \sigma \frac{m}{n}$$

By applying this technique to the coaxial cable, we found that for a cable of inner radius a and outer radius b,

$$\mathcal{L} = \frac{\mu}{2\pi} \ln \frac{b}{a}$$

$$\mathcal{C} = \frac{2\pi\epsilon}{\ln (b/a)}$$

$$\mathcal{G} = \frac{2\pi\sigma}{\ln (b/a)}$$

The general solutions to the transmission-line equations (6.83a) and (6.83b), expressed in phasor form, that is,

$$\frac{\partial \bar{V}}{\partial z} = -j\omega \mathcal{L}\bar{I} \tag{6.84a}$$

$$\frac{\partial \bar{I}}{\partial z} = -\mathcal{G}\bar{V} - j\omega \mathcal{C}\bar{V} \tag{6.84b}$$

are given by

$$\bar{V}(z) = \bar{A}e^{-\bar{\gamma}z} + \bar{B}e^{\bar{\gamma}z} \tag{6.85a}$$

$$\bar{I}(z) = \frac{1}{\bar{Z}_0}(\bar{A}e^{-\bar{\gamma}z} - \bar{B}e^{\bar{\gamma}z}) \tag{6.85b}$$

where

$$\bar{\gamma} = \sqrt{j\omega\mathcal{L}(\mathcal{G} + j\omega\mathcal{C})} \qquad [= \sqrt{j\omega\mu(\sigma + j\omega\epsilon)}]$$

$$\bar{Z}_0 = \sqrt{\frac{j\omega\mathcal{L}}{\mathcal{G} + j\omega\mathcal{C}}} \qquad \left[\neq \sqrt{\frac{j\omega\mu}{\sigma + j\omega\epsilon}}\right]$$

are the propagation constant and the characteristic impedance, respectively, of the line. For a lossless line ($\mathcal{G} = 0$), these reduce to

$$\bar{\gamma} = j\omega\sqrt{\mathcal{L}\mathcal{C}} \qquad (= j\omega\sqrt{\mu\epsilon})$$

$$\bar{Z}_0 = \sqrt{\frac{\mathcal{L}}{\mathcal{C}}} \qquad (\neq \sqrt{\mu/\epsilon})$$

The solutions given by (6.85a) and (6.85b) represent the superposition of ($+$) and ($-$) waves propagating in the medium between the conductors of the line, expressed in terms of the line voltage and current instead of in terms of the electric and magnetic fields. By applying these general solutions to the case of a lossless line short circuited at the far end and obtaining the particular solutions for that case, we discussed the standing wave phenomenon and the standing wave patterns resulting from the complete reflection of waves by the short circuit. We also examined the frequency behavior of the input impedance of a short-circuited line of length l, given by

$$\bar{Z}_{\text{in}} = jZ_0 \tan \beta l$$

and (a) illustrated its application in a technique for the location of short circuit in a line, and (b) learned that for a circuit element to behave as assumed by conventional (lumped) circuit theory, its dimensions must be a small fraction of the wavelength corresponding to the frequency of operation.

To extend the discussion of the reflection phenomenon to one of partial reflection and transmission, we first derived the boundary conditions at the interface between two dielectric media. These are given in vector form by

$$\mathbf{i}_n \times (\mathbf{E}_1 - \mathbf{E}_2) = 0 \qquad (6.86a)$$

$$\mathbf{i}_n \times (\mathbf{H}_1 - \mathbf{H}_2) = 0 \qquad (6.86b)$$

$$\mathbf{i}_n \cdot (\mathbf{D}_1 - \mathbf{D}_2) = 0 \qquad (6.86c)$$

$$\mathbf{i}_n \cdot (\mathbf{B}_1 - \mathbf{B}_2) = 0 \qquad (6.86d)$$

where $\mathbf{i}_n$ is the unit normal vector to the interface and directed into the medium having the subscript 1 for the fields. These boundary conditions point to the continuity of the tangential component of $\mathbf{E}$, the tangential component of $\mathbf{H}$, the normal component of $\mathbf{D}$, and the normal component of $\mathbf{B}$, at a point on the interface.

We used the boundary conditions (6.86a)–(6.86d) to investigate reflection and transmission of waves at a junction between two lossless lines. By applying them to the general solutions for the line voltage and current on either side of the junction, we deduced the ratio of the reflected wave voltage to the incident wave voltage, that is, the voltage reflection coefficient, to be

$$\Gamma_V = \frac{Z_{02} - Z_{01}}{Z_{02} + Z_{01}}$$

where Z_{01} is the characteristic impedance of the line from which the wave is incident and Z_{02} is the characteristic impedance of the line on which the wave is incident. The ratio of the transmitted wave voltage to the incident wave voltage, that is, the voltage transmission coefficient, is given by

$$\tau_V = 1 + \Gamma_V$$

The current reflection and transmission coefficients are given by

$$\Gamma_I = -\Gamma_V$$
$$\tau_I = 1 - \Gamma_V$$

Finally, we discussed the standing wave pattern resulting from the partial reflection of the wave at the junction and defined a quantity known as the standing wave ratio (SWR), which is a measure of the reflection phenomenon. In terms of Γ_V, it is given by

$$SWR = \frac{1 + |\Gamma_V|}{1 - |\Gamma_V|}$$

In retrospect, it can be seen that the discussion of the standing wave phenomenon and reflection and transmission at the junction of two lines is equally applicable to the solution of analogous uniform plane wave problems involving media unbounded in the two dimensions normal to the direction of propagation of the wave.

REVIEW QUESTIONS

6.1. What is a boundary condition? How do boundary conditions arise?

6.2. State the boundary conditions for the electric field components at the surface of a perfect conductor.

6.3. State the boundary conditions for the magnetic field components at the surface of a perfect conductor.

6.4. Summarize in vector form the boundary conditions at a perfect conductor surface, indicating correspondingly the Maxwell's equations in integral form from which they are derived.

6.5. Discuss the guiding of a uniform plane wave by a pair of parallel-plane, perfectly conducting sheets.

6.6. How is the voltage between the two conductors in a given cross-sectional plane of a parallel-plate transmission line related to the electric field in that plane?

6.7. How is the current flowing on the plates across a given cross-sectional plane of a parallel-plate transmission line related to the magnetic field in that plane?

6.8. What are transmission-line equations? How are they obtained from Maxwell's equations?

6.9. How is $\mathcal{L}$, the inductance per unit length of a transmission line, defined? What is it equal to for a parallel-plate transmission line?

6.10. How is $\mathcal{C}$, the capacitance per unit length of a transmission line, defined? What is it equal to for a parallel-plate transmission line?

6.11. How is $\mathcal{G}$, the conductance per unit length of a transmission line, defined? What is it equal to for a parallel-plate transmission line?

6.12. Are the three quantities $\mathcal{L}$, $\mathcal{C}$, and $\mathcal{G}$ independent? If not, how are they dependent on each other?

6.13. Draw the transmission-line equivalent circuit. How is it derived from the transmission-line equations?

6.14. Discuss the concept of the distributed circuit and compare it to a lumped circuit.

6.15. Discuss the physical phenomena associated with each of the elements in the transmission-line equivalent circuit.

6.16. What is a transverse electromagnetic wave?

6.17. What is a field map? Describe the procedure for drawing the field map for a transmission line of arbitrary cross section.

6.18. Draw a rough sketch of the field map for a line made up of two identical parallel cylindrical conductors with their axes separated by four times their radii.

6.19. Describe the procedure for computing the transmission line parameters $\mathcal{L}$, $\mathcal{C}$, and $\mathcal{G}$ from the field map.

6.20. How does a field map consisting of curvilinear squares simplify the computation of the line parameters?

6.21. Discuss the determination of $\mathcal{L}$, $\mathcal{C}$, and $\mathcal{G}$ for a coaxial cable by using the curvilinear squares technique.

6.22. By consulting an appropriate reference book, prepare a list of the expressions for $\mathcal{L}$, $\mathcal{C}$, and $\mathcal{G}$ for two or more transmission lines other than the parallel-plate and coaxial lines.

6.23. Discuss your understanding of the characteristic impedance of a transmission line. Why is it not equal to the intrinsic impedance of the medium between the conductors of the line?

6.24. What is the boundary condition to be satisfied at a short circuit on a line?

6.25. For an open-circuited line, what would be the boundary condition to be satisfied at the open circuit?

6.26. What is a standing wave? How do complete standing waves arise? Discuss their characteristics and give an example in mechanics.

6.27. What is a standing wave pattern? Discuss the voltage and current standing wave patterns for the short-circuited line.

6.28. What would be the voltage and current standing wave patterns for an open-circuited line?

6.29. Discuss the variation with frequency of the input reactance of a short-circuited line and its application in the determination of the location of a short circuit.

6.30. Can you suggest an alternative procedure to that described in Example 6.3 to locate a short circuit in a transmission line?

6.31. Under what condition do circuit elements behave as assumed by conventional (lumped) circuit theory?

6.32. State the boundary conditions for the electric field components at the interface between two dielectric media.

6.33. State the boundary conditions for the magnetic field components at the interface between two dielectric media.

6.34. Summarize in vector form the boundary conditions at the interface between two dielectric media, indicating correspondingly the Maxwell's equations in integral form from which they are derived.

6.35. What are the boundary conditions for the voltage and current at the junction between two transmission lines?

6.36. What is the voltage reflection coefficient at the junction between two transmission lines? How are the current reflection coefficient and the voltage and current transmission coefficients related to the voltage reflection coefficient?

6.37. What is the voltage reflection coefficient at the short circuit for a short-circuited line?

6.38. Can the transmitted wave current at the junction between two transmission lines be greater than the incident wave current? Explain.

6.39. What is a partial standing wave? Discuss the standing wave patterns corresponding to partial standing waves.

6.40. Define standing wave ratio (SWR). What are the standing wave ratios for (a) an infinitely long line, (b) a short-circuited line, (c) an open-circuited line, and (d) a line terminated by its characteristic impedance?

PROBLEMS

6.1. The plane $x + 2y + 3z = 5$ defines the surface of a perfect conductor. Find the possible direction(s) of the electric field intensity at a point on the conductor surface.

6.2. Given $\mathbf{E} = y\mathbf{i}_x + x\mathbf{i}_y$, determine if a perfect conductor can be placed in the surface $xy = 2$ without disturbing the field.

6.3. A perfect conductor occupies the region $x + 2y \leq 2$. Find the surface current density at a point on the conductor at which $\mathbf{H} = H_0\mathbf{i}_z$.

6.4. The displacement flux density at a point on the surface of a perfect conductor is given by $\mathbf{D} = D_0(\mathbf{i}_x + \sqrt{3}\,\mathbf{i}_y + 2\sqrt{3}\,\mathbf{i}_z)$. Find the magnitude of the surface charge density at that point.

6.5. It is known that at a point on the surface of a perfect conductor $\mathbf{D} = D_0(\mathbf{i}_x + 2\mathbf{i}_y + 2\mathbf{i}_z)$, $\mathbf{H} = H_0(2\mathbf{i}_x - 2\mathbf{i}_y + \mathbf{i}_z)$, and ρ_S is positive. Find ρ_S and $\mathbf{J}_S$ at that point.

6.6. Two infinite plane conducting sheets occupy the planes $x = 0$ and $x = 0.1$ m. An electric field given by

$$\mathbf{E} = E_0 \sin 10\pi x \cos 3\pi \times 10^9 t \, \mathbf{i}_z$$

where E_0 is a constant, exists in the region between the plates, which is free space. (a) Show that $\mathbf{E}$ satisfies the boundary condition on the sheets. (b) Obtain $\mathbf{H}$ associated with the given $\mathbf{E}$. (c) Find the surface current densities on the two sheets.

6.7. A parallel-plate transmission line is made up of perfect conductors of width $w = 0.1$ m and lying in the planes $x = 0$ and $x = 0.02$ m. The medium between the conductors is a perfect dielectric of $\mu = \mu_0$. For a uniform plane wave having the electric field

$$\mathbf{E} = 100\pi \cos (2\pi \times 10^6 t - 0.02\pi z) \, \mathbf{i}_x \text{ V/m}$$

propagating between the conductors, find (a) the voltage between the conductors, (b) the current along the conductors, and (c) the power flow along the line.

6.8. A parallel-plate transmission line made up of perfect conductors has $\mathcal{L}$ equal to 10^{-7} H/m. If the medium between the plates is characterized by $\sigma = 10^{-11}$ mho/m, $\epsilon = 6\epsilon_0$, and $\mu = \mu_0$, find $\mathcal{C}$ and $\mathcal{G}$ of the line.

6.9. If the conductors of a transmission line are imperfect, then the transmission-line equivalent circuit contains a resistance and additional inductance in the series branch. Assuming that the thickness of the (imperfect) conductors of a parallel-plate line is several skin depths at the frequency of interest, show from considerations of skin effect phenomenon in a good conductor medium that the resistance and inductance per unit length along the conductors are $2/\sigma_c \delta w$ and $2/\omega \sigma_c \delta w$, respectively, where σ_c is the conductivity of the (imperfect) conductors, w is the width and δ is the skin depth. The factor 2 arises because of two conductors.

6.10. Show that for a transverse electromagnetic wave, the voltage between the conductors and the current along the conductors in a given transverse plane are uniquely defined in terms of the electric and magnetic fields, respectively, in that plane.

6.11. By constructing a field map consisting of curvilinear squares for a coaxial cable having $b/a = 3.5$, obtain the approximate values of the line parameters $\mathcal{L}$, $\mathcal{C}$, and $\mathcal{G}$ in terms of μ, ϵ, and σ of the dielectric. Compare the approximate values with the exact values given by expressions derived in Example 6.2.

6.12. Figure 6.25 shows the cross section of a parallel-wire line, that is, a line having two cylindrical conductors of radii a and with their axes separated by 2d. For $d/a = 2$, construct a field map consisting of curvilinear squares and obtain approximate values for the line parameters $\mathcal{L}$, $\mathcal{C}$, and $\mathcal{G}$. Compare the

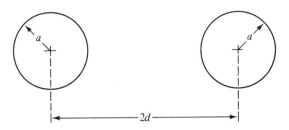

Figure 6.25. For Problem 6.12.

approximate values with the exact values given by expressions available from Sec. 10.6.

6.13. For a transmission line of arbitrary cross section and with the medium between the conductors characterized by $\sigma = 10^{-16}$ mho/m, $\epsilon = 2.5\epsilon_0$, and $\mu = \mu_0$, it is known that $\mathcal{C} = 10^{-10}$ F/m. (a) Find $\mathcal{L}$ and $\mathcal{G}$. (b) Find $\bar{Z}_0$ for $f = 10^6$ Hz.

6.14. For a coaxial cable employing air dielectric, find the ratio of the outer to the inner radii for which the characteristic impedance of the cable is 75 ohms.

6.15. Show that for the parallel-plate line, the characteristic impedance is d/w times the intrinsic impedance of the medium between the conductors of the line.

6.16. The strip line, employed in microwave integrated circuits, consists of a center conductor photoetched on the inner faces of two substrates sandwiched between two conductors, as shown by the cross-sectional view in Fig. 6.26. For the dimensions shown in the figure, construct a field map consisting of curvilinear squares and compute $\mathcal{L}$, $\mathcal{C}$, and Z_0, considering the substrate to be a perfect dielectric having $\epsilon = 9\epsilon_0$ and $\mu = \mu_0$. Assume for simplicity that the field is confined to the substrate region.

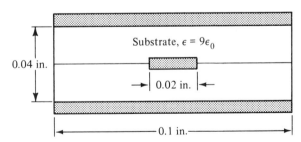

Figure 6.26. For Problem 6.16.

6.17. Consider a transmission-line equivalent circuit having impedance $\mathcal{Z}\, dz$ in the series branch and admittance $\mathcal{Y}\, dz$ in the shunt branch. (a) Write the transmission-line equations. (b) Show that $\bar{\gamma} = \sqrt{\mathcal{Z}\mathcal{Y}}$ and $\bar{Z}_0 = \sqrt{\mathcal{Z}/\mathcal{Y}}$. (c) If $\mathcal{Z}$ is the impedance of an inductor $\mathcal{L}_1$ and $\mathcal{Y}$ is the admittance of the parallel combination of an inductor $\mathcal{L}_2$ and a capacitor $\mathcal{C}$, find $\bar{\gamma}$ and discuss the propagation characteristics along the line.

6.18. Using the general solutions for the complex line voltage and current on a lossless line given by (6.50a) and (6.50b), respectively, obtain the particular solutions for the complex voltage and current on an open-circuited line. Then find the input impedance of an open-circuited line of length l.

6.19. Solve Example 6.3 by considering the standing wave patterns between the short circuit and the generator for the two frequencies of interest and by deducing the number of wavelengths at one of the two frequencies.

6.20. For an air dielectric short-circuited line of characteristic impedance 50 ohms, find the minimum values of the length for which its input impedance is equivalent to that of (a) an inductor of value 0.25×10^{-6} H at 100 MHz and (b) a capacitor of value 10^{-10} F at 100 MHz.

6.21. A transmission line of length 2 m having a nonmagnetic ($\mu = \mu_0$) perfect dielectric is short-circuited at the far end. A variable-frequency generator is connected at its input and the current drawn is monitored. It is found that the current reaches a maximum for $f = 500$ MHz and then a minimum for $f = 525$ MHz. Find the permittivity of the dielectric.

6.22. A voltage generator is connected to the input of a lossless line short circuited at the far end. The frequency of the generator is varied and the line voltage and line current at the input terminals are monitored. It is found that the voltage reaches a maximum value of 10 V at 405 MHz and the current reaches a maximum value of 0.2 amp at 410 MHz. (a) Find the characteristic impedance of the line. (b) Find the voltage and current values at 407 MHz.

6.23. Assuming that the criterion $f \ll v_p/2\pi l$ is satisfied for frequencies less than $0.1\, v_p/2\pi l$, compute the maximum length of an air dielectric short-circuited line for which the input impedance is approximately that of an inductor of value equal to the total inductance of the line for $f = 100$ MHz.

6.24. A lossless transmission line of length 2 m and having $\mathcal{L} = 0.5\mu_0$ and $\mathcal{C} = 18\epsilon_0$ is short circuited at the far end. (a) Find the phase velocity, v_p. (b) Find the wavelength, the length of the line in terms of the number of wavelengths, and the input impedance of the line for each of the following frequencies: 100 Hz; 100 MHz; and 12.5 MHz.

6.25. In Fig. 6.20, assume that medium 1 is characterized by $\epsilon = 12\epsilon_0$ and $\mu = 2\mu_0$ and that medium 2 is characterized by $\epsilon = 9\epsilon_0$ and $\mu = \mu_0$. If $\mathbf{E}_1 = E_0(3\mathbf{i}_x + 2\mathbf{i}_y - 6\mathbf{i}_z)$ and if $\mathbf{H}_1 = H_0(2\mathbf{i}_x - 3\mathbf{i}_y)$, find $\mathbf{E}_2$ and $\mathbf{H}_2$.

6.26. In Fig. 6.20, assume that medium 1 is characterized by $\epsilon = 4\epsilon_0$ and $\mu = 3\mu_0$ and that medium 2 is characterized by $\epsilon = 16\epsilon_0$ and $\mu = 9\mu_0$. If $\mathbf{D}_1 = D_0(\mathbf{i}_x - 2\mathbf{i}_y + \mathbf{i}_z)$ and if $\mathbf{B}_1 = B_0(\mathbf{i}_x + 2\mathbf{i}_y + 3\mathbf{i}_z)$, find $\mathbf{D}_2$ and $\mathbf{B}_2$.

6.27. Region 1 defined by $x + 2y < 2$ is free space and region 2 defined by $x + 2y > 2$ is a perfect dielectric medium having $\epsilon = 6\epsilon_0$ and $\mu = 2\mu_0$. Determine if the fields $\mathbf{E}_1 = E_0\mathbf{i}_y$ and $\mathbf{H}_1 = H_0\mathbf{i}_z$ and the fields $\mathbf{E}_2 = \dfrac{E_0}{3}(-\mathbf{i}_x + \mathbf{i}_y)$ and

$\mathbf{H}_2 = H_0\mathbf{i}_z$ at points 1 and 2, respectively, lying adjacent to and on either side of the boundary, satisfy the boundary conditions.

6.28. Repeat Example 6.6 with the values of Z_{01} and Z_{02} interchanged.

6.29. In the transmission-line system shown in Fig. 6.27, a power P_i is incident on the junction from line 1. Find (a) the power reflected into line 1, (b) the power transmitted into line 2, and (c) the power transmitted into line 3.

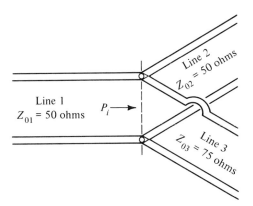

Figure 6.27. For Problem 6.29.

6.30. Show that the voltage minima of the standing wave pattern of Fig. 6.24 are sharper than the voltage maxima by computing the voltage amplitude halfway between the locations of voltage maxima and minima.

6.31. A line assumed to be infinitely long and of unknown characteristic impedance is connected to a line of characteristic impedance 50 ohms on which standing wave measurements are made. It is found that the standing wave ratio is 3 and that two consecutive voltage minima exist at 15 cm and 25 cm from the junction of the two lines. Find the unknown characteristic impedance.

6.32. A line assumed to be infinitely long and of unknown characteristic impedance when connected to a line of characteristic impedance 50 ohms produces a standing wave ratio of value 2 in the 50-ohm line. The same line when connected to a line of characteristic impedance 150 ohms produces a standing wave ratio of value 1.5 in the 150-ohm line. Find the unknown characteristic impedance.

7. WAVEGUIDES

In Chap. 6 we studied the principles of transmission lines, one of the two kinds of waveguiding systems. We learned that transmission lines are made up of two (or more) parallel conductors. The second kind of waveguiding system, namely, waveguides, generally consists of a single conductor. Guiding of waves in a waveguide is accomplished by the bouncing of the waves obliquely between the walls of the guide, as compared to the case of a transmission line in which the waves slide parallel to the conductors of the line. It is our goal in this chapter to learn the principles of waveguides.

We shall introduce the principle of waveguides by first considering a parallel-plate waveguide, that is, a waveguide consisting of two parallel, plane conductors and then extend it to the rectangular waveguide, which is a hollow metallic pipe of rectangular cross section, a common form of waveguide. We shall learn that waveguides are characterized by cutoff, which is the phenomenon of no propagation in a certain range of frequencies, and dispersion, which is the phenomenon of propagating waves of different frequencies possessing different phase velocities along the waveguide. In connection with the latter characteristic, we shall introduce the concept of group velocity. We shall also discuss the principles of cavity resonators, the microwave counterparts of resonant circuits, and of optical waveguides. To introduce the parallel-plate waveguide, we shall make use of the superposition of two uniform plane waves propagating at an angle to each other. Hence we shall begin the chapter with the discussion of uniform plane wave propagation in an arbitrary direction relative to the coordinate axes.

7.1 UNIFORM PLANE WAVE PROPAGATION IN AN ARBITRARY DIRECTION

In Chap. 4 we introduced the uniform plane wave propagating in the z direction by considering an infinite plane current sheet lying in the xy plane. If the current sheet lies in a plane making an angle to the xy plane, the uniform plane wave would then propagate in a direction different from the z direction. Thus let us consider a uniform plane wave propagating in the z' direction making an angle θ with the negative x axis as shown in Fig. 7.1. Let the electric field of the wave be entirely in the y direction. The magnetic field would then be directed as shown in the figure so that $\mathbf{E} \times \mathbf{H}$ points in the z' direction.

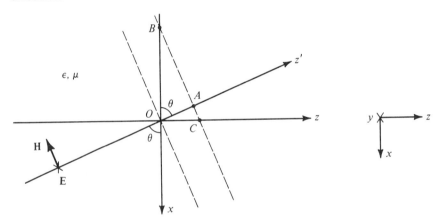

Figure 7.1. Uniform plane wave propagating in the z' direction lying in the xz plane and making an angle θ with the negative x axis.

We can write the expression for the electric field of the wave as

$$\mathbf{E} = E_0 \cos(\omega t - \beta z')\, \mathbf{i}_y \tag{7.1}$$

where $\beta = \omega\sqrt{\mu\epsilon}$ is the phase constant, that is, the rate of change of phase with distance along the z' direction for a fixed value of time. From the construction of Fig. 7.2(a), we, however, have

$$z' = -x \cos\theta + z \sin\theta \tag{7.2}$$

so that

$$\begin{aligned} \mathbf{E} &= E_0 \cos[\omega t - \beta(-x\cos\theta + z\sin\theta)]\, \mathbf{i}_y \\ &= E_0 \cos[\omega t - (-\beta\cos\theta)x - (\beta\sin\theta)z]\, \mathbf{i}_y \\ &= E_0 \cos(\omega t - \beta_x x - \beta_z z)\, \mathbf{i}_y \end{aligned} \tag{7.3}$$

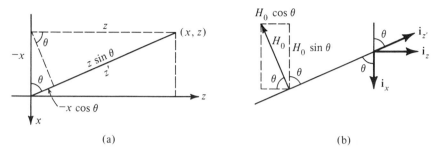

(a)   (b)

Figure 7.2. Constructions pertinent to the formulation of the expressions
for the fields of the uniform plane wave of Fig. 7.1.

where $\beta_x = -\beta \cos \theta$ and $\beta_z = \beta \sin \theta$ are the phase constants in the
positive x and positive z directions, respectively.

We note that $|\beta_x|$ and $|\beta_z|$ are less than β, the phase constant along the
direction of propagation of the wave. This can also be seen from Fig. 7.1 in
which two constant phase surfaces are shown by dashed lines passing through
the points O and A on the z' axis. Since the distance along the x direction
between the two constant phase surfaces, that is, the distance OB is equal to
$OA/\cos \theta$, the rate of change of phase with distance along the x direction is
equal to

$$\beta \frac{OA}{OB} = \frac{\beta(OA)}{OA/\cos \theta} = \beta \cos \theta$$

The minus sign for β_x simply signifies the fact that insofar as the x axis is
concerned, the wave is progressing in the negative x direction. Similarly,
since the distance along the z direction between the two constant phase surfaces, that is, the distance OC is equal to $OA/\sin \theta$, the rate of change of phase
with distance along the z direction is equal to

$$\beta \frac{OA}{OC} = \frac{\beta(OA)}{OA/\sin \theta} = \beta \sin \theta$$

Since the wave is progressing along the positive z direction, β_z is positive. We
further note that

$$\beta_x^2 + \beta_z^2 = (-\beta \cos \theta)^2 + (\beta \sin \theta)^2 = \beta^2 \tag{7.4}$$

and that

$$-\cos \theta \, \mathbf{i}_x + \sin \theta \, \mathbf{i}_z = \mathbf{i}_{z'} \tag{7.5}$$

where $\mathbf{i}_{z'}$ is the unit vector directed along z' direction, as shown in Fig. 7.2(b).
Thus the vector

$$\boldsymbol{\beta} = (-\beta \cos \theta)\mathbf{i}_x + (\beta \sin \theta)\mathbf{i}_z = \beta_x \mathbf{i}_x + \beta_z \mathbf{i}_z \tag{7.6}$$

defines completely the direction of propagation and the phase constant along the direction of propagation. Hence the vector $\boldsymbol{\beta}$ is known as the "propagation vector."

The expression for the magnetic field of the wave can be written as

$$\mathbf{H} = \mathbf{H}_0 \cos{(\omega t - \beta z')} \tag{7.7}$$

where

$$|\mathbf{H}_0| = \frac{E_0}{\sqrt{\mu/\epsilon}} = \frac{E_0}{\eta} \tag{7.8}$$

since the ratio of the electric field intensity to the magnetic field intensity of a uniform plane wave is equal to the intrinsic impedance of the medium. From the construction in Fig. 7.2(b), we observe that

$$\mathbf{H}_0 = H_0(-\sin\theta\,\mathbf{i}_x - \cos\theta\,\mathbf{i}_z) \tag{7.9}$$

Thus using (7.9) and substituting for z' from (7.2), we obtain

$$\mathbf{H} = H_0(-\sin\theta\,\mathbf{i}_x - \cos\theta\,\mathbf{i}_z)\cos{[\omega t - \beta(-x\cos\theta + z\sin\theta)]}$$

$$= -\frac{E_0}{\eta}(\sin\theta\,\mathbf{i}_x + \cos\theta\,\mathbf{i}_z)\cos{[\omega t - \beta_x x - \beta_z z]} \tag{7.10}$$

Generalizing the foregoing treatment to the case of a uniform plane wave propagating in a completely arbitrary direction in three dimensions, as shown in Fig. 7.3, and characterized by phase constants β_x, β_y, and β_z in the x, y, and z directions, respectively, we can write the expression for the electric field as

$$\mathbf{E} = \mathbf{E}_0 \cos{(\omega t - \beta_x x - \beta_y y - \beta_z z + \phi_0)}$$

$$= \mathbf{E}_0 \cos{[\omega t - (\beta_x\mathbf{i}_x + \beta_y\mathbf{i}_y + \beta_z\mathbf{i}_z)\cdot(x\mathbf{i}_x + y\mathbf{i}_y + z\mathbf{i}_z) + \phi_0]}$$

$$= \mathbf{E}_0 \cos{(\omega t - \boldsymbol{\beta}\cdot\mathbf{r} + \phi_0)} \tag{7.11}$$

where

$$\boldsymbol{\beta} = \beta_x\mathbf{i}_x + \beta_y\mathbf{i}_y + \beta_z\mathbf{i}_z \tag{7.12}$$

is the propagation vector,

$$\mathbf{r} = x\mathbf{i}_x + y\mathbf{i}_y + z\mathbf{i}_z \tag{7.13}$$

is the position vector, and ϕ_0 is the phase at the origin at $t = 0$. The position vector is the vector drawn from the origin to the point (x, y, z) and hence has components x, y, and z along the x, y, and z axes, respectively. The expression for the magnetic field of the wave is then given by

$$\mathbf{H} = \mathbf{H}_0 \cos(\omega t - \boldsymbol{\beta} \cdot \mathbf{r} + \phi_0) \tag{7.14}$$

where

$$|\mathbf{H}_0| = \frac{|\mathbf{E}_0|}{\eta} \tag{7.15}$$

Since **E**, **H**, and the direction of propagation are mutually perpendicular to each other, it follows that

$$\mathbf{E}_0 \cdot \boldsymbol{\beta} = 0 \tag{7.16a}$$

$$\mathbf{H}_0 \cdot \boldsymbol{\beta} = 0 \tag{7.16b}$$

$$\mathbf{E}_0 \cdot \mathbf{H}_0 = 0 \tag{7.16c}$$

In particular, $\mathbf{E} \times \mathbf{H}$ should be directed along the propagation vector $\boldsymbol{\beta}$ as illustrated in Fig. 7.3 so that $\boldsymbol{\beta} \times \mathbf{E}_0$ is directed along $\mathbf{H}_0$. We can therefore combine the facts (7.16) and (7.15) to obtain

$$\mathbf{H}_0 = \frac{\mathbf{i}_\beta \times \mathbf{E}_0}{\eta} = \frac{\mathbf{i}_\beta \times \mathbf{E}_0}{\sqrt{\mu/\epsilon}} = \frac{\omega\sqrt{\mu\epsilon}\,\mathbf{i}_\beta \times \mathbf{E}_0}{\omega\mu}$$

$$= \frac{\beta\mathbf{i}_\beta \times \mathbf{E}_0}{\omega\mu} = \frac{\boldsymbol{\beta} \times \mathbf{E}_0}{\omega\mu} \tag{7.17}$$

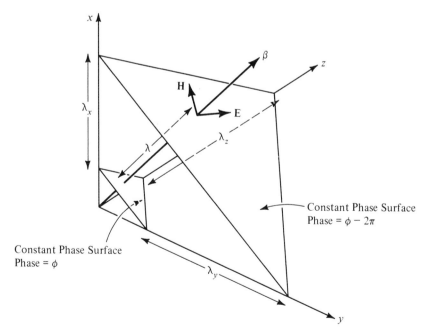

Figure 7.3. The various quantities associated with a uniform plane wave propagating in an arbitrary direction.

where $\mathbf{i}_\beta$ is the unit vector along $\boldsymbol{\beta}$. Thus

$$\mathbf{H} = \frac{1}{\omega\mu}\boldsymbol{\beta} \times \mathbf{E} \qquad (7.18)$$

Returning to Fig. 7.3, we can define several quantities pertinent to the uniform plane wave propagation in an arbitrary direction. The apparent wavelengths λ_x, λ_y, and λ_z along the coordinate axes x, y, and z, respectively, are the distances measured along those respective axes between two consecutive constant phase surfaces between which the phase difference is 2π, as shown in the figure, at a fixed time. From the interpretations of β_x, β_y, and β_z as being the phase constants along the x, y, and z axes, respectively, we have

$$\lambda_x = \frac{2\pi}{\beta_x} \qquad (7.19a)$$

$$\lambda_y = \frac{2\pi}{\beta_y} \qquad (7.19b)$$

$$\lambda_z = \frac{2\pi}{\beta_z} \qquad (7.19c)$$

We note that the wavelength λ along the direction of propagation is related to λ_x, λ_y, and λ_z in the manner

$$\frac{1}{\lambda^2} = \frac{1}{(2\pi/\beta)^2} = \frac{\beta^2}{4\pi^2} = \frac{\beta_x^2 + \beta_y^2 + \beta_z^2}{4\pi^2}$$

$$= \frac{1}{\lambda_x^2} + \frac{1}{\lambda_y^2} + \frac{1}{\lambda_z^2} \qquad (7.20)$$

The apparent phase velocities v_{px}, v_{py}, and v_{pz} along the x, y, and z axes, respectively, are the velocities with which the phase of the wave progresses with time along the respective axes. Thus

$$v_{px} = \frac{\omega}{\beta_x} \qquad (7.21a)$$

$$v_{py} = \frac{\omega}{\beta_y} \qquad (7.21b)$$

$$v_{pz} = \frac{\omega}{\beta_z} \qquad (7.21c)$$

The phase velocity v_p along the direction of propagation is related to v_{px}, v_{py}, and v_{pz} in the manner

$$\frac{1}{v_p^2} = \frac{1}{(\omega/\beta)^2} = \frac{\beta^2}{\omega^2} = \frac{\beta_x^2 + \beta_y^2 + \beta_z^2}{\omega^2}$$

$$= \frac{1}{v_{px}^2} + \frac{1}{v_{py}^2} + \frac{1}{v_{pz}^2} \tag{7.22}$$

The apparent wavelengths and phase velocities along the coordinate axes are greater than the actual wavelength and phase velocity, respectively, along the direction of propagation of the wave. This fact can be understood physically by considering, for example, water waves in an ocean striking the shore at an angle. The distance along the shoreline between two successive crests is greater than the distance between the same two crests measured along a line normal to the orientation of the crests. Also, an observer has to run faster along the shoreline in order to keep pace with a particular crest than he has to do in a direction normal to the orientation of the crests. We shall now consider an example.

Example 7.1.   Let us consider a 30 MHz uniform plane wave propagating in free space and given by the electric field vector

$$\mathbf{E} = 5(\mathbf{i}_x + \sqrt{3}\,\mathbf{i}_y) \cos\,[6\pi \times 10^7 t - 0.05\pi(3x - \sqrt{3}\,y + 2z)]\ \text{V/m}$$

Then comparing with the general expression for **E** given by (7.11), we have

$$\mathbf{E}_0 = 5(\mathbf{i}_x + \sqrt{3}\,\mathbf{i}_y)$$

$$\boldsymbol{\beta} \cdot \mathbf{r} = 0.05\pi(3x - \sqrt{3}\,y + 2z)$$

$$= 0.05\pi(3\mathbf{i}_x - \sqrt{3}\,\mathbf{i}_y + 2\mathbf{i}_z) \cdot (x\mathbf{i}_x + y\mathbf{i}_y + z\mathbf{i}_z)$$

$$\boldsymbol{\beta} = 0.05\pi(3\mathbf{i}_x - \sqrt{3}\,\mathbf{i}_y + 2\mathbf{i}_z)$$

$$\boldsymbol{\beta} \cdot \mathbf{E}_0 = 0.05\pi(3\mathbf{i}_x - \sqrt{3}\,\mathbf{i}_y + 2\mathbf{i}_z) \cdot 5(\mathbf{i}_x + \sqrt{3}\,\mathbf{i}_y)$$

$$= 0.25\pi(3 - 3) = 0$$

Hence (7.16a) is satisfied; $\mathbf{E}_0$ is perpendicular to $\boldsymbol{\beta}$.

$$\beta = |\boldsymbol{\beta}| = 0.05\pi\,|3\mathbf{i}_x - \sqrt{3}\,\mathbf{i}_y + 2\mathbf{i}_z| = 0.05\pi\sqrt{9 + 3 + 4} = 0.2\pi$$

$$\lambda = \frac{2\pi}{\beta} = \frac{2\pi}{0.2\pi} = 10\ \text{m}$$

This does correspond to a frequency of $\dfrac{3 \times 10^8}{10}$ Hz or 30 MHz in free space. The direction of propagation is along the unit vector

$$\mathbf{i}_\beta = \frac{\boldsymbol{\beta}}{|\boldsymbol{\beta}|} = \frac{3\mathbf{i}_x - \sqrt{3}\,\mathbf{i}_y + 2\mathbf{i}_z}{\sqrt{9 + 3 + 4}} = \frac{3}{4}\mathbf{i}_x - \frac{\sqrt{3}}{4}\mathbf{i}_y + \frac{1}{2}\mathbf{i}_z$$

From (7.17),

$$H_0 = \frac{1}{\omega\mu_0}\boldsymbol{\beta} \times \mathbf{E}_0$$

$$= \frac{0.05\pi \times 5}{6\pi \times 10^7 \times 4\pi \times 10^{-7}}(3\mathbf{i}_x - \sqrt{3}\,\mathbf{i}_y + 2\mathbf{i}_z) \times (\mathbf{i}_x + \sqrt{3}\,\mathbf{i}_y)$$

$$= \frac{1}{96\pi}\begin{vmatrix} \mathbf{i}_x & \mathbf{i}_y & \mathbf{i}_z \\ 3 & -\sqrt{3} & 2 \\ 1 & \sqrt{3} & 0 \end{vmatrix}$$

$$= \frac{1}{48\pi}(-\sqrt{3}\,\mathbf{i}_x + \mathbf{i}_y + 2\sqrt{3}\,\mathbf{i}_z)$$

Thus

$$\mathbf{H} = \frac{1}{48\pi}(-\sqrt{3}\,\mathbf{i}_x + \mathbf{i}_y + 2\sqrt{3}\,\mathbf{i}_z)\cos[6\pi \times 10^7 t$$
$$- 0.05\pi(3x - \sqrt{3}\,y + 2z)\;\text{amp/m}$$

To verify the expression for **H** just derived, we note that

$$\mathbf{H}_0 \cdot \boldsymbol{\beta} = \left[\frac{1}{48\pi}(-\sqrt{3}\,\mathbf{i}_x + \mathbf{i}_y + 2\sqrt{3}\,\mathbf{i}_z)\right] \cdot [0.05\pi(3\mathbf{i}_x - \sqrt{3}\,\mathbf{i}_y + 2\mathbf{i}_z)]$$

$$= \frac{0.05}{48}(-3\sqrt{3} - \sqrt{3} + 4\sqrt{3}) = 0$$

$$\mathbf{E}_0 \cdot \mathbf{H}_0 = 5(\mathbf{i}_x + \sqrt{3}\,\mathbf{i}_y) \cdot \frac{1}{48\pi}(-\sqrt{3}\,\mathbf{i}_x + \mathbf{i}_y + 2\sqrt{3}\,\mathbf{i}_z)$$

$$= \frac{5}{48\pi}(-\sqrt{3} + \sqrt{3}) = 0$$

$$\frac{|\mathbf{E}_0|}{|\mathbf{H}_0|} = \frac{5|\mathbf{i}_x + \sqrt{3}\,\mathbf{i}_y|}{(1/48\pi)|-\sqrt{3}\,\mathbf{i}_x + \mathbf{i}_y + 2\sqrt{3}\,\mathbf{i}_z|} = \frac{5\sqrt{1+3}}{(1/48\pi)\sqrt{3+1+12}}$$

$$= \frac{10}{1/12\pi} = 120\pi = \eta_0$$

Hence (7.16b), (7.16c), and (7.15) are satisfied.
Proceeding further, we find that

$$\beta_x = 0.05\pi \times 3 = 0.15\pi$$
$$\beta_y = -0.05\pi \times \sqrt{3} = -0.05\sqrt{3}\,\pi$$
$$\beta_z = 0.05\pi \times 2 = 0.1\pi$$

We then obtain

$$\lambda_x = \frac{2\pi}{\beta_x} = \frac{2\pi}{0.15\pi} = \frac{40}{3} \text{ m} = 13.333 \text{ m}$$

$$\lambda_y = \frac{2\pi}{|\beta_y|} = \frac{2\pi}{0.05\sqrt{3}\,\pi} = \frac{40}{\sqrt{3}} \text{ m} = 23.094 \text{ m}$$

$$\lambda_z = \frac{2\pi}{\beta_z} = \frac{2\pi}{0.1\pi} = 20 \text{ m}$$

$$v_{px} = \frac{\omega}{\beta_x} = \frac{6\pi \times 10^7}{0.15\pi} = 4 \times 10^8 \text{ m/s}$$

$$v_{py} = \frac{\omega}{|\beta_y|} = \frac{6\pi \times 10^7}{0.05\sqrt{3}\,\pi} = 4\sqrt{3} \times 10^8 \text{ m/s} = 6.928 \times 10^8 \text{ m/s}$$

$$v_{pz} = \frac{\omega}{\beta_z} = \frac{6\pi \times 10^7}{0.1\pi} = 6 \times 10^8 \text{ m/s}$$

Finally, to verify (7.20) and (7.22), we note that

$$\frac{1}{\lambda_x^2} + \frac{1}{\lambda_y^2} + \frac{1}{\lambda_z^2} = \frac{1}{(40/3)^2} + \frac{1}{(40/\sqrt{3})^2} + \frac{1}{20^2}$$

$$= \frac{9}{1600} + \frac{3}{1600} + \frac{4}{1600} = \frac{1}{100} = \frac{1}{10^2} = \frac{1}{\lambda^2}$$

and

$$\frac{1}{v_{px}^2} + \frac{1}{v_{py}^2} + \frac{1}{v_{pz}^2} = \frac{1}{(4 \times 10^8)^2} + \frac{1}{(4\sqrt{3} \times 10^8)^2} + \frac{1}{(6 \times 10^8)^2}$$

$$= \frac{1}{16 \times 10^{16}} + \frac{1}{48 \times 10^{16}} + \frac{1}{36 \times 10^{16}}$$

$$= \frac{1}{9 \times 10^{16}} = \frac{1}{(3 \times 10^8)^2} = \frac{1}{v_p^2} \qquad ∎$$

7.2   TRANSVERSE ELECTRIC WAVES IN A PARALLEL-PLATE WAVEGUIDE

Let us now consider the superposition of two uniform plane waves propagating symmetrically with respect to the z axis as shown in Fig. 7.4 and having the electric fields

$$\mathbf{E}_1 = E_0 \cos(\omega t - \boldsymbol{\beta}_1 \cdot \mathbf{r}) \mathbf{i}_y$$

$$= E_0 \cos(\omega t + \beta x \cos\theta - \beta z \sin\theta) \mathbf{i}_y \qquad (7.23a)$$

$$\mathbf{E}_2 = -E_0 \cos(\omega t - \boldsymbol{\beta}_2 \cdot \mathbf{r}) \mathbf{i}_y$$

$$= -E_0 \cos(\omega t - \beta x \cos\theta - \beta z \sin\theta) \mathbf{i}_y \qquad (7.23b)$$

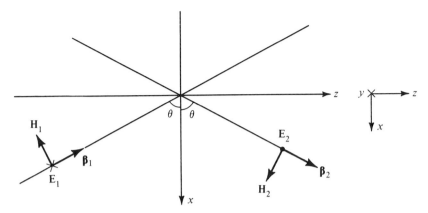

Figure 7.4. Superposition of two uniform plane waves propagating symmetrically with respect to the z axis.

where $\beta = \omega\sqrt{\mu\epsilon}$, with ϵ and μ being the permittivity and the permeability, respectively, of the medium. The corresponding magnetic fields are given by

$$\mathbf{H}_1 = \frac{E_0}{\eta}(-\sin\theta\,\mathbf{i}_x - \cos\theta\,\mathbf{i}_z)\cos(\omega t + \beta x\cos\theta - \beta z\sin\theta) \quad (7.24a)$$

$$\mathbf{H}_2 = \frac{E_0}{\eta}(\sin\theta\,\mathbf{i}_x - \cos\theta\,\mathbf{i}_z)\cos(\omega t - \beta x\cos\theta - \beta z\sin\theta) \quad (7.24b)$$

where $\eta = \sqrt{\mu/\epsilon}$. The electric and magnetic fields of the superposition of the two waves are given by

$$\begin{aligned}
\mathbf{E} &= \mathbf{E}_1 + \mathbf{E}_2 \\
&= E_0[\cos(\omega t - \beta z\sin\theta + \beta x\cos\theta) \\
&\quad - \cos(\omega t - \beta z\sin\theta - \beta x\cos\theta)]\mathbf{i}_y \\
&= -2E_0\sin(\beta x\cos\theta)\sin(\omega t - \beta z\sin\theta)\,\mathbf{i}_y \quad (7.25a)
\end{aligned}$$

$$\begin{aligned}
\mathbf{H} &= \mathbf{H}_1 + \mathbf{H}_2 \\
&= -\frac{E_0}{\eta}\sin\theta\,[\cos(\omega t - \beta z\sin\theta + \beta x\cos\theta) \\
&\quad - \cos(\omega t - \beta z\sin\theta - \beta x\cos\theta)]\mathbf{i}_x \\
&\quad -\frac{E_0}{\eta}\cos\theta\,[\cos(\omega t - \beta z\sin\theta + \beta x\cos\theta) \\
&\quad + \cos(\omega t - \beta z\sin\theta - \beta x\cos\theta)]\mathbf{i}_z \\
&= \frac{2E_0}{\eta}\sin\theta\sin(\beta x\cos\theta)\sin(\omega t - \beta z\sin\theta)\,\mathbf{i}_x \\
&\quad -\frac{2E_0}{\eta}\cos\theta\cos(\beta x\cos\theta)\cos(\omega t - \beta z\sin\theta)\,\mathbf{i}_z \quad (7.25b)
\end{aligned}$$

In view of the factors $\sin(\beta x \cos\theta)$ and $\cos(\beta x \cos\theta)$ for the x dependence and the factors $\sin(\omega t - \beta z \sin\theta)$ and $\cos(\omega t - \beta z \sin\theta)$ for the z dependence, the composite fields have standing wave character in the x direction and traveling wave character in the z direction. Thus we have standing waves in the x direction moving bodily in the z direction, as illustrated in Fig. 7.5, by considering the electric field for two different times. In fact, we find that the Poynting vector is given by

$$\mathbf{P} = \mathbf{E} \times \mathbf{H} = E_y \mathbf{i}_y \times (H_x \mathbf{i}_x + H_z \mathbf{i}_z)$$

$$= -E_y H_x \mathbf{i}_z + E_y H_z \mathbf{i}_x$$

$$= \frac{4E_0^2}{\eta} \sin\theta \sin^2(\beta x \cos\theta) \sin^2(\omega t - \beta z \sin\theta)\, \mathbf{i}_z$$

$$+ \frac{E_0^2}{\eta} \cos\theta \sin(2\beta x \cos\theta) \sin 2(\omega t - \beta z \sin\theta)\, \mathbf{i}_x \qquad (7.26)$$

The time-average Poynting vector is given by

$$\langle \mathbf{P} \rangle = \frac{4E_0^2}{\eta} \sin\theta \sin^2(\beta x \cos\theta) \langle \sin^2(\omega t - \beta z \sin\theta)\rangle\, \mathbf{i}_z$$

$$+ \frac{E_0^2}{\eta} \cos\theta \sin(2\beta x \cos\theta) \langle \sin 2(\omega t - \beta z \sin\theta)\rangle\, \mathbf{i}_x$$

$$= \frac{2E_0^2}{\eta} \sin\theta \sin^2(\beta x \cos\theta)\, \mathbf{i}_z \qquad (7.27)$$

Thus the time-average power flow is entirely in the z direction, thereby verifying our interpretation of the field expressions. Since the composite electric field is directed entirely transverse to the z direction, that is, the direction of time-average power flow, whereas the composite magnetic field is not, the composite wave is known as the "transverse electric," or TE wave.

From the expressions for the fields for the TE wave given by (7.25a) and (7.25b), we note that the electric field is zero for $\sin(\beta x \cos\theta)$ equal to zero, or

$$\beta x \cos\theta = \pm m\pi, \qquad m = 0, 1, 2, 3, \ldots$$

$$x = \pm \frac{m\pi}{\beta \cos\theta} = \pm \frac{m\lambda}{2\cos\theta}, \qquad m = 0, 1, 2, 3, \ldots \qquad (7.28)$$

where

$$\lambda = \frac{2\pi}{\beta} = \frac{2\pi}{\omega\sqrt{\mu\epsilon}} = \frac{1}{f\sqrt{\mu\epsilon}}$$

Thus if we place perfectly conducting sheets in these planes, the waves will propagate undisturbed, that is, as though the sheets were not present since the

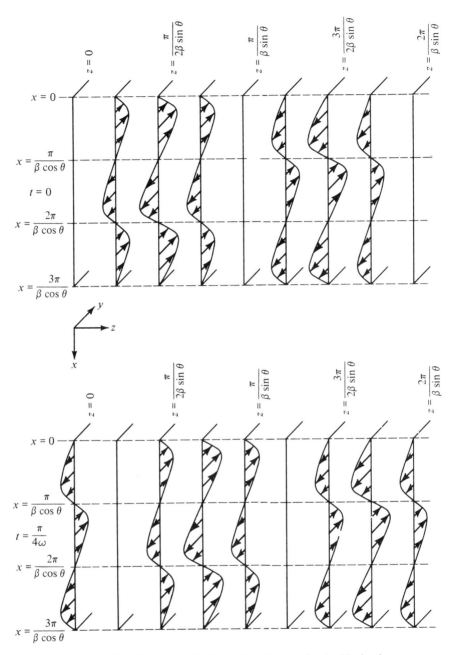

Figure 7.5. Standing waves in the x direction moving bodily in the z direction.

boundary condition that the tangential component of the electric field be zero on the surface of a perfect conductor is satisfied in these planes. The boundary condition that the normal component of the magnetic field be zero on the surface of a perfect conductor is also satisfied since H_x is zero in these planes.

If we consider any two adjacent sheets, the situation is actually one of uniform plane waves bouncing obliquely between the sheets, as illustrated in Fig. 7.6 for two sheets in the planes $x = 0$ and $x = \lambda/(2 \cos \theta)$, thereby guiding

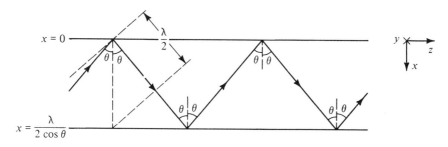

Figure 7.6. Uniform plane waves bouncing obliquely between two parallel plane perfectly conducting sheets.

the wave and hence the energy in the z direction, parallel to the plates. Thus we have a "parallel-plate waveguide," as compared to the parallel-plate transmission line in which the uniform plane wave slides parallel to the plates. We note from the constant phase surfaces of the obliquely bouncing wave shown in Fig. 7.6 that $\lambda/(2 \cos \theta)$ is simply one-half of the apparent wavelength of that wave in the x direction, that is, normal to the plates. Thus the fields have one-half apparent wavelength in the x direction. If we place the perfectly conducting sheets in the planes $x = 0$ and $x = m\lambda/(2 \cos \theta)$, the fields will then have m number of one-half apparent wavelengths in the x direction between the plates. The fields have no variations in the y direction. Thus the fields are said to correspond to "TE$_{m,0}$ modes" where the subscript m refers to the x direction, denoting m number of one-half apparent wavelengths in that direction and the subscript 0 refers to the y direction, denoting zero number of one-half apparent wavelengths in that direction.

Let us now consider a parallel-plate waveguide with perfectly conducting plates situated in the planes $x = 0$ and $x = a$, that is, having a fixed spacing a between them, as shown in Fig. 7.7(a). Then, for TE$_{m,0}$ waves guided by the plates, we have from (7.28),

$$a = \frac{m\lambda}{2 \cos \theta}$$

or

$$\cos \theta = \frac{m\lambda}{2a} = \frac{m}{2a} \frac{1}{f \sqrt{\mu\epsilon}} \tag{7.29}$$

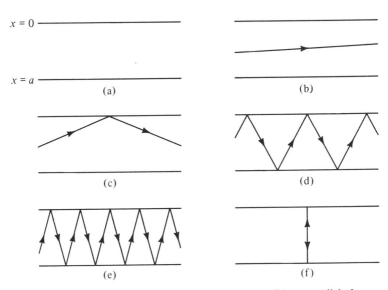

Figure 7.7. For illustrating the phenomenon of cutoff in a parallel-plate waveguide.

Thus waves of different wavelengths (or frequencies) bounce obliquely between the plates at different values of the angle θ. For very small wavelengths (very high frequencies), $m\lambda/2a$ is small, $\cos\theta \approx 0$, $\theta \approx 90°$, and the waves simply slide between the plates as in the case of the transmission line, as shown in Fig. 7.7(b). As λ increases (f decreases), $m\lambda/2a$ increases, θ decreases, and the waves bounce more and more obliquely, as shown in Fig. 7.7(c)–(e), until λ becomes equal to $2a/m$ for which $\cos\theta = 1$, $\theta = 0°$, and the waves simply bounce back and forth normally to the plates, as shown in Fig. 7.7(f), without any feeling of being guided parallel to the plates. For $\lambda > 2a/m$, $m\lambda/2a > 1$, $\cos\theta > 1$, and θ has no real solution, indicating that propagation does not occur for these wavelengths in the waveguide mode. This condition is known as the "cutoff" condition.

The cutoff wavelength, denoted by the symbol λ_c, is given by

$$\lambda_c = \frac{2a}{m} \tag{7.30}$$

This is simply the wavelength for which the spacing a is equal to m number of one-half wavelengths. Propagation of a particular mode is possible only if λ is less than the value of λ_c for that mode. The cutoff frequency is given by

$$f_c = \frac{m}{2a\sqrt{\mu\epsilon}} \tag{7.31}$$

Propagation of a particular mode is possible only if f is greater than the value of f_c for that mode. Consequently, waves of a given frequency f can propagate in all modes for which the cutoff wavelengths are greater than the wavelength or the cutoff frequencies are less than the frequency.

Substituting λ_c for $2a/m$ in (7.29), we have

$$\cos \theta = \frac{\lambda}{\lambda_c} = \frac{f_c}{f} \tag{7.32a}$$

$$\sin \theta = \sqrt{1 - \cos^2 \theta} = \sqrt{1 - \left(\frac{\lambda}{\lambda_c}\right)^2} = \sqrt{1 - \left(\frac{f_c}{f}\right)^2} \tag{7.32b}$$

$$\beta \cos \theta = \frac{2\pi}{\lambda} \frac{\lambda}{\lambda_c} = \frac{2\pi}{\lambda_c} = \frac{m\pi}{a} \tag{7.32c}$$

$$\beta \sin \theta = \frac{2\pi}{\lambda} \sqrt{1 - \left(\frac{\lambda}{\lambda_c}\right)^2} \tag{7.32d}$$

We see from (7.32d) that the phase constant along the z direction, that is, $\beta \sin \theta$, is real for $\lambda < \lambda_c$ and imaginary for $\lambda > \lambda_c$, thereby explaining once again the cutoff phenomenon. We now define the guide wavelength, λ_g, to be the wavelength in the z direction, that is, along the guide. This is given by

$$\lambda_g = \frac{2\pi}{\beta \sin \theta} = \frac{\lambda}{\sqrt{1 - (\lambda/\lambda_c)^2}} = \frac{\lambda}{\sqrt{1 - (f_c/f)^2}} \tag{7.33}$$

This is simply the apparent wavelength, in the z direction, of the obliquely bouncing uniform plane waves. The phase velocity along the guide axis, which is simply the apparent phase velocity, in the z direction, of the obliquely bouncing uniform plane waves, is

$$v_{pz} = \frac{\omega}{\beta \sin \theta} = \frac{v_p}{\sin \theta} = \frac{v_p}{\sqrt{1 - (\lambda/\lambda_c)^2}} = \frac{v_p}{\sqrt{1 - (f_c/f)^2}} \tag{7.34}$$

Finally, substituting (7.32a)–(7.32d) in the field expressions (7.25a) and (7.25b), we obtain

$$\mathbf{E} = -2E_0 \sin \left(\frac{m\pi x}{a}\right) \sin \left(\omega t - \frac{2\pi}{\lambda_g} z\right) \mathbf{i}_y \tag{7.35a}$$

$$\mathbf{H} = \frac{2E_0}{\eta} \frac{\lambda}{\lambda_g} \sin \left(\frac{m\pi x}{a}\right) \sin \left(\omega t - \frac{2\pi}{\lambda_g} z\right) \mathbf{i}_x$$

$$- \frac{2E_0}{\eta} \frac{\lambda}{\lambda_c} \cos \left(\frac{m\pi x}{a}\right) \cos \left(\omega t - \frac{2\pi}{\lambda_g} z\right) \mathbf{i}_z \tag{7.35b}$$

These expressions for the $TE_{m,0}$ mode fields in the parallel-plate waveguide do not contain the angle θ. They clearly indicate the standing wave character of the fields in the x direction, having m one-half sinusoidal variations between the plates. We shall now consider an example.

Example 7.2. Let us assume the spacing a between the plates of a parallel-plate waveguide to be 5 cm and investigate the propagating $TE_{m,0}$ modes for $f = 10,000$ MHz.

From (7.30), the cutoff wavelengths for $TE_{m,0}$ modes are given by

$$\lambda_c = \frac{2a}{m} = \frac{10}{m} \text{ cm} = \frac{0.1}{m} \text{ m}$$

This result is independent of the dielectric between the plates. If the medium between the plates is free space, then the cutoff frequencies for the $TE_{m,0}$ modes are

$$f_c = \frac{3 \times 10^8}{\lambda_c} = \frac{3 \times 10^8}{0.1/m} = 3m \times 10^9 \text{ Hz}$$

For $f = 10,000$ MHz $= 10^{10}$ Hz, the propagating modes are $TE_{1,0}(f_c = 3 \times 10^9$ Hz), $TE_{2,0}(f_c = 6 \times 10^9$ Hz), and $TE_{3,0}(f_c = 9 \times 10^9$ Hz).

For each propagating mode, we can find θ, λ_g, and v_{pz} by using (7.32a), (7.33), and (7.34), respectively. Values of these quantities are listed in the following:

Mode	λ_c, cm	f_c, MHz	θ, deg	λ_g, cm	v_{pz}, m/s
$TE_{1,0}$	10	3000	72.54	3.145	3.145×10^8
$TE_{2,0}$	5	6000	53.13	3.75	3.75×10^8
$TE_{3,0}$	3.33	9000	25.84	6.882	6.882×10^8

∎

7.3 PARALLEL-PLATE WAVEGUIDE DISCONTINUITY

In the previous section we introduced $TE_{m,0}$ waves in a parallel-plate waveguide. Let us now consider reflection and transmission at a dielectric discontinuity in a parallel-plate guide, as shown in Fig. 7.8. If a $TE_{m,0}$ wave is incident on the junction from section 1, then it will set up a reflected wave into

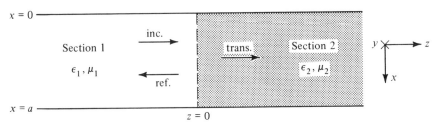

Figure 7.8. For consideration of reflection and transmission at a dielectric discontinuity in a parallel-plate waveguide.

section 1 and a transmitted wave into section 2, provided that mode propagates in that section. The fields corresponding to these incident, reflected, and transmitted waves must satisfy the boundary conditions at the dielectric discontinuity. These boundary conditions were derived in Sec. 6.5. Denoting the incident, reflected, and transmitted wave fields by the subscripts i, r, and t, respectively, we have from the continuity of the tangential component of $\mathbf{E}$ at a dielectric discontinuity,

$$E_{yi} + E_{yr} = E_{yt} \text{ at } z = 0 \qquad (7.36)$$

and from the continuity of the tangential component of $\mathbf{H}$ at a dielectric discontinuity,

$$H_{xi} + H_{xr} = H_{xt} \text{ at } z = 0 \qquad (7.37)$$

We now define the guide impedance, η_{g1}, of section 1 as

$$\eta_{g1} = \frac{E_{yi}}{-H_{xi}} \qquad (7.38)$$

Recognizing that $\mathbf{i}_y \times (-\mathbf{i}_x) = \mathbf{i}_z$, we note that η_{g1} is simply the ratio of the transverse components of the electric and magnetic fields of the $\mathrm{TE}_{m,0}$ wave which give rise to time-average power flow down the guide. From (7.35a) and (7.35b) applied to section 1, we have

$$\eta_{g1} = \eta_1 \frac{\lambda_{g1}}{\lambda_1} = \frac{\eta_1}{\sqrt{1 - (\lambda_1/\lambda_c)^2}} = \frac{\eta_1}{\sqrt{1 - (f_{c1}/f)^2}} \qquad (7.39)$$

The guide impedance is analogous to the characteristic impedance of a transmission line, if we recognize that E_{yi} and $-H_{xi}$ are analogous to V^+ and I^+, respectively. In terms of the reflected wave fields, it then follows that

$$\eta_{g1} = -\left(\frac{E_{yr}}{-H_{xr}}\right) = \frac{E_{yr}}{H_{xr}} \qquad (7.40)$$

This result can also be seen from the fact that for the reflected wave, the power flow is in the negative z direction and since $\mathbf{i}_y \times \mathbf{i}_x = -\mathbf{i}_z$, η_{g1} is equal to E_{yr}/H_{xr}. For the transmitted wave fields, we have

$$\frac{E_{yt}}{-H_{xt}} = \eta_{g2} \qquad (7.41)$$

where

$$\eta_{g2} = \eta_2 \frac{\lambda_{g2}}{\lambda_2} = \frac{\eta_2}{\sqrt{1 - (\lambda_2/\lambda_c)^2}} = \frac{\eta_2}{\sqrt{1 - (f_{c2}/f)^2}} \qquad (7.42)$$

is the guide impedance of section 2.

Using (7.38), (7.40), and (7.41), (7.37) can be written as

$$\frac{E_{yi}}{\eta_{g1}} - \frac{E_{yr}}{\eta_{g1}} = \frac{E_{yt}}{\eta_{g2}} \qquad (7.43)$$

Solving (7.36) and (7.43), we get

$$E_{yi}\left(1 - \frac{\eta_{g2}}{\eta_{g1}}\right) + E_{yr}\left(1 + \frac{\eta_{g2}}{\eta_{g1}}\right) = 0$$

or the reflection coefficient at the junction is given by

$$\Gamma = \frac{E_{yr}}{E_{yi}} = \frac{\eta_{g2} - \eta_{g1}}{\eta_{g2} + \eta_{g1}} \qquad (7.44)$$

and the transmission coefficient at the junction is given by

$$\tau = \frac{E_{yt}}{E_{yi}} = \frac{E_{yi} + E_{yr}}{E_{yi}} = 1 + \Gamma \qquad (7.45)$$

These expressions for Γ and τ are similar to those obtained in Sec. 6.6 for reflection and transmission at a transmission-line discontinuity. Hence insofar as reflection and transmission at the junction are concerned, we can replace the waveguide sections by transmission lines having characteristic impedances equal to the guide impedances, as shown in Fig. 7.9. It should be noted that unlike the characteristic impedance of a lossless line, which is a constant independent of frequency, the guide impedance of the lossless

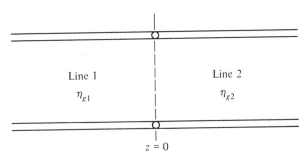

Figure 7.9. Transmission-line equivalent of parallel-plate waveguide discontinuity.

waveguide is a function of the frequency. We shall now consider an example.

Example 7.3. Let us consider the parallel-plate waveguide discontinuity shown in Fig. 7.10. For $TE_{1,0}$ waves of frequency $f = 5000$ MHz, incident on the junction from the free space side, we wish to find the reflection and transmission coefficients.

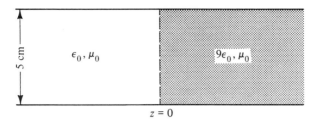

Figure 7.10. For illustrating the computation of reflection and transmission coefficients at a parallel-plate waveguide discontinuity.

For the $TE_{1,0}$ mode, $\lambda_c = 2a = 10$ cm, independent of the dielectric. For $f = 5000$ MHz,

$$\lambda_1 = \text{wavelength on the free space side} = \frac{3 \times 10^8}{5 \times 10^9} = 6 \text{ cm}$$

$$\lambda_2 = \text{wavelength on the dielectric side} = \frac{3 \times 10^8}{\sqrt{9} \times 5 \times 10^9} = \frac{6}{3} = 2 \text{ cm}$$

Since $\lambda < \lambda_c$ in both sections, $TE_{1,0}$ mode propagates in both sections. Thus

$$\eta_{g1} = \frac{\eta_1}{\sqrt{1 - (\lambda_1/\lambda_c)^2}} = \frac{120\pi}{\sqrt{1 - (6/10)^2}} = 471.24 \text{ ohms}$$

$$\eta_{g2} = \frac{\eta_2}{\sqrt{1 - (\lambda_2/\lambda_c)^2}} = \frac{120\pi/\sqrt{9}}{\sqrt{1 - (2/10)^2}} = \frac{40\pi}{\sqrt{1 - 0.04}} = 128.25 \text{ ohms}$$

$$\Gamma = \frac{\eta_{g2} - \eta_{g1}}{\eta_{g2} + \eta_{g1}} = \frac{128.25 - 471.24}{128.25 + 471.24} = -0.572$$

$$\tau = 1 + \Gamma = 1 - 0.572 = 0.428$$

For $f = 4000$ MHz, we would obtain $\Gamma = -0.629$ and $\tau = 0.371$. ∎

7.4 DISPERSION AND GROUP VELOCITY*

In Sec. 7.2 we learned that for the propagating range of frequencies, the phase velocity and the wavelength along the axis of the parallel-plate waveguide are given by

$$v_{pz} = \frac{v_p}{\sqrt{1 - (f_c/f)^2}} \tag{7.46}$$

and

$$\lambda_g = \frac{\lambda}{\sqrt{1 - (f_c/f)^2}} \tag{7.47}$$

---

*This section may be omitted without loss of continuity.

where $v_p = 1/\sqrt{\mu\epsilon}$, $\lambda = v_p/f = 1/f\sqrt{\mu\epsilon}$, and f_c is the cutoff frequency. We note that for a particular mode, the phase velocity of propagation along the guide axis varies with the frequency. As a consequence of this characteristic of the guided wave propagation, the field patterns of the different frequency components of a signal comprising a band of frequencies do not maintain the same phase relationships as they propagate down the guide. This phenomenon is known as "dispersion," so termed after the phenomenon of dispersion of colors by a prism.

To discuss dispersion, let us consider a simple example of two infinitely long trains A and B traveling in parallel, one below the other, with each train made up of boxcars of identical size and having wavy tops, as shown in Fig. 7.11. Let the spacings between the peaks (centers) of successive boxcars be 50 m and 90 m, and let the speeds of the trains be 20 m/s and 30 m/s, for trains A and B, respectively. Let the peaks of the cars numbered 0 for the two trains be aligned at time $t = 0$, as shown in Fig. 7.11(a). Now, as time progresses, the two peaks get out of alignment as shown, for example, for $t = 1$ s in Fig. 7.11(b), since train B is traveling faster than train A. But at the same time, the gap between the peaks of cars numbered -1 decreases. This continues until at $t = 4$ s, the peak of car "-1" of train A having moved by a distance of 80 m aligns with the peak of car "-1" of train B, which will have moved by a distance of 120 m, as shown in Fig. 7.11(c). For an observer following the movement of the two trains as a group, the group appears to have moved by a distance of 30 m although the individual trains will have moved by 80 m and 120 m, respectively. Thus we can talk of a "group velocity," that is, the velocity with which the group as a whole is moving. In this case, the group velocity is 30 m/4 s or 7.5 m/s.

The situation in the case of the guided wave propagation of two different frequencies in the parallel-plate waveguide is exactly similar to the two-train example just discussed. The distance between the peaks of two successive cars is analogous to the guide wavelength, and the speed of the train is analogous to the phase velocity along the guide axis. Thus let us consider the field patterns corresponding to two waves of frequencies f_A and f_B propagating in the same mode, having guide wavelengths λ_{gA} and λ_{gB}, and phase velocities along the guide axis v_{pzA} and v_{pzB}, respectively, as shown, for example, for the electric field of the $TE_{1,0}$ mode in Fig. 7.12. Let the positive peaks numbered 0 of the two patterns be aligned at $t = 0$, as shown in Fig. 7.12(a). As the individual waves travel with their respective phase velocities along the guide, these two peaks get out of alignment but some time later, say Δt, the positive peaks numbered -1 will align at some distance, say Δz, from the location of the alignment of the "0" peaks, as shown in Fig. 7.12(b). Since the "-1"th peak of wave A will have traveled a distance $\lambda_{gA} + \Delta z$ with a phase velocity v_{pzA} and the "-1"th peak of wave B will have traveled a distance $\lambda_{gB} + \Delta z$

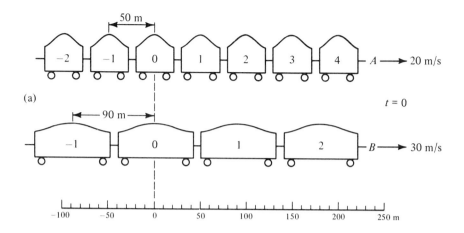

(a)

$t = 0$

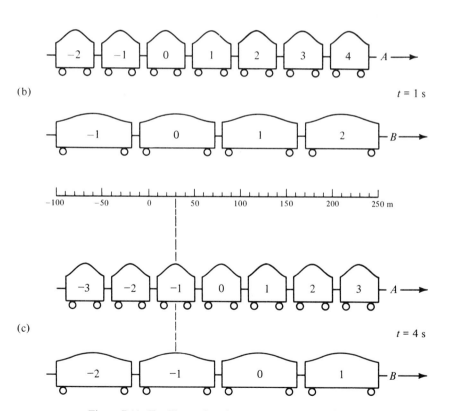

(b)

$t = 1$ s

(c)

$t = 4$ s

Figure 7.11. For illustrating the concept of group velocity.

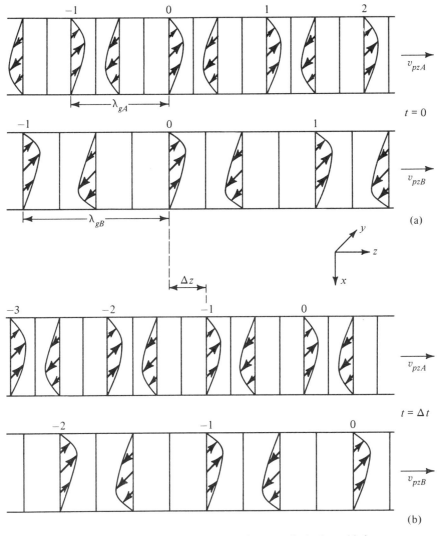

Figure 7.12. For illustrating the concept of group velocity for guided wave propagation.

with a phase velocity v_{pzB} in this time Δt, we have

$$\lambda_{gA} + \Delta z = v_{pzA} \Delta t \qquad (7.48a)$$

$$\lambda_{gB} + \Delta z = v_{pzB} \Delta t \qquad (7.48b)$$

Solving (7.48a) and (7.48b) for Δt and Δz, we obtain

$$\Delta t = \frac{\lambda_{gA} - \lambda_{gB}}{v_{pzA} - v_{pzB}} \qquad (7.49a)$$

and

$$\Delta z = \frac{\lambda_{gA} v_{pzB} - \lambda_{gB} v_{pzA}}{v_{pzA} - v_{pzB}} \tag{7.49b}$$

The group velocity, v_g, is then given by

$$v_g = \frac{\Delta z}{\Delta t} = \frac{\lambda_{gA} v_{pzB} - \lambda_{gB} v_{pzA}}{\lambda_{gA} - \lambda_{gB}} = \frac{\lambda_{gA} \lambda_{gB} f_B - \lambda_{gB} \lambda_{gA} f_A}{\lambda_{gA} \lambda_{gB} \left(\dfrac{1}{\lambda_{gB}} - \dfrac{1}{\lambda_{gA}} \right)}$$

$$= \frac{f_B - f_A}{\dfrac{1}{\lambda_{gB}} - \dfrac{1}{\lambda_{gA}}} = \frac{\omega_B - \omega_A}{\beta_{zB} - \beta_{zA}} \tag{7.50}$$

where β_{zA} and β_{zB} are the phase constants along the guide axis, corresponding to f_A and f_B, respectively. Thus the group velocity of a signal comprised of two frequencies is the ratio of the difference between the two radian frequencies to the difference between the corresponding phase constants along the guide axis.

If we now have a signal comprised of a number of frequencies, then a value of group velocity can be obtained for each pair of these frequencies in accordance with (7.50). In general, these values of group velocity will all be different. In fact, this is the case for wave propagation in the parallel-plate guide, as can be seen from Fig. 7.13, which is a plot of ω versus β_z corresponding to the parallel-plate guide for which

$$\beta_z = \frac{2\pi}{\lambda_g} = \frac{2\pi}{\lambda} \sqrt{1 - \left(\frac{\lambda}{\lambda_c} \right)^2} = \omega \sqrt{\mu \epsilon} \sqrt{1 - \left(\frac{f_c}{f} \right)^2} \tag{7.51}$$

Such a plot is known as the "ω–β_z diagram" or the "dispersion diagram."

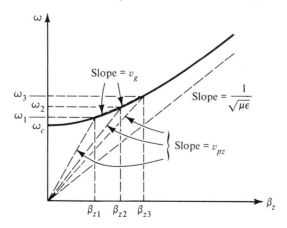

Figure 7.13. Dispersion diagram for the parallel-plate waveguide.

The phase velocity, ω/β_z, for a particular frequency is given by the slope of the line drawn from the origin to the point, on the dispersion curve, corresponding to that frequency as shown in the figure for the three frequencies ω_1, ω_2, and ω_3. The group velocity for a particular pair of frequencies is given by the slope of the line joining the two points, on the curve, corresponding to the two frequencies as shown in the figure for the two pairs ω_1, ω_2 and ω_2, ω_3. Since the curve is nonlinear, it can be seen that the two group velocities are not equal. We cannot then attribute a particular value of group velocity for the group of the three frequencies ω_1, ω_2, and ω_3.

If, however, the three frequencies are very close, as in the case of a narrow-band signal, it is meaningful to assign a group velocity to the entire group having a value equal to the slope of the tangent to the dispersion curve at the center frequency. Thus the group velocity corresponding to a narrow band of frequencies centered around a predominant frequency ω is given by

$$v_g = \frac{d\omega}{d\beta_z} \qquad (7.52)$$

For the parallel-plate waveguide under consideration, we have from (7.51),

$$\frac{d\beta_z}{d\omega} = \sqrt{\mu\epsilon}\sqrt{1 - \left(\frac{f_c}{f}\right)^2} + \omega\sqrt{\mu\epsilon} \cdot \frac{1}{2}\left(1 - \frac{f_c^2}{f^2}\right)^{-1/2}\frac{f_c^2}{\pi f^3}$$

$$= \sqrt{\mu\epsilon}\left(1 - \frac{f_c^2}{f^2} + \frac{\omega}{2\pi}\frac{f_c^2}{f^3}\right)\left(1 - \frac{f_c^2}{f^2}\right)^{-1/2}$$

$$= \sqrt{\mu\epsilon}\left(1 - \frac{f_c^2}{f^2}\right)^{-1/2}$$

and

$$v_g = \frac{d\omega}{d\beta_z} = \frac{1}{\sqrt{\mu\epsilon}}\sqrt{1 - \frac{f_c^2}{f^2}} = v_p\sqrt{1 - \left(\frac{f_c}{f}\right)^2} \qquad (7.53)$$

As a numerical example, for the case of Example 7.2, the group velocities for $f = 10,000$ MHz for the three propagating modes $TE_{1,0}$, $TE_{2,0}$, and $TE_{3,0}$ are 2.862×10^8 m/s, 2.40×10^8 m/s, and 1.308×10^8 m/s, respectively. From (7.46) and (7.53), we note that

$$v_{pz}v_g = v_p^2 \qquad (7.54)$$

An example of a narrow-band signal is an amplitude modulated signal, having a carrier frequency ω modulated by a low frequency $\Delta\omega \ll \omega$ as given by

$$E_x(t) = E_{x0}(1 + m\cos\Delta\omega\cdot t)\cos\omega t \qquad (7.55)$$

where m is the percentage modulation. Such a signal is actually equivalent to a superposition of unmodulated signals of three frequencies $\omega - \Delta\omega$, ω, and

$\omega + \Delta\omega$, as can be seen by expanding the right side of (7.55). Thus

$$E_x(t) = E_{x0} \cos \omega t + m E_{x0} \cos \omega t \cos \Delta\omega \cdot t$$

$$= E_{x0} \cos \omega t + \frac{m E_{x0}}{2} [\cos (\omega - \Delta\omega)t + \cos (\omega + \Delta\omega)t] \quad (7.56)$$

The frequencies $\omega - \Delta\omega$ and $\omega + \Delta\omega$ are the side frequencies. When the amplitude modulated signal propagates in a dispersive channel such as the parallel-plate waveguide under consideration, the different frequency components undergo phase changes in accordance with their respective phase constants. Thus if $\beta_z - \Delta\beta_z$, β_z, and $\beta_z + \Delta\beta_z$ are the phase constants corresponding to $\omega - \Delta\omega$, ω, and $\omega + \Delta\omega$, respectively, assuming linearity of the dispersion curve within the narrow band, the amplitude modulated wave is given by

$$E_x(z, t) = E_{x0} \cos (\omega t - \beta_z z)$$

$$+ \frac{m E_{x0}}{2} \{\cos [(\omega - \Delta\omega)t - (\beta_z - \Delta\beta_z)z]$$

$$+ \cos [(\omega + \Delta\omega)t - (\beta_z + \Delta\beta_z)z]\}$$

$$= E_{x0} \cos (\omega t - \beta_z z)$$

$$+ \frac{m E_{x0}}{2} \{\cos [(\omega t - \beta_z z) - (\Delta\omega \cdot t - \Delta\beta_z \cdot z)]$$

$$+ \cos [(\omega t - \beta_z z) + (\Delta\omega \cdot t - \Delta\beta_z \cdot z)]\}$$

$$= E_{x0} \cos (\omega t - \beta_z z) + m E_{x0} \cos (\omega t - \beta_z z) \cos (\Delta\omega \cdot t - \Delta\beta_z \cdot z)$$

$$= E_{x0}[1 + m \cos (\Delta\omega \cdot t - \Delta\beta_z \cdot z)] \cos (\omega t - \beta_z z) \quad (7.57)$$

This indicates that although the carrier frequency phase changes in accordance with the phase constant β_z, the modulation envelope and hence the information travels with the group velocity $\Delta\omega/\Delta\beta_z$, as shown in Fig. 7.14. In

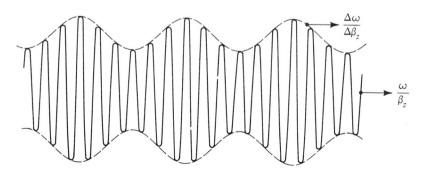

Figure 7.14. For illustrating that the modulation envelope travels with the group velocity.

view of this and since v_g is less than v_p, the fact that v_{pz} is greater than v_p is not a violation of the theory of relativity. Since it is always necessary to use some modulation technique to convey information from one point to another, the information always takes more time to reach from one point to another in a dispersive channel than in the corresponding nondispersive medium.

7.5 RECTANGULAR WAVEGUIDE AND CAVITY RESONATOR

Thus far, we have restricted our discussion to $\text{TE}_{m,0}$ wave propagation in a parallel-plate waveguide. From Sec. 7.2, we recall that the parallel-plate waveguide is made up of two perfectly conducting sheets in the planes $x = 0$ and $x = a$ and that the electric field of the $\text{TE}_{m,0}$ mode has only a y component with m number of one-half sinusoidal variations in the x direction and no variations in the y direction. If we now introduce two perfectly conducting sheets in two constant y planes, say, $y = 0$ and $y = b$, the field distribution will remain unaltered since the electric field is entirely normal to the plates, and hence the boundary condition of zero tangential electric field is satisfied for both sheets. We then have a metallic pipe with rectangular cross section in the xy plane, as shown in Fig. 7.15. Such a structure is known as the "rectangular waveguide" and is, in fact, a common form of waveguide.

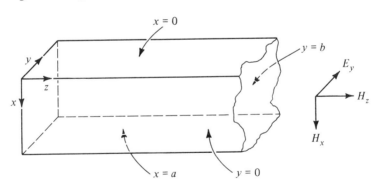

Figure 7.15. A rectangular waveguide.

Since the $\text{TE}_{m,0}$ mode field expressions derived for the parallel-plate waveguide satisfy the boundary conditions for the rectangular waveguide, those expressions as well as the entire discussion of the parallel-plate waveguide case hold also for $\text{TE}_{m,0}$ mode propagation in the rectangular waveguide case. We learned that the $\text{TE}_{m,0}$ modes can be interpreted as due to uniform plane waves having electric field in the y direction and bouncing obliquely between the conducting walls $x = 0$ and $x = a$, and with the associated cutoff condition characterized by bouncing of the waves back and forth normally to these walls, as shown in Fig. 7.16(a). For the cutoff condition, the dimension a is equal to m number of one-half wavelengths such that

$$[\lambda_c]_{\text{TE}_{m,0}} = \frac{2a}{m} \tag{7.58}$$

In a similar manner, we can have uniform plane waves having electric field in the x direction and bouncing obliquely between the walls $y = 0$ and $y = b$, and with the associated cutoff condition characterized by bouncing of the waves back and forth normally to these walls, as shown in Fig. 7.16(b), thereby resulting in $\text{TE}_{0,n}$ modes having no variations in the x direction and n number of one-half sinusoidal variations in the y direction. For the cutoff

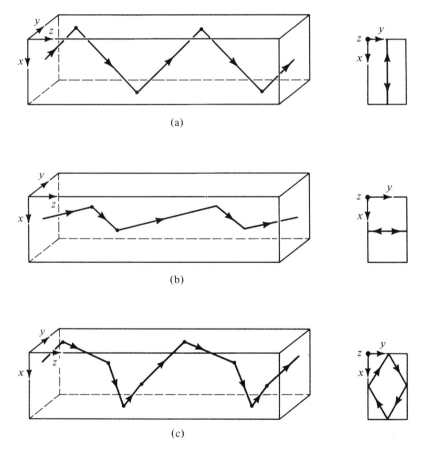

(a)

(b)

(c)

Figure 7.16. Propagation and cutoff of (a) $\text{TE}_{m,0}$, (b) $\text{TE}_{0,n}$, and (c) $\text{TE}_{m,n}$ modes in a rectangular waveguide.

condition, the dimension b is equal to n number of one-half wavelengths such that

$$[\lambda_c]_{\mathrm{TE}_{0,n}} = \frac{2b}{n} \tag{7.59}$$

We can even have $\mathrm{TE}_{m,n}$ modes having m number of one-half sinusoidal variations in the x direction and n number of one-half sinusoidal variations in the y direction due to uniform plane waves having both x and y components of the electric field and bouncing obliquely between all four walls of the guide and with the associated cutoff condition characterized by bouncing of the waves back and forth obliquely between the four walls as shown, for example, in Fig. 7.16(c). For the cutoff condition, the dimension a must be equal to m number of one-half apparent wavelengths in the x direction and the dimension b must be equal to n number of one-half apparent wavelengths in the y direction such that

$$\frac{1}{[\lambda_c]^2_{\mathrm{TE}_{m,n}}} = \frac{1}{(2a/m)^2} + \frac{1}{(2b/n)^2} \tag{7.60}$$

or

$$[\lambda_c]_{\mathrm{TE}_{m,n}} = \frac{1}{\sqrt{(m/2a)^2 + (n/2b)^2}} \tag{7.61}$$

*▪▪▪▪▪   At this point, it may be of interest to obtain the field expressions for the $\mathrm{TE}_{m,n}$ modes. To do this, we shall first show, by making use of the expansions for the Maxwell's curl equations in Cartesian coordinates given by (3.12a)–(3.12c) and (3.26a)–(3.26c), that all transverse (x and y) field components are derivable from the longitudinal field component H_z. It is convenient to use the phasor forms of the field components and the differential equations. Since all components of the fields are then dependent on t and z in the manner $e^{j[\omega t-(2\pi/\lambda_g)z]}$, we can replace $\partial/\partial t$ by $j\omega$ and $\partial/\partial z$ by $-j(2\pi/\lambda_g)$. Furthermore, $E_z = 0$ in view of TE modes and J_x, J_y, and J_z are all zero since the medium inside the waveguide is a perfect dielectric. Thus the phasor forms of (3.12a)–(3.12c) and (3.26a)–(3.26c) pertinent to the discussion here are

$$j\frac{2\pi}{\lambda_g}\bar{E}_y = -j\omega\mu\bar{H}_x \tag{7.62a}$$

$$-j\frac{2\pi}{\lambda_g}\bar{E}_x = -j\omega\mu\bar{H}_y \tag{7.62b}$$

$$\frac{\partial\bar{E}_y}{\partial x} - \frac{\partial\bar{E}_x}{\partial y} = -j\omega\mu\bar{H}_z \tag{7.62c}$$

*The portion between the symbols ▪▪▪▪▪▪▪ may be omitted without loss of continuity.

$$\frac{\partial \bar{H}_z}{\partial y} + j\frac{2\pi}{\lambda_g}\bar{H}_y = j\omega\epsilon\bar{E}_x \tag{7.62d}$$

$$-j\frac{2\pi}{\lambda_g}\bar{H}_x - \frac{\partial \bar{H}_z}{\partial x} = j\omega\epsilon\bar{E}_y \tag{7.62e}$$

$$\frac{\partial \bar{H}_y}{\partial x} - \frac{\partial \bar{H}_x}{\partial y} = 0 \tag{7.62f}$$

Solving (7.62a), (7.62b), (7.62d), and (7.62e), for $\bar{E}_x$, $\bar{E}_y$, $\bar{H}_x$, and $\bar{H}_y$ in terms of $\bar{H}_z$, we obtain

$$\bar{E}_x = \frac{j\omega\mu}{(2\pi/\lambda_g)^2 - \omega^2\mu\epsilon}\frac{\partial \bar{H}_z}{\partial y} \tag{7.63a}$$

$$\bar{E}_y = -\frac{j\omega\mu}{(2\pi/\lambda_g)^2 - \omega^2\mu\epsilon}\frac{\partial \bar{H}_z}{\partial x} \tag{7.63b}$$

$$\bar{H}_x = j\frac{2\pi/\lambda_g}{(2\pi/\lambda_g)^2 - \omega^2\mu\epsilon}\frac{\partial \bar{H}_z}{\partial x} \tag{7.63c}$$

$$\bar{H}_y = j\frac{2\pi/\lambda_g}{(2\pi/\lambda_g)^2 - \omega^2\mu\epsilon}\frac{\partial \bar{H}_z}{\partial y} \tag{7.63d}$$

Furthermore by substituting (7.63a) and (7.63b) into (7.62c) and rearranging, we obtain a differential equation for $\bar{H}_z$ as given by

$$\frac{\partial^2 \bar{H}_z}{\partial x^2} + \frac{\partial^2 H_z}{\partial y^2} + \left[-\left(\frac{2\pi}{\lambda_g}\right)^2 + \omega^2\mu\epsilon\right]\bar{H}_z = 0 \tag{7.64}$$

When the differential equation (7.64) is solved by using the separation of variables technique and subject to appropriate boundary conditions, the solution for $\bar{H}_z$ is obtained, which can then be put into (7.63a)–(7.63d) to obtain the transverse field components. We shall, however, not pursue this approach but shall write the solution for H_z from our knowledge of H_z for $TE_{m,0}$ modes and the subsequent discussion of $TE_{0,n}$ and $TE_{m,n}$ modes. To do this, we first note from (7.35b) that for $TE_{m,0}$ modes,

$$H_z = H_0 \cos\left(\frac{m\pi x}{a}\right)\cos\left(\omega t - \frac{2\pi}{\lambda_g}z\right) \tag{7.65a}$$

where we have replaced the amplitude factor by H_0. The expression for H_z for $TE_{0,n}$ modes can then be obtained by letting $x \longrightarrow y$, $m \longrightarrow n$, and $a \longrightarrow b$ in (7.65a). Thus for $TE_{0,n}$ modes

$$H_z = H_0 \cos\left(\frac{n\pi y}{b}\right)\cos\left(\omega t - \frac{2\pi}{\lambda_g}z\right) \tag{7.65b}$$

Combining (7.65a) and (7.65b), we have for TE$_{m,n}$ modes,

$$H_z = H_0 \cos\left(\frac{m\pi x}{a}\right) \cos\left(\frac{n\pi y}{b}\right) \cos\left(\omega t - \frac{2\pi}{\lambda_g}z\right) \qquad (7.66)$$

Note that (7.66) reduces to (7.65a) for $n = 0$ and to (7.65b) for $m = 0$. Writing H_z in phasor form, that is,

$$\bar{H}_z = H_0 \cos\left(\frac{m\pi x}{a}\right) \cos\left(\frac{n\pi y}{b}\right) e^{-j(2\pi/\lambda_g)z} \qquad (7.67)$$

and substituting into (7.64), we obtain

$$-\left(\frac{2\pi}{\lambda_g}\right)^2 + \omega^2 \mu\epsilon = \left(\frac{m\pi}{a}\right)^2 + \left(\frac{n\pi}{b}\right)^2$$

$$= (2\pi)^2\left[\frac{1}{(2a/m)^2} + \frac{1}{(2b/m)^2}\right]$$

$$= \left(\frac{2\pi}{\lambda_c}\right)^2 \qquad (7.68)$$

Substituting (7.68) and (7.67) into (7.63a)–(7.63d), we finally obtain the expressions for the transverse field components:

$$\bar{E}_x = j\frac{\omega\mu\lambda_c^2}{4\pi^2}\frac{n\pi}{b}H_0 \cos\left(\frac{m\pi x}{a}\right) \sin\left(\frac{n\pi y}{b}\right) e^{-j(2\pi/\lambda_g)z} \qquad (7.69a)$$

$$\bar{E}_y = -j\frac{\omega\mu\lambda_c^2}{4\pi^2}\frac{m\pi}{a}H_0 \sin\left(\frac{m\pi x}{a}\right) \cos\left(\frac{n\pi y}{b}\right) e^{-j(2\pi/\lambda_g)z} \qquad (7.69b)$$

$$\bar{H}_x = j\frac{\lambda_c^2}{2\pi\lambda_g}\frac{m\pi}{a}H_0 \sin\left(\frac{m\pi x}{a}\right) \cos\left(\frac{n\pi y}{b}\right) e^{-j(2\pi/\lambda_g)z} \qquad (7.69c)$$

$$\bar{H}_y = j\frac{\lambda_c^2}{2\pi\lambda_g}\frac{n\pi}{b}H_0 \cos\left(\frac{m\pi x}{a}\right) \sin\left(\frac{n\pi y}{b}\right) e^{-j(2\pi/\lambda_g)z} \qquad (7.69d)$$

Note that the sine terms in these field expressions satisfy the boundary conditions of zero tangential electric field and zero normal magnetic field at the walls of the waveguide. ▪▪▪▪▪

The entire treatment of guided waves in Sec. 7.2 can be repeated starting with the superposition of two uniform plane waves having their magnetic fields entirely in the y direction, thereby leading to "transverse magnetic waves," or "TM waves," so termed because the magnetic field for these waves has no z component, whereas the electric field has. Insofar as the cutoff phenomenon is concerned, these modes are obviously governed by the same condition as the corresponding TE modes. There cannot, however, be any

$TM_{m,0}$ or $TM_{0,n}$ modes in a rectangular waveguide since the z component of the electric field, being tangential to all four walls of the guide, requires sinusoidal variations in both x and y directions in order that the boundary condition of zero tangential component of electric field is satisfied on all four walls. Thus for $TM_{m,n}$ modes in a rectangular waveguide, both m and n must be nonzero and the cutoff wavelengths are the same as for the $TE_{m,n}$ modes, that is,

$$[\lambda_c]_{TM_{m,n}} = \frac{1}{\sqrt{(m/2a)^2 + (n/2b)^2}} \qquad (7.70)$$

The foregoing discussion of the modes of propagation in a rectangular waveguide points out that a signal of given frequency can propagate in several modes, namely, all modes for which the cutoff frequencies are less than the signal frequency or the cutoff wavelengths are greater than the signal wavelength. Waveguides are, however, designed so that only one mode, the mode with the lowest cutoff frequency (or the largest cutoff wavelength), propagates. This is known as the "dominant mode." From (7.58), (7.59), (7.61), and (7.70), we can see that the dominant mode is the $TE_{1,0}$ mode or the $TE_{0,1}$ mode, depending on whether the dimension a or the dimension b is the larger of the two. By convention, the larger dimension is designated to be a, and hence the $TE_{1,0}$ mode is the dominant mode. We shall now consider an example.

Example 7.4. It is desired to determine the lowest four cutoff frequencies referred to the cutoff frequency of the dominant mode for three cases of rectangular waveguide dimensions: (i) $b/a = 1$, (ii) $b/a = 1/2$, and (iii) $b/a = 1/3$. Given $a = 3$ cm, it is then desired to find the propagating mode(s) for $f = 9000$ MHz for each of the three cases.

From (7.61) and (7.70), the expression for the cutoff wavelength for a $TE_{m,n}$ mode where $m = 0, 1, 2, 3, \ldots$ and $n = 0, 1, 2, 3, \ldots$ but not both m and n equal to zero and for a $TM_{m,n}$ mode where $m = 1, 2, 3, \ldots$ and $n = 1, 2, 3, \ldots$ is given by

$$\lambda_c = \frac{1}{\sqrt{(m/2a)^2 + (n/2b)^2}}$$

The corresponding expression for the cutoff frequency is

$$f_c = \frac{v_p}{\lambda_c} = \frac{1}{\sqrt{\mu\epsilon}}\sqrt{\left(\frac{m}{2a}\right)^2 + \left(\frac{n}{2b}\right)^2}$$

$$= \frac{1}{2a\sqrt{\mu\epsilon}}\sqrt{m^2 + \left(n\frac{a}{b}\right)^2}$$

The cutoff frequency of the dominant mode $TE_{1,0}$ is $1/2a\sqrt{\mu\epsilon}$. Hence

$$\frac{f_c}{[f_c]_{TE_{1,0}}} = \sqrt{m^2 + \left(n\frac{a}{b}\right)^2}$$

By assigning different pairs of values for m and n, the lowest four values of $f_c/[f_c]_{TE_{1,0}}$ can be computed for each of the three specified values of b/a. These computed values and the corresponding modes are shown in Fig. 7.17.

For $a = 3$ cm, and assuming free space for the dielectric in the waveguide,

$$[f_c]_{TE_{1,0}} = \frac{1}{2a\sqrt{\mu\epsilon}} = \frac{3 \times 10^8}{2 \times 0.03} = 5000 \text{ MHz}$$

Hence for a signal of frequency $f = 9000$ MHz, all the modes for which $f_c/[f_c]_{TE_{1,0}}$ is less than 1.8 propagate. From Fig. 7.17, these are

$$TE_{1,0}, TE_{0,1}, TM_{1,1}, TE_{1,1} \quad \text{for } b/a = 1$$
$$TE_{1,0} \qquad\qquad\qquad\qquad \text{for } b/a = 1/2$$
$$TE_{1,0} \qquad\qquad\qquad\qquad \text{for } b/a = 1/3$$

It can be seen from Fig. 7.17 that for $b/a \leq 1/2$, the second lowest cutoff

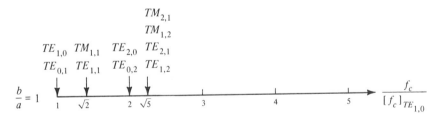

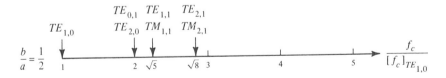

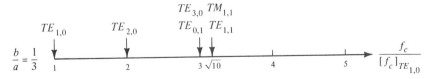

Figure 7.17. Lowest four cutoff frequencies referred to the cutoff frequency of the dominant mode for three cases of rectangular waveguide dimensions.

frequency that corresponds to that of the $TE_{2,0}$ mode is twice the cutoff frequency of the dominant mode $TE_{1,0}$. For this reason, the dimension b of a rectangular waveguide is generally chosen to be less than or equal to $a/2$ in order to achieve single-mode transmission over a complete octave (factor of two) range of frequencies.   ■

Let us now consider guided waves of equal magnitude propagating in the positive z and negative z directions in a rectangular waveguide. This can be achieved by terminating the guide by a perfectly conducting sheet in a constant z plane, that is, a transverse plane of the guide. Due to perfect reflection from the sheet, the fields will then be characterized by standing wave nature along the guide axis, that is, in the z direction, in addition to the standing wave nature in the x and y directions. The standing wave pattern along the guide axis will have nulls of transverse electric field on the terminating sheet and in planes parallel to it at distances of integer multiples of $\lambda_g/2$ from that sheet. Placing of perfect conductors in these planes will not disturb the fields since the boundary condition of zero tangential electric field is satisfied in those planes.

Conversely, if we place two perfectly conducting sheets in two constant z planes separated by a distance d, then, in order for the boundary conditions to be satisfied, d must be equal to an integer multiple of $\lambda_g/2$. We then have a rectangular box of dimensions a, b, and d in the x, y, and z directions, respectively, as shown in Fig. 7.18. Such a structure is known as a "cavity

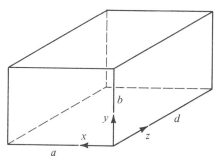

Figure 7.18. A rectangular cavity resonator.

resonator" and is the counterpart of the low-frequency lumped parameter resonant circuit at microwave frequencies since it supports oscillations at frequencies for which the above condition, that is,

$$d = l\frac{\lambda_g}{2}, \qquad l = 1, 2, 3, \ldots \tag{7.71}$$

is satisfied. Recalling that λ_g is simply the apparent wavelength of the obliquely bouncing uniform plane wave along the z direction, we find that the wavelength corresponding to the mode of oscillation for which the fields have m number of one-half sinusoidal variations in the x direction, n number of one-half sinusoidal variations in the y direction, and l number of one-half sinusoidal variations in the z direction is given by

$$\frac{1}{\lambda_{osc}^2} = \frac{1}{(2a/m)^2} + \frac{1}{(2b/n)^2} + \frac{1}{(2d/l)^2} \tag{7.72}$$

or

$$\lambda_{osc} = \frac{1}{\sqrt{(m/2a)^2 + (n/2b)^2 + (l/2d)^2}} \tag{7.73}$$

The expression for the frequency of oscillation is then given by

$$f_{osc} = \frac{v_p}{\lambda_{osc}} = \frac{1}{\sqrt{\mu\epsilon}}\sqrt{\left(\frac{m}{2a}\right)^2 + \left(\frac{n}{2b}\right)^2 + \left(\frac{l}{2d}\right)^2} \tag{7.74}$$

The modes are designated by three subscripts in the manner $TE_{m,n,l}$ and $TM_{m,n,l}$. Since m, n, and l can assume combinations of integer values, an infinite number of frequencies of oscillation are possible for a given set of dimensions for the cavity resonator. We shall now consider an example.

Example 7.5. The dimensions of a rectangular cavity resonator with air dielectric are $a = 4$ cm, $b = 2$ cm, and $d = 4$ cm. It is desired to determine the three lowest frequencies of oscillation and specify the mode(s) of oscillation, transverse with respect to the z direction, for each frequency.

By substituting $\mu = \mu_0$, $\epsilon = \epsilon_0$, and the given dimensions for a, b, and d in (7.74), we obtain

$$f_{osc} = 3 \times 10^8 \sqrt{\left(\frac{m}{0.08}\right)^2 + \left(\frac{n}{0.04}\right)^2 + \left(\frac{l}{0.08}\right)^2}$$

$$= 3750\sqrt{m^2 + 4n^2 + l^2}\ \text{MHz}$$

By assigning combinations of integer values for m, n, and l and recalling that both m and n must be nonzero for TM modes, we obtain the three lowest frequencies of oscillation to be

$3750 \times \sqrt{2} = 5303$ MHz for $TE_{1,0,1}$ mode

$3750 \times \sqrt{5} = 8385$ MHz for $TE_{0,1,1}$, $TE_{2,0,1}$, and $TE_{1,0,2}$ modes

$3750 \times \sqrt{6} = 9186$ MHz for $TE_{1,1,1}$ and $TM_{1,1,1}$ modes   ∎

7.6   OPTICAL WAVEGUIDES

Thus far we have been concerned with waveguides that have conductors as boundaries. In this section we shall briefly consider another class of waveguides. These waveguides, having dielectrics as their boundaries, form the basis for waveguiding at optical frequencies. The principle of optical wave-guides suggests itself from the phenomenon of guiding of waves by means of oblique reflections at the boundaries of the guide. Thus let us consider a uniform plane wave incident obliquely on a plane boundary between two different perfect dielectric media at an angle of incidence θ_i to the normal to the boundary, as shown in Fig. 7.19. To satisfy the boundary conditions at

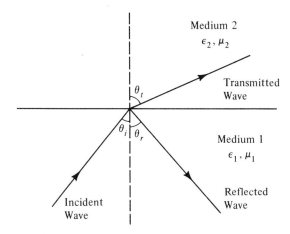

Figure 7.19. Reflection and transmission of an obliquely incident uniform plane wave on a plane boundary between two different perfect dielectric media.

the interface between the two media, a reflected wave and a transmitted wave will be set up. Let θ_r be the angle of reflection and θ_t be the angle of transmission. Then without writing the expressions for the fields, we can find the relationship between θ_i, θ_r, and θ_t by noting that in order for the incident, reflected, and transmitted waves to be in step at the boundary, their apparent phase velocities parallel to the boundary must be equal, that is

$$\frac{v_{p1}}{\sin \theta_i} = \frac{v_{p1}}{\sin \theta_r} = \frac{v_{p2}}{\sin \theta_t} \tag{7.75}$$

where v_{p1} $(= 1/\sqrt{\mu_1 \epsilon_1})$ and v_{p2} $(= 1/\sqrt{\mu_2 \epsilon_2})$ are the phase velocities along the directions of propagation of the waves in medium 1 and medium 2, respectively.

From (7.75), we have

$$\sin \theta_r = \sin \theta_i \tag{7.76a}$$

$$\sin \theta_t = \frac{v_{p2}}{v_{p1}} \sin \theta_i = \sqrt{\frac{\mu_1 \epsilon_1}{\mu_2 \epsilon_2}} \sin \theta_i \tag{7.76b}$$

or

$$\theta_r = \theta_i \qquad (7.77a)$$

$$\theta_t = \sin^{-1} \left(\sqrt{\frac{\mu_1 \epsilon_1}{\mu_2 \epsilon_2}} \sin \theta_i \right) \qquad (7.77b)$$

Equation (7.77a) is known as the "law of reflection" and (7.77b) is known as the "law of refraction," or "Snell's law." Snell's law is commonly cast in terms of the refractive index, denoted by the symbol n and defined as the ratio of the velocity of light in free space to the phase velocity in the medium. Thus if $n_1 \ (= c/v_{p1})$ and $n_2 \ (= c/v_{p2})$ are the refractive indices for media 1 and 2, respectively, then

$$\theta_t = \sin^{-1} \left(\frac{n_1}{n_2} \sin \theta_i \right) \qquad (7.78)$$

Assuming that $\mu_1 = \mu_2 = \mu_0$, which is generally the case, we note from (7.76b) that for $\epsilon_2 > \epsilon_1$, $\sin \theta_t < \sin \theta_i$ and $\theta_t < \theta_i$ so that the transmitted wave is refracted toward the normal to the boundary. For $\epsilon_2 < \epsilon_1$, $\sin \theta_t > \sin \theta_i$ and $\theta_t > \theta_i$ so that the transmitted wave is refracted away from the normal to the boundary. Hence for this case there exists a value of θ_i for which $\theta_t = 90°$. Denoting this "critical angle" of incidence to be θ_c, we have from (7.76b).

$$\sqrt{\frac{\epsilon_1}{\epsilon_2}} \sin \theta_c = \sin 90° = 1$$

or

$$\theta_c = \sin^{-1} \sqrt{\frac{\epsilon_2}{\epsilon_1}} = \sin^{-1} \frac{n_2}{n_1} \qquad (7.79)$$

For $\theta_i > \theta_c$, there is no real solution for θ_t and "total internal reflection" occurs, that is, the incident wave is entirely reflected. Hence if we have a dielectric slab of permittivity ϵ_1, sandwiched between two dielectric media of permittivity $\epsilon_2 < \epsilon_1$, then by launching waves at an angle of incidence greater than the critical angle, it is possible to achieve guided wave propagation, as shown in Fig. 7.20. This is the principle of optical waveguides. As in the case of metallic waveguides, a given frequency signal may propagate in several modes for which the cutoff frequencies are less than the wave frequency. We shall, however, not pursue a discussion of these modes; instead, we shall conclude this section with a brief description of an optical fiber, which is a common form of optical waveguide.

An optical fiber, so termed because of its filamentary appearance, consists typically of a core and a cladding, having cylindrical cross sections as shown in Fig. 7.21(a). The core is made up of a material of permittivity greater than

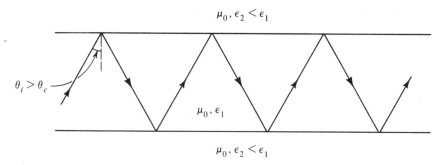

Figure 7.20. Total internal reflection in a dielectric slab waveguide.

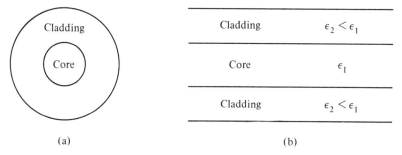

(a)   (b)

Figure 7.21. (a) Transverse and (b) longitudinal cross sections of an optical fiber.

that of the cladding so that a critical angle exists for waves inside the core incident on the interface between the core and the cladding, and hence waveguiding is made possible in the core by total internal reflection. The phenomenon may be visualized by considering a longitudinal cross section of the fiber through its axis, shown in Fig. 7.21(b), and comparing it with that of the slab waveguide shown in Fig. 7.20. Although the cladding is not essential for the purpose of waveguiding in the core since the permittivity of the core material is greater than that of free space, the cladding serves two useful purposes: (a) It avoids scattering and field distortion by the supporting structure of the fiber since the field decays exponentially outside the core and hence is negligible outside the cladding. (b) It allows single-mode propagation for a larger value of the radius of the core than permitted in the absence of the cladding.

7.7 SUMMARY

In this chapter we studied the principles of waveguides. To introduce the waveguiding phenomenon, we first learned how to write the expressions for the electric and magnetic fields of a uniform plane wave propagating in an arbitrary direction with respect to the coordinate axes. These expressions are given by

$$\mathbf{E} = \mathbf{E}_0 \cos (\omega t - \boldsymbol{\beta} \cdot \mathbf{r} + \phi_0)$$

$$\mathbf{H} = \mathbf{H}_0 \cos (\omega t - \boldsymbol{\beta} \cdot \mathbf{r} + \phi_0)$$

where $\boldsymbol{\beta}$ and $\mathbf{r}$ are the propagation and position vectors given by

$$\boldsymbol{\beta} = \beta_x \mathbf{i}_x + \beta_y \mathbf{i}_y + \beta_z \mathbf{i}_z$$

$$\mathbf{r} = x\mathbf{i}_x + y\mathbf{i}_y + z\mathbf{i}_z$$

and ϕ_0 is the phase of the wave at the origin at $t = 0$. The magnitude of $\boldsymbol{\beta}$ is equal to $\omega\sqrt{\mu\epsilon}$, the phase constant along the direction of propagation of the wave. The direction of $\boldsymbol{\beta}$ is the direction of propagation of the wave. We learned that

$$\mathbf{E}_0 \cdot \boldsymbol{\beta} = 0$$

$$\mathbf{H}_0 \cdot \boldsymbol{\beta} = 0$$

$$\mathbf{E}_0 \cdot \mathbf{H}_0 = 0$$

that is, $\mathbf{E}_0$, $\mathbf{H}_0$, and $\boldsymbol{\beta}$ are mutually perpendicular, and that

$$\frac{|\mathbf{E}_0|}{|\mathbf{H}_0|} = \eta = \sqrt{\frac{\mu}{\epsilon}}$$

Also, since $\mathbf{E} \times \mathbf{H}$ should be directed along the propagation vector $\boldsymbol{\beta}$, it then follows that

$$\mathbf{H} = \frac{1}{\omega\mu} \boldsymbol{\beta} \times \mathbf{E}$$

The quantities β_x, β_y, and β_z are the phase constants along the x, y, and z axes, respectively. The apparent wavelengths and the apparent phase velocities along the coordinate axes are given, respectively, by

$$\lambda_i = \frac{2\pi}{\beta_i}, \qquad i = x, y, z$$

$$v_{pi} = \frac{\omega}{\beta_i}, \qquad i = x, y, z$$

By considering the superposition of two uniform plane waves propagating at an angle to each other and placing two perfect conductors in appropriate planes such that the boundary condition of zero tangential electric field is satisfied, we introduced the parallel-plate waveguide. We learned that the composite wave is a transverse electric, or TE wave since the electric field is entirely transverse to the direction of time-average power flow, that is, the guide axis, but the magnetic field is not. In terms of the uniform plane wave propagation, the phenomenon is one of waves bouncing obliquely between the conductors as they progress down the guide. For a fixed spacing a between the conductors of the guide, waves of different frequencies bounce obliquely at different angles such that the spacing a is equal to an integer, say, m number of one-half apparent wavelengths normal to the plates and hence the fields have m number of one-half-sinusoidal variations normal to the plates. These are said to correspond to $TE_{m,0}$ modes where the subscript 0 implies no variations of the fields in the direction parallel to the plates and transverse to the guide axis. When the frequency is such that the spacing a is equal to m one-half wavelengths, the waves bounce normally to the plates without the feeling of being guided along the axis, thereby leading to the cutoff condition. Thus the cutoff wavelengths corresponding to $TE_{m,0}$ modes are given by

$$\lambda_c = \frac{2a}{m}$$

and the cutoff frequencies are given by

$$f_c = \frac{v_p}{\lambda_c} = \frac{m}{2a\sqrt{\mu\epsilon}}$$

A given frequency signal can propagate in all modes for which $\lambda < \lambda_c$ or $f > f_c$. For the propagating range of frequencies, the wavelength along the guide axis, that is, the guide wavelength, and the phase velocity along the guide axis are given, respectively, by

$$\lambda_g = \frac{\lambda}{\sqrt{1 - (\lambda/\lambda_c)^2}} = \frac{\lambda}{\sqrt{1 - (f_c/f)^2}}$$

$$v_{pz} = \frac{v_p}{\sqrt{1 - (\lambda/\lambda_c)^2}} = \frac{v_p}{\sqrt{1 - (f_c/f)^2}}$$

We discussed the solution of problems involving reflection and transmission at a discontinuity in a waveguide by using the transmission-line analogy. This consists of replacing each section of the waveguide by a transmission line whose characteristic impedance is equal to the guide impedance and then computing the reflection and transmission coefficients as in the transmission-line case. The guide impedance, η_g, which is the ratio of the transverse electric field to the transverse magnetic field, is given for the TE modes by

$$\eta_g = \frac{\eta}{\sqrt{1 - (\lambda/\lambda_c)^2}} = \frac{\eta}{\sqrt{1 - (f_c/f)^2}}$$

We discussed the phenomenon of dispersion arising from the frequency dependence of the phase velocity along the guide axis, and we introduced the concept of group velocity. Group velocity is the velocity with which the envelope of a narrow-band modulated signal travels in the dispersive channel and hence it is the velocity with which the information is transmitted. It is given by

$$v_g = \frac{d\omega}{d\beta_z} = v_p \sqrt{1 - \left(\frac{f_c}{f}\right)^2}$$

where β_z is the phase constant along the guide axis.

We extended the treatment of the parallel-plate waveguide to the rectangular waveguide, which is a metallic pipe of rectangular cross section. By considering a rectangular waveguide of cross-sectional dimensions a and b, we discussed transverse electric or TE modes as well as transverse magnetic or TM modes, and learned that while $TE_{m,n}$ modes can include values of m or n equal to zero, $TM_{m,n}$ modes require that both m and n be nonzero, where m and n refer to the number of one-half sinusoidal variations of the fields along the dimensions a and b, respectively. The cutoff wavelengths for the $TE_{m,n}$ or $TM_{m,n}$ modes are given by

$$\lambda_c = \frac{1}{\sqrt{(m/2a)^2 + (n/2b)^2}}$$

The mode that has the largest cutoff wavelength or the lowest cutoff frequency is the dominant mode, which here is the $TE_{1,0}$ mode. Waveguides are generally designed to transmit only the dominant mode.

By placing perfect conductors in two transverse planes of a rectangular waveguide separated by an integer multiple of one-half the guide wavelength, we introduced the cavity resonator, which is the microwave counterpart of the lumped parameter resonant circuit encountered in low-frequency circuit theory. For a rectangular cavity resonator having dimensions a, b, and d, the frequencies of oscillation for the $TE_{m,n,l}$ or $TM_{m,n,l}$ modes are given by

$$f_{osc} = \frac{1}{\sqrt{\mu\epsilon}} \sqrt{\left(\frac{m}{2a}\right)^2 + \left(\frac{n}{2b}\right)^2 + \left(\frac{l}{2d}\right)^2}$$

where l refers to the number of one-half sinusoidal variations of the fields along the dimension d.

Finally, we discussed the principle of optical waveguides. By considering a uniform plane wave incident at an angle θ_i from medium 1 of permittivity ϵ_1 and permeability μ_1 onto medium 2 of permittivity ϵ_2 and permeability μ_2, we derived Snell's law of refraction

$$\theta_t = \sin^{-1}\left(\sqrt{\frac{\mu_1\epsilon_1}{\mu_2\epsilon_2}} \sin \theta_i\right)$$

where θ_t is the angle of transmission into medium 2. For $\mu_1 = \mu_2$ and for $\epsilon_2 < \epsilon_1$, there exists a critical angle of incidence θ_c given by

$$\theta_c = \sin^{-1} \sqrt{\frac{\epsilon_2}{\epsilon_1}}$$

above which total internal reflection of the wave occurs into medium 1. Thus optical waveguides consist of a dielectric medium sandwiched between two dielectric media of lesser permittivity so as to permit waveguiding by means of total internal reflection.

REVIEW QUESTIONS

7.1. What is the propagation vector? Interpret the significance of its magnitude and direction.

7.2. Discuss how the phase constants along the coordinate axes are less than the phase constant along the direction of propagation of a uniform plane wave propagating in an arbitrary direction.

7.3. Write the expressions for the electric and magnetic fields of a uniform plane wave propagating in an arbitrary direction and list all the conditions to be satisfied by the electric field, magnetic field, and propagation vectors.

7.4. What are apparent wavelengths? Why are they longer than the wavelength along the direction of propagation?

7.5. What are apparent phase velocities? Why are they greater than the phase velocity along the direction of propagation?

7.6. Discuss how the superposition of two uniform plane waves propagating at an angle to each other gives rise to a composite wave consisting of standing waves traveling bodily transverse to the standing waves.

7.7. What is a transverse electric wave? Discuss the reasoning behind the nomenclature $TE_{m,0}$ modes.

7.8. How would you characterize a transverse magnetic wave?

7.9. Compare the phenomenon of guiding of uniform plane waves in a parallel-plate waveguide with that in a parallel-plate transmission line.

7.10. Discuss how the cutoff condition arises in a waveguide.

7.11. Explain the relationship between the cutoff wavelength and the spacing between the plates of a parallel-plate waveguide based on the phenomenon at cutoff.

7.12. Is the cutoff wavelength dependent on the dielectric in the waveguide? Is the cutoff frequency dependent on the dielectric in the waveguide?

7.13. What is guide wavelength?

7.14. Provide a physical explanation for the frequency dependence of the phase velocity along the guide axis.

7.15. Define guide impedance.

7.16. Discuss the use of the transmission-line analogy for solving problems involving reflection and transmission at a waveguide discontinuity.

7.17. Why are the reflection and transmission coefficients for a given mode at a lossless waveguide discontinuity dependent on frequency whereas the reflection and transmission coefficients at the junction of two lossless lines are independent of frequency?

7.18. Discuss the phenomenon of dispersion.

7.19. Discuss the concept of group velocity with the aid of an example.

7.20. What is a dispersion diagram? Explain how the phase and group velocities can be determined from a dispersion diagram.

7.21. When is it meaningful to attribute a group velocity to a signal comprised of more than two frequencies? Why?

7.22. Discuss the propagation of a narrow-band amplitude modulated signal in a dispersive channel.

7.23. Discuss the nomenclature associated with the modes of propagation in a rectangular waveguide.

7.24. Explain the relationship between the cutoff wavelength and the dimensions of a rectangular waveguide based on the phenomenon at cutoff.

7.25. Discuss the reasoning behind the formulation of the expression for H_z for $TE_{m,n}$ modes in a rectangular waveguide.

7.26. Briefly outline the procedure for deriving the transverse field components in a rectangular waveguide from the longitudinal field component.

7.27. Why can there be no transverse magnetic modes having no variations for the fields along one of the dimensions of a rectangular waveguide?

7.28. What is meant by the dominant mode? Why are waveguides designed so that they propagate only the dominant mode?

7.29. Why is the dimension b of a rectangular waveguide generally chosen to be less than or equal to one-half the dimension a?

7.30. Explain why, when driving through a mountain tunnel or under a road bridge, you are able to receive signals in the FM band but not in the AM band of an AM–FM radio.

7.31. What is a cavity resonator?

7.32. How do the dimensions of a rectangular cavity resonator determine the frequencies of oscillation of the resonator?

7.33. Discuss the condition required to be satisfied by the incident, reflected, and transmitted waves at the interface between two dielectric media.

7.34. What is Snell's law?

7.35. What is total internal reflection? What are the requirements for total internal reflection?

7.36. Discuss the principle of optical waveguides.

7.37. Compare the phenomenon at cutoff in a metallic waveguide with that at cutoff in an optical waveguide.

7.38. Provide a brief description of an optical fiber.

PROBLEMS

7.1. Assuming the x and y axes to be directed eastward and northward, respectively, find the expression for the propagation vector of a uniform plane wave of frequency 15 MHz in free space propagating in the direction 30° north of east.

7.2. The propagation vector of a uniform plane wave in a perfect dielectric medium having $\epsilon = 4.5\epsilon_0$ and $\mu = \mu_0$ is given by

$$\boldsymbol{\beta} = 2\pi(3\mathbf{i}_x + 4\mathbf{i}_y + 5\mathbf{i}_z)$$

Find (a) the apparent wavelengths and (b) the apparent phase velocities, along the coordinate axes.

7.3. For a uniform plane wave propagating in free space, the apparent phase velocities along the x and y directions are found to be $6\sqrt{2} \times 10^8$ m/s and $2\sqrt{3} \times 10^8$ m/s, respectively. Find the direction of propagation of the wave.

7.4. The electric field vector of a uniform plane wave propagating in a perfect dielectric medium having $\epsilon = 9\epsilon_0$ and $\mu = \mu_0$ is given by

$$\mathbf{E} = 10(-\mathbf{i}_x - 2\sqrt{3}\,\mathbf{i}_y + \sqrt{3}\,\mathbf{i}_z) \cos{[16\pi \times 10^6 t}$$
$$- 0.04\pi(\sqrt{3}\,x - 2y - 3z)]$$

Find (a) the frequency, (b) the direction of propagation, (c) the wavelength along the direction of propagation, (d) the apparent wavelengths along the x, y, and z axes, and (e) the apparent phase velocities along the x, y, and z axes.

7.5. Given

$$\mathbf{E} = 10\mathbf{i}_x \cos{[6\pi \times 10^7 t - 0.1\pi(y + \sqrt{3}\,z)]}$$

(a) Determine if the given $\mathbf{E}$ represents the electric field of a uniform plane wave propagating in free space. (b) If the answer to part (a) is "yes," find the corresponding magnetic field vector $\mathbf{H}$.

7.6. Given

$$\mathbf{E} = (\mathbf{i}_x - 2\mathbf{i}_y - \sqrt{3}\,\mathbf{i}_z) \cos{[15\pi \times 10^6 t - 0.05\pi(\sqrt{3}\,x + z)]}$$

$$\mathbf{H} = \frac{1}{60\pi}(\mathbf{i}_x + 2\mathbf{i}_y - \sqrt{3}\,\mathbf{i}_z) \cos{[15\pi \times 10^6 t - 0.05\pi(\sqrt{3}\,x + z)]}$$

(a) Perform all the necessary tests and determine if these fields represent a uniform plane wave propagating in a perfect dielectric medium. (b) Find the permittivity and the permeability of the medium.

7.7. Two equal-amplitude uniform plane waves of frequency 25 MHz and having their electric fields along the y direction propagate along the directions i_z and $\frac{1}{2}(\sqrt{3}\,i_x + i_z)$ in free space. (a) Find the direction of propagation of the composite wave. (b) Find the wavelength along the direction of propagation and the wavelength transverse to the direction of propagation of the composite wave.

7.8. Show that $\langle\sin^2 (\omega t - \beta z \sin \theta)\rangle$ and $\langle\sin 2(\omega t - \beta z \sin \theta)\rangle$ are equal to zero and $1/2$, respectively.

7.9. Find the spacing a for a parallel-plate waveguide having a dielectric of $\epsilon = 9\epsilon_0$ and $\mu = \mu_0$ such that 6000 MHz is 20 percent above the cutoff frequency of the dominant mode, that is, the mode with the lowest cutoff frequency.

7.10. The dimension a of a parallel-plate waveguide filled with a dielectric having $\epsilon = 4\epsilon_0$ and $\mu = \mu_0$ is 4 cm. Determine the propagating $TE_{m,0}$ modes for a wave of frequency 6000 MHz. For each propagating mode, find f_c, θ, and λ_g.

7.11. The spacing a between the plates of a parallel-plate waveguide is equal to 5 cm. The dielectric between the plates is free space. If a generator of fundamental frequency 1800 MHz and rich in harmonics excites the waveguide, find all frequencies that propagate in $TE_{1,0}$ mode only.

7.12. The electric and magnetic fields of the composite wave resulting from the superposition of two uniform plane waves are given by

$$\mathbf{E} = E_{x0} \cos \beta_x x \cos (\omega t - \beta_z z)\,i_x$$
$$+ E_{z0} \sin \beta_x x \sin (\omega t - \beta_z z)\,i_z$$
$$\mathbf{H} = H_{y0} \cos \beta_x x \cos (\omega t - \beta_z z)\,i_y$$

(a) Find the time-average Poynting vector. (b) Discuss the nature of the composite wave.

7.13. Transverse electric modes are excited in an air dielectric parallel-plate waveguide of dimension $a = 5$ cm by setting up at its mouth a field distribution having

$$\mathbf{E} = 10(\sin 20\pi x + 0.5 \sin 60\pi x) \sin 10^{10}\pi t\,i_y$$

Determine the propagating mode(s) and obtain the expression for the electric field of the propagating wave.

7.14. For the parallel-plate waveguide discontinuity of Example 7.3, find the reflection and transmission coefficients for $f = 7500$ MHz propagating in (a) $TE_{1,0}$ mode and (b) $TE_{2,0}$ mode.

7.15. The left half of a parallel-plate waveguide of dimension $a = 4$ cm is filled with a dielectric of $\epsilon = 4\epsilon_0$ and $\mu = \mu_0$. The right half is filled with a dielectric of $\epsilon = 9\epsilon_0$ and $\mu = \mu_0$. For $TE_{1,0}$ wave of frequency 2500 MHz incident on the discontinuity from the left, find the reflection and transmission coefficients.

7.16. Assume that the permittivity of the dielectric to the right side of the parallel-plate waveguide discontinuity of Fig. 7.10 is unknown. If the reflection coefficient for $TE_{1,0}$ waves of frequency 5000 MHz incident on the junction from the free space side is -0.2643, find the permittivity of the dielectric.

7.17. For the two-train example of Fig. 7.11, find the group velocity if the speed of train numbered B is (a) 36 m/s and (b) 40 m/s, instead of 30 m/s. Discuss your results with the aid of sketches.

7.18. Find the velocity with which the group of two frequencies 2400 MHz and 2500 MHz travels in a parallel-plate waveguide of dimension $a = 2.5$ cm and having a perfect dielectric of $\epsilon = 9\epsilon_0$ and $\mu = \mu_0$.

7.19. For a narrow-band amplitude modulated signal having the carrier frequency 5000 MHz propagating in an air dielectric parallel-plate waveguide of dimension $a = 5$ cm, find the velocity with which the modulation envelope travels.

7.20. For an $\omega - \beta_z$ relationship given by

$$\omega = \omega_0 + k\beta_z^2$$

where ω_0 and k are positive constants, find the phase and group velocities for (a) $\omega = 1.5\omega_0$, (b) $\omega = 2\omega_0$, and (c) $\omega = 3\omega_0$.

7.21. By considering the parallel-plate waveguide, show that a point on the obliquely bouncing wavefront, traveling with the phase velocity along the oblique direction, progresses parallel to the guide axis with the group velocity.

7.22. Write the expression for $\bar{E}_z$ for TM modes in a rectangular waveguide. Then obtain the transverse field components by following a procedure similar to that used in the text for TE modes.

7.23. For an air dielectric rectangular waveguide of dimensions $a = 3$ cm and $b = 1.5$ cm, find all propagating modes for $f = 12,000$ MHz.

7.24. For a rectangular waveguide of dimensions $a = 5$ cm and $b = 5/3$ cm, and having a dielectric of $\epsilon = 9\epsilon_0$ and $\mu = \mu_0$, find all propagating modes for $f = 2500$ MHz.

7.25. For $f = 3000$ MHz, find the dimensions a and b of an air dielectric rectangular waveguide such that $TE_{1,0}$ mode propagates with a 30 percent safety factor ($f = 1.30f_c$) but also such that the frequency is 30 percent below the cutoff frequency of the next higher order mode.

7.26. For an air dielectric rectangular cavity resonator having the dimensions $a = 2.5$ cm, $b = 2$ cm, and $d = 5$ cm, find the five lowest frequencies of oscillation. Identify the mode(s) for each frequency.

7.27. For a rectangular cavity resonator having the dimensions $a = b = d = 2$ cm, and filled with a dielectric of $\epsilon = 9\epsilon_0$ and $\mu = \mu_0$, find the three lowest frequencies of oscillation. Identify the mode(s) for each frequency.

7.28. In Fig. 7.19, let $\epsilon_1 = 4\epsilon_0$, $\epsilon_2 = 9\epsilon_0$, and $\mu_1 = \mu_2 = \mu_0$. (a) For $\theta_i = 30°$, find θ_t. (b) Is there a critical angle of incidence for which $\theta_t = 90°$?

7.29. In Fig. 7.19, let $\epsilon_1 = 4\epsilon_0$, $\epsilon_2 = 2.25\epsilon_0$, and $\mu_1 = \mu_2 = \mu_0$. (a) For $\theta_i = 30°$, find θ_t. (b) Find the value of the critical angle of incidence θ_c, for which $\theta_t = 90°$.

7.30. A thin-film waveguide employed in integrated optics circuits consists of a substrate upon which a thin film of refractive index greater than that of the substrate is deposited. The medium above the thin film is air. For refractive indices of the substrate and the film equal to 1.51 and 1.53, respectively, find the minimum bouncing angle of the total internally reflected waves in the film.

8. ANTENNAS

In the preceding four chapters we studied the principles of propagation and transmission of electromagnetic waves. The remaining important topic pertinent to electromangetic wave phenomena is radiation of electromagnetic waves. We have, in fact, touched on the principle of radiation of electromagnetic waves in Chap. 4 when we derived the electromagnetic field due to the infinite plane sheet of sinusoidally time-varying, spatially uniform current density. We learned that the current sheet gives rise to uniform plane waves "radiating" away from the sheet to either side of it. We pointed out at that time that the infinite plane current sheet is, however, an idealized, hypothetical source. With the experience gained thus far in our study of the elements of engineering electromagnetics, we are now in a position to learn the principles of radiation from physical antennas, which is our goal in this chapter.

We shall begin the chapter with the derivation of the electromagnetic field due to an elemental wire antenna, known as the "Hertzian dipole." After studying the radiation characteristics of the Hertzian dipole, we shall consider the example of a half-wave dipole to illustrate the use of superposition to represent an arbitrary wire antenna as a series of Hertzian dipoles in order to determine its radiation fields. We shall also discuss the principles of arrays of physical antennas and the concept of image antennas to take into account ground effects. Finally, we shall briefly consider the receiving properties of antennas and learn of their reciprocity with the radiating properties.

8.1 HERTZIAN DIPOLE

The Hertzian dipole is an elemental antenna consisting of an infinitesimally long piece of wire carrying an alternating current $I(t)$, as shown in Fig. 8.1. To maintain the current flow in the wire, we postulate two point charges $Q_1(t)$ and $Q_2(t)$ terminating the wire at its two ends, so that the law of conservation of charge is satisfied. Thus if

$$I(t) = I_0 \cos \omega t \tag{8.1}$$

then

$$\frac{dQ_1}{dt} = I(t) = I_0 \cos \omega t \tag{8.2a}$$

$$\frac{dQ_2}{dt} = -I(t) = -I_0 \cos \omega t \tag{8.2b}$$

and

$$Q_1(t) = \frac{I_0}{\omega} \sin \omega t \tag{8.3a}$$

$$Q_2(t) = -\frac{I_0}{\omega} \sin \omega t = -Q_1(t) \tag{8.3b}$$

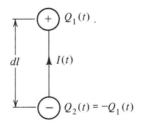

Figure 8.1. Hertzian dipole.

The time variations of I, Q_1, and Q_2, given by (8.1), (8.3a) and (8.3b), respectively, are illustrated by the curves and the series of sketches for the dipoles in Fig. 8.2, corresponding to one complete period. The different sizes of the arrows associated with the dipoles denote the different strengths of the current whereas the number of the plus or minus signs is indicative of the strength of the charges.

To determine the electromagnetic field due to the Hertzian dipole, we shall employ an intuitive approach based upon the knowledge gained in the previous chapters as follows: From the application of what we have learned in Chap. 1, we can obtain the expressions for the electric and magnetic fields

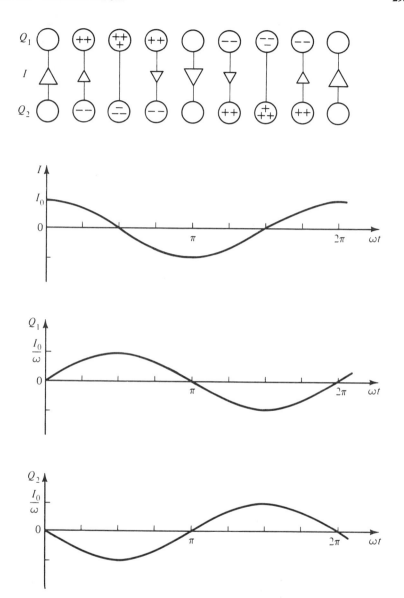

Figure 8.2. Time variations of charges and current associated with the Hertzian dipole.

due to the point charges and the current element, respectively, associated with the Hertzian dipole, assuming that the fields follow exactly the time-variations of the charges and the current. These expressions do not, however, take into account the fact that time-varying electric and magnetic fields give rise to wave propagation. Hence we shall extend them from considerations of our knowledge of wave propagation and then check if the resulting solutions satisfy Maxwell's equations. If they do not, we will then have to modify them so that they do satisfy Maxwell's equations and at the same time reduce to the originally derived expressions in the region where wave propagation effects are small, that is, at distances from the dipole small compared to a wavelength.

To follow the approach outlined in the preceding paragraph, we locate the dipole at the origin with the current directed along the z axis, as shown in Fig. 8.3, and derive first the expressions for the fields by applying the simple laws learned in Secs. 1.5 and 1.6. The symmetry associated with the problem is such that it is simpler to use a spherical coordinate system. Hence if the reader is not already familiar with the spherical coordinate system, it is suggested that Appendix A be read at this stage. To review briefly, a point in the spherical coordinate system is defined by the intersection of a sphere centered at the origin, a cone having its apex at the origin and its surface symmetrical about the z axis, and a plane containing the z axis. Thus the coordinates for a given point, say P, are r, the radial distance from the origin, θ, the angle which the radial line from the origin to the point makes with the z axis, and ϕ, the angle which the line drawn from the origin to the projection of the point onto the xy plane makes with the x axis, as shown in Fig. 8.3. A vector drawn at a given point is represented in terms of the unit vectors

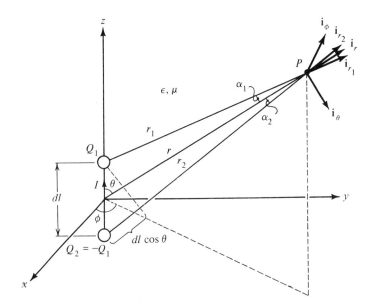

Figure 8.3. For the determination of the electromagnetic field due to the Hertzian dipole.

i_r, i_θ, and i_ϕ directed in the increasing r, θ, and ϕ directions respectively, at that point. It is important to note that all three of these unit vectors are not uniform unlike the unit vectors i_x, i_y, and i_z in the Cartesian coordinate system.

Now using the expression for the electric field due to a point charge given by (1.52), we can write the electric field at point P due to the arrangement of the two point charges Q_1 and $-Q_1$ in Fig. 8.3 to be

$$\mathbf{E} = \frac{Q_1}{4\pi\epsilon r_1^2}\mathbf{i}_{r_1} - \frac{Q_1}{4\pi\epsilon r_2^2}\mathbf{i}_{r_2} \tag{8.4}$$

where r_1 and r_2 are the distances from Q_1 to P and Q_2 $(= -Q_1)$ to P, respectively, and $\mathbf{i}_{r_1}$ and $\mathbf{i}_{r_2}$ are unit vectors directed along the lines from Q_1 to P and Q_2 to P, respectively, as shown in Fig. 8.3. Noting that

$$\mathbf{i}_{r_1} = \cos\alpha_1\,\mathbf{i}_r + \sin\alpha_1\,\mathbf{i}_\theta \tag{8.5a}$$

$$\mathbf{i}_{r_2} = \cos\alpha_2\,\mathbf{i}_r - \sin\alpha_2\,\mathbf{i}_\theta \tag{8.5b}$$

we obtain the r and θ components of the electric field at P to be

$$E_r = \frac{Q_1}{4\pi\epsilon}\left(\frac{\cos\alpha_1}{r_1^2} - \frac{\cos\alpha_2}{r_2^2}\right) \tag{8.6a}$$

$$E_\theta = \frac{Q_1}{4\pi\epsilon}\left(\frac{\sin\alpha_1}{r_1^2} + \frac{\sin\alpha_2}{r_2^2}\right) \tag{8.6b}$$

For infinitesimal value of the length dl of the current element, that is, for $dl \ll r$,

$$\left(\frac{\cos\alpha_1}{r_1^2} - \frac{\cos\alpha_2}{r_2^2}\right) \approx \frac{1}{r_1^2} - \frac{1}{r_2^2}$$

$$= \frac{(r_2 - r_1)(r_2 + r_1)}{r_1^2 r_2^2} \approx \frac{(dl\cos\theta)\,2r}{r^4}$$

$$= \frac{2\,dl\cos\theta}{r^3} \tag{8.7a}$$

and

$$\left(\frac{\sin\alpha_1}{r_1^2} + \frac{\sin\alpha_2}{r_2^2}\right) \approx \frac{2\sin\alpha_1}{r^2}$$

$$\approx \frac{dl\sin\theta}{r^3} \tag{8.7b}$$

where we have also used the approximations that for $dl \ll r$, $(r_2 - r_1) \approx dl\cos\theta$ and $\sin\alpha_1 \approx \dfrac{(dl/2)\sin\theta}{r}$. These are, of course, exact in the limit that

$dl \rightarrow 0$. Substituting (8.7a) and (8.7b) in (8.6a) and (8.6b), respectively, we obtain the electric field at point P due to the arrangement of the two point charges to be given by

$$E = \frac{Q_1\,dl}{4\pi\epsilon r^3}(2\cos\theta\,\mathbf{i}_r + \sin\theta\,\mathbf{i}_\theta) \tag{8.8}$$

Note that $Q_1\,dl$ is the electric dipole moment associated with the Hertzian dipole.

Using the Biot–Savart law given by (1.68), we can write the magnetic field at point P due to the infinitesimal current element in Fig. 8.3 to be

$$H = \frac{B}{\mu} = \frac{I\,dl\,\mathbf{i}_z \times \mathbf{i}_r}{4\pi r^2}$$

$$= \frac{I\,dl\sin\theta}{4\pi r^2}\mathbf{i}_\phi \tag{8.9}$$

To extend the expressions for E and H given by (8.8) and (8.9), respectively, we observe that when the charges and current vary with time, the fields also vary with time giving rise to wave propagation. The effect of a given time-variation of the source quantity is therefore felt at a point in space not instantaneously but only after a time lag. This time lag is equal to the time it takes for the wave to propagate from the source point to the observation point, that is, r/v_p, or $\beta r/\omega$, where $v_p\ (= 1/\sqrt{\mu\epsilon})$ and $\beta(= \omega\sqrt{\mu\epsilon})$ are the phase velocity and the phase constant, respectively. Thus for

$$Q_1 = \frac{I_0}{\omega}\sin\omega t \tag{8.10}$$

$$I = I_0\cos\omega t \tag{8.11}$$

we would intuitively expect the fields at point P to be given by

$$E = \frac{[(I_0/\omega)\sin\omega(t - \beta r/\omega)]\,dl}{4\pi\epsilon r^3}(2\cos\theta\,\mathbf{i}_r + \sin\theta\,\mathbf{i}_\theta)$$

$$= \frac{I_0\,dl\sin(\omega t - \beta r)}{4\pi\epsilon\omega r^3}(2\cos\theta\,\mathbf{i}_r + \sin\theta\,\mathbf{i}_\theta) \tag{8.12a}$$

$$H = \frac{[I_0\cos\omega(t - \beta r/\omega)]\,dl}{4\pi r^2}\sin\theta\,\mathbf{i}_\phi$$

$$= \frac{I_0\,dl\cos(\omega t - \beta r)}{4\pi r^2}\sin\theta\,\mathbf{i}_\phi \tag{8.12b}$$

There is, however, one thing wrong with our intuitive expectation of the fields due to the Hertzian dipole! The fields do not satisfy Maxwell's curl equations

$$\mathbf{V} \times \mathbf{E} = -\frac{\partial \mathbf{B}}{\partial t} = -\mu \frac{\partial \mathbf{H}}{\partial t} \tag{8.13a}$$

$$\mathbf{V} \times \mathbf{H} = \mathbf{J} + \frac{\partial \mathbf{D}}{\partial t} = \epsilon \frac{\partial \mathbf{E}}{\partial t} \tag{8.13b}$$

(where we have set $\mathbf{J} = 0$ in view of the perfect dielectric medium). For example, let us try the curl equation for $\mathbf{H}$. First we note from Appendix B that the expansion for the curl of a vector in spherical coordinates is

$$\mathbf{V} \times \mathbf{A} = \frac{1}{r \sin \theta}\left[\frac{\partial}{\partial \theta}(A_\phi \sin \theta) - \frac{\partial A_\theta}{\partial \phi}\right]\mathbf{i}_r$$

$$+ \frac{1}{r}\left[\frac{1}{\sin \theta}\frac{\partial A_r}{\partial \phi} - \frac{\partial}{\partial r}(r A_\phi)\right]\mathbf{i}_\theta$$

$$+ \frac{1}{r}\left[\frac{\partial}{\partial r}(r A_\theta) - \frac{\partial A_r}{\partial \theta}\right]\mathbf{i}_\phi \tag{8.14}$$

Thus

$$\mathbf{V} \times \mathbf{H} = \frac{1}{r \sin \theta}\frac{\partial}{\partial \theta}\left[\frac{I_0\, dl \cos(\omega t - \beta r)}{4\pi r^2}\sin^2 \theta\right]\mathbf{i}_r$$

$$- \frac{1}{r}\frac{\partial}{\partial r}\left[\frac{I_0\, dl \cos(\omega t - \beta r)}{4\pi r}\sin \theta\right]\mathbf{i}_\theta$$

$$= \frac{I_0\, dl \cos(\omega t - \beta r)}{4\pi r^3}(2\cos \theta\, \mathbf{i}_r + \sin \theta\, \mathbf{i}_\theta)$$

$$- \frac{\beta I_0\, dl \sin(\omega t - \beta r)}{4\pi r^2}\sin \theta\, \mathbf{i}_\theta$$

$$= \epsilon \frac{\partial \mathbf{E}}{\partial t} - \frac{\beta I_0\, dl \sin(\omega t - \beta r)}{4\pi r^2}\sin \theta\, \mathbf{i}_\theta$$

$$\neq \epsilon \frac{\partial \mathbf{E}}{\partial t} \tag{8.15}$$

The reason behind the discrepancy associated with the expressions for the fields due to the Hertzian dipole can be understood by recalling that in Sec. 4.6 we learned from considerations of the Poynting vector that the fields far from a physical antenna vary inversely with the radial distance away from the antenna. The expressions we have derived do not contain inverse distance dependent terms and hence they are not complete, thereby causing the discrepancy. The complete field expressions must contain terms involving $1/r$ in addition to those in (8.12a) and (8.12b). Since for small r, $1/r \ll 1/r^2 \ll 1/r^3$, the addition of terms involving $1/r$ and containing $\sin \theta$ to (8.12a) and (8.12b) would still maintain the fields in the region close to the dipole to be predominantly the same as those given by (8.12a) and (8.12b), while making the $1/r$ terms predominant for large r since for large r, $1/r \gg 1/r^2 \gg 1/r^3$.

Thus let us modify the expression for **H** given by (8.12b) by adding a second term containing $1/r$ in the manner

$$\mathbf{H} = \frac{I_0 \, dl \sin \theta}{4\pi} \left[\frac{\cos(\omega t - \beta r)}{r^2} + \frac{A \cos(\omega t - \beta r + \delta)}{r} \right] \mathbf{i}_\phi \qquad (8.16)$$

where A and δ are constants to be determined. Then from Maxwell's curl equation for **H**, given by (8.13b), we have

$$\epsilon \frac{\partial \mathbf{E}}{\partial t} = \nabla \times \mathbf{H} = \frac{1}{r \sin \theta} \frac{\partial}{\partial \theta} (H_\phi \sin \theta) \, \mathbf{i}_r - \frac{1}{r} \frac{\partial}{\partial r} (rH_\phi) \, \mathbf{i}_\theta$$

$$= \frac{2I_0 \, dl \cos \theta}{4\pi} \left[\frac{\cos(\omega t - \beta r)}{r^3} + \frac{A \cos(\omega t - \beta r + \delta)}{r^2} \right] \mathbf{i}_r$$

$$+ \frac{I_0 \, dl \sin \theta}{4\pi} \left[\frac{\cos(\omega t - \beta r)}{r^3} - \frac{\beta \sin(\omega t - \beta r)}{r^2} \right.$$

$$\left. - \frac{A\beta \sin(\omega t - \beta r + \delta)}{r} \right] \mathbf{i}_\theta \qquad (8.17)$$

$$\mathbf{E} = \frac{2I_0 \, dl \cos \theta}{4\pi\epsilon\omega} \left[\frac{\sin(\omega t - \beta r)}{r^3} + \frac{A \sin(\omega t - \beta r + \delta)}{r^2} \right] \mathbf{i}_r$$

$$+ \frac{I_0 \, dl \sin \theta}{4\pi\epsilon\omega} \left[\frac{\sin(\omega t - \beta r)}{r^3} + \frac{\beta \cos(\omega t - \beta r)}{r^2} \right.$$

$$\left. + \frac{A\beta \cos(\omega t - \beta r + \delta)}{r} \right] \mathbf{i}_\theta \qquad (8.18)$$

Now, from Maxwell's curl equation for **E** given by (8.13a), we have

$$\mu \frac{\partial \mathbf{H}}{\partial t} = -\nabla \times \mathbf{E} = -\frac{1}{r} \left[\frac{\partial}{\partial r} (rE_\theta) - \frac{\partial E_r}{\partial \theta} \right] \mathbf{i}_\phi$$

$$= \frac{I_0 \, dl \sin \theta}{4\pi\epsilon\omega} \left[\frac{2\beta \cos(\omega t - \beta r)}{r^3} - \frac{2A \sin(\omega t - \beta r + \delta)}{r^3} \right.$$

$$\left. - \frac{\beta^2 \sin(\omega t - \beta r)}{r^2} - \frac{A\beta^2 \sin(\omega t - \beta r + \delta)}{r} \right] \mathbf{i}_\phi \qquad (8.19)$$

$$\mathbf{H} = \frac{I_0 \, dl \sin \theta}{4\pi} \left[\frac{2 \sin(\omega t - \beta r)}{\beta r^3} + \frac{2A \cos(\omega t - \beta r + \delta)}{\beta^2 r^3} \right.$$

$$\left. + \frac{\cos(\omega t - \beta r)}{r^2} + \frac{A \cos(\omega t - \beta r + \delta)}{r} \right] \mathbf{i}_\phi \qquad (8.20)$$

We, however, have to rule out the $1/r^3$ terms in (8.20) since for small r, these terms are more predominant than the $1/r^2$ dependence required by (8.12b).

Equation (8.20) will then also be consistent with (8.16) from which we derived (8.18) and then (8.20). Thus we set

$$\frac{2 \sin (\omega t - \beta r)}{\beta r^3} + \frac{2A \cos (\omega t - \beta r + \delta)}{\beta^2 r^3} = 0 \tag{8.21}$$

which gives us

$$\delta = \frac{\pi}{2} \tag{8.22a}$$

$$A = \beta \tag{8.22b}$$

Substituting (8.22a) and (8.22b) in (8.18) and (8.20), we then have

$$\mathbf{E} = \frac{2I_0\, dl \cos \theta}{4\pi\epsilon\omega} \left[\frac{\sin (\omega t - \beta r)}{r^3} + \frac{\beta \cos (\omega t - \beta r)}{r^2} \right] \mathbf{i}_r$$

$$+ \frac{I_0\, dl \sin \theta}{4\pi\epsilon\omega} \left[\frac{\sin (\omega t - \beta r)}{r^3} + \frac{\beta \cos (\omega t - \beta r)}{r^2} \right.$$

$$\left. - \frac{\beta^2 \sin (\omega t - \beta r)}{r} \right] \mathbf{i}_\theta \tag{8.23a}$$

$$\mathbf{H} = \frac{I_0\, dl \sin \theta}{4\pi} \left[\frac{\cos (\omega t - \beta r)}{r^2} - \frac{\beta \sin (\omega t - \beta r)}{r} \right] \mathbf{i}_\phi \tag{8.23b}$$

These expressions for $\mathbf{E}$ and $\mathbf{H}$ satisfy both of Maxwell's curl equations, reduce to (8.12a) and (8.12b), respectively, for small r ($\beta r \ll 1$), and they vary inversely with r for large r ($\beta r \gg 1$). They represent the complete electromagnetic field due to the Hertzian dipole.

8.2 RADIATION RESISTANCE AND DIRECTIVITY

In the previous section we derived the expressions for the complete electromagnetic field due to the Hertzian dipole. These expressions look very complicated. Fortunately, it is seldom necessary to work with the complete field expressions because one is often interested in the field far from the dipole which is governed predominantly by the terms involving $1/r$. We, however, had to derive the complete field in order to obtain the amplitude and phase of these $1/r$ terms relative to the amplitude and phase of the current in the Hertzian dipole, since these terms alone do not satisfy Maxwell's equations. Furthermore, by going through the exercise, we learned how to solve a difficult problem through intuitive extension and reasoning based on previously gained knowledge.

Thus from (8.23a) and (8.23b), we find that for a Hertzian dipole of length dl oriented along the z axis and carrying current

$$I = I_0 \cos \omega t \qquad (8.24)$$

the electric and magnetic fields at values of r far from the dipole are given by

$$\mathbf{E} = -\frac{\beta^2 I_0 \, dl \sin \theta}{4\pi\epsilon\omega r} \sin (\omega t - \beta r) \, \mathbf{i}_\theta$$

$$= -\frac{\eta\beta I_0 \, dl \sin \theta}{4\pi r} \sin (\omega t - \beta r) \, \mathbf{i}_\theta \qquad (8.25a)$$

$$\mathbf{H} = -\frac{\beta I_0 \, dl \sin \theta}{4\pi r} \sin (\omega t - \beta r) \, \mathbf{i}_\phi \qquad (8.25b)$$

These fields are known as the "radiation fields," since they are the components of the total fields that contribute to the time-average radiated power away from the dipole (see Problem 8.6). Before we discuss the nature of these fields, let us find out quantitatively what we mean by "far from the dipole." To do this, we look at the expression for the complete magnetic field given by (8.23b) and note that the ratio of the amplitudes of the $1/r^2$ and $1/r$ terms is equal to $1/\beta r$. Hence for $\beta r \gg 1$, or $r \gg \lambda/2\pi$, the $1/r^2$ term is negligible compared to the $1/r$ term. Thus even at a distance of a few wavelengths from the dipole, the fields are predominantly radiation fields.

Returning now to the expressions for the radiation fields given by (8.25a) and (8.25b), we note that at any given point, (a) the electric field (E_θ), the magnetic field (H_ϕ), and the direction of propagation (r) are mutually perpendicular and (b) the ratio of E_θ to H_ϕ is equal to η, the intrinsic impedance of the medium, which are characteristic of uniform plane waves. The phase of the field, however, is uniform over the surfaces $r = $ constant, that is, spherical surfaces centered at the dipole, whereas the amplitude of the field is uniform over surfaces $(\sin \theta)/r = $ constant. Hence the fields are only locally uniform plane waves, that is, over any infinitesimal area normal to the r direction at a given point.

The Poynting vector due to the radiation fields is given by

$$\mathbf{P} = \mathbf{E} \times \mathbf{H}$$

$$= E_\theta \mathbf{i}_\theta \times H_\phi \mathbf{i}_\phi = E_\theta H_\phi \mathbf{i}_r$$

$$= \frac{\eta\beta^2 I_0^2 (dl)^2 \sin^2 \theta}{16\pi^2 r^2} \sin^2 (\omega t - \beta r) \, \mathbf{i}_r \qquad (8.26)$$

By evaluating the surface integral of the Poynting vector over any surface enclosing the dipole, we can find the power flow out of that surface, that is, the power "radiated" by the dipole. For convenience in evaluating the

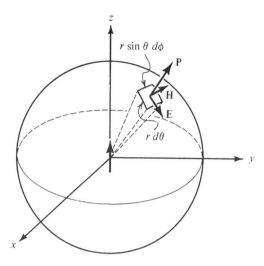

Figure 8.4. For computing the power radiated by the Hertzian dipole.

surface integral, we choose the spherical surface of radius r and centered at the dipole, as shown in Fig. 8.4. Thus noting that the differential surface area on the spherical surface is $(r\,d\theta)(r\sin\theta\,d\phi)\,\mathbf{i}_r$ or $r^2\sin\theta\,d\theta\,d\phi\,\mathbf{i}_r$, we obtain the instantaneous power radiated to be

$$
\begin{aligned}
P_{\text{rad}} &= \int_{\theta=0}^{\pi}\int_{\phi=0}^{2\pi}\mathbf{P}\cdot r^2\sin\theta\,d\theta\,d\phi\,\mathbf{i}_r \\
&= \int_{\theta=0}^{\pi}\int_{\phi=0}^{2\pi}\frac{\eta\beta^2 I_0^2(dl)^2\sin^3\theta}{16\pi^2}\sin^2(\omega t-\beta r)\,d\theta\,d\phi \\
&= \frac{\eta\beta^2 I_0^2(dl)^2}{8\pi}\sin^2(\omega t-\beta r)\int_{\theta=0}^{\pi}\sin^3\theta\,d\theta \\
&= \frac{\eta\beta^2 I_0^2(dl)^2}{6\pi}\sin^2(\omega t-\beta r) \\
&= \frac{2\pi\eta I_0^2}{3}\left(\frac{dl}{\lambda}\right)^2\sin^2(\omega t-\beta r)
\end{aligned}
\tag{8.27}
$$

The time-average power radiated by the dipole, that is, the average of P_{rad} over one period of the current variation, is

$$
\begin{aligned}
\langle P_{\text{rad}}\rangle &= \frac{2\pi\eta I_0^2}{3}\left(\frac{dl}{\lambda}\right)^2\langle\sin^2(\omega t-\beta r)\rangle \\
&= \frac{\pi\eta I_0^2}{3}\left(\frac{dl}{\lambda}\right)^2 \\
&= \frac{1}{2}I_0^2\left[\frac{2\pi\eta}{3}\left(\frac{dl}{\lambda}\right)^2\right]
\end{aligned}
\tag{8.28}
$$

We now define a quantity known as the "radiation resistance" of the antenna, denoted by the symbol R_{rad}, as the value of a fictitious resistor that dissipates the same amount of time-average power as that radiated by the antenna when a current of the same peak amplitude as that in the antenna is passed through it. Recalling that the average power dissipated in a resistor R when a current $I_0 \cos \omega t$ is passed through it is $\frac{1}{2} I_0^2 R$, we note from (8.28) that the radiation resistance of the Hertzian dipole is

$$R_{\text{rad}} = \frac{2\pi\eta}{3} \left(\frac{dl}{\lambda}\right)^2 \text{ ohms} \tag{8.29}$$

For free space, $\eta = \eta_0 = 120\pi$ ohms, and

$$R_{\text{rad}} = 80\pi^2 \left(\frac{dl}{\lambda}\right)^2 \text{ ohms} \tag{8.30}$$

As a numerical example, for (dl/λ) equal to 0.01, $R_{\text{rad}} = 80\pi^2(0.01)^2 = 0.08$ ohms. Thus for a current of peak amplitude 1 amp, the time-average radiated power is equal to 0.04 W. This indicates that a Hertzian dipole of length 0.01λ is not a very effective radiator.

We note from (8.29) that the radiation resistance and hence the radiated power are proportional to the square of the electrical length, that is, the physical length expressed in terms of wavelength, of the dipole. The result given by (8.29) is, however, valid only for small values of dl/λ since if dl/λ is not small, the amplitude of the current along the antenna can no longer be uniform and its variation must be taken into account in deriving the radiation fields and hence the radiation resistance. We shall do this in the following section for a half-wave dipole, that is, for a dipole of length equal to $\lambda/2$.

Let us now examine the directional characteristics of the radiation from the Hertzian dipole. We note from (8.25a) and (8.25b) that, for a constant r, the amplitude of the fields is proportional to $\sin \theta$. Similarly, we note from (8.26) that, for a constant r, the power density is proportional to $\sin^2 \theta$. Thus an observer wandering on the surface of an imaginary sphere centered at the dipole views different amplitudes of the fields and of the power density at different points on the surface. The situation is illustrated in Fig. 8.5(a) for the power density by attaching to different points on the spherical surface vectors having lengths proportional to the Poynting vectors at those points. It can be seen that the power density is largest for $\theta = \pi/2$, that is, in the plane normal to the axis of the dipole, and decreases continuously toward the axis of the dipole, becoming zero along the axis.

It is customary to depict the radiation characteristic by means of a "radiation pattern," as shown in Fig. 8.5(b), which can be imagined to be

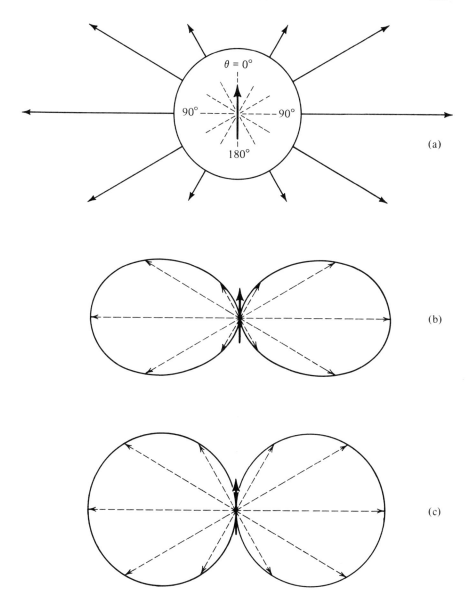

Figure 8.5. The directional characteristics of radiation from the Hertzian dipole.

obtained by shrinking the radius of the spherical surface in Fig. 8.5(a) to zero with the Poynting vectors attached to it and then joining the tips of the Poynting vectors. Thus the distance from the dipole point to a point on the radiation pattern is proportional to the power density in the direction of that point. Similarly, the radiation pattern for the fields can be drawn as shown in Fig. 8.5(c), based upon the sin θ dependence of the fields. In view of the independence of the fields from ϕ, the patterns of Fig. 8.5(b)–(c) are valid for any plane containing the axis of the dipole. In fact, the three-dimensional radiation patterns can be imagined to be the figures obtained by revolving these patterns about the dipole axis. For a general case, the radiation may also depend on ϕ and hence it will be necessary to draw a radiation pattern for the $\theta = \pi/2$ plane. Here, this pattern is merely a circle centered at the dipole.

We now define a parameter known as the "directivity" of the antenna, denoted by the symbol D, as the ratio of the maximum power density radiated by the antenna to the average power density. To elaborate on the definition of D, imagine that we take the power radiated by the antenna and distribute it equally in all directions by shortening some of the vectors in Fig. 8.5(a) and lengthening the others so that they all have equal lengths. The pattern then becomes nondirectional and the power density, which is the same in all directions, will be less than the maximum power density of the original pattern. Obviously, the more directional the radiation pattern of an antenna is, the greater is the directivity.

From (8.26), we obtain the maximum power density radiated by the Hertzian dipole to be

$$[P_r]_{\max} = \frac{\eta \beta^2 I_0^2 (dl)^2 [\sin^2 \theta]_{\max}}{16\pi^2 r^2} \sin^2 (\omega t - \beta r)$$

$$= \frac{\eta \beta^2 I_0^2 (dl)^2}{16\pi^2 r^2} \sin^2 (\omega t - \beta r) \tag{8.31}$$

By dividing the radiated power given by (8.27) by the surface area $4\pi r^2$ of the sphere of radius r, we obtain the average power density to be

$$[P_r]_{\text{av}} = \frac{P_{\text{rad}}}{4\pi r^2} = \frac{\eta \beta^2 I_0^2 (dl)^2}{24\pi^2 r^2} \sin^2 (\omega t - \beta r) \tag{8.32}$$

Thus the directivity of the Hertzian dipole is given by

$$D = \frac{[P_r]_{\max}}{[P_r]_{\text{av}}} = 1.5 \tag{8.33}$$

8.3 HALF-WAVE DIPOLE

In the previous section we found the radiation fields due to a Hertzian dipole, which is an elemental antenna of infinitesimal length. If we now have an antenna of any length having a specified current distribution, we can divide it into a series of Hertzian dipoles and by applying superposition we can find the radiation fields for that antenna. We shall illustrate this procedure in this section by considering the half-wave dipole, which is a commonly used form of antenna.

The half-wave dipole is a center-fed, straight wire antenna of length L equal to $\lambda/2$ and having the current distribution

$$I(z) = I_0 \cos \frac{\pi z}{L} \cos \omega t \qquad \text{for} - \frac{L}{2} < z < \frac{L}{2} \qquad (8.34)$$

where the dipole is assumed to be oriented along the z axis with its center at the origin, as shown in Fig. 8.6(a). As can be seen from Fig. 8.6(a), the amplitude of the current distribution varies cosinusoidally along the antenna with zeros at the ends and maximum at the center. To see how this distribution comes about, the half-wave dipole may be imagined to be the evolution of an open-circuited transmission line with the conductors folded perpendicularly to the line at points $\lambda/4$ from the end of the line. The current standing wave pattern for an open-circuited line is shown in Fig. 8.6(b). It consists of zero current at the open circuit and maximum current at $\lambda/4$ from the open circuit, that is, at points a and a'. Hence it can be seen that when the conductors are folded perpendicularly to the line at a and a', the half-wave dipole shown in Fig. 8.6(a) results.

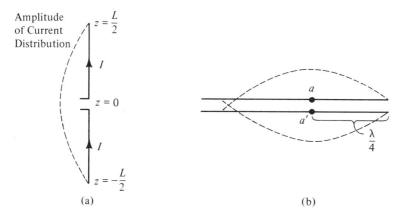

(a)   (b)

Figure 8.6. (a) Half-wave dipole. (b) Open-circuited transmission line for illustrating the evolution of the half-wave dipole.

Now to find the radiation field due to the half-wave dipole, we divide it into a number of Hertzian dipoles, each of length dz' as shown in Fig. 8.7. If we consider one of these dipoles situated at distance z' from the origin, then from (8.34) the current in this dipole is $I_0 \cos (\pi z'/L) \cos \omega t$. From (8.25a) and (8.25b), the radiation fields due to this dipole at point P situated at dis-

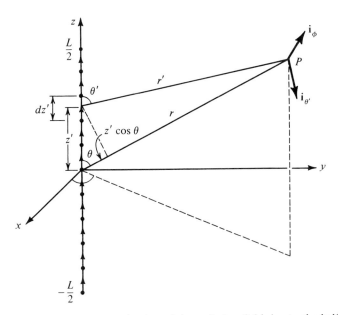

Figure 8.7. For the determination of the radiation field due to the half-wave dipole.

tance r' from it are given by

$$d\mathbf{E} = -\frac{\eta \beta I_0 \cos (\pi z'/L) \, dz' \sin \theta'}{4\pi r'} \sin (\omega t - \beta r') \, \mathbf{i}_{\theta'} \qquad (8.35a)$$

$$d\mathbf{H} = -\frac{\beta I_0 \cos (\pi z'/L) \, dz' \sin \theta'}{4\pi r'} \sin (\omega t - \beta r') \, \mathbf{i}_{\phi} \qquad (8.35b)$$

where θ' is the angle between the z axis and the line from the current element to the point P and $\mathbf{i}_{\theta'}$ is the unit vector perpendicular to that line, as shown in Fig. 8.7. The fields due to the entire current distribution of the half-wave dipole are then given by

$$\mathbf{E} - \int_{z'=-L/2}^{L/2} d\mathbf{E}$$

$$= -\int_{z'=-L/2}^{L/2} \frac{\eta \beta I_0 \cos (\pi z'/L) \sin \theta' \, dz'}{4\pi r'} \sin (\omega t - \beta r') \, \mathbf{i}_{\theta'} \qquad (8.36a)$$

$$\mathbf{H} = \int_{z'=-L/2}^{L/2} d\mathbf{H}$$

$$= -\int_{z'=-L/2}^{L/2} \frac{\beta I_0 \cos (\pi z'/L) \sin \theta' \, dz'}{4\pi r'} \sin (\omega t - \beta r') \, \mathbf{i}_{\phi} \qquad (8.36b)$$

where r', θ' and $\mathbf{i}_{\theta'}$ are functions of z'.

For radiation fields, r' is at least equal to several wavelengths and hence $\gg L$. We can therefore set $i_{\theta'} \approx i_\theta$ and $\theta' \approx \theta$ since they do not vary significantly for $-L/2 < z' < L/2$. We can also set $r' \approx r$ in the amplitude factors for the same reason, but for r' in the phase factors we substitute $r - z' \cos\theta$ since $\sin(\omega t - \beta r') = \sin(\omega t - \pi r'/L)$ can vary appreciably over the range $-L/2 < z' < L/2$. Thus we have

$$\mathbf{E} = E_\theta \mathbf{i}_\theta$$

where

$$
\begin{aligned}
E_\theta &= -\int_{z'=-L/2}^{L/2} \frac{\eta \beta I_0 \cos(\pi z'/L) \sin\theta}{4\pi r} \sin(\omega t - \beta r + \beta z' \cos\theta)\, dz' \\
&= -\frac{\eta(\pi/L)I_0 \sin\theta}{4\pi r} \int_{z'=-L/2}^{L/2} \cos\frac{\pi z'}{L} \sin\left(\omega t - \frac{\pi}{L}r + \frac{\pi}{L}z' \cos\theta\right) dz' \\
&= -\frac{\eta I_0}{2\pi r} \frac{\cos[(\pi/2)\cos\theta]}{\sin\theta} \sin\left(\omega t - \frac{\pi}{L}r\right)
\end{aligned}
\tag{8.37a}
$$

Similarly,

$$\mathbf{H} = H_\phi \mathbf{i}_\phi$$

where

$$H_\phi = -\frac{I_0}{2\pi r} \frac{\cos[(\pi/2)\cos\theta]}{\sin\theta} \sin\left(\omega t - \frac{\pi}{L}r\right) \tag{8.37b}$$

The Poynting vector due to the radiation fields of the half-wave dipole is given by

$$
\begin{aligned}
\mathbf{P} &= \mathbf{E} \times \mathbf{H} = E_\theta H_\phi \mathbf{i}_r \\
&= \frac{\eta I_0^2}{4\pi^2 r^2} \frac{\cos^2[(\pi/2)\cos\theta]}{\sin^2\theta} \sin^2\left(\omega t - \frac{\pi}{L}r\right) \mathbf{i}_r
\end{aligned}
\tag{8.38}
$$

The power radiated by the half-wave dipole is given by

$$
\begin{aligned}
P_{\text{rad}} &= \int_{\theta=0}^{\pi} \int_{\phi=0}^{2\pi} \mathbf{P} \cdot r^2 \sin\theta\, d\theta\, d\phi\, \mathbf{i}_r \\
&= \int_{\theta=0}^{\pi} \int_{\phi=0}^{2\pi} \frac{\eta I_0^2}{4\pi^2} \frac{\cos^2[(\pi/2)\cos\theta]}{\sin\theta} \sin^2\left(\omega t - \frac{\pi}{L}r\right) d\theta\, d\phi \\
&= \frac{\eta I_0^2}{\pi} \sin^2\left(\omega t - \frac{\pi}{L}r\right) \int_{\theta=0}^{\pi/2} \frac{\cos^2[(\pi/2)\cos\theta]}{\sin\theta} d\theta \\
&= \frac{0.609 \eta I_0^2}{\pi} \sin^2\left(\omega t - \frac{\pi}{L}r\right)
\end{aligned}
\tag{8.39}
$$

The time-average radiated power is

$$\langle P_{\text{rad}} \rangle = \frac{0.609 \eta I_0^2}{\pi} \left\langle \sin^2 \left(\omega t - \frac{\pi}{L} r \right) \right\rangle$$

$$= \frac{1}{2} I_0^2 \left(\frac{0.609 \eta}{\pi} \right) \tag{8.40}$$

Thus the radiation resistance of the half-wave dipole is

$$R_{\text{rad}} = \frac{0.609 \eta}{\pi} \text{ ohms} \tag{8.41}$$

For free space, $\eta = \eta_0 = 120\pi$ ohms, and

$$R_{\text{rad}} = 0.609 \times 120 = 73 \text{ ohms} \tag{8.42}$$

Turning our attention now to the directional characteristics of the half-wave dipole, we note from (8.37a) and (8.37b) that the radiation pattern for the fields is $\left[\cos \left(\frac{\pi}{2} \cos \theta \right) \right] / \sin \theta$ whereas for the power density, it is $\left[\cos^2 \left(\frac{\pi}{2} \cos \theta \right) \right] / \sin^2 \theta$. These patterns, which are sketched in Fig. 8.8(a)–(b), are slightly more directional than the corresponding patterns for the Hertzian dipole. To find the directivity of the half-wave dipole, we note from (8.38)

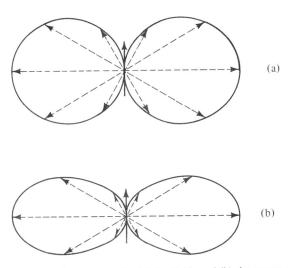

Figure 8.8. Radiation patterns for (a) the fields and (b) the power density due to the half-wave dipole.

that the maximum power density is

$$[P_r]_{max} = \frac{\eta I_0^2}{4\pi^2 r^2} \left\{ \frac{\cos^2 [(\pi/2)\cos\theta]}{\sin^2 \theta} \right\}_{max} \sin^2 \left(\omega t - \frac{\pi}{L} r \right)$$

$$= \frac{\eta I_0^2}{4\pi^2 r^2} \sin^2 \left(\omega t - \frac{\pi}{L} r \right) \tag{8.43}$$

The average power density obtained by dividing P_{rad} by $4\pi r^2$ is

$$[P_r]_{av} = \frac{0.609 \eta I_0^2}{4\pi^2 r^2} \sin^2 \left(\omega t - \frac{\pi}{L} r \right) \tag{8.44}$$

Thus the directivity of the half-wave dipole is given by

$$D = \frac{[P_r]_{max}}{[P_r]_{av}} = \frac{1}{0.609} = 1.642 \tag{8.45}$$

8.4 ANTENNA ARRAYS

In Sec. 4.5 we illustrated the principle of an antenna array by considering an array of two parallel, infinite plane current sheets of uniform densities. We learned that by appropriately choosing the spacing between the current sheets and the amplitudes and phases of the current densities, a desired radiation characteristic can be obtained. The infinite plane current sheet is, however, a hypothetical antenna for which the fields are truly uniform plane waves propagating in the one dimension normal to the sheet. Now that we have gained some knowledge of physical antennas, in this section we shall consider arrays of such antennas.

The simplest array we can consider consists of two Hertzian dipoles, oriented parallel to the z axis and situated at points on the x axis on either side of and equidistant from the origin, as shown in Fig. 8.9. We shall consider the amplitudes of the currents in the two dipoles to be equal, but we shall allow a phase difference α between them. Thus if $I_1(t)$ and $I_2(t)$ are the currents in the dipoles situated at $(d/2, 0, 0)$ and $(-d/2, 0, 0)$, respectively, then

$$I_1 = I_0 \cos \left(\omega t + \frac{\alpha}{2} \right) \tag{8.46a}$$

$$I_2 = I_0 \cos \left(\omega t - \frac{\alpha}{2} \right) \tag{8.46b}$$

For simplicity, we shall consider a point P in the xz plane and compute the field at that point due to the array of the two dipoles. To do this, we note

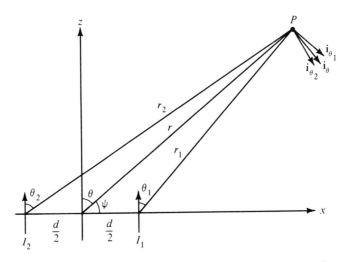

Figure 8.9. For computing the radiation field due to an array of two Hertzian dipoles.

from (8.25a) that the electric field intensities at the point P due to the individual dipoles are given by

$$\mathbf{E}_1 = -\frac{\eta\beta I_0\,dl\,\sin\theta_1}{4\pi r_1}\sin\left(\omega t - \beta r_1 + \frac{\alpha}{2}\right)\mathbf{i}_{\theta_1} \qquad (8.47a)$$

$$\mathbf{E}_2 = -\frac{\eta\beta I_0\,dl\,\sin\theta_2}{4\pi r_2}\sin\left(\omega t - \beta r_2 - \frac{\alpha}{2}\right)\mathbf{i}_{\theta_2} \qquad (8.47b)$$

where $\theta_1, \theta_2, r_1, r_2, \mathbf{i}_{\theta_1}$, and $\mathbf{i}_{\theta_2}$ are as shown in Fig. 8.9.

For $r \gg d$, that is, for points far from the array, which is the region of interest, we can set $\theta_1 \approx \theta_2 \approx \theta$ and $\mathbf{i}_{\theta_1} \approx \mathbf{i}_{\theta_2} \approx \mathbf{i}_\theta$. Also, we can set $r_1 \approx r_2 \approx r$ in the amplitude factors, but for r_1 and r_2 in the phase factors, we substitute

$$r_1 \approx r - \frac{d}{2}\cos\psi \qquad (8.48a)$$

$$r_2 \approx r + \frac{d}{2}\cos\psi \qquad (8.48b)$$

where ψ is the angle made by the line form the origin to P with the axis of the array, that is, the x axis, as shown in Fig. 8.9. Thus we obtain the resultant field to be

$$\mathbf{E} = \mathbf{E}_1 + \mathbf{E}_2$$

$$= -\frac{\eta \beta I_0 \, dl \sin \theta}{4\pi r} \left[\sin \left(\omega t - \beta r + \frac{\beta d}{2} \cos \psi + \frac{\alpha}{2} \right) \right.$$

$$\left. + \sin \left(\omega t - \beta r - \frac{\beta d}{2} \cos \psi - \frac{\alpha}{2} \right) \right] \mathbf{i}_\theta$$

$$= -\frac{2\eta \beta I_0 \, dl \sin \theta}{4\pi r} \cos \left(\frac{\beta d \cos \psi + \alpha}{2} \right) \sin \left(\omega t - \beta r \right) \mathbf{i}_\theta \qquad (8.49)$$

Comparing (8.49) with the expression for the electric field at P due to a single dipole situated at the origin, we note that the resultant field of the array is simply equal to the single dipole field multiplied by the factor $2 \cos \left(\frac{\beta d \cos \psi + \alpha}{2} \right)$, known as the "array factor." Thus the radiation pattern of the resultant field is given by the product of $\sin \theta$, which is the radiation pattern of the single dipole field, and $\left| \cos \left(\frac{\beta d \cos \psi + \alpha}{2} \right) \right|$, which is the radiation pattern of the array if the antennas were isotropic. We shall call these three patterns the "resultant pattern," the "unit pattern," and the "group pattern," respectively. It is apparent that the group pattern is independent of the nature of the individual antennas as long as they have the same spacing and carry currents having the same relative amplitudes and phase differences. It can also be seen that the group pattern is the same in any plane containing the axis of the array. In other words, the three-dimensional group pattern is simply the pattern obtained by revolving the group pattern in the xz plane about the x axis, that is, the axis of the array.

Example 8.1.   For the array of two antennas carrying currents having equal amplitudes, let us consider several pairs of d and α and investigate the group patterns.

Case 1: $d = \lambda/2$, $\alpha = 0$. The group pattern is

$$\left| \cos \left(\frac{\beta \lambda}{4} \cos \psi \right) \right| = \cos \left(\frac{\pi}{2} \cos \psi \right)$$

This is shown sketched in Fig. 8.10(a). It has maxima perpendicular to the axis of the array and nulls along the axis of the array. Such a pattern is known as a "broadside pattern."

Case 2: $d = \lambda/2$, $\alpha = \pi$. The group pattern is

$$\left| \cos \left(\frac{\beta \lambda}{4} \cos \psi + \frac{\pi}{2} \right) \right| = \left| \sin \left(\frac{\pi}{2} \cos \psi \right) \right|$$

This is shown sketched in Fig. 8.10(b). It has maxima along the axis of the array and nulls perpendicular to the axis of the array. Such a pattern is known as an "endfire pattern."

Case 3: $d = \lambda/4$, $\alpha = -\pi/2$. The group pattern is

$$\left| \cos \left(\frac{\beta\lambda}{8} \cos \psi - \frac{\pi}{4} \right) \right| = \cos \left(\frac{\pi}{4} \cos \psi - \frac{\pi}{4} \right)$$

This is shown sketched in Fig. 8.10(c). It has a maximum along $\psi = 0$ and null along $\psi = \pi$. Again, this is an endfire pattern, but directed to one side. This case is the same as the one considered in Sec. 4.5.

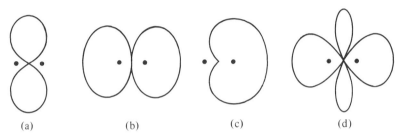

(a)   (b)   (c)   (d)

Figure 8.10. Group patterns for an array of two antennas carrying currents of equal amplitude for (a) $d = \lambda/2$, $\alpha = 0$, (b) $d = \lambda/2$, $\alpha = \pi$, (c) $d = \lambda/4$, $\alpha = -\pi/2$, and (d) $d = \lambda$, $\alpha = 0$.

Case 4: $d = \lambda$, $\alpha = 0$. The group pattern is

$$\left| \cos \left(\frac{\beta\lambda}{2} \cos \psi \right) \right| = |\cos (\pi \cos \psi)|$$

This is shown sketched in Fig. 8.10(d). It has maxima along $\psi = 0°$, $90°$, $180°$, and $270°$ and nulls along $\psi = 60°$, $120°$, $240°$, and $300°$.

Proceeding further, we can obtain the resultant pattern for an array of two Hertzian dipoles by multiplying the unit pattern by the group pattern. Thus recalling that the unit pattern for the Hertzian dipole is $\sin \theta$ in the plane of the dipole and considering values of $\lambda/2$ and 0 for d and α, respectively, for which the group pattern is given in Fig. 8.10(a), we obtain the resultant pattern in the xz plane, as shown in Fig. 8.11(a). In the xy plane, that is, the plane normal to the axis of the dipole, the unit pattern is a circle and hence the resultant pattern is the same as the group pattern, as illustrated in Fig. 8.11(b). ∎

Example 8.2. The procedure of multiplication of the unit and group patterns to obtain the resultant pattern illustrated in Example 8.1 can be extended to an array containing any number of antennas. Let us, for example, consider

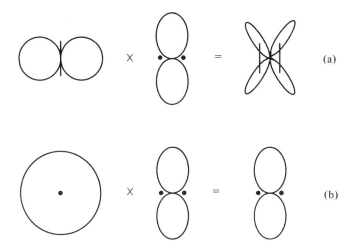

Figure 8.11. Determination of the resultant pattern of an antenna array by multiplication of unit and group patterns.

a linear array of four isotropic antennas spaced $\lambda/2$ apart and fed in phase, as shown in Fig. 8.12(a), and obtain the resultant pattern.

To obtain the resultant pattern of the four-element array, we replace it by a two-element array of spacing λ, as shown in Fig. 8.12(b), in which each element forms a unit representing a two-element array of spacing $\lambda/2$. The

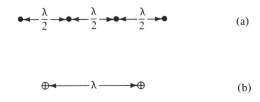

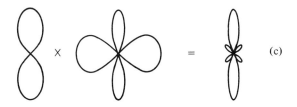

Figure 8.12. Determination of the resultant pattern for a linear array of four isotropic antennas.

unit pattern is then the pattern shown in Fig. 8.10(a). The group pattern, which is the pattern of two isotropic radiators having $d = \lambda$ and $\alpha = 0$, is the pattern given in Fig. 8.10(d). The resultant pattern of the four-element array is the product of these two patterns, as illustrated in Fig. 8.12(c). If the individual elements of the four-element array are not isotropic, then this pattern becomes the group pattern for the determination of the new resultant pattern.   ∎

8.5 IMAGE ANTENNAS

Thus far we have considered the antennas to be situated in an unbounded medium so that the waves radiate in all directions from the antenna without giving rise to reflections from any obstacles. In practice, however, we have to consider the effect of the ground even if no other obstacles are present. To do this, it is reasonable to assume that the ground is a perfect conductor. Hence in this section we shall consider antennas situated above an infinite plane, perfect conductor surface and introduce the concept of image sources, a technique that is also useful in solving static field problems.

Thus let us consider a Hertzian dipole oriented vertically and located at a height h above a plane, perfect conductor surface, as shown in Fig. 8.13(a). Since no waves can penetrate into the perfect conductor, as we learned in Sec. 5.6, the waves radiated from the dipole onto the conductor give rise to reflected waves, as shown in Fig. 8.13(a) for two directions of incidence. For a given incident wave onto the conductor surface, the angle of reflection is equal to the angle of incidence, as can be seen intuitively from the following reasons: (a) the reflected wave must propagate away from the conductor surface, (b) the apparent wavelengths of the incident and reflected waves parallel to the conductor surface must be equal, and (c) the tangential component of the resultant electric field on the conductor surface must be zero, which also determines the polarity of the reflected wave electric field.

If we now produce the directions of propagation of the two reflected waves backward, they meet at a point which is directly beneath the dipole and at the same distance h below the conductor surface as the dipole is above it. Thus the reflected waves appear to be originating from an antenna, which is the "image" of the actual antenna about the conductor surface. This image antenna must also be a vertical antenna since in order for the boundary condition of zero tangential electric field to be satisfied at all points on the conductor surface, the image antenna must have the same radiation pattern as that of the actual antenna, as shown in Fig. 8.13(a). In particular, the current in the image antenna must be directed in the same sense as that in the actual antenna to be consistent with the polarity of the reflected wave electric field. It can therefore be seen that the charges associated with the image dipole

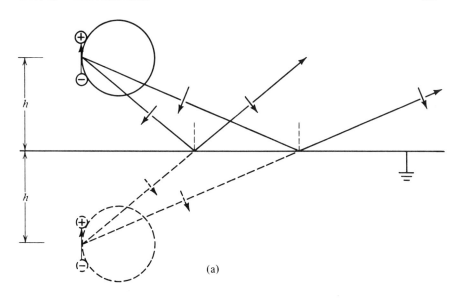

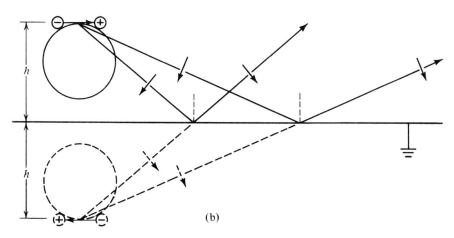

Figure 8.13. For illustrating the concept of image antennas. (a) Vertical Hertzian dipole and (b) horizontal Hertzian dipole above a plane perfect-conductor surface.

have signs opposite to those of the corresponding charges associated with the actual dipole.

A similar reasoning can be applied to the case of a horizontal dipole above a perfect conductor surface, as shown in Fig. 8.13(b). Here it can be seen that the current in the image antenna is directed in the opposite sense to that in the actual antenna. This again results in charges associated with the image dipole having signs opposite to those of the corresponding charges associated with the actual dipole. In fact, this is always the case.

From the foregoing discussion it can be seen that the field due to an antenna in the presence of the conductor is the same as the resultant field of the array formed by the actual antenna and the image antenna. There is, of course, no field inside the conductor. The image antenna is only a virtual antenna that seves to simplify the field determination outside the conductor. The simplicity arises from the fact that we can use the knowledge gained on antenna arrays in the previous section to determine the radiation pattern. Thus, for example, for a vertical Hertzian dipole at a height of $\lambda/2$ above the conductor surface, the radiation pattern in the vertical plane is the product of the unit pattern, which is the radiation pattern of the single dipole in the plane of its axis, and the group pattern corresponding to an array of two isotropic radiators spaced λ apart and fed in phase. This multiplication and the resultant pattern are illustrated in Fig. 8.14. The radiation patterns for the case of the horizontal dipole can be obtained in a similar manner.

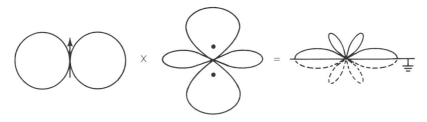

Figure 8.14. Determination of radiation pattern in the vertical plane for a vertical Hertzian dipole above a plane perfect-conductor surface.

8.6   RECEIVING ANTENNAS

Thus far we have considered the radiating, or transmitting, properties of antennas. Fortunately, it is not necessary to repeat all the derivations for the discussion of the receiving properties of antennas since reciprocity dictates that the receiving pattern of an antenna be the same as its transmitting pattern. To illustrate this in simple terms without going through the general proof of reciprocity, let us consider a Hertzian dipole situated at the origin and directed along the z axis, as shown in Fig. 8.15. We know that the radiation pattern is then given by $\sin \theta$ and that the polarization of the radiated field is such that the electric field is in the plane of the dipole axis.

To investigate the receiving properties of the Hertzian dipole, we assume that it is situated in the radiation field of a second antenna so that the incoming waves are essentially| uniform plane waves. Thus let us consider a uni-

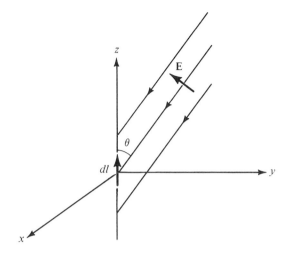

Figure 8.15. For investigating the receiving properties of a Hertzian dipole.

form plane wave with its electric field **E** in the plane of the dipole and incident
on the dipole at an angle θ with its axis, as shown in Fig. 8.15. Then the component of the incident electric field parallel to the dipole is $E \sin \theta$. Since the
dipole is infinitesimal in length, the voltage induced in the dipole, which is
the line integral of the electric field intensity along the length of the dipole,
is simply equal to $(E \sin \theta)\, dl$ or to $E\, dl \sin \theta$. This indicates that for a given
amplitude of the incident wave field, the induced voltage in the dipole is
proportional to $\sin \theta$. Furthermore, for an incident uniform plane wave
having its electric field normal to the dipole axis, the voltage induced in the
dipole is zero, that is, the dipole does not respond to polarization with electric field normal to the plane of its axis. These properties are reciprocal to
the transmitting properties of the dipole. Since an arbitrary antenna can be
decomposed into a series of Hertzian dipoles, it then follows that reciprocity
holds for an arbitrary antenna. Thus any transmitting antenna can be used
as a receiving antenna and vice versa.

    We shall now briefly consider the loop antenna, a common type of
receiving antenna. A simple form of loop antenna consists of a circular
loop of wire with a pair of terminals. We shall orient the circular loop antenna
with its axis aligned with the z axis, as shown in Fig. 8.16, and we shall assume
that it is electrically short, that is, its dimensions are small compared to the
wavelength of the incident wave, so that the spatial variation of the field
over the area of the loop is negligible. For a uniform plane wave incident on
the loop, we can find the voltage induced in the loop, that is, the line integral
of the electric field intensity around the loop, by using Faraday's law. Thus
if **H** is the magnetic field intensity associated with the wave, the magnitude

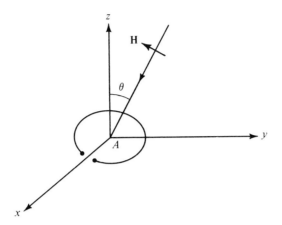

Figure 8.16. A circular loop antenna.

of the induced voltage is given by

$$
|V| = \left| -\frac{d}{dt} \int_{\substack{\text{area of} \\ \text{the loop}}} \mathbf{B} \cdot d\mathbf{S} \right|
$$

$$
= \left| -\mu \frac{d}{dt} \int_{\substack{\text{area of} \\ \text{the loop}}} \mathbf{H} \cdot d\mathbf{S}\, \mathbf{i}_z \right|
$$

$$
= \mu A \left| \frac{\partial H_z}{\partial t} \right| \tag{8.50}
$$

where A is the area of the loop. Hence the loop does not respond to a wave having its magnetic field entirely parallel to the plane of the loop, that is, normal to the axis of the loop.

For a wave having its magnetic field in the plane of the axis of the loop, and incident on the loop at an angle θ with its axis, as shown in Fig. 8.16, $H_z = H \sin \theta$ and hence the induced voltage has a magnitude

$$
|V| = \mu A \left| \frac{\partial H}{\partial t} \right| \sin \theta \tag{8.51}
$$

Thus the receiving pattern of the loop antenna is given by $\sin \theta$, same as that of a Hertzian dipole aligned with the axis of the loop antenna. The loop antenna, however, responds best to polarization with magnetic field in the plane of its axis, whereas the Hertzian dipole responds best to polarization with electric field in the plane of its axis.

Example 8.3. The directional properties of a receiving antenna can be used to locate the source of an incident signal. To illustrate the principle, let us consider two vertical loop antennas, numbered 1 and 2, situated on the x axis at $x = 0$ m and $x = 200$ m, respectively. By rotating the loop antennas about the vertical (z axis), it is found that no (or minimum) signal is induced in antenna 1 when it is in the xz plane and in antenna 2 when it is in a plane making an angle of 5° with the axis, as shown by the top view in Fig. 8.17. Let us find the location of the source of the signal.

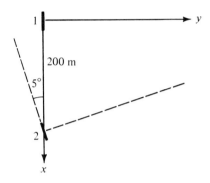

Figure 8.17. Top view of two loop antennas used to locate the source of an incident signal.

Since the receiving properties of a loop antenna are such that no signal is induced for a wave arriving along its axis, the source of the signal is located at the intersection of the axes of the two loops when they are oriented so as to receive no (or minimum) signal. From simple geometrical considerations, the source of the signal is therefore located on the y axis at $y = 200/\tan 5°$ or 2.286 km. ∎

8.7 SUMMARY

In this chapter we studied the principles of antennas. We first introduced the Hertzian dipole, which is an elemental wire antenna, and derived the complete electromagnetic field due to the Hertzian dipole by employing an intuitive approach based on the knowledge gained in the previous chapters. For a Hertzian dipole of length dl, oriented along the z axis at the origin, and carrying current

$$I(t) = I_0 \cos \omega t$$

we found the complete electromagnetic field to be given by

$$E = \frac{2I_0 \, dl \cos \theta}{4\pi\epsilon\omega}\left[\frac{\sin(\omega t - \beta r)}{r^3} + \frac{\beta \cos(\omega t - \beta r)}{r^2}\right]i_r$$

$$+ \frac{I_0 \, dl \sin \theta}{4\pi\epsilon\omega}\left[\frac{\sin(\omega t - \beta r)}{r^3} + \frac{\beta \cos(\omega t - \beta r)}{r^2}\right.$$

$$\left. - \frac{\beta^2 \sin(\omega t - \beta r)}{r}\right]i_\theta$$

$$H = \frac{I_0 \, dl \sin \theta}{4\pi}\left[\frac{\cos(\omega t - \beta r)}{r^2} - \frac{\beta \sin(\omega t - \beta r)}{r}\right]i_\phi$$

where $\beta = \omega\sqrt{\mu\epsilon}$ is the phase constant.

For $\beta r \gg 1$ or for $r \gg \lambda/2\pi$, the only important terms in the complete field expressions are the $1/r$ terms since the remaining terms are negligible compared to these terms. Thus for $r \gg \lambda/2\pi$, the Hertzian dipole fields are given by

$$E = -\frac{\eta\beta I_0 \, dl \sin \theta}{4\pi r}\sin(\omega t - \beta r)\, i_\theta$$

$$H = -\frac{\beta I_0 \, dl \sin \theta}{4\pi r}\sin(\omega t - \beta r)\, i_\phi$$

where $\eta = \sqrt{\mu/\epsilon}$ is the intrinsic impedance of the medium. These fields, known as the radiation fields, correspond to locally uniform plane waves radiating away from the dipole and, in fact, are the only components of the complete fields contributing to the time-average radiated power. We found the time-average power radiated by the Hertzian dipole to be given by

$$\langle P_{\rm rad}\rangle = \frac{1}{2}I_0^2\left[\frac{2\pi\eta}{3}\left(\frac{dl}{\lambda}\right)^2\right]$$

and identified the quantity inside the brackets to be its radiation resistance. The radiation resistance, $R_{\rm rad}$, of an antenna is the value of a fictitious resistor that will dissipate the same amount of time-average power as that radiated by the antenna when a current of the same peak amplitude as that in the antenna is passed through it. Thus for the Hertzian dipole,

$$R_{\rm rad} = \frac{2\pi\eta}{3}\left(\frac{dl}{\lambda}\right)^2$$

We then examined the directional characteristics of the radiation fields of the Hertzian dipole, as indicated by the factor $\sin \theta$ in the field expressions and hence by the factor $\sin^2 \theta$ for the power density. We discussed the radiation patterns and introduced the concept of the directivity of an antenna. The directivity, D, of an antenna is defined as the ratio of the maximum power density radiated by the antenna to the average power density. For the

Hertzian dipole,

$$D = 1.5$$

As an illustration of obtaining the radiation fields due to a wire antenna of arbitrary length and arbitrary current distribution by representing it as a series of Hertzian dipoles and using superposition, we considered the example of a half-wave dipole and derived its radiation fields. We found that for a center-fed half-wave dipole of length $L \, (= \lambda/2)$, oriented along the z axis with its center at the origin, and having the current distribution given by

$$I(z) = I_0 \cos \frac{\pi z}{L} \cos \omega t \qquad \text{for} -\frac{L}{2} < z < \frac{L}{2}$$

the radiation fields are

$$\mathbf{E} = -\frac{\eta I_0}{2\pi r} \frac{\cos\left[(\pi/2)\cos\theta\right]}{\sin\theta} \sin\left(\omega t - \frac{\pi}{L}r\right) \mathbf{i}_\theta$$

$$\mathbf{H} = -\frac{I_0}{2\pi r} \frac{\cos\left[(\pi/2)\cos\theta\right]}{\sin\theta} \sin\left(\omega t - \frac{\pi}{L}r\right) \mathbf{i}_\phi$$

From these, we sketched the radiation patterns and computed the radiation resistance and the directivity of the half-wave dipole to be

$$R_{\text{rad}} = 73 \text{ ohms} \quad \text{for free space}$$

$$D = 1.642$$

We discussed antenna arrays and introduced the technique of obtaining the resultant radiation pattern of an array by multiplication of the unit and the group patterns. For an array of two antennas having the spacing d and fed with currents of equal amplitude but differing in phase by α, we found the group pattern for the fields to be $\left|\cos\dfrac{(\beta d \cos\psi + \alpha)}{2}\right|$, where ψ is the angle measured from the axis of the array, and we investigated the group patterns for several pairs of values of d and α. For example, for $d = \lambda/2$ and $\alpha = 0$, the pattern corresponds to maximum radiation broadside to the axis of the array, whereas for $d = \lambda/2$ and $\alpha = \pi$, the pattern corresponds to maximum radiation endfire to the axis of the array.

To take into account the effect of ground on antennas, we introduced the concept of an image antenna in a perfect conductor and discussed the application of the array techniques in conjunction with the actual and the image antennas to obtain the radiation pattern of the actual antenna in the presence of the ground.

Finally, we discussed the reciprocity between the receiving and radiating properties of an antenna by considering the simple case of the Hertzian dipole.

We introduced the loop antenna and illustrated the application of its directional properties for locating the source of an incident signal.

REVIEW QUESTIONS

8.1. What is a Hertzian dipole?

8.2. Discuss the time-variations of the current and charges associated with the Hertzian dipole.

8.3. Briefly describe the spherical coordinate system.

8.4. Explain why it is simpler to use the spherical coordinate system to find the fields due to the Hertzian dipole.

8.5. Discuss the reasoning associated with the intuitive extension of the fields due to the time-varying current and charges of the Hertzian dipole based on time-varying electromagnetic phenomena.

8.6. Explain the reason for the inconsistency with Maxwell's equations of the intuitively derived fields due to the time-varying current and charges of the Hertzian dipole.

8.7. Briefly outline the reasoning used for the removal of the inconsistency with Maxwell's equations of the intuitively derived fields due to the Hertzian dipole.

8.8. Discuss the characteristics of the complete electromagnetic field due to the Hertzian dipole.

8.9. Consult an appropriate reference book and compare the procedure used for obtaining the electromagnetic field due to the Hertzian dipole with the procedure used here.

8.10. What are radiation fields? Why are they important?

8.11. Discuss the characteristics of the radiation fields.

8.12. Define the radiation resistance of an antenna.

8.13. Why is the expression for the radiation resistance of a Hertzian dipole not valid for a linear antenna of any length?

8.14. Explain why power lines are not effective radiators.

8.15. What is a radiation pattern?

8.16. Discuss the radiation pattern for the power density due to the Hertzian dipole.

8.17. Define the directivity of an antenna. What is the directivity of a Hertzian dipole?

8.18. What is the directivity of a fictitious antenna that radiates equally in all directions into one hemisphere?

8.19. How do you find the radiation fields due to an antenna of arbitrary length and arbitrary current distribution?

8.20. Discuss the evolution of the half-wave dipole from an open-circuited transmission line.

8.21. Justify the approximations involved in evaluating the integrals in the determination of the radiation fields due to the half-wave dipole.

8.22. What are the values of the radiation resistance and the directivity for a half-wave dipole?

8.23. What is an antenna array?

8.24. Justify the approximations involved in the determination of the resultant field of an array of two antennas.

8.25. Why is it that the distances r_1 and r_2 in the phase factors in Eqs. (8.47a) and (8.47b) cannot be set equal to r, but the same quantities in the amplitude factors can be set equal to r?

8.26. What is an array factor? Provide a physical explanation for the array factor.

8.27. Discuss the concept of unit and group patterns and their multiplication to obtain the resultant pattern of an array.

8.28. Distinguish between broadside and endfire radiation patterns.

8.29. Discuss the concept of an image antenna to find the field of an antenna in the vicinity of a perfect conductor.

8.30. What determines the sense of the current flow in an image antenna relative to that in the actual antenna?

8.31. How does the concept of an image antenna simplify the determination of the radiation pattern of an antenna above a perfect conductor surface?

8.32. Discuss the reciprocity associated with the transmitting and receiving properties of an antenna. Can you think of a situation in which reciprocity does not hold?

8.33. What is the receiving pattern of a loop antenna?

8.34. How should you orient a loop antenna to receive (a) a maximum signal and (b) a minimum signal?

8.35. Discuss the application of the directional receiving properties of a loop antenna in the location of the source of a radio signal.

8.36. How would you determine the direction of arrival of a radio signal by employing an array of two antennas located in the plane of incidence of the signal?

PROBLEMS

8.1. The electric dipole moment associated with a Hertzian dipole of length 0.1 m is given by

$$\mathbf{p} = 10^{-9} \sin 2\pi \times 10^7 t \, \mathbf{i}_z \text{ C-m}$$

Find the current in the dipole.

8.2. Evaluate the curl of **E** given by Eq. (8.12a) and show that it is not equal to $-\mu \dfrac{\partial \mathbf{H}}{\partial t}$ where **H** is given by Eq. (8.12b).

8.3. Show that in the limit $\omega \longrightarrow 0$, the complete field expressions given by Eqs. (8.23a) and (8.23b) tend to Eqs. (8.12a) and (8.12b), respectively.

8.4. Show that the radiation fields given by Eqs. (8.25a) and (8.25b) do not by themselves satisfy both of Maxwell's curl equations.

8.5. Find the value of r at which the amplitude of the radiation field term in Eq. (8.23a) is equal to the resultant amplitude of the remaining two terms in the θ component.

8.6. Obtain the Poynting vector corresponding to the complete electromagnetic field due to the Hertzian dipole and show that the $1/r^3$ and $1/r^2$ terms do not contribute to the time-average power flow from the dipole.

8.7. A straight wire of length 1 m situated in free space carries a uniform current $10 \cos 4\pi \times 10^6 t$ amp. (a) Calculate the amplitude of the electric field intensity at a distance of 10 km in a direction at right angle to the wire. (b) Calculate the radiation resistance and the time-average power radiated by the wire.

8.8. Compute the radiation resistance per kilometer length of a straight power-line wire. Comment on the effectiveness of the power line as a radiator.

8.9. Find the time-average power required to be radiated by a Hertzian dipole in order to produce an electric field intensity of peak amplitude 0.01 V/m at a distance of 1 km broadside to the dipole.

8.10. A Hertzian dipole situated at the origin and oriented along the x axis carries a current $I_1 = I_0 \cos \omega t$. A second Hertzian dipole, having the same length and also situated at the origin but oriented along the z axis, carries a current $I_2 = I_0 \sin \omega t$. Find the polarization of the radiated electric field at (a) a point on the x axis, (b) a point on the z axis, (c) a point on the y axis, and (d) a point on the line $x = y$, $z = 0$.

8.11. Find the ratio of the currents in two antennas having directivities D_1 and D_2 and radiation resistances $R_{\text{rad } 1}$ and $R_{\text{rad } 2}$ for which the maximum radiated power densities are equal.

8.12. The radiation pattern for the power density of an antenna located at the origin is dependent on θ in the manner $\sin^4 \theta$. Find the directivity of the antenna.

8.13. The radiation pattern for the power density of an antenna located at the origin is dependent on θ in the manner $\sin^2 \theta \cos^2 \theta$. Find the directivity of the antenna.

8.14. In Fig. 8.7, let $L = 2$ m, and investigate the variations of r' and $\pi r'/L$ for $-L/2 < z' < L/2$ for (a) a point in the xy plane at $r = 1$ km and (b) a point on the z axis at $r = 1$ km.

8.15. By dividing the interval $0 < \theta < \pi/2$ into nine equal parts, numerically compute the value of

$$\int_{\theta=0}^{\pi/2} \frac{\cos^2\left[(\pi/2)\cos\theta\right]}{\sin\theta}\, d\theta$$

8.16. Complete the missing steps in the evaluation of the integral in Eq. (8.37a).

8.17. Find the time-average power required to be radiated by a half-wave dipole in order to produce an electric field intensity of peak amplitude 0.01 V/m at a distance of 1 km broadside to the dipole.

8.18. Compare the correct value of the radiation resistance of the half-wave dipole with the incorrect value that would result from using the expression for the radiation resistance of the Hertzian dipole.

8.19. A short dipole is a center-fed straight wire antenna having a length small compared to a wavelength. The amplitude of the current distribution can then be approximated as decreasing linearly from a maximum at the center to zero at the ends. Thus for a short dipole of length L lying along the z axis between $z = -L/2$ and $z = L/2$, the current distribution is given by

$$I(z) = \begin{cases} I_0\left(1 + \dfrac{2z}{L}\right)\cos\omega t & \text{for } -\dfrac{L}{2} < z < 0 \\[2ex] I_0\left(1 - \dfrac{2z}{L}\right)\cos\omega t & \text{for } 0 < z < \dfrac{L}{2} \end{cases}$$

(a) Obtain the radiation fields of the short dipole. (b) Find the radiation resistance and the directivity of the short dipole.

8.20. For the array of two antennas of Example 8.1, find and sketch the group patterns for (a) $d = \lambda/4$, $\alpha = \pi/2$ and (b) $d = 2\lambda$, $\alpha = 0$.

8.21. For the array of two antennas of Example 8.1, having $d = \lambda/4$, find the value of α for which the maxima of the group pattern are directed along $\psi = \pm 60°$, and then sketch the group pattern.

8.22. Obtain the resultant pattern for a linear array of eight isotropic antennas, spaced $\lambda/2$ apart, carrying equal currents, and fed in phase.

8.23. Obtain the resultant pattern for a linear array of three isotropic antennas, spaced $\lambda/2$ apart, carrying unequal currents in the ratio 1:2:1, and fed in phase.

8.24. For the array of two Hertzian dipoles of Fig. 8.9, find and sketch the resultant pattern in the xz plane for $d = \lambda/2$ and $\alpha = \pi$.

8.25. For the array of two Hertzian dipoles of Fig. 8.9, find and sketch the resultant pattern in the xz plane for $d = \lambda/4$ and $\alpha = -\pi/2$.

8.26. For a horizontal Hertzian dipole at a height $\lambda/4$ above a plane, perfect conductor surface, find and sketch the radiation pattern in (a) the vertical plane perpendicular to the axis of the antenna and (b) the vertical plane containing the axis of the antenna.

8.27. For a vertical antenna of length $\lambda/4$ above a plane, perfect conductor surface, find (a) the radiation pattern in the vertical plane and (b) the directivity.

8.28. A Hertzian dipole is situated parallel to a corner reflector, which is an arrangement of two plane, perfect conductors at right angles to each other, as shown by the cross-sectional view in Fig. 8.18. (a) Locate the image antennas required to satisfy the boundary conditions on the corner reflector surface. (b) Find and sketch the radiation pattern in the cross-sectional plane.

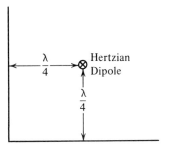

Figure 8.18. For Problem 8.28.

8.29. If the Hertzian dipole in Fig. 8.18 is situated at a distance $\lambda/2$ from the corner and equidistant from the two planes, find the ratio of the radiation field at a point broadside to the dipole and away from the corner to the radiation field in the absence of the corner reflector.

8.30. An arrangement of two identical Hertzian dipoles situated at the origin and oriented along the x and y axes, known as the turnstile antenna, is used for receiving circularly polarized signals arriving along the z axis. Determine how you would combine the voltages induced in the two dipoles so that the turnstile antenna is responsive to circular polarization rotating in the clockwise sense as viewed by the antenna but not to that of the counterclockwise sense of rotation.

8.31. A vertical loop antenna of area 1 m² is situated at a distance of 10 km from a vertical wire antenna of length $\lambda/4$ above a perfectly conducting ground (directivity = 3.28; see Problem 8.27) radiating at 2 MHz. The loop antenna is oriented so as to maximize the signal induced in it. For a time-average radiated power of 10 kW, find the amplitude of the voltage induced in the loop antenna.

8.32. An interferometer consists of an array of two identical antennas with spacing d. Show that for a uniform plane wave incident on the array at an angle ψ to the axis of the array, as shown in Fig. 8.19, the phase difference $\Delta\phi$ between the voltage induced in antenna 1 and the voltage induced in antenna 2 is

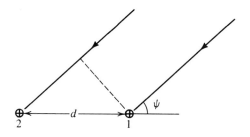

Figure 8.19. For Problem 8.32.

$\dfrac{2\pi}{\lambda}d\cos\psi$, where λ is the wavelength of the incident wave. For $d = 2\lambda$ and for $\Delta\phi = 30°$, find all possible values of ψ. Take into account the fact that the phase measurement is ambiguous by the amount $\pm 2n\pi$ where n is an integer.

9. STATIC AND QUASISTATIC FIELDS

In the preceding five chapters we studied the principles of propagation, transmission, and radiation of electromagnetic waves. These phenomena are based upon the interaction between the time-varying electric and magnetic fields as indicated by Maxwell's equations that we introduced in Chaps. 2 and 3. To conclude our study of the elements of engineering electromagnetics, we shall devote this chapter to static fields, that is, fields independent of time, and quasistatic fields, which are low-frequency extensions of static fields. Since we have already built up many of the concepts and tools of engineering electromagnetics in the previous chapters, our goal in this chapter will be to start with Maxwell's equations, set the time variations equal to zero, and proceed with a logical development of the topics.

Perhaps the most important quantity in the study of static fields is the electric potential, a scalar that is related to the static electric field intensity through a vector operation known as the "gradient." We shall introduce the gradient and the electric potential at the outset and illustrate the computation of the static electric field through the use of the potential concept. We shall then consider the solution of two important differential equations involving the potential, known as "Poisson's equation" and "Laplace's equation," which have applications in electronic devices, among others. We shall then extend our study to the quasistatic case, illustrating the determination of low-frequency behavior of physical structures via the quasistatic field approach, and we shall finally conclude the chapter with a discussion of magnetic circuits.

9.1 GRADIENT AND ELECTRIC POTENTIAL

For static fields, $\partial/\partial t = 0$, and Maxwell's curl equations given for time-varying fields by

$$\nabla \times \mathbf{E} = -\frac{\partial \mathbf{B}}{\partial t} \tag{9.1}$$

$$\nabla \times \mathbf{H} = \mathbf{J} + \frac{\partial \mathbf{D}}{\partial t} \tag{9.2}$$

reduce to

$$\nabla \times \mathbf{E} = 0 \tag{9.3}$$

$$\nabla \times \mathbf{H} = \mathbf{J} \tag{9.4}$$

respectively. Equation (9.3) states that the curl of the static electric field is equal to zero. If the curl of a vector is zero, then that vector can be expressed as the "gradient" of a scalar, since the curl of the gradient of a scalar is identically equal to zero. The gradient of a scalar, say Φ, denoted $\nabla\Phi$ (del Φ) is given in Cartesian coordinates by

$$\nabla\Phi = \left(\mathbf{i}_x \frac{\partial}{\partial x} + \mathbf{i}_y \frac{\partial}{\partial y} + \mathbf{i}_z \frac{\partial}{\partial z}\right)\Phi$$

$$= \frac{\partial\Phi}{\partial x}\mathbf{i}_x + \frac{\partial\Phi}{\partial y}\mathbf{i}_y + \frac{\partial\Phi}{\partial z}\mathbf{i}_z \tag{9.5}$$

The curl of $\nabla\Phi$ is then given by

$$\nabla \times \nabla\Phi = \begin{vmatrix} \mathbf{i}_x & \mathbf{i}_y & \mathbf{i}_z \\ \dfrac{\partial}{\partial x} & \dfrac{\partial}{\partial y} & \dfrac{\partial}{\partial z} \\ (\nabla\Phi)_x & (\nabla\Phi)_y & (\nabla\Phi)_z \end{vmatrix}$$

$$= \begin{vmatrix} \mathbf{i}_x & \mathbf{i}_y & \mathbf{i}_z \\ \dfrac{\partial}{\partial x} & \dfrac{\partial}{\partial y} & \dfrac{\partial}{\partial z} \\ \dfrac{\partial\Phi}{\partial x} & \dfrac{\partial\Phi}{\partial y} & \dfrac{\partial\Phi}{\partial z} \end{vmatrix}$$

$$= 0 \tag{9.6}$$

To discuss the physical interpretation of the gradient, we note that

$$\nabla\Phi \cdot d\mathbf{l} = \left(\frac{\partial\Phi}{\partial x}\mathbf{i}_x + \frac{\partial\Phi}{\partial y}\mathbf{i}_y + \frac{\partial\Phi}{\partial z}\mathbf{i}_z\right) \cdot (dx\,\mathbf{i}_x + dy\,\mathbf{i}_y + dz\,\mathbf{i}_z)$$

$$= \frac{\partial\Phi}{\partial x}dx + \frac{\partial\Phi}{\partial y}dy + \frac{\partial\Phi}{\partial z}dz$$

$$= d\Phi \tag{9.7}$$

Let us consider a surface on which Φ is equal to a constant, say Φ_0, and a point P on that surface as shown in Fig. 9.1(a). If we now consider another point Q_1 on the same surface and an infinitesimal distance away from P, $d\Phi$ between these two points is zero since Φ is constant on the surface. Thus for the vector $d\mathbf{l}_1$ drawn from P to Q_1, $[\nabla\Phi]_P \cdot d\mathbf{l}_1 = 0$ and hence $[\nabla\Phi]_P$ is perpendicular to $d\mathbf{l}_1$. Since this is true for all points Q_1, Q_2, Q_3, . . . on the

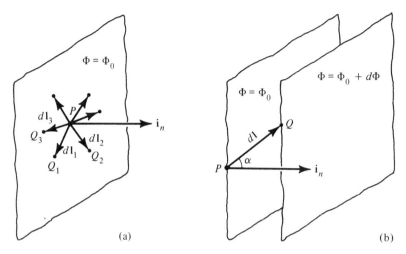

Figure 9.1. For discussing the physical interpretation of the gradient of a scalar function.

constant Φ surface, it follows that $[\nabla\Phi]_P$ must be normal to all possible infinitesimal displacement vectors $d\mathbf{l}_1$, $d\mathbf{l}_2$, $d\mathbf{l}_3$, . . . drawn at P and hence is normal to the surface. Denoting $\mathbf{i}_n$ to be the unit normal vector to the surface at P, we then have

$$[\nabla\Phi]_P = |\nabla\Phi|_P \, \mathbf{i}_n \qquad (9.8)$$

Let us now consider two surfaces on which Φ is constant, having values Φ_0 and $\Phi_0 + d\Phi$, as shown in Fig. 9.1(b). Let P and Q be points on the $\Phi = \Phi_0$ and $\Phi = \Phi_0 + d\Phi$ surfaces, respectively, and $d\mathbf{l}$ be the vector drawn from P to Q. Then from (9.7) and (9.8),

$$d\Phi = [\nabla\Phi]_P \cdot d\mathbf{l}$$
$$= |\nabla\Phi|_P \, \mathbf{i}_n \cdot d\mathbf{l}$$
$$= |\nabla\Phi|_P \, dl \cos \alpha \qquad (9.9)$$

where α is the angle between $\mathbf{i}_n$ at P and $d\mathbf{l}$. Thus

$$|\nabla\Phi|_P = \frac{d\Phi}{dl \cos \alpha} \qquad (9.10)$$

Since $dl \cos \alpha$ is the distance between the two surfaces along $\mathbf{i}_n$ and hence is the shortest distance between them, it follows that $|\nabla \Phi|_P$ is the maximum rate of increase of Φ at the point P. Thus the gradient of a scalar function Φ at a point is a vector having magnitude equal to the maximum rate of increase of Φ at that point and is directed along the direction of the maximum rate of increase, which is normal to the constant Φ surface passing through that point. This concept of the gradient of a scalar function is often utilized to find a unit vector normal to a given surface. We shall illustrate this by means of an example.

Example 9.1.   Let us find the unit vector normal to the surface $y = x^2$ at the point $(2, 4, 1)$ by using the concept of the gradient of a scalar.

Writing the equation for the surface as

$$x^2 - y = 0$$

we note that the scalar function that is constant on the surface is given by

$$\Phi(x, y, z) = x^2 - y$$

The gradient of the scalar function is then given by

$$
\begin{aligned}
\nabla \Phi &= \nabla(x^2 - y) \\
&= \frac{\partial(x^2 - y)}{\partial x}\mathbf{i}_x + \frac{\partial(x^2 - y)}{\partial y}\mathbf{i}_y + \frac{\partial(x^2 - y)}{\partial z}\mathbf{i}_z \\
&= 2x\mathbf{i}_x - \mathbf{i}_y
\end{aligned}
$$

The value of the gradient at the point $(2, 4, 1)$ is $2(2)\mathbf{i}_x - \mathbf{i}_y = 4\mathbf{i}_x - \mathbf{i}_y$. Thus the required unit vector is

$$\mathbf{i}_n = \pm \frac{4\mathbf{i}_x - \mathbf{i}_y}{|4\mathbf{i}_x - \mathbf{i}_y|} = \pm\left(\frac{4}{\sqrt{17}}\mathbf{i}_x - \frac{1}{\sqrt{17}}\mathbf{i}_y\right) \qquad \blacksquare$$

Returning to Maxwell's curl equation for the static electric field given by (9.3), we can now express $\mathbf{E}$ as the gradient of a scalar function, say, Φ. The question then arises as to what this scalar function is. To obtain the answer, let us consider a region of static electric field. Then we can draw a set of surfaces orthogonal everywhere to the field lines, as shown in Fig. 9.2. These surfaces correspond to the constant Φ surfaces. Since on any such surface $\mathbf{E} \cdot d\mathbf{l} = 0$, no work is involved in the movement of a test charge from one point to another on the surface. Such surfaces are known as the "equipotential surfaces." Since they are orthogonal to the field lines, they may physically be occupied by conductors without affecting the field distribution.

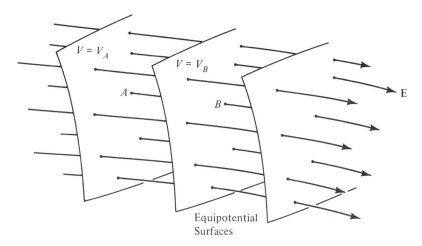

Figure 9.2. A set of equipotential surfaces in a region of static electric field.

Movement of a test charge from a point, say A, on one equipotential surface to a point, say B, on another equipotential surface involves an amount of work per unit charge equal to $\int_A^B \mathbf{E} \cdot d\mathbf{l}$ to be done by the field. This quantity is known as the "electric potential difference" between the points A and B and is denoted by the symbol $[V]_A^B$. It has the units of volts. There is a potential drop from A to B if work is done by the field and a potential rise if work is done against the field by an external agent. The situation is similar to that in the earth's gravitational field for which there is a potential drop associated with the movement of a mass from a point of higher elevation to a point of lower elevation and a potential rise for just the opposite case.

It is convenient to define an "electric potential" associated with each point. The potential at point A, denoted V_A, is simply the potential difference between point A and a reference point, say O. It is the amount of work per unit charge done by the field in connection with the movement of a test charge from A to O, or the amount of work per unit charge done against the field by an external agent in moving the test charge from O to A. Thus

$$V_A = \int_A^O \mathbf{E} \cdot d\mathbf{l} = -\int_O^A \mathbf{E} \cdot d\mathbf{l} \qquad (9.11)$$

and

$$[V]_A^B = \int_A^B \mathbf{E} \cdot d\mathbf{l} = \int_A^O \mathbf{E} \cdot d\mathbf{l} + \int_O^B \mathbf{E} \cdot d\mathbf{l}$$

$$= \int_A^O \mathbf{E} \cdot d\mathbf{l} - \int_B^O \mathbf{E} \cdot d\mathbf{l}$$

$$= V_A - V_B \qquad (9.12)$$

If we now consider points A and B to be separated by infinitesimal length $d\mathbf{l}$ from A to B, then the incremental potential drop from A to B is $\mathbf{E}_A \cdot d\mathbf{l}$, or the incremental potential rise dV along the length $d\mathbf{l}$ is given by

$$dV = -\mathbf{E}_A \cdot d\mathbf{l} \tag{9.13}$$

Writing

$$dV = [\boldsymbol{\nabla} V]_A \cdot d\mathbf{l} \tag{9.14}$$

in accordance with (9.7), we then have

$$[\boldsymbol{\nabla} V]_A \cdot d\mathbf{l} = -\mathbf{E}_A \cdot d\mathbf{l} \tag{9.15}$$

Since (9.15) is true at any point A in the static electric field, it follows that

$$\mathbf{E} = -\boldsymbol{\nabla} V \tag{9.16}$$

Thus we have obtained the result that the static electric field is the negative of the gradient of the electric potential.

Before proceeding further we note that the potential difference we have defined here has the same meaning as the voltage between two points, defined in Sec. 2.1. We, however, recall that the voltage between two points A and B in a time-varying field is in general dependent on the path followed from A to B to evaluate $\int_A^B \mathbf{E} \cdot d\mathbf{l}$ since according to Faraday's law

$$\oint_C \mathbf{E} \cdot d\mathbf{l} = -\frac{d}{dt} \int_S \mathbf{B} \cdot d\mathbf{S} \tag{9.17}$$

is not in general equal to zero. On the other hand, the potential difference (or voltage) between two points A and B in a static electric field is independent of the path followed from A to B to evaluate $\int_A^B \mathbf{E} \cdot d\mathbf{l}$ since for static fields, $\partial/\partial t = 0$, and (9.17) reduces to

$$\oint_C \mathbf{E} \cdot d\mathbf{l} = 0 \tag{9.18}$$

Thus the potential difference between two points in a static electric field has a unique value. Fields for which the line integral around a closed path is zero are known as "conservative" fields. The static electric field is a conservative field. The earth's gravitational field is another example of a conservative field since the work done in moving a mass around a closed path is equal to zero.

Returning now to the discussion of electric potential, let us consider the electric field of a point charge and investigate the electric potential due to the point charge. To do this, we recall from Sec. 1.5 that the electric field intensity due to a point charge Q is directed radially away from the point charge and its magnitude is $Q/4\pi\epsilon_0 R^2$ where R is the radial distance from the point charge. Since the equipotential surfaces are everywhere orthogonal to the field lines, it then follows that they are spherical surfaces centered at the point charge, as shown by the cross-sectional view in Fig. 9.3. If we now consider two equipotential surfaces of radii R and $R + dR$, the potential drop from the surface of radius R to the surface of radius $R + dR$ is $\dfrac{Q}{4\pi\epsilon_0 R^2} dR$ or, the incremental potential rise dV is given by

$$dV = -\frac{Q}{4\pi\epsilon_0 R^2} dR$$

$$= d\left(\frac{Q}{4\pi\epsilon_0 R} + C\right) \tag{9.19}$$

where C is a constant. Thus

$$V(R) = \frac{Q}{4\pi\epsilon_0 R} + C \tag{9.20}$$

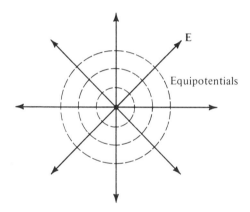

Figure 9.3. Cross-sectional view of equipotential surfaces and electric field lines for a point charge.

We can conveniently set C equal to zero by noting that it is equal to $V(\infty)$ and by choosing $R = \infty$ for the reference point. Thus we obtain the electric potential due to a point charge Q to be

$$V = \frac{Q}{4\pi\epsilon_0 R} \tag{9.21}$$

We note that the potential drops off inversely with the radial distance away from the point charge.

Equation (9.21) is often the starting point for the computation of the potential field due to static charge distributions and the subsequent determination of the electric field by using (9.16). We shall illustrate this by considering the case of the electric dipole in the following example and we shall include a few other cases in the problems.

Example 9.2.   As we have learned in Sec. 5.2, the electric dipole consists of two equal and opposite point charges. Let us consider a static electric dipole consisting of point charges Q and $-Q$ situated on the z axis at $z = d/2$ and $z = -d/2$, respectively, as shown in Fig. 9.4(a) and find the potential and hence the electric field at distances far from the dipole.

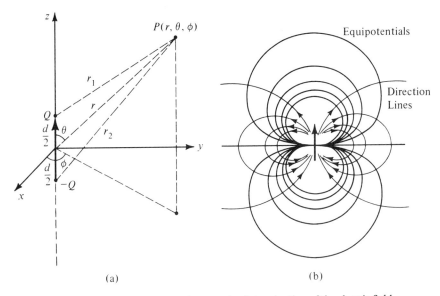

(a)   (b)

Figure 9.4. (a) Geometry pertinent to the determination of the electric field due to an electric dipole. (b) Cross sections of equipotential surfaces and direction lines of the electric field for the electric dipole.

First we note that in view of the symmetry associated with the dipole around the z axis, it is convenient to use the spherical coordinate system. As discussed in Appendix A, the spherical coordinates of a point P are the distance r from the origin O to the point P, the angle θ which the line OP makes with the z axis, and the angle ϕ which the line from the origin to the projection of P onto the xy plane makes with the x axis as shown in Fig. 9.4(a). Denoting the distance from the point charge Q to P to be r_1 and the distance from the point charge $-Q$ to P to be r_2, we write the expression for the electric potential at P due to the electric dipole as

$$V = \frac{Q}{4\pi\epsilon_0 r_1} + \frac{-Q}{4\pi\epsilon_0 r_2}$$

$$= \frac{Q}{4\pi\epsilon_0}\left(\frac{1}{r_1} - \frac{1}{r_2}\right)$$

For a point P far from the dipole, that is, for $r \gg d$, the lines drawn from the two charges to the point are almost parallel. Hence

$$r_1 \approx r - \frac{d}{2}\cos\theta$$

$$r_2 \approx r + \frac{d}{2}\cos\theta$$

and

$$\frac{1}{r_1} - \frac{1}{r_2} = \frac{r_2 - r_1}{r_1 r_2} \approx \frac{d\cos\theta}{r^2}$$

so that

$$V \approx \frac{Qd\cos\theta}{4\pi\epsilon_0 r^2} = \frac{\mathbf{p} \cdot \mathbf{i}_r}{4\pi\epsilon_0 r^2}$$

where $\mathbf{p} = Qd\mathbf{i}_z$ is the dipole moment of the electric dipole. Thus the potential field of the electric dipole drops off inversely with the square of the distance from the dipole.

Now, from (9.16) and noting from Appendix B that the gradient of a scalar in spherical coordinates is given by

$$\nabla\Phi = \frac{\partial\Phi}{\partial r}\mathbf{i}_r + \frac{1}{r}\frac{\partial\Phi}{\partial\theta}\mathbf{i}_\theta + \frac{1}{r\sin\theta}\frac{\partial\Phi}{\partial\phi}\mathbf{i}_\phi$$

we obtain the electric field intensity due to the dipole to be

$$\mathbf{E} = -\nabla V = -\frac{\partial}{\partial r}\left(\frac{Qd\cos\theta}{4\pi\epsilon_0 r^2}\right)\mathbf{i}_r - \frac{1}{r}\frac{\partial}{\partial\theta}\left(\frac{Qd\cos\theta}{4\pi\epsilon_0 r^2}\right)\mathbf{i}_\theta$$

$$= \frac{Qd}{4\pi\epsilon_0 r^3}(2\cos\theta\,\mathbf{i}_r + \sin\theta\,\mathbf{i}_\theta)$$

We note that this result agrees with the one obtained directly in (8.8) in Sec. 8.1.

Finally, a sketch of the direction lines of the electric field and of the cross sections of the equipotential surfaces ($\cos\theta/r^2 = \text{constant}$) is shown in Fig. 9.4(b). Although it is possible to derive the equation for the direction lines, it is not essential to do so since they can be sketched by recognizing that (a) they must originate from the positive charge and end on the negative charge and (b) they must be everywhere perpendicular to the equipotential surfaces. ∎

9.2 POISSON'S EQUATION

In the previous section we learned that for the static electric field, $\nabla \times \mathbf{E}$ is equal to zero, and hence

$$\mathbf{E} = -\nabla V$$

Substituting this result into Maxwell's divergence equation for **D**, and assuming ϵ to be uniform, we obtain

$$\nabla \cdot \mathbf{D} = \nabla \cdot \epsilon \mathbf{E} = \epsilon \nabla \cdot \mathbf{E}$$
$$= \epsilon \nabla \cdot (-\nabla V) = \rho$$

or

$$\nabla \cdot \nabla V = -\frac{\rho}{\epsilon}$$

The quantity $\nabla \cdot \nabla V$ is known as the "Laplacian" of V, denoted $\nabla^2 V$ (del squared V). Thus we have

$$\nabla^2 V = -\frac{\rho}{\epsilon} \qquad (9.22)$$

This equation is known as "Poisson's equation." It governs the relationship between the volume charge density ρ in a region and the potential in that region. In Cartesian coordinates,

$$\nabla^2 V = \nabla \cdot \nabla V$$
$$= \left(\mathbf{i}_x \frac{\partial}{\partial x} + \mathbf{i}_y \frac{\partial}{\partial y} + \mathbf{i}_z \frac{\partial}{\partial z} \right) \cdot \left(\frac{\partial V}{\partial x} \mathbf{i}_x + \frac{\partial V}{\partial y} \mathbf{i}_y + \frac{\partial V}{\partial z} \mathbf{i}_z \right)$$
$$= \frac{\partial^2 V}{\partial x^2} + \frac{\partial^2 V}{\partial y^2} + \frac{\partial^2 V}{\partial z^2} \qquad (9.23)$$

and Poisson's equation becomes

$$\frac{\partial^2 V}{\partial x^2} + \frac{\partial^2 V}{\partial y^2} + \frac{\partial^2 V}{\partial z^2} = -\frac{\rho}{\epsilon} \qquad (9.24)$$

For the one-dimensional case in which V varies with x only, $\partial^2 V/\partial y^2$ and $\partial^2 V/\partial z^2$ are both equal to zero, and (9.24) reduces to

$$\frac{\partial^2 V}{\partial x^2} = \frac{d^2 V}{dx^2} = -\frac{\rho}{\epsilon} \qquad (9.25)$$

To illustrate an example of the application of (9.25), let us consider the space charge layer in a p–n junction semiconductor with zero bias, as shown in Fig. 9.5(a), in which the region $x < 0$ is doped p-type and the region $x > 0$ is doped n-type. To review briefly the formation of the space charge layer, we note that since the density of the holes on the p side is larger than that on the n side, there is a tendency for the holes to diffuse to the n side and recombine with the electrons. Similarly, there is a tendency for the electrons on the n side to diffuse to the p side and recombine with the holes. The diffusion of holes leaves behind negatively charged acceptor atoms and the

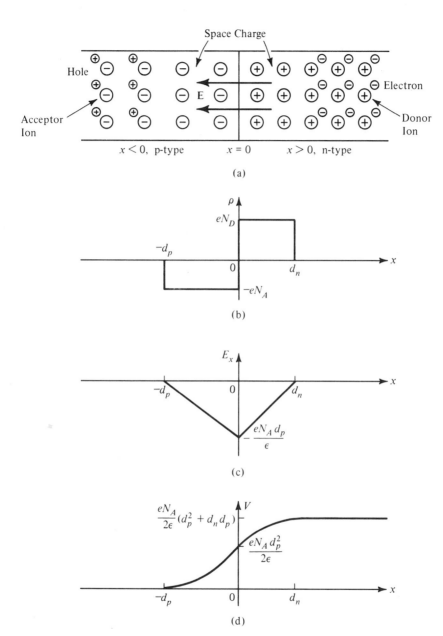

Figure 9.5. For illustrating the application of Poisson's equation for the determination of the potential distribution for a p–n junction semiconductor.

diffusion of electrons leaves behind positively charged donor atoms. Since these acceptor and donor atoms are immobile, a space charge layer, also known as the "depletion layer," is formed in the region of the junction with negative charges on the p side and positive charges on the n side. This space charge gives rise to an electric field directed from the n side of the junction to the p side so that it opposes diffusion of the mobile carriers across the junction thereby resulting in an equilibrium.

For simplicity, let us consider an abrupt junction, that is, a junction in which the impurity concentration is constant on either side of the junction. Let N_A and N_D be the acceptor and donor ion concentrations, respectively, and d_p and d_n be the widths in the p and n regions, respectively, of the depletion layer. The space charge density ρ is then given by

$$\rho = \begin{cases} -eN_A & \text{for } -d_p < x < 0 \\ eN_D & \text{for } \quad 0 < x < d_n \end{cases} \tag{9.26}$$

as shown in Fig. 9.5(b), where e is the electronic charge. Since the semiconductor is electrically neutral, the total acceptor charge must be equal to the total donor charge, that is,

$$eN_A d_p = eN_D d_n \tag{9.27}$$

Substituting (9.26) into (9.25), we obtain

$$\frac{d^2V}{dx^2} = \begin{cases} \dfrac{eN_A}{\epsilon} & \text{for } -d_p < x < 0 \\[2mm] -\dfrac{eN_D}{\epsilon} & \text{for } \quad 0 < x < d_n \end{cases} \tag{9.28}$$

This equation governs the potential distribution in the depletion layer.

To solve (9.28) for V, we integrate it once and obtain

$$\frac{dV}{dx} = \begin{cases} \dfrac{eN_A}{\epsilon} x + C_1 & \text{for } -d_p < x < 0 \\[2mm] -\dfrac{eN_D}{\epsilon} x + C_2 & \text{for } \quad 0 < x < d_n \end{cases}$$

where C_1 and C_2 are constants of integration. To evaluate C_1 and C_2, we note that since $\mathbf{E} = -\nabla V = -(\partial V/\partial x)\mathbf{i}_x$, $\partial V/\partial x$ is simply equal to $-E_x$. Since the electric field lines begin on the positive charges and end on the negative charges, the field and hence $\partial V/\partial x$ must vanish at $x = -d_p$ and $x = d_n$, giving us

$$\frac{dV}{dx} = \begin{cases} \dfrac{eN_A}{\epsilon}(x + d_p) & \text{for } -d_p < x < 0 \\[2mm] -\dfrac{eN_D}{\epsilon}(x - d_n) & \text{for } \quad 0 < x < d_n \end{cases} \tag{9.29}$$

The field intensity, that is, $-(dV/dx)$, may now be sketched as a function of x as shown in Fig. 9.5(c).

Proceeding further, we integrate (9.29) and obtain

$$V = \begin{cases} \dfrac{eN_A}{2\epsilon}(x + d_p)^2 + C_3 & \text{for } -d_p < x < 0 \\[3mm] -\dfrac{eN_D}{2\epsilon}(x - d_n)^2 + C_4 & \text{for } \quad 0 < x < d_n \end{cases}$$

where C_3 and C_4 are constants of integration. To evaluate C_3 and C_4, we first set the potential at $x = -d_p$ arbitrarily equal to zero to obtain C_3 equal to zero. Then we make use of the condition that the potential be continuous at $x = 0$, since the discontinuity in dV/dx at $x = 0$ is finite, to obtain

$$\frac{eN_A}{2\epsilon} d_p^2 = -\frac{eN_D}{2\epsilon} d_n^2 + C_4$$

or

$$C_4 = \frac{e}{2\epsilon}(N_A d_p^2 + N_D d_n^2)$$

Substituting this value for C_4 and setting C_3 equal to zero in the expression for V, we get the required solution

$$V = \begin{cases} \dfrac{eN_A}{2\epsilon}(x + d_p)^2 & \text{for } -d_p < x < 0 \\[3mm] -\dfrac{eN_D}{2\epsilon}(x^2 - 2xd_n) + \dfrac{eN_A}{2\epsilon}d_p^2 & \text{for } \quad 0 < x < d_n \end{cases} \qquad (9.30)$$

The variation of potential with x as given by (9.30) is shown in Fig. 9.5(d).

We can proceed further and find the width $d = d_p + d_n$ of the depletion layer by setting $V(d_n)$ equal to the contact potential, V_0, that is, the potential difference across the depletion layer resulting from the electric field in the layer. Thus

$$V_0 = V(d_n) = \frac{eN_D}{2\epsilon} d_n^2 + \frac{eN_A}{2\epsilon} d_p^2$$

$$= \frac{e}{2\epsilon} \frac{N_D(N_A + N_D)}{N_A + N_D} d_n^2 + \frac{e}{2\epsilon} \frac{N_A(N_A + N_D)}{N_A + N_D} d_p^2$$

$$= \frac{e}{2\epsilon} \frac{N_A N_D}{N_A + N_D}(d_n^2 + d_p^2 + 2d_n d_p)$$

$$= \frac{e}{2\epsilon} \frac{N_A N_D}{N_A + N_D} d^2$$

where we have used (9.27). Finally, we obtain the result that

$$d = \sqrt{\frac{2\epsilon V_0}{e}\left(\frac{1}{N_A} + \frac{1}{N_D}\right)}$$

which tells us that the depletion layer width is smaller the heavier the doping is. This property is used in tunnel diodes to achieve layer widths on the order of 10^{-6} cm by heavy doping as compared to widths on the order of 10^{-4} cm in ordinary p–n junctions.

We have just illustrated an example of the application of Poisson's equation involving the solution for the potential distribution for a given charge distribution. Poisson's equation is even more useful for the solution of problems in which the charge distribution is the quantity to be determined given the functional dependence of the charge density on the potential. We shall, however, not pursue this topic any further.

9.3 LAPLACE'S EQUATION

In the previous section we derived Poisson's equation

$$\nabla^2 V = -\frac{\rho}{\epsilon}$$

If the charge density in a region is zero, then Poisson's equation reduces to

$$\nabla^2 V = 0 \qquad (9.31)$$

This equation is known as "Laplace's equation." It governs the behavior of the potential in a charge-free region. In Cartesian coordinates, it is given by

$$\frac{\partial^2 V}{\partial x^2} + \frac{\partial^2 V}{\partial y^2} + \frac{\partial^2 V}{\partial z^2} = 0 \qquad (9.32)$$

Laplace's equation is also satisfied by the potential in conductors under steady current condition. For the steady current condition, $\partial \rho / \partial t = 0$ and the continuity equation given for the time-varying case by

$$\nabla \cdot \mathbf{J}_c + \frac{\partial \rho}{\partial t} = 0$$

reduces to

$$\nabla \cdot \mathbf{J}_c = 0 \qquad (9.33)$$

Replacing $\mathbf{J}_c$ by $\sigma \mathbf{E} = -\sigma \nabla V$ where σ is the conductivity of the conductor and assuming σ to be constant, we obtain

$$\nabla \cdot \sigma \mathbf{E} = \sigma \nabla \cdot \mathbf{E} = -\sigma \nabla \cdot \nabla V = -\sigma \nabla^2 V = 0$$

or

$$\nabla^2 V = 0$$

The problems for which Laplace's equation is applicable consist of finding the potential distribution in the region between two conductors given the charge distribution on the surfaces of the conductors or the potentials of the conductors or a combination of the two. The procedure involves the solving of Laplace's equation subject to the boundary conditions on the surfaces of the conductors. The electric field intensity between the conductors is then found by using $\mathbf{E} = -\nabla V$, from which the conduction current density is obtained by using $\mathbf{J}_c = \sigma\mathbf{E}$, if the medium is a conductor. We shall illustrate this by means of an example involving variation of V in one dimension.

Example 9.3.   Let us consider two infinite, plane, parallel, perfectly conducting plates occupying the planes $x = 0$ and $x = d$ and kept at potentials $V = 0$ and $V = V_0$, respectively, as shown by the cross-sectional view in Fig. 9.6, and find the solution for Laplace's equation in the region between the plates. The arrangement may be considered an idealization of two parallel plates having dimensions very large compared to the spacing between them.

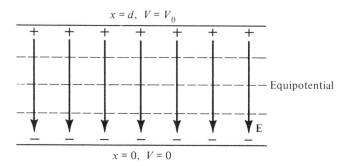

Figure 9.6. For illustrating the solution of Laplace's equation in one dimension.

The potential is obviously a function of x only and hence (9.32) reduces to

$$\frac{\partial^2 V}{\partial x^2} = \frac{d^2 V}{dx^2} = 0$$

Integrating this equation twice, we obtain

$$V(x) = Ax + B$$

where A and B are constants of integration. To determine the values of A and B, we make use of the boundary conditions for V, that is,

$$V = 0 \qquad \text{for } x = 0$$
$$V = V_0 \qquad \text{for } x = d$$

giving us

$$0 = A(0) + B \quad \text{or} \quad B = 0$$

$$V_0 = A(d) + B = Ad \quad \text{or} \quad A = \frac{V_0}{d}$$

Thus the required solution for the potential is given by

$$V = \frac{V_0}{d}x \quad \text{for } 0 < x < d$$

which tells us that the equipotentials are planes parallel to the conductors, as shown in Fig. 9.6.

Proceeding further, we obtain

$$\mathbf{E} = -\nabla V = -\frac{\partial V}{\partial x}\mathbf{i}_x = -\frac{V_0}{d}\mathbf{i}_x \quad \text{for } 0 < x < d$$

This field is uniform and directed from the higher potential plate to the lower potential plate, as shown in Fig. 9.6. The surface charge densities on the two plates are given by

$$[\rho_S]_{x=0} = [\mathbf{D}]_{x=0} \cdot \mathbf{i}_x = -\frac{\epsilon V_0}{d}\mathbf{i}_x \cdot \mathbf{i}_x = -\frac{\epsilon V_0}{d}$$

$$[\rho_S]_{x=d} = [\mathbf{D}]_{x=d} \cdot (-\mathbf{i}_x) = -\frac{\epsilon V_0}{d}\mathbf{i}_x \cdot (-\mathbf{i}_x) = \frac{\epsilon V_0}{d}$$

The magnitude of the surface charge per unit area on either plate is $Q = |\rho_S|(1) = \epsilon V_0/d$, and the capacitance per unit area of the plates, that is, the ratio of Q to V_0, is equal to ϵ/d.

If the medium between the plates is a conductor, then the conduction current density is given by

$$\mathbf{J}_c = \sigma\mathbf{E} = -\frac{\sigma V_0}{d}\mathbf{i}_x$$

The conduction current from the higher potential plate to the lower potential plate per unit area of the plates is $I_c = |\mathbf{J}_c|(1) = \sigma V_0/d$, and the conductance per unit area of the plates, that is, the ratio of I_c to V_0, is equal to σ/d. ∎

We have just illustrated the solution of Laplace's equation by considering an example involving the variation of V in one dimension only. Before going on to the solution of Laplace's equation in two dimensions, a brief discussion of the applicability of Laplace's equation in the determination of transmission-line parameters and field maps is in order. To do this, we recall that a transmission line is characterized by fields that are entirely transverse to its axis. Hence in any given transverse plane, that is, cross-sectional plane, $\oint_c \mathbf{E} \cdot d\mathbf{l}$ $= 0$ and $\mathbf{E}$ possesses the same spatial characteristics in the transverse dimensions as those of a static field although it is time-varying. Thus by solving Laplace's equation in the cross-sectional plane, subject to the boundary conditions at the conductors of the line, we can obtain the field map consisting of equipotential "lines" and electric field lines. The equipotential lines, being everywhere orthogonal to the electric field lines, are identical to the magnetic field lines. Conversely, the graphical field mapping technique discussed in Sec. 6.3 is equally applicable to the solution of Laplace's equation if we recognize that the magnetic field lines are equivalent to equipotential lines. A comparison of the results of Example 9.3 with the parallel-plate transmission line case in Sec. 6.2 serves as an example for this discussion.

Returning to the solution of Laplace's equation, we now consider its solution in two dimensions, say x and y. The potential, being independent of z, then satisfies the equation

$$\frac{\partial^2 V}{\partial x^2} + \frac{\partial^2 V}{\partial y^2} = 0 \tag{9.34}$$

Equation (9.34) is a partial differential equation in two dimensions. As we have already discussed in Sec. 4.4, the technique by means of which it is solved is the "separation of variables" technique. It consists of assuming that the solution for the potential is the product of two functions, one of which is a function of x only and the second is a function of y only. Denoting these functions to be X and Y, respectively, we have

$$V(x, y) = X(x) Y(y) \tag{9.35}$$

Substituting this assumed solution into the differential equation, we obtain

$$Y\frac{d^2X}{dx^2} + X\frac{d^2Y}{dy^2} = 0$$

Dividing both sides by XY and rearranging, we get

$$\frac{1}{X}\frac{d^2X}{dx^2} = -\frac{1}{Y}\frac{d^2Y}{dy^2} \tag{9.36}$$

The left side of (9.36) is a function of x only; the right side is a function of y only. Thus (9.36) states that a function of x only is equal to a function of y only. A function of x only other than a constant cannot be equal to a function of y only other than the same constant for all values of x and y. For example,

$2x$ is equal to $4y$ for only those pairs of values of x and y for which $x = 2y$. Since we are seeking a solution that is good for all pairs of x and y, the only solution that satisfies (9.36) is that for which each side of (9.36) is equal to a constant. Denoting this constant to be α^2, we have

$$\frac{d^2X}{dx^2} = \alpha^2 X \qquad (9.37a)$$

and

$$\frac{d^2Y}{dy^2} = -\alpha^2 Y \qquad (9.37b)$$

Thus we have obtained two ordinary differential equations involving separately the variables x and y, starting with the partial differential equation involving both of the variables x and y. It is for this reason that the method is known as the separation of variables technique.

The solutions for (9.37a) and (9.37b) are given by

$$X(x) = \begin{cases} Ae^{\alpha x} + Be^{-\alpha x} & \text{for } \alpha \neq 0 \\ A_0 x + B_0 & \text{for } \alpha = 0 \end{cases} \qquad (9.38a)$$

where A, B, A_0, and B_0 are arbitrary constants, and

$$Y(y) = \begin{cases} C \cos \alpha y + D \sin \alpha y & \text{for } \alpha \neq 0 \\ C_0 y + D_0 & \text{for } \alpha = 0 \end{cases} \qquad (9.38b)$$

where C, D, C_0, and D_0 are arbitrary constants. Substituting (9.38a) and (9.38b) into (9.35), we obtain

$$V(x, y) = \begin{cases} (Ae^{\alpha x} + Be^{-\alpha x})(C \cos \alpha y + D \sin \alpha y) & \text{for } \alpha \neq 0 \\ (A_0 x + B_0)(C_0 y + D_0) & \text{for } \alpha = 0 \end{cases} \qquad (9.39)$$

Equation (9.39) is the general solution for Laplace's equation in the two dimensions x and y. The arbitrary constants are evaluated from the boundary conditions specified for a given problem. We shall now consider two examples.

Example 9.4. Let us consider an infinitely long rectangular slot cut in a semi-infinite plane conducting slab held at zero potential, as shown by the cross-sectional view, transverse to the slot, in Fig. 9.7. With reference to the coordinate system shown in the figure, assume that a potential distribution $V = V_0 \sin (\pi y/b)$, where V_0 is a constant, is created at the mouth $x = a$ of the slot by the application of a potential to an appropriately shaped conductor away from the mouth of the slot not shown in the figure. We wish to find the potential distribution in the slot.

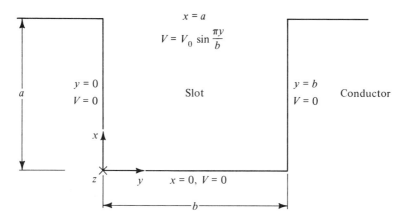

Figure 9.7. Cross-sectional view of a rectangular slot cut in a semi-infinite plane conducting slab at zero potential. The potential at the mouth of the slot is $V_0 \sin (\pi y/b)$ volts.

Since the slot is infinitely long in the z direction with uniform cross section, the problem is two dimensional in x and y and the general solution for V given by (9.39) is applicable. The boundary conditions are

$$V = 0 \qquad \text{for } y = 0, \, 0 < x < a \qquad (9.40a)$$

$$V = 0 \qquad \text{for } y = b, \, 0 < x < a \qquad (9.40b)$$

$$V = 0 \qquad \text{for } x = 0, \, 0 < y < b \qquad (9.40c)$$

$$V = V_0 \sin \frac{\pi y}{b} \qquad \text{for } x = a, \, 0 < y < b \qquad (9.40d)$$

The solution corresponding to $\alpha = 0$ does not fit the boundary conditions since V is required to be zero for two values of y and in the range $0 < x < a$. Hence we can ignore that solution and consider only the solution for $\alpha \neq 0$.

Applying the boundary condition (9.40a), we have

$$0 = (Ae^{\alpha x} + Be^{-\alpha x})(C) \qquad \text{for } 0 < x < a$$

The only way of satisfying this equation for a range of values of x is by setting $C = 0$. Next, applying the boundary condition (9.40c), we have

$$0 = (A + B)D \sin \alpha y \qquad \text{for } 0 < y < b$$

This requires that $(A + B)D = 0$, which can be satisfied by either $A + B = 0$ or $D = 0$. We, however, rule out $D = 0$ since it results in a trivial solution of zero for the potential. Hence we set

$$A + B = 0 \qquad \text{or} \qquad B = -A$$

Thus the solution for V reduces to

$$V(x, y) = (Ae^{\alpha x} - Ae^{-\alpha x})D \sin \alpha y$$
$$= A' \sinh \alpha x \sin \alpha y \qquad (9.41)$$

where $A' = 2AD$.

Next, applying boundary condition (9.40b) to (9.41), we obtain

$$0 = A' \sinh \alpha x \sin \alpha b \qquad \text{for } 0 < x < a$$

To satisfy this equation without obtaining a trivial solution of zero for the potential, we set

$$\sin \alpha b = 0$$

or

$$\alpha b = n\pi \qquad n = 1, 2, 3, \dots$$

$$\alpha = \frac{n\pi}{b} \qquad n = 1, 2, 3, \dots$$

Since several values of α satisfy the boundary condition, several solutions are possible for the potential. To take this into account, we write the solution as the superposition of all these solutions multiplied by different arbitrary constants. In this manner, we obtain

$$V(x, y) = \sum_{n=1,2,3,\dots}^{\infty} A'_n \sinh \frac{n\pi x}{b} \sin \frac{n\pi y}{b} \qquad \text{for } 0 < y < b \qquad (9.42)$$

Finally, applying the boundary condition (9.40d) to (9.42), we get

$$V_0 \sin \frac{\pi y}{b} = \sum_{n=1,2,3,\dots}^{\infty} A'_n \sinh \frac{n\pi a}{b} \sin \frac{n\pi y}{b} \qquad \text{for } 0 < y < b \qquad (9.43)$$

On the right side of (9.43) we have an infinite series of sine terms in y, but on the left side we have only one sine term in y. Equating the coefficients of the sine terms having the same arguments, we obtain

$$A'_n \sinh \frac{n\pi a}{b} = \begin{cases} V_0 & \text{for } n = 1 \\ 0 & \text{for } n \neq 1 \end{cases}$$

or

$$A'_1 = \frac{V_0}{\sinh (\pi a/b)}$$

$$A'_n = 0 \qquad \text{for } n \neq 1$$

Substituting this result in (9.42), we obtain the required solution for V as

$$V(x, y) = V_0 \frac{\sinh (\pi x/b)}{\sinh (\pi a/b)} \sin \frac{\pi y}{b} \qquad (9.44)$$

We may now compute the potential at any point inside the slot given the values of a, b, and V_0. For example, for $a = b$, that is, for a square slot, (9.44) gives the potential at the center of the slot to be $0.1993V_0$. ∎

Example 9.5. Let us assume that the rectangular slot of Fig. 9.7 is covered at the mouth $x = a$ by a conducting plate that is kept at a potential $V = V_0$, making sure that the edges touching the corners of the slot are insulated, as shown in Fig. 9.8(a), and find the solution for the potential in the slot for this new boundary condition.

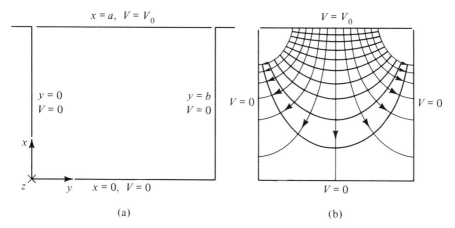

Figure 9.8. (a) Cross-sectional view of a rectangular slot in a semi-infinite plane conducting slab at zero potential and covered at the mouth by a conducting plate kept at a potential V_0. (b) Equipotentials and direction lines of electric field in the slot for the case $b/a = 1$.

Since the boundary conditions (9.40a)–(9.40c) remain the same, all we need to do to find the required solution for the potential is to substitute the new boundary condition

$$V = V_0 \qquad \text{for } x = a, \ 0 < y < b$$

in (9.42) and evaluate the coefficients A'_n. Thus we have

$$V_0 = \sum_{n=1,2,3,\ldots}^{\infty} A'_n \sinh \frac{n\pi a}{b} \sin \frac{n\pi y}{b} \qquad \text{for } 0 < y < b \qquad (9.45)$$

In this equation we have an infinite series on the right side, but the left side is a constant. Thus we cannot hope to obtain A'_n by simply comparing the coefficients of the sine terms having like arguments as in Example 9.4. If we do so, we get the ridiculous answer of $V_0 = 0$ and all $A'_n = 0$ since there is no constant term on the right side and there are no sine terms on the left side.

The way out of the dilemma is to make use of the so-called orthogonality property of sine functions, given by

$$\int_{y=0}^{p} \sin \frac{n\pi y}{p} \sin \frac{m\pi y}{p} \, dy = \begin{cases} 0 & n \neq m \\ \dfrac{p}{2} & n = m \end{cases}$$

where m and n are integers. Multiplying both sides of (9.45) by $\sin \dfrac{m\pi y}{b} \, dy$ and integrating between the limits 0 and b, we have

$$\int_{y=0}^{b} V_0 \sin \frac{m\pi y}{b} \, dy = \int_{y=0}^{b} \sum_{n=1,2,3,\ldots}^{\infty} A'_n \sinh \frac{n\pi a}{b} \sin \frac{n\pi y}{b} \sin \frac{m\pi y}{b} \, dy$$

The integration and summation on the right side can be interchanged, giving us

$$\int_{y=0}^{b} V_0 \sin \frac{m\pi y}{b} \, dy = \sum_{n=1,2,3,\ldots}^{\infty} A'_n \sinh \frac{n\pi a}{b} \int_{y=0}^{b} \sin \frac{n\pi y}{b} \sin \frac{m\pi y}{b} \, dy$$

or

$$\frac{V_0 b}{m\pi}(1 - \cos m\pi) = \left(A'_m \sinh \frac{m\pi a}{b} \right) \frac{b}{2}$$

$$A'_m = \begin{cases} \dfrac{4V_0}{m\pi} \dfrac{1}{\sinh (m\pi a/b)} & \text{for } m \text{ odd} \\ 0 & \text{for } m \text{ even} \end{cases}$$

Substituting this result in (9.42), we obtain the required solution for the potential inside the slot as

$$V = \sum_{n=1,3,5,\ldots}^{\infty} \frac{4V_0}{n\pi} \frac{\sinh (n\pi x/b)}{\sinh (n\pi a/b)} \sin \frac{n\pi y}{b} \qquad (9.46)$$

The numerical values of potentials may now be computed for points inside the slot for given values of a, b, and V_0, and equipotentials may be sketched by joining points having approximately the same potential values. The electric field lines can then be drawn orthogonal to the equipotentials. The resulting sketches for a square slot are shown in Fig. 9.8(b). ∎

9.4 COMPUTER SOLUTION OF LAPLACE'S EQUATION*

In the previous section we illustrated the solution of the two-dimensional Laplace's equation in Cartesian coordinates x and y. In this section we shall discuss the approximate solution of the two-dimensional Laplace's equation

---

*This section may be omitted without loss of continuity.

which forms the basis for adaptation to digital computers. To illustrate the principle behind the approximate solution, let us suppose that we know the potentials V_1, V_2, V_3, and V_4 at four points equidistant from a point $P(0, 0, 0)$ and lying on mutually perpendicular axes, which we call x and y, passing through P as shown in Fig. 9.9. We wish to find the potential V_0 at P in terms of V_1, V_2, V_3, and V_4.

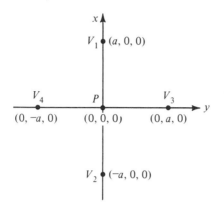

Figure 9.9. For illustrating the principle behind the approximate solution of Laplace's equation in two dimensions.

Assuming no variation of V in the z direction, we require that

$$[\nabla^2 V]_P = \left[\frac{\partial^2 V}{\partial x^2} + \frac{\partial^2 V}{\partial y^2}\right]_{(0,0,0)} = 0 \tag{9.47}$$

To solve this equation approximately for V_0, we note that

$$\left[\frac{\partial^2 V}{\partial x^2}\right]_{(0,0,0)} = \left[\frac{\partial}{\partial x}\left(\frac{\partial V}{\partial x}\right)\right]_{(0,0,0)}$$

$$\approx \frac{1}{a}\left\{\left[\frac{\partial V}{\partial x}\right]_{(a/2,0,0)} - \left[\frac{\partial V}{\partial x}\right]_{(-a/2,0,0)}\right\}$$

$$\approx \frac{1}{a}\left\{\frac{[V]_{(a,0,0)} - [V]_{(0,0,0)}}{a} - \frac{[V]_{(0,0,0)} - [V]_{(-a,0,0)}}{a}\right\}$$

$$= \frac{1}{a^2}[(V_1 - V_0) - (V_0 - V_2)]$$

$$= \frac{1}{a^2}(V_1 + V_2 - 2V_0) \tag{9.48a}$$

Similarly,

$$\left[\frac{\partial^2 V}{\partial y^2}\right]_{(0,0,0)} \approx \frac{1}{a^2}(V_3 + V_4 - 2V_0) \tag{9.48b}$$

Substituting (9.48a) and (9.48b) into (9.47) and rearranging, we obtain

$$V_0 \approx \frac{1}{4}(V_1 + V_2 + V_3 + V_4) \qquad (9.49)$$

Thus the potential at P is approximately equal to the average of the potentials at the four equidistant points lying along mutually perpendicular axes through P. The result becomes more and more accurate as the spacing a becomes less and less.

Equation (9.49) forms the basis for the computer solution of Laplace's equation. To illustrate the technique, let us consider the problem of Example 9.4 and assume $a = b$, that is, a square slot. We can then divide the area of the slot into a 4×4 grid of squares, as shown in Fig. 9.10. If we assume V_0 to be 100 V, then the potentials at the five grid points along the mouth of the slot are $100 \sin 0$, $100 \sin \frac{\pi}{4}$, $100 \sin \frac{\pi}{2}$, $100 \sin \frac{3\pi}{4}$, and $100 \sin \pi$ or 0, 70.71, 100, 70.71, and 0 V, respectively, as shown in the figure. The potentials at the grid points along the remaining three sides of the slot are all zero. The exact values of potentials at the grid points inside the slot computed from the

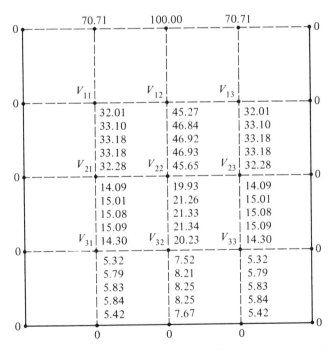

Figure 9.10. For illustrating the computer solution of Laplace's equation in two dimensions.

analytical solution given by (9.44) are shown by the upper rows of numbers beside the grid points for later comparison with those obtained by the computer solution technique.

The computer solution consists of finding the potentials at the nine grid points inside the slot from the given values at the grid points on the boundaries of the slot. Irrespective of how this is achieved, we must obtain a final set of values such that the potential at each grid point inside the slot is the average of the potentials at the neighboring four grid points, or the sum of the four neighboring potentials is equal to four times the potential at the grid point, in accordance with (9.49). The simplest technique adaptable to computer solution is to start with values of zero for all unknown potentials. Each unknown potential is then replaced by the average of the four neighboring potentials by traversing the grid in a systematic manner and by replacing in this process old values with new values as they are computed, until a set of values satisfying (9.49) at each grid point, to within a specified error, is obtained. Any symmetry associated with the problem, as in the present case, can be utilized to advantage for achieving a reduction in the number of computations.

The method we just discussed is known as the "iteration" technique since it involves the iterative process of converging an initially assumed solution to a final one consistent with Laplace's equation in the approximate sense given by (9.49). There are several variations of the iteration technique. For example, by employing an initial guess other than zeros, a faster convergence may be achieved. The end result will, however, still be only to within the specified accuracy.

The values of potentials obtained by the iteration technique for a specified maximum allowable value of 0.1 V for the error

$$\Delta = \left[V_0 - \frac{1}{4}(V_1 + V_2 + V_3 + V_4) \right]$$

are shown by the second rows of numbers beside the grid points in Fig. 9.10. When the specified maximum error is decreased to 0.01 V, thereby demanding a more accurate solution, the values of potentials obtained are shown by the third rows of numbers beside the grid points in Fig. 9.10. When these two rows are compared with the upper rows, it appears that the specification of a greater required accuracy in the iteration leads to a less accurate end result. This is, however, not the case since the iteration method can only converge to a solution that is consistent with (9.49) and not to the analytical solution.

Hence let us find the exact values of the unknown potentials consistent with (9.49). To do this, we write a set of simultaneous equations for these potentials by applying (9.49) at each grid point inside the slot. Thus denoting the unknown potentials to be V_{ij}, $i, j = 1, 2, 3$, as shown in Fig. 9.10, we obtain a set of nine equations given in matrix form by

$$
\begin{bmatrix}
4 & -1 & 0 & -1 & 0 & 0 & 0 & 0 & 0 \\
-1 & 4 & -1 & 0 & -1 & 0 & 0 & 0 & 0 \\
0 & -1 & 4 & 0 & 0 & -1 & 0 & 0 & 0 \\
-1 & 0 & 0 & 4 & -1 & 0 & -1 & 0 & 0 \\
0 & -1 & 0 & -1 & 4 & -1 & 0 & -1 & 0 \\
0 & 0 & -1 & 0 & -1 & 4 & 0 & 0 & -1 \\
0 & 0 & 0 & -1 & 0 & 0 & 4 & -1 & 0 \\
0 & 0 & 0 & 0 & -1 & 0 & -1 & 4 & -1 \\
0 & 0 & 0 & 0 & 0 & -1 & 0 & -1 & 4
\end{bmatrix}
\begin{bmatrix}
V_{11} \\ V_{12} \\ V_{13} \\ V_{21} \\ V_{22} \\ V_{23} \\ V_{31} \\ V_{32} \\ V_{33}
\end{bmatrix}
=
\begin{bmatrix}
70.71 \\ 100.00 \\ 70.71 \\ 0 \\ 0 \\ 0 \\ 0 \\ 0 \\ 0
\end{bmatrix}
\tag{9.50}
$$

The matrix equation (9.50) can be inverted directly, since it involves only a 9×9 matrix. Imagine, however, the situation if the number of grid points is large. For example, even for a 16×16 grid of squares, it will be necessary to invert a 225×225 matrix! Fortunately, however, it is not necessary to directly invert the matrix. To illustrate this we see from the partitionings in (9.50) that it can be written in compact form as

$$
\begin{bmatrix}
M & -U & 0 \\
-U & M & -U \\
0 & -U & M
\end{bmatrix}
\begin{bmatrix}
V_1 \\ V_2 \\ V_3
\end{bmatrix}
=
\begin{bmatrix}
V_g \\ 0 \\ 0
\end{bmatrix}
\tag{9.51}
$$

where

$$
M =
\begin{bmatrix}
4 & -1 & 0 \\
-1 & 4 & -1 \\
0 & -1 & 4
\end{bmatrix}
\tag{9.52a}
$$

$$
U =
\begin{bmatrix}
1 & 0 & 0 \\
0 & 1 & 0 \\
0 & 0 & 1
\end{bmatrix}
\tag{9.52b}
$$

$$
V_i =
\begin{bmatrix}
V_{i1} \\ V_{i2} \\ V_{i3}
\end{bmatrix}
\qquad i = 1, 2, 3
\tag{9.52c}
$$

$$
V_g =
\begin{bmatrix}
70.71 \\ 100.00 \\ 70.71
\end{bmatrix}
\tag{9.52d}
$$

From (9.51), we can write the following equations successively:

$$-UV_2 + MV_3 = 0$$

$$V_2 = MV_3 \tag{9.53a}$$

$$-UV_1 + MV_2 - UV_3 = 0$$

$$V_1 = MV_2 - UV_3 = (M^2 - U)V_3 \tag{9.53b}$$

$$MV_1 - UV_2 = V_g$$

$$(M^3 - 2M)V_3 = V_g$$

$$V_3 = (M^3 - 2M)^{-1}V_g \tag{9.53c}$$

Substituting for M and V_g in (9.53c) from (9.52a) and (9.52d), respectively, and simplifying, we get

$$\begin{bmatrix} V_{31} \\ V_{32} \\ V_{33} \end{bmatrix} = \begin{bmatrix} 68 & -48 & 12 \\ -48 & 80 & -48 \\ 12 & -48 & 68 \end{bmatrix}^{-1} \begin{bmatrix} 70.71 \\ 100.00 \\ 70.71 \end{bmatrix} \tag{9.54}$$

Thus, we have simplified the problem into one of inversion of a 3×3 matrix. Inverting the 3×3 matrix and performing the matrix multiplication on the right side of (9.54), we obtain the values of V_{31}, V_{32}, and V_{33}. The remaining values can then be found from (9.53a) and (9.53b). The results are shown by the fourth rows of numbers beside the grid points in Fig. 9.10.

It can now be seen by comparing the second and third rows of values with the fourth rows of values that the iteration method does converge closer to the exact solution consistent with (9.49) as the specified allowable error is decreased. To obtain a solution closer to the exact analytical solution, we must decrease the spacing between the grid points. For example, for an 8×8 grid of squares, the solution obtained by the iteration method for a specified maximum error of 0.01 V is shown by the set of numbers in the last rows in Fig. 9.10.

9.5 LOW-FREQUENCY BEHAVIOR VIA QUASISTATICS

In Example 6.4 in Sec. 6.4 we illustrated the determination of the low-frequency behavior of a physical structure from its input impedance by considering the example of the short-circuited line. We expressed the input impedance of the short-circuited line as an infinite series involving powers of the frequency ω and by considering the term proportional to ω we found that for a line of length l, the input impedance is equivalent to that of a single inductor for frequencies low enough such that $l \ll \lambda$, the wavelength corresponding to the frequency. In this section we shall illustrate the determination of the low-frequency behavior by a quasistatic extension of the static field existing in the structure when the frequency of the source driving the structure is zero. The quasistatic extension consists of starting with a time-varying field having the same spatial characteristics as that of the static field and obtaining the field solutions containing terms up to and including the first power in ω.

To introduce the quasistatic field approach, we shall first consider the same physical structure as a short-circuited parallel-plate line, that is, an arrangement of two parallel, plane, perfect conductors joined at one end by another perfectly conducting sheet, as shown in Fig. 9.11(a). We shall neglect fringing of the fields by assuming that the spacing d between the plates is very small compared to the dimensions of the plates or that the structure is part of a structure of much larger extent in the y and z directions. For a constant current source of value I_0 driving the structure at the end $z = -l$, as shown in the figure, such that the surface current densities on the two plates are given by

$$\mathbf{J}_S = \begin{cases} \dfrac{I_0}{w}\mathbf{i}_z & \text{for } x = 0 \\[2mm] -\dfrac{I_0}{w}\mathbf{i}_z & \text{for } x = d \end{cases} \tag{9.55}$$

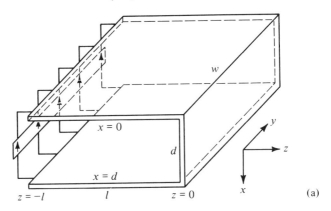

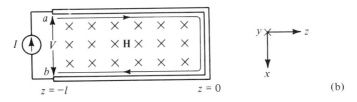

Figure 9.11. (a) A parallel-plate structure short-circuited at one end and driven by a current source at the other end. (b) Magnetic field between the plates for a constant current source.

the medium between the plates is characterized by a uniform y-directed magnetic field as shown by the cross-sectional view in Fig. 9.11(b). The field is zero outside the plates. From the boundary condition for the tangential magnetic field intensity at the surface of a perfect conductor, the magnitude of this field is I_0/w. Thus we obtain the static magnetic field intensity between the plates to be

$$\mathbf{H} = \frac{I_0}{w}\mathbf{i}_y \quad \text{for } 0 < x < d \tag{9.56}$$

The corresponding magnetic flux density is given by

$$\mathbf{B} = \mu\mathbf{H} = \frac{\mu I_0}{w}\mathbf{i}_y \quad \text{for } 0 < x < d \tag{9.57}$$

The magnetic flux ψ linking the current is simply the flux crossing the cross-sectional plane of the structure. Since $\mathbf{B}$ is uniform in the cross-sectional plane and normal to it,

$$\psi = B_y(dl) = \frac{\mu dl}{w}I_0 \tag{9.58}$$

The ratio of this magnetic flux to the current, that is, the inductance of the structure, is

$$L = \frac{\psi}{I_0} = \frac{\mu dl}{w} \tag{9.59}$$

To discuss the quasistatic behavior of the structure, we now let the current source be varying sinusoidally with time at a frequency ω and assume that the magnetic field between the plates varies accordingly. Thus for

$$I(t) = I_0 \cos \omega t \tag{9.60}$$

we have

$$\mathbf{H}_0 = \frac{I_0}{w} \cos \omega t \, \mathbf{i}_y \tag{9.61}$$

where the subscript 0 denotes that the field is of the zeroth power in ω. In terms of phasor notation, we have for

$$\bar{I} = I_0 \tag{9.62}$$

$$\bar{H}_{y0} = \frac{I_0}{w} \tag{9.63}$$

The time-varying magnetic field (9.61) gives rise to an electric field in accordance with Maxwell's curl equation for $\mathbf{E}$. Expansion of the curl equation for the case under consideration gives

$$\frac{\partial E_x}{\partial z} = -\frac{\partial B_{y0}}{\partial t} = -\mu\frac{\partial H_{y0}}{\partial t}$$

or, in phasor form

$$\frac{\partial \bar{E}_x}{\partial z} = -j\omega\mu\bar{H}_{y0} \tag{9.64}$$

Substituting for $\bar{H}_{y0}$ from (9.63), we have

$$\frac{\partial \bar{E}_x}{\partial z} = -j\omega\mu\frac{\bar{I}_0}{w}$$

or

$$\bar{E}_x = -j\omega\mu\frac{\bar{I}_0}{w}z + \bar{C} \tag{9.65}$$

The constant $\bar{C}$ is, however, equal to zero since $[\bar{E}_x]_{z=0} = 0$ to satisfy the boundary condition of zero tangential electric field on the perfect conductor surface. Thus we obtain the quasistatic electric field in the structure to be

$$\bar{E}_{x1} = -j\omega\frac{\mu z}{w}\bar{I}_0 \tag{9.66}$$

where the subscript 1 denotes that the field is of the first power in ω.

The voltage developed across the current source is now given by

$$\begin{aligned}
\bar{V} &= \int_a^b [\bar{E}_{x1}]_{z=-l}\, dx \\
&= j\omega\frac{\mu dl}{w}\bar{I}_0 \\
&= j\omega L\bar{I}_0 \tag{9.67}
\end{aligned}$$

Thus the quasistatic extension of the static field in the structure of Fig. 9.11 illustrates that its input behavior for low frequencies is equivalent to that of a single inductor as we found in Example 6.4.

Example 9.6.   Let us consider the case of two parallel perfectly conducting plates separated by a lossy medium characterized by conductivity σ, permittivity ϵ, and permeability μ and driven by a voltage source at one end, as shown in Fig. 9.12(a). We wish to determine its low-frequency behavior by using the quasistatic field approach.

Assuming the voltage source to be a constant voltage source, we first obtain the static electric field in the medium between the plates to be

$$\mathbf{E} = \frac{V_0}{d}\mathbf{i}_x$$

following the procedure of Example 9.3. The conduction current density in

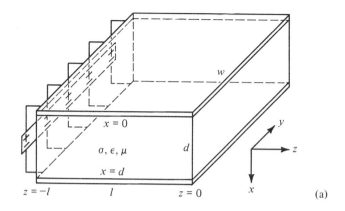

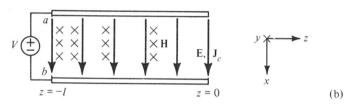

Figure 9.12. (a) A parallel-plate structure with lossy medium between the plates and driven by a voltage source. (b) Electric and magnetic fields between the plates for a constant voltage source.

the medium is then given by

$$J_c = \sigma E = \frac{\sigma V_0}{d} i_x$$

The conduction current gives rise to a static magnetic field in accordance with Maxwell's curl equation for **H** given for static fields by

$$\nabla \times H = J_c = \sigma E$$

For the case under consideration, this reduces to

$$\frac{\partial H_y}{\partial z} = -\sigma E_x = -\frac{\sigma V_0}{d}$$

giving us

$$H_y = -\frac{\sigma V_0 z}{d} + C_1$$

The constant C_1 is, however, equal to zero since $[H_y]_{z=0} = 0$ in view of the boundary condition that the surface current density on the plates must be zero at $z = 0$. Thus the static magnetic field in the medium between the plates is given by

$$\mathbf{H} = -\frac{\sigma V_0 z}{d}\mathbf{i}_y$$

The static electric and magnetic field distributions are shown by the cross-sectional view of the structure in Fig. 9.12(b).

To determine the quasistatic behavior of the structure, we now let the voltage source be varying sinusoidally with time at a frequency ω and assume that the electric and magnetic fields vary with time accordingly. Thus for

$$V = V_0 \cos \omega t$$

we have

$$\mathbf{E}_0 = \frac{V_0}{d} \cos \omega t \, \mathbf{i}_x \qquad (9.68a)$$

$$\mathbf{H}_0 = -\frac{\sigma V_0 z}{d} \cos \omega t \, \mathbf{i}_y \qquad (9.68b)$$

where the subscript 0 denotes that the fields are of the zeroth power in ω. In terms of phasor notation, we have for $\bar{V} = V_0$,

$$\bar{E}_{x0} = \frac{V_0}{d} \qquad (9.69a)$$

$$\bar{H}_{y0} = -\frac{\sigma V_0 z}{d} \qquad (9.69b)$$

The time-varying electric field (9.68a) gives rise to a magnetic field in accordance with

$$\nabla \times \mathbf{H} = \frac{\partial \mathbf{D}_0}{\partial t} = \epsilon \frac{\partial \mathbf{E}_0}{\partial t}$$

and the time-varying magnetic field (9.68b) gives rise to an electric field in accordance with

$$\nabla \times \mathbf{E} = -\frac{\partial \mathbf{B}_0}{\partial t} = -\mu \frac{\partial \mathbf{H}_0}{\partial t}$$

For the case under consideration and using phasor notation, these equations reduce to

$$\frac{\partial \bar{H}_y}{\partial z} = -j\omega\epsilon \bar{E}_{x0} = -j\omega\frac{\epsilon V_0}{d}$$

$$\frac{\partial \bar{E}_x}{\partial z} = -j\omega\mu \bar{H}_{y0} = j\omega\frac{\mu\sigma V_0 z}{d}$$

giving us

$$\bar{H}_{y1} = -j\omega \frac{\epsilon V_0 z}{d} + \bar{C}_2$$

$$\bar{E}_{x1} = j\omega \frac{\mu\sigma V_0 z^2}{2d} + \bar{C}_3$$

where the subscript 1 denotes that the fields are of the first power in ω. The constant $\bar{C}_2$ is, however, equal to zero in view of the boundary condition that the surface current density on the plates must be zero at $z = 0$. To evaluate the constant $\bar{C}_3$, we note that $[\bar{E}_{x1}]_{z=-l} = 0$ since the boundary condition at the source end, that is,

$$\bar{V} = \int_a^b [\bar{E}_x]_{z=-l} \, dx$$

is satisfied by $\bar{E}_{x0}$ alone. Thus we have

$$j\omega \frac{\mu\sigma V_0 (-l)^2}{2d} + \bar{C}_3 = 0$$

or

$$\bar{C}_3 = -j\omega \frac{\mu\sigma V_0 l^2}{2d}$$

Substituting for $\bar{C}_3$ and $\bar{C}_2$ in the expressions for $\bar{E}_{x1}$ and $\bar{H}_{y1}$, respectively, we get

$$\bar{E}_{x1} = j\omega \frac{\mu\sigma V_0 (z^2 - l^2)}{2d} \qquad (9.70a)$$

$$\bar{H}_{y1} = -j\omega \frac{\epsilon V_0 z}{d}$$

The result for $\bar{H}_{y1}$ is, however, not complete since $\bar{E}_{x1}$ gives rise to a conduction current of density proportional to ω which in turn provides an additional contribution to $\bar{H}_{y1}$. Denoting this contribution to be H_{y1}^c, we have

$$\frac{\partial \bar{H}_{y1}^c}{\partial z} = -\sigma \bar{E}_{x1} = -j\omega \frac{\mu\sigma^2 V_0 (z^2 - l^2)}{2d}$$

$$\bar{H}_{y1}^c = -j\omega \frac{\mu\sigma^2 V_0 (z^3 - 3zl^2)}{6d} + \bar{C}_4$$

The constant $\bar{C}_4$ is zero for the same reason that $\bar{C}_2$ is zero. Hence setting $\bar{C}_4$ equal to zero and adding the resulting expression for $\bar{H}_{y1}^c$ to the right side of the expression for $\bar{H}_{y1}$, we obtain the complete expression for $\bar{H}_{y1}$ as

$$\bar{H}_{y1} = -j\omega \frac{\epsilon V_0 z}{d} - j\omega \frac{\mu\sigma^2 V_0 (z^3 - 3zl^2)}{6d} \qquad (9.70b)$$

The total field components correct to the first power in ω are then given by

$$\bar{E}_x = \bar{E}_{x0} + \bar{E}_{x1}$$
$$= \frac{V_0}{d} + j\omega\frac{\mu\sigma V_0(z^2 - l^2)}{2d} \tag{9.71a}$$

$$\bar{H}_y = \bar{H}_{y0} + \bar{H}_{y1}$$
$$= -\frac{\sigma V_0 z}{d} - j\omega\frac{\epsilon V_0 z}{d} - j\omega\frac{\mu\sigma^2 V_0(z^3 - 3zl^2)}{6d} \tag{9.71b}$$

The current drawn from the voltage source is

$$\bar{I} = w[\bar{H}_y]_{z=-l}$$
$$= \left(\frac{\sigma wl}{d} + j\omega\frac{\epsilon wl}{d} - j\omega\frac{\mu\sigma^2 wl^3}{3d}\right)\bar{V} \tag{9.72}$$

Finally, the input admittance of the structure is given by

$$\bar{Y} = \frac{\bar{I}}{\bar{V}} = j\omega\frac{\epsilon wl}{d} + \frac{\sigma wl}{d}\left(1 - j\omega\frac{\mu\sigma l^2}{3}\right)$$

$$\approx j\omega\frac{\epsilon wl}{d} + \frac{1}{\dfrac{d}{\sigma wl}\left(1 + j\omega\dfrac{\mu\sigma l^2}{3}\right)}$$

$$= j\omega\frac{\epsilon wl}{d} + \frac{1}{\dfrac{d}{\sigma wl} + j\omega\dfrac{\mu dl}{3w}}$$

$$= j\omega C + \frac{1}{R + (j\omega L/3)} \tag{9.73}$$

where $C = \epsilon wl/d$ is the capacitance of the structure if the material is a perfect dielectric, $R = d/\sigma wl$ is the d.c. resistance (reciprocal of the conductance) of the structure, and $L = \mu dl/w$ is the inductance of the structure if the material is lossless and the two plates are short-circuited at $z = 0$. The equivalent circuit corresponding to (9.73) consists of capacitance C in parallel with the series combination of resistance R and inductance $L/3$, as shown in Fig. 9.13. ∎

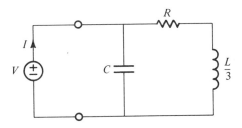

Figure 9.13. Equivalent circuit for the low-frequency input behavior of the structure of Fig. 9.12.

9.6 MAGNETIC CIRCUITS

In this section we shall introduce the principle of magnetic circuits. A simple example of magnetic circuit is the toroidal magnetic core of uniform permeability μ and having a uniform, circular cross-sectional area A and mean circumference l, as shown in Fig. 9.14. A current I_0 amp is passed through a filamentary wire of N turns wound around the toroid. Because of this current, a magnetic field is established in the core in the direction of advance of a right-hand screw as it is turned in the sense of the current, as shown in Fig. 9.14.

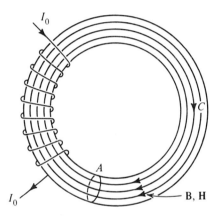

Figure 9.14. A toroidal magnetic circuit.

If the permeability of the core material is very large compared to the permeability of the surrounding medium, which is free space, the magnetic flux is confined almost entirely to the core in a manner similar to conduction current flow in wires or fluid flow in pipes. To illustrate this, let us consider lines of magnetic flux density on either side of a plane interface between a magnetic material of $\mu \gg \mu_0$ and free space, as shown in Fig. 9.15. Then from the boundary conditions for the magnetic field components, we have

$$B_1 \sin \alpha_1 = B_2 \sin \alpha_2 \qquad (9.74a)$$

$$H_1 \cos \alpha_1 = H_2 \cos \alpha_2 \qquad (9.74b)$$

Dividing (9.74a) by (9.74b), we get

$$\frac{B_1}{H_1} \tan \alpha_1 = \frac{B_2}{H_2} \tan \alpha_2$$

$$\mu_1 \tan \alpha_1 = \mu_2 \tan \alpha_2$$

$$\frac{\tan \alpha_1}{\tan \alpha_2} = \frac{\mu_2}{\mu_1} = \frac{\mu_0}{\mu_1} \ll 1$$

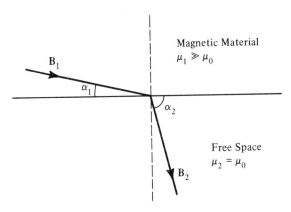

Figure 9.15. Lines of magnetic flux density at the boundary between free space and a magnetic material of $\mu \gg \mu_0$.

Thus $\alpha_1 \ll \alpha_2$, and

$$\frac{B_2}{B_1} = \frac{\sin \alpha_1}{\sin \alpha_2} \ll 1$$

For example, if the values of μ_1 and α_1 are $1000\,\mu_0$ and $5°$, respectively, then $\alpha_2 = 89.35°$ and $B_2/B_1 = 0.087$. For $\alpha_1 = 3°$, $\alpha_2 = 88.9°$ and $B_2/B_1 = 0.052$. The magnetic flux is for all practical purposes confined entirely to the core and very little flux appears as leakage flux outside the core.

If we assume that the magnetic flux ψ over the cross-sectional area of the toroid is equal to the flux density B_m at the mean radius of the toroid times the cross-sectional area of the toroid, we can then write

$$B_m = \frac{\psi}{A} \tag{9.75}$$

$$H_m = \frac{B_m}{\mu} = \frac{\psi}{\mu A} \tag{9.76}$$

From Ampere's circuital law, the magnetomotive force around the closed path C along the mean circumference of the toroid is equal to the current enclosed by that path. This current is equal to NI_0 since the filamentary wire penetrates the surface bounded by the path N times. Thus

$$\oint_C \mathbf{H} \cdot d\mathbf{l} = NI_0$$

$$H_m l = NI_0 \tag{9.77}$$

Substituting for H_m from (9.76) and rearranging, we obtain

$$\psi = \frac{\mu NI_0 A}{l}$$

We now define the "reluctance" of the magnetic circuit, denoted by the symbol $\mathfrak{R}$, as the ratio of the ampere turns NI_0 applied to the magnetic circuit to the magnetic flux ψ. Thus

$$\mathfrak{R} = \frac{NI_0}{\psi} = \frac{l}{\mu A} \tag{9.78}$$

The reluctance of the magnetic circuit is analogous to the electric circuit quantity resistance and has the units of ampere turns per weber. We note from (9.78) that for a given magnetic material, the reluctance appears to be purely a function of the dimensions of the circuit. This is, however, not true since for the ferromagnetic materials used for the cores, μ is a function of the magnetic flux density in the material, as we learned in Sec. 5.3.

As a numerical example of computations involving the magnetic circuit of Fig. 9.14, let us consider a core of cross-sectional area 2 cm² and mean circumference 20 cm. Let the material of the core be annealed sheet steel for which the B versus H relationship is shown by the curve of Fig. 9.16. Then to establish a magnetic flux of 3×10^{-4} Wb in the core, the mean flux density must be $(3 \times 10^{-4})/(2 \times 10^{-4})$ or 1.5 Wb/m². From Fig. 9.16, the corresponding value of H is 1000 amp/m. The number of ampere turns required to establish the flux is then equal to $1000 \times 20 \times 10^{-2}$, or, 200, and the reluctance of the core is $200/(3 \times 10^{-4})$, or $(2/3) \times 10^6$ amp-turns/Wb. We shall now consider a more detailed example.

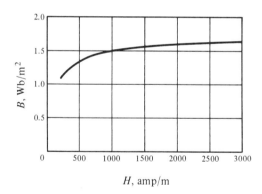

Figure 9.16. B versus H curve for annealed sheet steel.

Example 9.7.   A magnetic circuit containing three legs and with an air gap in the right leg is shown in Fig. 9.17(a). A filamentary wire of N turns carrying current I is wound around the center leg. The core material is annealed sheet steel, for which the B versus H relationship is shown in Fig. 9.16. The dimensions of the magnetic circuit are

$$A_1 = A_3 = 3 \text{ cm}^2, \qquad A_2 = 6 \text{ cm}^2$$
$$l_1 = l_3 = 20 \text{ cm}, \qquad l_2 = 10 \text{ cm}, \qquad l_g = 0.2 \text{ mm}$$

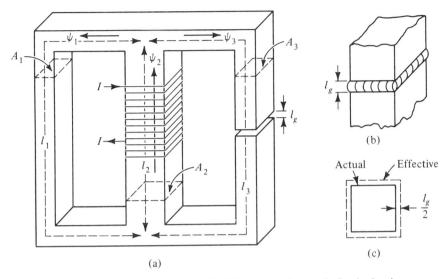

Figure 9.17. (a) A magnetic circuit. (b) Fringing of magnetic flux in the air gap of the magnetic circuit. (c) Effective and actual cross sections for the air gap.

Let us determine the value of NI required to establish a magnetic flux of 4×10^{-4} Wb in the air gap.

The current in the winding establishes a magnetic flux in the center leg which divides between the right and left legs. Fringing of the flux occurs in the air gap, as shown in Fig. 9.17(b). This is taken into account by using an effective cross section larger than the actual cross section, as shown in Fig. 9.17(c). Using subscripts 1, 2, 3, and g for the quantities associated with the left, center, and right legs, and the air gap, respectively, we can write

$$\psi_3 = \psi_g$$
$$\psi_2 = \psi_1 + \psi_3$$

Also, applying Ampere's circuital law to the right and left loops of the magnetic circuit, we obtain, respectively,

$$NI = H_2 l_2 + H_3 l_3 + H_g l_g$$
$$NI = H_2 l_2 + H_1 l_1$$

It follows from these two equations that

$$H_1 l_1 = H_3 l_3 + H_g l_g$$

which can also be written directly from a consideration of the outer loop of the magnetic circuit.

Noting from Fig. 9.17(c) that the effective cross section of the air gap is $(\sqrt{3} + l_g)^2 = 3.07 \text{ cm}^2$, we find the required magnetic flux density in the air gap to be

$$B_g = \frac{\psi_g}{(A_g)_{\text{eff}}} = \frac{4 \times 10^{-4}}{3.07 \times 10^{-4}} = 1.303 \text{ Wb/m}^2$$

The magnetic field intensity in the air gap is

$$H_g = \frac{B_g}{\mu_0} = \frac{1.303}{4\pi \times 10^{-7}} = 0.1037 \times 10^7 \text{ amp/m}$$

The flux density in leg 3 is

$$B_3 = \frac{\psi_3}{A_3} = \frac{\psi_g}{A_3} = \frac{4 \times 10^{-4}}{3 \times 10^{-4}} = 1.333 \text{ Wb/m}^2$$

From Fig. 9.16, the value of H_3 is 475 amp/m.

Knowing the values of H_g and H_3, we then obtain

$$\begin{aligned} H_1 l_1 &= H_3 l_3 + H_g l_g \\ &= 475 \times 0.2 + 0.1037 \times 10^7 \times 0.2 \times 10^{-3} \\ &= 302.4 \text{ amp} \end{aligned}$$

$$H_1 = \frac{302.4}{0.2} = 1512 \text{ amp/m}$$

From Fig. 9.16, the value of B_1 is 1.56 Wb/m^2 and hence the flux in leg 1 is

$$\psi_1 = B_1 A_1 = 1.56 \times 3 \times 10^{-4} = 4.68 \times 10^{-4} \text{ Wb}$$

Thus

$$\begin{aligned} \psi_2 &= \psi_1 + \psi_3 \\ &= 4.68 \times 10^{-4} + 4 \times 10^{-4} = 8.68 \times 10^{-4} \text{ Wb} \end{aligned}$$

$$B_2 = \frac{\psi_2}{A_2} = \frac{8.68 \times 10^{-4}}{6 \times 10^{-4}} = 1.447 \text{ Wb/m}^2$$

From Fig. 9.16, the value of H_2 is 750 amp/m. Finally, we obtain the required number of ampere turns to be

$$\begin{aligned} NI &= H_2 l_2 + H_1 l_1 \\ &= 750 \times 0.2 + 302.4 \\ &= 452.4 \end{aligned}$$

■

9.7 SUMMARY

In this chapter we learned that Maxwell's equations for static fields are given by

$$\nabla \times \mathbf{E} = 0 \qquad (9.79a)$$

$$\nabla \times \mathbf{H} = \mathbf{J} \qquad (9.79b)$$

$$\nabla \cdot \mathbf{D} = \rho \qquad (9.79c)$$

$$\nabla \cdot \mathbf{B} = 0 \qquad (9.79d)$$

whereas the continuity equation is

$$\nabla \cdot \mathbf{J} = 0 \qquad (9.80)$$

These equations together with the constitutive relations

$$\mathbf{D} = \epsilon \mathbf{E} \qquad (9.81a)$$

$$\mathbf{H} = \frac{\mathbf{B}}{\mu} \qquad (9.81b)$$

$$\mathbf{J} = \mathbf{J}_c = \sigma \mathbf{E} \qquad (9.81c)$$

govern the behavior of static fields.

First we learned from (9.79a) that, since the curl of the gradient of a scalar function is identically zero, $\mathbf{E}$ can be expressed as the gradient of a scalar function. The gradient of a scalar function Φ is given in Cartesian coordinates by

$$\nabla \Phi = \frac{\partial \Phi}{\partial x} \mathbf{i}_x + \frac{\partial \Phi}{\partial y} \mathbf{i}_y + \frac{\partial \Phi}{\partial z} \mathbf{i}_z$$

The magnitude of $\nabla \Phi$ at a given point is the maximum rate of increase of Φ at that point, and its direction is the direction in which the maximum rate of increase occurs, that is, normal to the constant Φ surface passing through that point.

From considerations of work associated with the movement of a test charge in the static electric field, we found that

$$\mathbf{E} = -\nabla V \qquad (9.82)$$

where V is the electric potential. The electric potential V_A at a point A is the amount of work per unit charge done by the field in the movement of a test charge from the point A to a reference point O. It is the potential difference between A and O. Thus

370

$$V_A = [V]_A^O = \int_A^O \mathbf{E} \cdot d\mathbf{l} = -\int_O^A \mathbf{E} \cdot d\mathbf{l}$$

The potential difference between two points has the same physical meaning as the voltage between the two points. The voltage is, however, not a unique quantity since it depends on the path employed for evaluating it, whereas the potential difference, being independent of the path, has a unique value.

We considered the potential field of a point charge and found that for the point charge

$$V = \frac{Q}{4\pi\epsilon R}$$

where R is the radial distance away from the point charge. The equipotential surfaces for the point charge are thus spherical surfaces centered at the point charge. We illustrated the application of the potential concept in the determination of electric field due to charge distributions by considering the example of an electric dipole.

Substituting (9.82) into (9.79c), we derived Poisson's equation

$$\nabla^2 V = -\frac{\rho}{\epsilon} \qquad (9.83)$$

which states that the Laplacian of the electric potential at a point is equal to $-1/\epsilon$ times the volume charge density at that point. In Cartesian coordinates,

$$\nabla^2 V = \frac{\partial^2 V}{\partial x^2} + \frac{\partial^2 V}{\partial y^2} + \frac{\partial^2 V}{\partial z^2}$$

For the one-dimensional case in which the charge density is a function of x only, (9.83) reduces to

$$\frac{\partial^2 V}{\partial x^2} = \frac{d^2 V}{dx^2} = -\frac{\rho}{\epsilon}$$

We illustrated the solution of this equation by considering the example of a p–n junction diode.

If $\rho = 0$, Poisson's equation reduces to Laplace's equation

$$\nabla^2 V = 0 \qquad (9.84)$$

This equation is applicable for a charge-free dielectric region as well as for a conducting medium. We illustrated the solution of the one-dimensional Laplace's equation

$$\frac{\partial^2 V}{\partial x^2} = \frac{d^2 V}{dx^2} = 0$$

by considering a parallel-plate arrangement. By using the separation of variables technique, we obtained the general solution to Laplace's equation in two dimensions

$$\frac{\partial^2 V}{\partial x^2} + \frac{\partial^2 V}{\partial y^2} = 0$$

and illustrated its application by considering two examples. We also discussed the applicability of Laplace's equation for the determination of transmission-line parameters and field maps.

To illustrate the computer solution of Laplace's equation, we derived the approximate solution to Laplace's equation in two dimensions. This solution states that the potential V_0 at a point P is given by

$$V_0 \approx \frac{1}{4}(V_1 + V_2 + V_3 + V_4) \tag{9.85}$$

where V_1, V_2, V_3, and V_4 are the potentials at four equidistant points lying along mutually perpendicular axes through P. By means of an example, we discussed the iteration technique of computer solution based on the repeated application of (9.85) to a set of grid points in the region of interest until a solution that converges to within a specified error is obtained. We also discussed the direct solution for the potentials at the grid points consistent with (9.85) using matrix inversion techniques.

After having considered the solution of static field problems, we then turned to the quasistatic extension of the static field solution as a means of obtaining the low-frequency behavior of a physical structure. The quasistatic field approach involves starting with a time-varying field having the same spatial characteristics as the static field in the physical structure and then obtaining field solutions containing terms up to and including the first power in frequency by using Maxwell's curl equations for time-varying fields. We illustrated this approach by considering two examples, one of them involving a lossy medium.

Finally, we introduced the magnetic circuit, which is essentially an arrangement of closed paths for magnetic flux to flow around just as current in electric circuits. The closed paths are provided by ferromagnetic cores which because of their high permeability relative to that of the surrounding medium confine the flux almost entirely to within the core regions. We illustrated the analysis of magnetic circuits by considering two examples, one of them including an air gap in one of the legs.

REVIEW QUESTIONS

9.1. State Maxwell's curl equations for static fields.

9.2. What is the expansion for the gradient of a scalar in Cartesian coordinates? When can a vector be expressed as the gradient of a scalar?

9.3. Discuss the physical interpretation for the gradient of a scalar function.

9.4. Discuss the application of the gradient concept for the determination of unit vector normal to a surface.

9.5. How would you find the rate of increase of a scalar function along a specified direction by using the gradient concept?

9.6. Define electric potential. What is its relationship to the static electric field intensity?

9.7. Distinguish between voltage, as applied to time-varying fields, and potential difference.

9.8. What is a conservative field? Give two examples of conservative fields.

9.9. Describe the equipotential surfaces for a point charge.

9.10. Discuss the determination of the electric field intensity due to a charge distribution by using the potential concept.

9.11. What is the Laplacian of a scalar? What is its expansion in Cartesian coordinates?

9.12. State Poisson's equation.

9.13. Outline the solution of Poisson's equation for the potential in a region of known charge density varying in one dimension.

9.14. State Laplace's equation. In what regions is it valid?

9.15. Discuss the application of Laplace's equation for a conducting medium.

9.16. Outline the solution of Laplace's equation in one dimension.

9.17. Why is Laplace's equation applicable to the determination of transmission-line parameters and field maps?

9.18. Outline the solution of Laplace's equation in two dimensions by the separation of variables technique.

9.19. What is the principle behind the approximate solution of Laplace's equation in two dimensions?

9.20. Discuss the iteration technique for the computer solution of Laplace's equation in two dimensions.

9.21. By consulting appropriate reference books, discuss two variations of the iteration technique for the computer solution of Laplace's equation.

9.22. How would you apply the iteration technique for the computer solution of Laplace's equation in three dimensions?

9.23. What is meant by the quasistatic extension of the static field in a physical structure?

9.24. Outline the steps involved in the determination of the quasistatic electric field in a parallel-plate structure short circuited at one end.

9.25. Why must the surface current density on the plates of the structure of Fig. 9.12 be zero at $z = 0$?

9.26. Discuss the quasistatic behavior of the structure of Fig. 9.12 for $\sigma \approx 0$.

9.27. What is a magnetic circuit? Why is the magnetic flux in a magnetic circuit confined almost entirely to the core?

9.28. Define the reluctance of a magnetic circuit. What is the analogous electric circuit quantity?

9.29. Why is the reluctance for a given set of dimensions of a magnetic circuit not a constant?

9.30. How is the fringing of the magnetic flux in an air gap in a magnetic circuit taken into account?

PROBLEMS

9.1. Find the gradients of the following scalar functions: (a) $\sqrt{x^2 + y^2 + z^2}$; (b) xyz.

9.2. Determine which of the following vectors can be expressed as the gradient of a scalar function: (a) $y\mathbf{i}_x - x\mathbf{i}_y$; (b) $x\mathbf{i}_x + y\mathbf{i}_y + z\mathbf{i}_z$; (c) $2xy^3z\mathbf{i}_x + 3x^2y^2z\mathbf{i}_y + x^2y^3\mathbf{i}_z$.

9.3. Find the unit vector normal to the plane surface $5x + 2y + 4z = 20$.

9.4. Find a unit vector normal to the surface $x^2 - y^2 = 5$ at the point $(3, 2, 1)$.

9.5. Find the rate of increase of the scalar function x^2y at the point $(1, 2, 1)$ in the direction of the vector $\mathbf{i}_x - \mathbf{i}_y$.

9.6. For the static electric field given by $\mathbf{E} = y\mathbf{i}_x + x\mathbf{i}_y$, find the potential difference between points $A(1, 1, 1)$ and $B(2, 2, 2)$.

9.7. For a point charge Q situated at the point $(1, 2, 0)$, find the potential difference between the point $A(3, 4, 1)$ and the point $B(5, 5, 0)$.

9.8. An arrangement of point charges known as the linear quadrupole consists of point charges Q, $-2Q$, and Q at the points $(0, 0, d)$, $(0, 0, 0)$, and $(0, 0, -d)$, respectively. Obtain the expression for the electric potential and hence for the electric field intensity at distances from the quadrupole large compared to d.

9.9. For a line charge of uniform density 10^{-3} C/m situated along the z axis between $(0, 0, -1)$ and $(0, 0, 1)$, obtain the series expression for the electric potential at the point $(0, y, 0)$ by dividing the line charge into 100 equal segments and considering the charge in each segment to be a point charge located at the center of the segment. Then find the series expression for the electric field intensity at the point $(0, 1, 0)$.

9.10. Repeat Problem 9.9, assuming the line charge density to be $10^{-3}|z|$ C/m.

9.11. The potential distribution in a simplified model of a vacuum diode consisting of cathode in the plane $x = 0$ and anode in the plane $x = d$ and held at a potential V_0 relative to the cathode is given by

$$V = V_0\left(\frac{x}{d}\right)^{4/3} \qquad \text{for } 0 < x < d$$

(a) Find the space charge density distribution in the region $0 < x < d$.
(b) Find the surface charge densities on the cathode and the anode.

9.12. Show that for the p–n junction diode of Fig. 9.5(a), the boundary condition of the continuity of the normal component of displacement flux density at $x = 0$ is automatically satisfied by Eq. (9.29).

9.13. Assume that the impurity concentration for the p–n junction diode of Fig. 9.5(a) is a linear function of distance across the junction. The space charge density distribution is then given by

$$\rho = kx \qquad \text{for } -d/2 < x < d/2$$

where d is the width of the space charge region and k is the proportionality constant. Find the solution for the potential in the space charge region.

9.14. Two infinitely long cylindrical conductors having radii a and b and coaxial with the z axis are held at a potential difference of V_0. Using the cylindrical coordinate system, obtain the solution for the potential and hence for the electric field intensity in the charge-free dielectric region between the cylinders. Find the expression for the capacitance per unit length of the cylinders.

9.15. The region between the two plates of Fig. 9.6 is filled with two perfect dielectric media having permittivities ϵ_1 for $0 < x < t$ and ϵ_2 for $t < x < d$. (a) Find the solutions for the potentials in the two regions $0 < x < t$ and $t < x < d$. (b) Find the potential at the interface $x = t$.

9.16. Repeat Problem 9.15 if the two media are imperfect dielectrics having conductivities σ_1 and σ_2.

9.17. The potential distribution at the mouth of the slot of Fig. 9.7 is given by

$$V = V_0 \sin \frac{\pi y}{b} + \frac{1}{3} V_0 \sin \frac{3\pi y}{b}$$

(a) Find the solution for the potential distribution inside the slot. (b) Compute the value of the potential at the center of the slot, assuming the slot to be square.

9.18. Repeat Problem 9.17 for the potential distribution at the mouth of the slot given by

$$V = V_0 \sin^3 \frac{\pi y}{b}$$

9.19. Assume that the rectangular slot of Fig. 9.7 is covered at the mouth by conducting plates such that the potential distribution is given by

$$V = \begin{cases} 0 & \text{for } 0 < y < b/4 \\ V_0 & \text{for } b/4 < y < 3b/4 \\ 0 & \text{for } 3b/4 < y < b \end{cases}$$

Find the solution for the potential inside the slot.

9.20. For the rectangular slot of Example 9.4, (a) find the expression for the electric field intensity inside the slot and (b) find the electric field intensity at the center of the slot, assuming the slot to be square.

9.21. For the slot of Example 9.4, assume $a = b$ and the potential distribution at the mouth to be $100 \sin^3 (\pi y/b)$ V. Compute the value of the potential at the center of the slot by (a) applying the iteration method to a 4×4 grid of squares, (b) using the 4×4 grid of squares to obtain the exact solution consistent with Eq. (9.49), and (c) applying the iteration method to an 8×8 grid of squares. Compare the results with the exact value given by the analytical solution found in Problem 9.18. Use a value of $\Delta = 0.01$ V for the iteration methods.

9.22. For the example of Fig. 9.10, divide the slot into a 16×16 grid of squares and by computing the potentials at the grid points surrounding the center of the slot by using the iteration technique and $\Delta = 0.01$ V, estimate the value of the electric field intensity at the center of the slot. Compare the estimated value with the exact value obtained in Problem 9.20.

9.23. The cross section of a structure that repeats endlessly in the plane of the paper is shown in Fig. 9.18. For the grid of points shown in the figure, compute the exact value of the potential at point A consistent with Eq. (9.49).

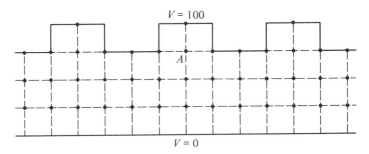

Figure 9.18. For Problem 9.23.

9.24. By considering Laplace's equation in three dimensions, show that the potential at a given point P in a charge-free region is approximately equal to the average of the potentials at the six equidistant points lying along mutually perpendicular axes through P. Then compute by the iteration method the potential at the center of the cubical box shown in Fig. 9.19 in which the top face is kept at 100 V relative to the other five faces. Use a $4 \times 4 \times 4$ grid of cubes and a value of 0.01 V for Δ.

9.25. For the structure of Fig. 9.11, assume that the medium between the plates is an imperfect dielectric of conductivity σ. (a) Show that the input impedance correct to the first power in ω is the same as if σ were zero. (b) Obtain the input impedance correct to the second power in ω and determine the equivalent circuit.

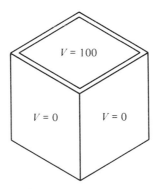

Figure 9.19. For problem 9.24.

9.26. For the structure of Fig. 9.11, continue the analysis beyond the quasistatic extension and obtain the input impedance correct to the third power in ω. Determine the equivalent circuit.

9.27. For the structure of Fig. 9.12, assume $\sigma = 0$ and continue the analysis beyond the quasistatic extension to obtain the input admittance correct to the third power in ω. Determine the equivalent circuit.

9.28. Find the condition(s) under which the quasistatic input behavior of the structure of Fig. 9.12 is essentially equivalent to (a) a capacitor in parallel with a resistor and (b) a resistor in series with an inductor.

9.29. For the toroidal magnetic circuit of Fig. 9.14, assume $A = 5$ cm^2, $l = 20$ cm, and annealed sheet steel for the material of the core. Find the reluctance of the circuit for an applied NI equal to 400 amp-turns.

9.30. For the magnetic circuit of Fig. 9.17, assume the air gap to be in the center leg. Find the required NI to establish a magnetic flux of 9×10^{-4} Wb in the air gap.

9.31. For the magnetic circuit of Fig. 9.17, assume that there is no air gap. Find the magnetic flux established in the center leg for an applied NI equal to 180 amp-turns.

9.32. For the magnetic circuit of Fig. 9.17, assume no air gap and $A_1 = 5$ cm^2 with all other dimensions remaining as specified in Example 9.7. Find the magnetic flux density in the center leg for an applied NI equal to 150 amp-turns.

10. SPECIAL TOPICS

In Chap. 1 we learned the basic mathematical tools and physical concepts of vectors and fields. In Chaps. 2 and 3 we learned the fundamental laws of electromagnetics, namely, Maxwell's equations, first in integral form and then in differential form. Then in Chaps. 4 through 9 we studied the elements of their engineering applications which comprised the topics of propagation, transmission, and radiation of electromagnetic waves, and static and quasi-static fields.

This final chapter is devoted to seven independent topics that are based on Chaps. 4 through 9, in that order. The first six topics can be studied separately following the respective chapters. The seventh topic can be studied following Chaps. 8 and 9. These special topics, although independent of each other, have the common goal of extending the knowledge gained in the corresponding previous chapter for the purpose of illustrating a related phenomenon, or application, or technique.

10.1 WAVE PROPAGATION IN IONIZED MEDIUM

In Chap. 4 we studied uniform plane wave propagation in free space. In this section we shall extend the discussion to wave propagation in ionized medium. An example of ionized medium is the earth's ionosphere which is a region of the upper atmosphere extending from approximately 50 km to more than 1000 km above the earth. In this region the constituent gases are ionized, mostly because of ultraviolet radiation from the sun, thereby resulting in the production of positive ions and electrons that are free to move under the influence of the fields of a wave incident upon the medium. The positive ions are, however, heavy compared to electrons and hence they are relatively immobile. The electron motion produces a current that influences the wave propagation.

In fact, in Sec. 1.5 we considered the motion of a cloud of electrons of uniform density N under the influence of a time-varying electric field

$$\mathbf{E} = E_0 \cos \omega t \, \mathbf{i}_x \tag{10.1}$$

and found that the resulting current density is given by

$$\mathbf{J} = \frac{Ne^2}{m\omega} E_0 \sin \omega t \, \mathbf{i}_x = \frac{Ne^2}{m} \int \mathbf{E} \, dt \tag{10.2}$$

where e and m are the electronic charge and mass, respectively. This result is based on the mechanism of continuous acceleration of the electrons by the force due to the applied electric field. In the case of the ionized medium, the electron motion is, however, impeded by the collisions of the electrons with the heavy particles and other electrons. We shall ignore these collisions as well as the negligible influence of the magnetic field associated with the wave.

Considering uniform plane wave propagation in the z direction in an unbounded ionized medium, and with the electric field oriented in the x direction, we then have

$$\frac{\partial E_x}{\partial z} = -\frac{\partial B_y}{\partial t} = -\mu_0 \frac{\partial H_y}{\partial t} \tag{10.3a}$$

$$\frac{\partial H_y}{\partial z} = -J_x - \frac{\partial D_x}{\partial t} = -\frac{Ne^2}{m} \int E_x \, dt - \epsilon_0 \frac{\partial E_x}{\partial t} \tag{10.3b}$$

Differentiating (10.3a) with respect to z and then substituting for $\partial H_y/\partial z$ from (10.3b), we obtain the wave equation

$$\frac{\partial^2 E_x}{\partial z^2} = -\mu_0 \frac{\partial}{\partial t}\left[-\frac{Ne^2}{m}\int E_x \, dt - \epsilon_0 \frac{\partial E_x}{\partial t}\right]$$

$$= \frac{\mu_0 Ne^2}{m} E_x + \mu_0 \epsilon_0 \frac{\partial^2 E_x}{\partial t^2} \qquad (10.4)$$

Substituting

$$E_x = E_0 \cos(\omega t - \beta z) \qquad (10.5)$$

corresponding to the uniform plane wave solution into (10.4) and simplifying, we get

$$\beta^2 = \omega^2 \mu_0 \epsilon_0 - \frac{\mu_0 Ne^2}{m}$$

$$= \omega^2 \mu_0 \epsilon_0 \left(1 - \frac{Ne^2}{m\epsilon_0 \omega^2}\right) \qquad (10.6)$$

Thus the phase constant for propagation in the ionized medium is given by

$$\beta = \omega\sqrt{\mu_0 \epsilon_0 \left(1 - \frac{Ne^2}{m\epsilon_0 \omega^2}\right)} \qquad (10.7)$$

This result indicates that the ionized medium behaves as through the permittivity of free space is modified by the multiplying factor $\left(1 - \dfrac{Ne^2}{m\epsilon_0 \omega^2}\right)$. We may therefore write

$$\beta = \omega\sqrt{\mu_0 \epsilon_{\text{eff}}} \qquad (10.8)$$

where

$$\epsilon_{\text{eff}} = \epsilon_0 \left(1 - \frac{Ne^2}{m\epsilon_0 \omega^2}\right) \qquad (10.9)$$

is the "effective permittivity" of the ionized medium. We note that for $\omega \rightarrow \infty$, $\epsilon_{\text{eff}} \rightarrow \epsilon_0$ and the medium behaves just as free space. This is to be expected since (10.2) indicates that for $\omega \rightarrow \infty$, $\mathbf{J} \rightarrow 0$. As ω decreases from ∞, ϵ_{eff} becomes less and less until for ω equal to $\sqrt{Ne^2/m\epsilon_0}$, ϵ_{eff} becomes zero. Hence for $\omega > \sqrt{Ne^2/m\epsilon_0}$, ϵ_{eff} is positive, β is real, and the solution for the electric field remains to be that of a propagating wave. For $\omega < \sqrt{Ne^2/m\epsilon_0}$, ϵ_{eff} is negative, β becomes imaginary, and the solution for the electric field corresponds to no propagation.

Thus waves of frequency $f > \sqrt{Ne^2/4\pi^2 m\epsilon_0}$ propagate in the ionized medium and waves of frequency $f < \sqrt{Ne^2/4\pi^2 m\epsilon_0}$ do not propagate. The quantity $\sqrt{Ne^2/4\pi^2 m\epsilon_0}$ is known as the "plasma frequency" and is denoted by the symbol, f_N. Substituting values for e, m, and ϵ_0, we get

$$f_N = \sqrt{80.6N} \text{ Hz} \qquad (10.10)$$

where N is in electrons per meter cubed. We can now write ϵ_{eff} as

$$\epsilon_{\text{eff}} = \epsilon_0\left(1 - \frac{f_N^2}{f^2}\right) \tag{10.11}$$

Proceeding further, we obtain the phase velocity for the propagating range of frequencies, that is, for $f > f_N$, to be

$$v_p = \frac{1}{\sqrt{\mu_0\epsilon_{\text{eff}}}} = \frac{1}{\sqrt{\mu_0\epsilon_0(1 - f_N^2/f^2)}}$$

$$= \frac{c}{\sqrt{1 - f_N^2/f^2}} \tag{10.12}$$

where $c = 1/\sqrt{\mu_0\epsilon_0}$ is the velocity of light in free space. From (10.12), we observe that $v_p > c$ and is a function of the wave frequency. The fact that $v_p > c$ is not a violation of the principle of relativity since the dispersive nature of the medium resulting from the dependence of v_p upon f ensures that information always travels with a velocity less than c (see Sec. 7.4).

To apply what we have learned above concerning propagation in an ionized medium to the case of the earth's ionosphere, we first provide a brief description of the ionosphere. A typical distribution of the ionospheric electron density versus height above the earth is shown in Fig. 10.1. The electron

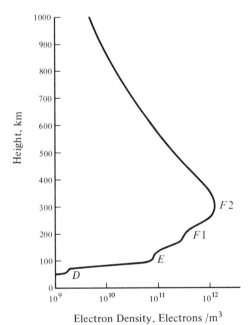

Electron Density, Electrons $/\text{m}^3$

Figure 10.1. A typical distribution of ionospheric electron density versus height above the earth.

density exists in the form of several layers known as D, E, and F layers in which the ionization changes with the hour of the day, the season, the sunspot cycle, and the geographic location. The nomenclature behind the designation of the letters for the layers is due to Appleton in England who in 1925 and at about the same time as Breit and Tuve in the United States demonstrated experimentally the reflection of radio waves by the ionosphere. In his early work, Appleton was accustomed to writing E for the electric field of the wave reflected from the first layer he recognized. Later, when he recognized a second layer, at a greater height, he wrote F for the field of the wave reflected from it. Still later he conjectured that there might be a third layer lower from either of the first two and thus he decided to name the possible lower layer D, thereby leaving earlier letters of the alphabet for other possible undiscovered, still lower layers. Electrons were indeed detected later in the D region.

The D region extends over the altitude range of about 50 km to about 90 km. Since collisions between electrons and heavy particles cannot be neglected in this region, it is mainly an absorbing region. The E region extends from about 90 km to about 150 km. Diurnal and seasonal variations of the E layer electron density are strongly correlated with the zenith angle of the sun. In the F region the lower of the two strata is designated as the $F1$ layer and the higher, more intense ionized stratum is designated as the $F2$ layer. The $F1$ ledge is usually located between 160 km and 200 km. Above this region the $F2$ layer electron density increases with altitude, reaching a peak at a height generally lying between 250 km and 400 km. Above this peak the electron density decreases monotonically with altitude. The $F1$ ledge is present only during the day. During the night the $F1$ and $F2$ layers are identified as a single F layer. The $F2$ layer is the most important from the point of view of radio communication since it contains the greatest concentration of electrons. Paradoxically, it also exhibits several anomalies.

Wave propagation in the ionosphere is complicated by the presence of the earth's magnetic field. If we neglect the earth's magnetic field, then for a wave of frequency f incident vertically on the ionosphere from a transmitter on the ground, it is evident from the propagation condition $f > f_N$ that the wave propagates up to the height at which $f = f_N$ and since it cannot propagate beyond that height, it gets reflected at that height. Thus waves of frequencies less than the maximum plasma frequency corresponding to the peak of the $F2$ layer cannot penetrate the ionosphere. Hence for communication with satellites orbiting above the peak of the ionosphere, frequencies greater than this maximum plasma frequency, also known as the "critical frequency," must be employed. While this critical frequency is a function of the time of day, the season, the sunspot cycle, and the geographic location, it is not greater than about 15 MHz and can be as low as a few megahertz. For a wave incident obliquely on the ionosphere, reflection is possible for frequencies greater than the critical frequency, up to about three times its value. Hence for earth-to-satellite communication, frequencies generally exceeding about 40 MHz are employed. Lower frequencies permit long-distance, ground-to-ground communication via reflections from the ionospheric layers. This mode of propagation is familiarly known as the "sky wave mode" of propagation. For very low frequencies of the order of several kilohertz and less, the lower boundary of the ionosphere and the earth form a waveguide, thereby permitting waveguide mode of propagation.

In Sec. 4.5 we learned that Doppler shift of frequency occurs when the source or the observer is in motion. Doppler shift can also occur for the case of fixed source and observer if the medium in which the wave propagates is changing with time. The ionosphere provides an example of this phenomenon. For simplicity, let us consider a hypothetical plane slab ionosphere of thickness s and having uniform electron density N. Then for a uniform plane wave of frequency ω propagating normal to the slab, the phase shift undergone by the wave in the thickness of the slab is given by

$$
\phi = \omega t - \beta s = \omega t - \omega \sqrt{\mu_0 \epsilon_0 \left(1 - \frac{Ne^2}{m\epsilon_0 \omega^2}\right)} s
$$

$$
= \omega t - \frac{\omega s}{c} \sqrt{1 - \frac{80.6N}{f^2}} \qquad (10.13)
$$

If the electron density is now varying with time, the rate of change of phase with time is given by

$$
\frac{d\phi}{dt} = \omega + \frac{40.3\,\omega s}{cf^2}\left(1 - \frac{80.6N}{f^2}\right)^{-1/2} \frac{dN}{dt} \qquad (10.14)
$$

Thus the Doppler shift in the frequency is

$$
\omega_D = \frac{40.3\,\omega s}{cf^2}\left(1 - \frac{80.6N}{f^2}\right)^{-1/2} \frac{dN}{dt}
$$

or

$$
f_D = \frac{40.3s}{cf}\left(1 - \frac{80.6N}{f^2}\right)^{-1/2} \frac{dN}{dt} \qquad (10.15)
$$

The Doppler shift introduced by the changing ionosphere can be a source of error in satellite navigational systems based on the Doppler shift due to the moving satellite. It is, however, one of the tools for studying the ionosphere.

In this section we learned that in an ionized medium, wave propagation occurs only for frequencies exceeding the plasma frequency corresponding to the electron density. Applying this to the case of the earth's ionosphere, we found that this imposes a lower limit in frequency for communication with satellites. We also extended the discussion of Doppler shift to the case of a time-varying propagation medium.

REVIEW QUESTIONS

10.1. What is an ionized medium? What influences wave propagation in an ionized medium?

10.2. Provide physical explanation for the frequency dependence of the effective permittivity of an ionized medium.

10.3. Discuss the condition for propagation in an ionized medium.

10.4. What is plasma frequency? How is it related to the electron density?

10.5. Provide a brief description of the earth's ionosphere and discuss how it affects communication.

10.6. Discuss the phenomenon of Doppler shift due to a time-varying medium.

PROBLEMS

10.1. Assume the ionosphere to be represented by a parabolic distribution of electron density as given by

$$N(h) = \frac{10^{14}}{80.6}\left[1 - \left(\frac{h - 300}{100}\right)^2\right] \text{el/m}^3 \qquad \text{for } 200 < h < 400$$

where h is the height above the ground in kilometers. (a) Find the height at which a vertically incident wave of frequency 8 MHz is reflected. (b) Find the frequency of a vertically incident wave which gets reflected at a height of 220 km. (c) What is the lowest frequency below which communication is not possible across the peak of the layer?

10.2. For a uniform plane wave of frequency 10 MHz propagating normal to a slab of ionized medium of thickness 50 km and uniform plasma frequency 8 MHz, find (a) the phase velocity in the slab, (b) the wavelength in the slab, and (c) the number of wavelengths undergone by the wave in the slab.

10.3. For a uniform plane wave propagating normal to a hypothetical slab ionosphere of thickness 100 km and uniform electron density $(10^{14}/80.6)$ el/m^3, changing with time at the rate of 10^8 el/m^3/s, find the Doppler shift in frequency for (a) $f = 10.1$ MHz and (b) $f = 40$ MHz.

10.4. If you have studied Sec. 7.4, you should be able to show that the group velocity for propagation in the ionized medium is given by

$$v_g = c\sqrt{1 - \frac{f_N^2}{f^2}}$$

Show that for a hypothetical slab ionosphere of thickness s and uniform plasma frequency f_N, a narrow-band modulated signal of carrier frequency $f \gg f_N$ propagating normal to the slab undergoes a time delay in excess of that of the free space value by the amount $f_N^2 s / 2cf^2$.

10.2 WAVE PROPAGATION IN ANISOTROPIC MEDIUM

In Sec. 5.2 we learned that for certain dielectric materials known as "anisotropic dielectric materials," $\mathbf{D}$ is not in general parallel to $\mathbf{E}$ and the relationship between $\mathbf{D}$ and $\mathbf{E}$ is expressed by means of a permittivity tensor consisting of a 3×3 matrix. Similarly, in Sec. 5.3 we learned of the anisotropic property of certain magnetic materials. There are several important applications based on wave propagation in anisotropic materials. A general treatment is, however, very complicated. Hence we shall consider two simple cases.

For the first example, we consider an anisotropic dielectric medium characterized by the $\mathbf{D}$ to $\mathbf{E}$ relationship given by

$$\begin{bmatrix} D_x \\ D_y \\ D_z \end{bmatrix} = \begin{bmatrix} \epsilon_{xx} & 0 & 0 \\ 0 & \epsilon_{yy} & 0 \\ 0 & 0 & \epsilon_{zz} \end{bmatrix} \begin{bmatrix} E_x \\ E_y \\ E_z \end{bmatrix} \qquad (10.16)$$

and having the permeability μ_0. This simple form of permittivity tensor can be achieved in certain anisotropic liquids and crystals by an appropriate choice of the coordinate system. It is easy to see that the characteristic polarizations for this case are all linear directed along the coordinate axes and having the effective permittivities ϵ_{xx}, ϵ_{yy}, and ϵ_{zz} for the x-, y-, and z-directed polarizations, respectively. Let us consider a uniform plane wave propagating in the z direction. The wave will generally contain both x and y components of the fields. It can be decomposed into two waves, one having an x-directed electric field and the other having a y-directed electric field. These component waves travel individually in the anisotropic medium as though it is isotropic but with different phase velocities since the effective permittivities are different. In view of this, the phase relationship between the two waves, and hence the polarization of the composite wave, changes with distance along the direction of propagation.

To illustrate the foregoing discussion quantitatively, let us consider the electric field of the wave to be linearly polarized at $z = 0$ as given by

$$\mathbf{E}(0) = (E_{x0}\mathbf{i}_x + E_{y0}\mathbf{i}_y)\cos \omega t \qquad (10.17)$$

Then assuming $(+)$ wave only, the electric field at an arbitrary value of z is given by

$$\mathbf{E}(z) = E_{x0} \cos(\omega t - \beta_1 z)\,\mathbf{i}_x + E_{y0} \cos(\omega t - \beta_2 z)\,\mathbf{i}_y \qquad (10.18)$$

where

$$\beta_1 = \omega\sqrt{\mu_0 \epsilon_{xx}} \qquad (10.19a)$$

$$\beta_2 = \omega\sqrt{\mu_0 \epsilon_{yy}} \qquad (10.19b)$$

are the phase constants corresponding to the x-polarized and y-polarized component waves, respectively. Thus the phase difference between the x and y components of the field is given by

$$\Delta\phi = (\beta_2 - \beta_1)z \qquad (10.20)$$

As the composite wave progresses along the z direction, $\Delta\phi$ changes from zero at $z = 0$ to $\pi/2$ at $z = \pi/2(\beta_2 - \beta_1)$ to π at $z = \pi/(\beta_2 - \beta_1)$, and so on. The polarization of the composite wave thus changes from linear at $z = 0$ to elliptical for $z > 0$, becoming linear again at $z = \pi/(\beta_2 - \beta_1)$, but rotated by an angle of $2\tan^{-1}(E_{y0}/E_{x0})$, as shown in Fig. 10.2. Thereafter, it becomes elliptical again, returning back to the original linear polarization at $z = 2\pi/(\beta_2 - \beta_1)$, and so on.

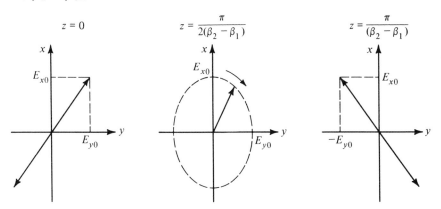

Figure 10.2. The change in polarization of the field of a wave propagating in the anisotropic dielectric medium characterized by Eq. (10.16).

For the second example, we consider propagation in a ferrite medium. Ferrites are a class of magnetic materials which when subject to a d.c. magnetizing field exhibit anisotropic magnetic properties. Since there are phase differences associated with the relationships between the components of **B** and the components of **H** due to this anisotropy, it is convenient to use the phasor notation and write the relationship in terms of the phasor components. For an applied d.c. magnetic field along the direction of propagation of the wave, which we assume to be the z direction, this relationship is given by

$$\begin{bmatrix} \bar{B}_x \\ \bar{B}_y \\ \bar{B}_z \end{bmatrix} = \begin{bmatrix} \mu & -j\kappa & 0 \\ j\kappa & \mu & 0 \\ 0 & 0 & \mu_0 \end{bmatrix} \begin{bmatrix} \bar{H}_x \\ \bar{H}_y \\ \bar{H}_z \end{bmatrix} \tag{10.21}$$

where μ and κ depend upon the material, the strength of the d.c. magnetic field, and the wave frequency.

To find the characteristic polarizations, we first note from (10.21) that

$$\bar{B}_x = \mu\bar{H}_x - j\kappa\bar{H}_y \tag{10.22a}$$

$$\bar{B}_y = j\kappa\bar{H}_x + \mu\bar{H}_y \tag{10.22b}$$

Setting $\bar{B}_x/\bar{B}_y$ equal to $\bar{H}_x/\bar{H}_y$, we then have

$$\frac{\mu\bar{H}_x - j\kappa\bar{H}_y}{j\kappa\bar{H}_x + \mu\bar{H}_y} = \frac{\bar{H}_x}{\bar{H}_y}$$

which upon solution for $\bar{H}_x/\bar{H}_y$ gives

$$\frac{\bar{H}_x}{\bar{H}_y} = \pm j \tag{10.23}$$

This result corresponds to equal amplitudes of H_x and H_y and phase difference of $\pm 90°$. Thus the characteristic polarizations are both circular, rotating in opposite senses as viewed along the z direction.

The effective permeabilities of the ferrite medium corresponding to the characteristic polarizations are

$$\frac{\bar{B}_x}{\bar{H}_x} = \frac{\mu\bar{H}_x - j\kappa\bar{H}_y}{\bar{H}_x}$$

$$= \mu - j\kappa\frac{\bar{H}_y}{\bar{H}_x}$$

$$= \mu \mp \kappa \qquad \text{for } \frac{\bar{H}_x}{\bar{H}_y} = \pm j \tag{10.24}$$

The phase constants associated with the propagation of the characteristic waves are

$$\beta_\pm = \omega\sqrt{\epsilon(\mu \mp \kappa)} \tag{10.25}$$

where the subscripts $+$ and $-$ refer to $\bar{H}_x/\bar{H}_y = +j$ and $\bar{H}_x/\bar{H}_y = -j$, respectively. We note from (10.25) that β_+ can become imaginary if $(\mu - \kappa) < 0$. When this happens, wave propagation does not occur for that characteristic polarization. We shall hereafter assume that the wave frequency is such that both characteristic waves propagate.

Let us now consider the magnetic field of the wave to be linearly polarized in the x direction at $z = 0$, that is,

$$\mathbf{H}(0) = H_0 \cos \omega t \, \mathbf{i}_x \tag{10.26}$$

Then we can express (10.26) as the superposition of two circularly polarized fields having opposite senses of rotation in the xy plane in the manner

$$\mathbf{H}(0) = \left(\frac{H_0}{2} \cos \omega t \, \mathbf{i}_x + \frac{H_0}{2} \sin \omega t \, \mathbf{i}_y \right)$$
$$+ \left(\frac{H_0}{2} \cos \omega t \, \mathbf{i}_x - \frac{H_0}{2} \sin \omega t \, \mathbf{i}_y \right) \tag{10.27}$$

The circularly polarized field inside the first pair of parentheses on the right side of (10.27) corresponds to

$$\frac{\bar{H}_x}{\bar{H}_y} = \frac{H_0/2}{-jH_0/2} = +j$$

whereas that inside the second pair of parentheses corresponds to

$$\frac{\bar{H}_x}{\bar{H}_y} = \frac{H_0/2}{jH_0/2} = -j$$

Assuming propagation in the positive z direction, the field at an arbitrary value of z is then given by

$$\mathbf{H}(z) = \left[\frac{H_0}{2} \cos(\omega t - \beta_+ z) \, \mathbf{i}_x + \frac{H_0}{2} \sin(\omega t - \beta_+ z) \, \mathbf{i}_y \right]$$
$$+ \left[\frac{H_0}{2} \cos(\omega t - \beta_- z) \, \mathbf{i}_x - \frac{H_0}{2} \sin(\omega t - \beta_- z) \, \mathbf{i}_y \right]$$
$$= \left[\frac{H_0}{2} \cos\left(\omega t - \frac{\beta_+ + \beta_-}{2} z - \frac{\beta_+ - \beta_-}{2} z \right) \mathbf{i}_x \right.$$
$$\left. + \frac{H_0}{2} \sin\left(\omega t - \frac{\beta_+ + \beta_-}{2} z - \frac{\beta_+ - \beta_-}{2} z \right) \mathbf{i}_y \right]$$
$$+ \left[\frac{H_0}{2} \cos\left(\omega t - \frac{\beta_+ + \beta_-}{2} z + \frac{\beta_+ - \beta_-}{2} z \right) \mathbf{i}_x \right.$$
$$\left. - \frac{H_0}{2} \sin\left(\omega t - \frac{\beta_+ + \beta_-}{2} z + \frac{\beta_+ - \beta_-}{2} z \right) \mathbf{i}_y \right]$$
$$= \left[H_0 \cos\left(\frac{\beta_- - \beta_+}{2} z \right) \mathbf{i}_x + H_0 \sin\left(\frac{\beta_- - \beta_+}{2} z \right) \mathbf{i}_y \right]$$
$$\cdot \cos\left(\omega t - \frac{\beta_+ + \beta_-}{2} z \right) \tag{10.28}$$

The result given by (10.28) indicates that the x and y components of the field are in phase at any given value of z. Hence the field is linearly polarized for all values of z. The direction of polarization is, however, a function of z since

$$\frac{H_y}{H_x} = \frac{H_0 \sin [(\beta_- - \beta_+)/2]z}{H_0 \cos [(\beta_- - \beta_+)/2]z} = \tan \frac{\beta_- - \beta_+}{2}z \qquad (10.29)$$

and hence the angle made by the field vector with the x axis is $\frac{\beta_- - \beta_+}{2}z$.

Thus the direction of polarization rotates linearly with z at a rate of $\frac{\beta_- - \beta_+}{2}$.

This phenomenon is known as "Faraday rotation" and is illustrated with the aid of the sketches in Fig. 10.3. The sketches in any given column correspond to a fixed value of z whereas the sketches in a given row correspond to a fixed value of t. At $z = 0$, the field is linearly polarized in the x direction and is the superposition of two counter-rotating circularly polarized fields as shown by the time series of sketches in the first column. If the medium is isotropic, the two counter-rotating circularly polarized fields undergo the same amount of phase lag with z and the field remains linearly polarized in the x direction as shown by the dashed lines in the second and third columns. For the case of the anisotropic medium, the two circularly polarized fields undergo different amounts of phase lag with z. Hence their superposition results in a linear polarization making an angle with the x direction and increasing linearly with z as shown by the solid lines in the second and third columns.

The phenomenon of Faraday rotation in a ferrite medium that we have just discussed forms the basis for a number of devices in the microwave field. The phenomenon itself is not restricted to ferrites. For example, an ionized medium immersed in a d.c. magnetic field possesses anisotropic properties which give rise to Faraday rotation of a linearly polarized wave propagating along the d.c. magnetic field. A natural example of this is propagation along the earth's magnetic field in the ionosphere. A simple modern example of the application of Faraday rotation is, however, illustrated by the magneto-optical switch. In fact, Faraday rotation was originally discovered in the optics regime.

The magneto-optical switch is a device for modulating a laser beam by switching on and off an electric current. The electric current generates a magnetic field that rotates the magnetization vector in a magnetic iron-garnet film on a substrate of garnet, in the plane of the film through which a light wave passes. When it enters the film, the light wave field is linearly polarized normal to the plane of the film. If the current in the electric circuit is off, the

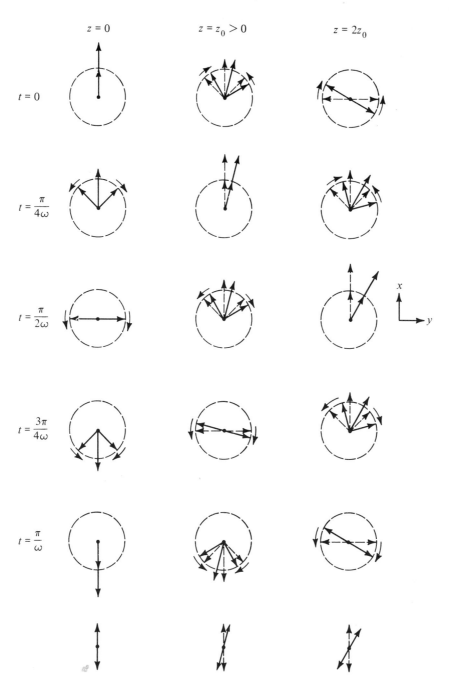

Figure 10.3. For illustrating the phenomenon of Faraday rotation.

magnetization vector is normal to the direction of propagation of the wave and the wave emerges out of the film without change of polarization, as shown in Fig. 10.4(a). If the current in the electric circuit is on, the magnetization vector is parallel to the direction of propagation of the wave, the light wave undergoes Faraday rotation and emerges out of the film with its polarization rotated by 90°, as shown in Fig. 10.4(b). After it emerges out of the film, the light beam is passed through a polarizer which has the property of absorbing light of the original polarization but passing through the light of

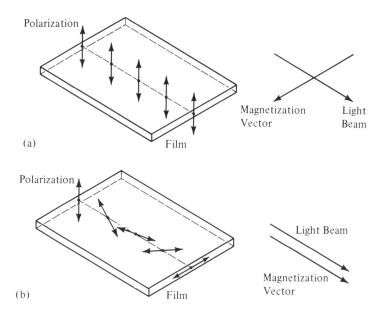

Figure 10.4. For illustrating the principle of operation of a magneto-optical switch.

the 90°-rotated polarization. Thus the beam is made to turn on and off by the switching on and off of the current in the electric circuit. In this manner, any coded message can be made to be carried by the light beam.

In this section we discussed wave propagation in an anisotropic medium. In particular, we learned that in a ferrite medium, a linearly polarized wave propagating along the direction of an applied d.c. magnetic field undergoes Faraday rotation. We then briefly mentioned other examples of media in which Faraday rotation takes place and finally discussed the operation of the magneto-optical switch, a device employing Faraday rotation for modulating a light beam.

REVIEW QUESTIONS

10.7. Discuss the principle behind wave propagation in an anisotropic medium based on the decomposition of the wave into characteristic waves.

10.8. When does a wave propagate in an anisotropic medium without change in polarization?

10.9. What is Faraday rotation? When does Faraday rotation take place in an anisotropic medium?

10.10. Consult appropriate reference books and list three applications of Faraday rotation.

10.11. What is a magneto-optical switch? Discuss its operation.

PROBLEMS

10.5. For the anisotropic medium characterized by the **D** to **E** relationship given by Eq. (10.16), assume $\epsilon_{xx} = 4\epsilon_0$, $\epsilon_{yy} = 9\epsilon_0$, and $\epsilon_{zz} = 2\epsilon_0$, and find the distance in which the phase difference between the x and y components of a plane wave of frequency 10^9 Hz propagating in the z direction changes by the amount π.

10.6. Show that for plane wave propagation in an anisotropic medium, the angle between **E** and **H** is not in general equal to $90°$. For the anisotropic dielectric medium of Problem 10.5, find the angle between **E** and **H** for **E** linearly polarized along the bisector of the angle between the x and y axes.

10.7. For a wave of frequency ω, the quantities μ and κ in the permeability matrix of Eq. (10.21) are given by

$$\mu = \mu_0 \left[1 + \frac{\omega_0 \omega_M}{\omega_0^2 - \omega^2} \right]$$

$$\kappa = -\mu_0 \frac{\omega \omega_M}{\omega_0^2 - \omega^2}$$

where $\omega_0 = \mu_0 |e| H_0/m$, $\omega_M = \mu_0 |e| M_0/m$, H_0 is the d.c. magnetizing field, M_0 is the magnetic dipole moment per unit volume in the material in the absence of the wave, e is the charge of an electron, and m is the mass of an electron. (a) Show that the effective permeabilities corresponding to the characteristic polarizations are $\mu_0 \left[1 + \dfrac{\omega_M}{\omega_0 \mp \omega} \right]$ for $\bar{H}_x/\bar{H}_y = \pm j$. (b) Compute the Faraday rotation angle in degrees per centimeter along the z direction for $\omega = 10^{10}$ rad/s, if $\omega_M = 5 \times 10^{10}$ rad/s, $\omega_0 = 1.5 \times 10^{10}$ rad/s, and $\epsilon = 9\epsilon_0$.

10.8. For the quantities defined in Problem 10.7 for the ferrite medium, show that for $\omega_0 \ll \omega$ and $\omega_M \ll \omega$, the Faraday rotation per unit distance along the z direction is $\frac{\omega_M}{2}\sqrt{\mu_0 \epsilon}$. Compute its value in degrees per centimeter if $\omega_M = 5 \times 10^{10}$ rad/s and $\epsilon = 9\epsilon_0$.

10.3   THE SMITH CHART

In Sec. 6.6 we studied reflection and transmission at the junction of two transmission lines. We found that when a line of certain characteristic impedance is terminated by another line of different characteristic impedance, as shown in Fig. 10.5, standing waves result on the first line. The degree of existence of the standing waves was defined by the standing wave ratio (SWR) which is the ratio of the voltage maximum to the voltage minimum of the standing wave pattern. In this section we shall proceed further and introduce the Smith Chart, which is a useful graphical aid in the solution of transmission-line and many other problems.

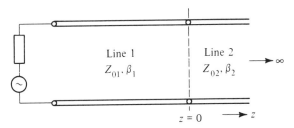

Figure 10.5. A transmission line terminated by another infinitely long transmission line.

First we define the line impedance $\bar{Z}(z)$ at a given value of z on the line as the ratio of the complex line voltage to the complex line current at that value of z, that is,

$$\bar{Z}(z) = \frac{\bar{V}(z)}{\bar{I}(z)} \qquad (10.30)$$

From the solutions for the line voltage and line current on line 2 given by (6.71a) and (6.71b), respectively, the line impedance in line 2 is given by

$$\bar{Z}_2(z) = \frac{\bar{V}_2(z)}{\bar{I}_2(z)} = Z_{02}$$

Thus the line impedance at all points on line 2 is simply equal to the characteristic impedance of that line. This is because the line is infinitely long and hence there is only a $(+)$ wave on the line. From the solutions for the line voltage and line current in line 1 given by (6.70a) and (6.70b), respectively, the line impedance for that line is given by

$$\bar{Z}_1(z) = \frac{\bar{V}_1(z)}{\bar{I}_1(z)} = Z_{01}\frac{\bar{V}_1^+ e^{-j\beta_1 z} + \bar{V}_1^- e^{j\beta_1 z}}{\bar{V}_1^+ e^{-j\beta_1 z} - \bar{V}_1^+ e^{j\beta_1 z}}$$

$$= Z_{01}\frac{1 + \bar{\Gamma}_V(z)}{1 - \bar{\Gamma}_V(z)} \qquad (10.31)$$

where

$$\bar{\Gamma}_V(z) = \frac{\bar{V}_1^- e^{j\beta_1 z}}{\bar{V}_1^+ e^{-j\beta_1 z}} = \bar{\Gamma}_V(0)e^{j2\beta_1 z} \qquad (10.32)$$

$$\bar{\Gamma}_V(0) = \frac{\bar{V}_1^-}{\bar{V}_1^+} = \frac{Z_{02} - Z_{01}}{Z_{02} + Z_{01}} \qquad (10.33)$$

The quantity $\bar{\Gamma}_V(0)$ is the voltage reflection coefficient at the junction $z = 0$ and $\bar{\Gamma}_V(z)$ is the voltage reflection coefficient at any value of z.

To compute the line impedance at a particular value of z, we first compute $\bar{\Gamma}_V(0)$ from a knowledge of Z_{02} which is the terminating impedance to line 1. We then compute $\bar{\Gamma}_V(z) = \bar{\Gamma}_V(0)e^{j2\beta_1 z}$ which is a complex number having the same magnitude as that of $\bar{\Gamma}_V(0)$ but a phase angle equal to $2\beta_1 z$ plus the phase angle of $\bar{\Gamma}_V(0)$. The computed value of $\bar{\Gamma}_V(z)$ is then substituted in (10.31) to find $\bar{Z}_1(z)$. All of this complex algebra is eliminated through the use of the Smith Chart.

The Smith Chart is a mapping of the values of normalized line impedance onto the reflection coefficient $(\bar{\Gamma}_V)$ plane. The normalized line impedance $\bar{Z}_n(z)$ is the ratio of the line impedance to the characteristic impedance of the line. From (10.31), and omitting the subscript 1 for the sake of generality, we have

$$\bar{Z}_n(z) = \frac{\bar{Z}(z)}{Z_0} = \frac{1 + \bar{\Gamma}_V(z)}{1 - \bar{\Gamma}_V(z)} \qquad (10.34)$$

Conversely,

$$\bar{\Gamma}_V(z) = \frac{\bar{Z}_n(z) - 1}{\bar{Z}_n(z) + 1} \qquad (10.35)$$

Writing $\bar{Z}_n = r + jx$ and substituting into (10.35), we find that

$$|\bar{\Gamma}_V| = \left|\frac{r + jx - 1}{r + jx + 1}\right| = \frac{\sqrt{(r-1)^2 + x^2}}{\sqrt{(r+1)^2 + x^2}} \le 1 \qquad \text{for } r \ge 0$$

Thus, we note that all passive values of normalized line impedances, that is, points in the right half of the complex $\bar{Z}_n$ plane shown in Fig. 10.6(a) are mapped onto the region within the circle of radius unity in the complex $\bar{\Gamma}_V$ plane shown in Fig. 10.6(b).

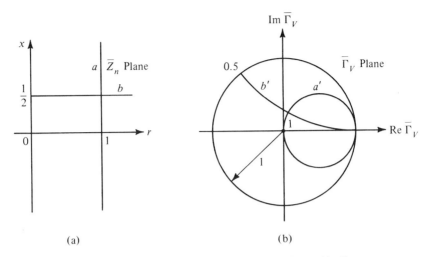

Figure 10.6. For illustrating the development of the Smith Chart.

We can now assign values for $\bar{Z}_n$, compute the corresponding values of $\bar{\Gamma}_V$, and plot them on the $\bar{\Gamma}_V$ plane but indicating the values of $\bar{Z}_n$ instead of the values of $\bar{\Gamma}_V$. To do this in a systematic manner, we consider contours in the $\bar{Z}_n$ plane corresponding to constant values of r, as shown for example by the line marked a for $r = 1$, and corresponding to constant values of x, as shown for example by the line marked b for $x = \frac{1}{2}$ in Fig. 10.6(a).

By considering several points along line a, computing the corresponding values of $\bar{\Gamma}_V$, plotting them on the $\bar{\Gamma}_V$ plane, and joining them, we obtain the contour marked a' in Fig. 10.6(b). Although it can be shown analytically that this contour is a circle of radius $\frac{1}{2}$ and centered at $(1/2, 0)$, it is a simple task to write a computer program to perform this operation, including the plotting. Similarly, by considering several points along line b and following the same procedure, we obtain the contour marked b' in Fig. 10.6(b). Again, it can be shown analytically that this contour is a portion of a circle of radius 2 and centered at $(1, 2)$. We can now identify the points on contour a' as corresponding to $r = 1$ by placing the number 1 beside it and the points on contour b' as corresponding to $x = \frac{1}{2}$ by placing the number 0.5 beside it. The point of intersection of contours a' and b' then corresponds to $\bar{Z}_n = 1 + j0.5$.

When the procedure discussed above is applied to many lines of constant r and constant x covering the entire right half of the $\bar{Z}_n$ plane, we obtain the Smith Chart. In a commercially available form shown in Fig. 10.7, the Smith Chart contains contours of constant r and constant x at appropriate increments of r and x in the range $0 < r < \infty$ and $-\infty < x < \infty$ so that interpolation between the contours can be carried out to a good degree of accuracy.

Let us now consider the transmission line system shown in Fig. 10.8,

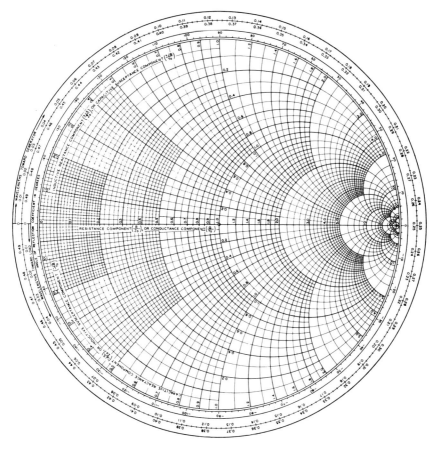

Figure 10.7. A commercially available form of the Smith Chart (copyrighted by and reproduced with the permission of Kay Elemetrics Corp., Pine Brook, N.J.).

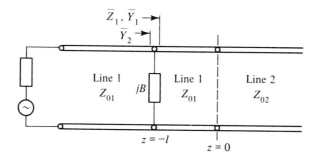

Figure 10.8. A transmission-line system for illustrating the computation of several quantities by using the Smith Chart.

which is the same as that in Fig. 10.5 except that a reactive element having susceptance (reciprocal of reactance) B is connected in parallel with line 1 at a distance l from the junction. Let us assume $Z_{01} = 150$ ohms, $Z_{02} = 50$ ohms, $B = -0.003$ mho, and $l = 0.375\lambda_1$, where λ_1 is the wavelength in line 1 corresponding to the source frequency, and find the following quantities by using the Smith Chart, as shown in Fig. 10.9:

1. $\bar{Z}_1$, line impedance just to the right of jB: First we note that since line 2 is infinitely long, the load for line 1 is simply 50 ohms. Normalizing this with respect to the characteristic impedance of line 1, we obtain the normalized load impedance for line 1 to be

$$\bar{Z}_n(0) = \frac{50}{150} = \frac{1}{3}$$

Locating this on the Smith Chart at point A in Fig. 10.9 amounts to computing the reflection coefficient at the junction, that is, $\bar{\Gamma}_V(0)$. Now the reflection coefficient at $z = -l = -0.375\lambda_1$, being equal to $\bar{\Gamma}_V(0)e^{-j2\beta_1 l} = \bar{\Gamma}_V(0)e^{-j1.5\pi}$, can be located on the Smith Chart by moving A such that the magnitude remains constant but the phase angle decreases by 1.5π. This is equivalent to moving it on a circle with

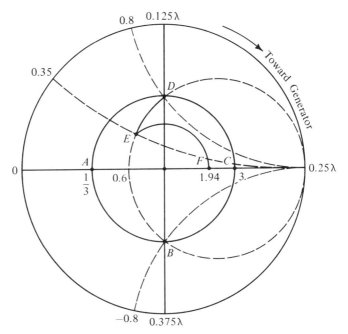

Figure 10.9. For illustrating the use of the Smith Chart in the computation of several quantities for the transmission-line system of Fig. 10.8.

its center at the center of the Smith Chart and in the clockwise direction by 1.5π or $270°$ so that point B is reached. Actually, it is not necessary to compute this angle since the Smith Chart contains a distance scale in terms of λ along its periphery for movement from load toward generator and vice versa, based on a complete revolution for one-half wavelength. The normalized impedance at point B can now be read off the chart and multiplied by the characteristic impedance of the line to obtain the required impedance value. Thus

$$\bar{Z}_1 = (0.6 - j0.8)150 = (90 - j120) \text{ ohms.}$$

2. SWR on line 1 to the right of jB: From (6.81)

$$\text{SWR} = \frac{1 + |\Gamma_V|}{1 - |\Gamma_V|} = \frac{1 + |\bar{\Gamma}_V|e^{j0}}{1 - |\bar{\Gamma}_V|e^{j0}} \tag{10.36}$$

Comparing the right side of (10.36) with the expression for $\bar{Z}_n$ given by (10.34), we note that it is simply equal to $\bar{Z}_n$ corresponding to phase angle of $\bar{\Gamma}_V$ equal to zero. Thus, to find the SWR, we locate the point on the Smith Chart having the same $|\bar{\Gamma}_V|$ as that for $z = 0$, but having a phase angle equal to zero, that is, the point C in Fig. 10.9, and then read off the normalized resistance value at that point. Here, it is equal to 3 and hence the required SWR is equal to 3. In fact, the circle passing through C and having its center at the center of the Smith Chart is known as the "constant SWR ($= 3$) circle" since for any normalized load impedance to line 1 lying on that circle, the SWR is the same (and equal to 3).

3. $\bar{Y}_1$, line admittance just to the right of jB: To find this, we note that the normalized line admittance $\bar{Y}_n$ at any value of z, that is, the line admittance normalized with respect to the line characteristic admittance Y_0 (reciprocal of Z_0) is given by

$$\begin{aligned}
\bar{Y}_n(z) &= \frac{\bar{Y}(z)}{Y_0} = \frac{Z_0}{\bar{Z}(z)} = \frac{1}{\bar{Z}_n(z)} \\
&= \frac{1 - \bar{\Gamma}_V(z)}{1 + \bar{\Gamma}_V(z)} = \frac{1 + \bar{\Gamma}_V(z)e^{\pm j\pi}}{1 - \bar{\Gamma}_V(z)e^{\pm j\pi}} \\
&= \frac{1 + \bar{\Gamma}_V(z)e^{\pm j2\beta\lambda/4}}{1 - \bar{\Gamma}_V(z)e^{\pm j2\beta\lambda/4}} = \frac{1 + \bar{\Gamma}_V(z \pm \lambda/4)}{1 - \bar{\Gamma}_V(z \pm \lambda/4)} \\
&= \bar{Z}_n\left(z \pm \frac{\lambda}{4}\right)
\end{aligned} \tag{10.37}$$

Thus $\bar{Y}_n$ at a given value of z is equal to $\bar{Z}_n$ at a value of z located $\lambda/4$ from it. On the Smith Chart this corresponds to the point on the constant SWR circle passing through B and diametrically opposite to it, that is, the point D. Thus,

$$\bar{Y}_{n1} = 0.6 + j0.8$$

and

$$\bar{Y}_1 = Y_{01}\bar{Y}_{n1} = \frac{1}{150}(0.6 + j0.8)$$

$$= (0.004 + j0.0053) \text{ mho}$$

In fact, the Smith Chart can be used as an admittance chart instead of as an impedance chart, that is, by knowing the line admittance at one point on the line, the line admittance at another point on the line can be found by proceeding in the same manner as for impedances. As an example, to find $\bar{Y}_1$, we can first find the normalized line admittance at $z = 0$ by locating the point C diametrically opposite to point A on the constant SWR cirlce. Then we find $\bar{Y}_{n1}$ by simply going on the constant SWR circle by the distance $l (= 0.375\lambda_1)$ toward the generator. This leads to point D, thereby giving us the same result for $\bar{Y}_1$ as found above.

4. SWR on line 1 to the left of jB: To find this, we first locate the normalized line admittance just to the left of jB, which then determines the constant SWR circle corresponding to the portion of line 1 to the left of jB. Thus, noting that $\bar{Y}_2 = \bar{Y}_1 + jB$, or $\bar{Y}_{n2} = \bar{Y}_{n1} + jB/Y_{01}$, and hence

$$\text{Re}[\bar{Y}_{n2}] = \text{Re}[\bar{Y}_{n1}] \qquad (10.38a)$$

$$\text{Im}[\bar{Y}_{n2}] = \text{Im}[\bar{Y}_{n1}] + \frac{B}{Y_{01}} \qquad (10.38b)$$

we start at point D and go along the constant real part (conductance) circle to reach point E for which the imaginary part differs from the imaginary part at D by the amount B/Y_{01}, that is, $-0.003 \big/ \frac{1}{150}$, or -0.45. We then draw the constant SWR circle passing through E and then read off the required SWR value at point F. This value is equal to 1.94.

The steps outlined above in part 4 can be applied in reverse to determine the location and the value of the susceptance required to achieve an SWR of unity to the left of it, that is, a condition of no standing waves. This procedure is known as transmission-line "matching." It is important from the point of view of eliminating or minimizing certain undesirable effects of standing waves in electromagnetic energy transmission.

To illustrate the solution to the matching problem, we first recognize that an SWR of unity is represented by the center point of the Smith Chart. Hence matching is achieved if $\bar{Y}_{n2}$ falls at the center of the Smith Chart. Now since the difference between $\bar{Y}_{n1}$ and $\bar{Y}_{n2}$ is only in the imaginary part as indicated by (10.38a) and (10.38b), $\bar{Y}_{n1}$ must lie on the constant conductance circle passing through the center of the Smith Chart (this circle is known as the "unit conductance circle" since it corresponds to normalized real part equal to unity). $\bar{Y}_{n1}$ must also lie on the constant SWR circle corresponding to the portion of the line to the right of jB. Hence it is given by the point(s) of intersection of this constant SWR circle and the unit conductance circle. There are two such points G and H, as shown in Fig. 10.10, in which the points A and

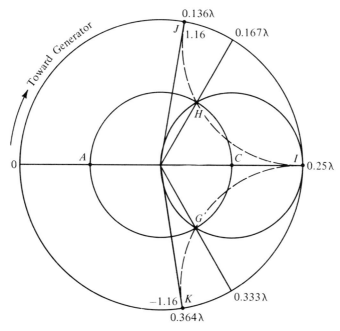

Figure 10.10. Solution of transmission-line matching problem by using the Smith Chart.

C are repeated from Fig. 10.9. There are thus two solutions to the matching problem. If we choose G to correspond to $\bar{Y}_{n1}$, then since the distance from C to G is $(0.333 - 0.250)\lambda_1$, or $0.083\lambda_1$, jB must be located at $z = -0.083\lambda_1$. To find the value of jB, we note that the normalized susceptance value corresponding to G is -1.16 and hence $B/Y_{01} = 1.16$, or $jB = j1.16 Y_{01} = j0.00773$ mho. If, however, we choose the point H to correspond to $\bar{Y}_{n1}$, then we find in a similar manner that jB must be located at $z = (0.250 + 0.167)\lambda_1$ or $0.417\lambda_1$ and its value must be $-j0.00773$ mho.

The reactive element jB used to achieve the matching is commonly realized by means of a short-circuited section of line, known as a "stub." This is based on the fact that the input impedance of a short-circuited line is purely

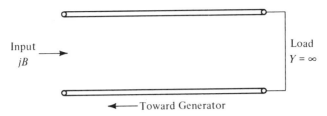

Figure 10.11. A short-circuited stub.

reactive, as shown in Sec. 6.4. The length of the stub for a required input susceptance can be found by considering the short circuit as the load, as shown in Fig. 10.11, and using the Smith Chart. The admittance corresponding to a short circuit is infinity and hence the load admittance normalized with respect to the characteristic admittance of the stub is also equal to infinity. This is located on the Smith Chart at point I in Fig. 10.10. We then go along the constant SWR circle passing through I (the outermost circle) toward the generator (input) until we reach the point corresponding to the required input susceptance of the stub normalized with respect to the characteristic admittance of the stub. Assuming the characteristic impedance of the stub to be the same as that of the line, this quantity is here equal to $j1.16$ or $-j1.16$, depending on whether point G or point H is chosen for the location of the stub. This leads us to point J or point K, and hence the stub length is $(0.25 + 0.136)\lambda_1$, or $0.386\lambda_1$, for $jB = j1.16$, and $(0.364 - 0.25)\lambda_1$, or $0.114\lambda_1$, for $jB = -j1.16$. The arrangement of the stub corresponding to the solution for which the stub location is at $z = -0.083\lambda_1$, and the stub length is $0.386\lambda_1$, is shown in Fig. 10.12.

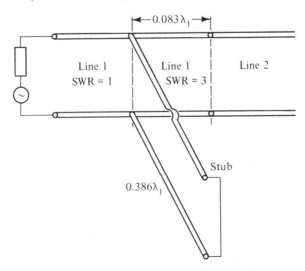

Figure 10.12. A solution to the matching problem for the transmission-line system of Fig. 10.5.

In this section we introduced the Smith Chart, which is a graphical aid in the solution of transmission-line problems. After first discussing the basis behind the construction of the Smith Chart, we illustrated its use by considering a transmission-line system and computing several quantities of interest. We concluded the section with the solution of a transmission-line matching problem.

REVIEW QUESTIONS

10.12. Define line impedance. What is its value for an infinitely long line?

10.13. What is the basis behind the construction of the Smith Chart? How does the Smith Chart simplify the solution of transmission-line problems?

10.14. Briefly discuss the mapping of the normalized line impedances from the complex $\bar{Z}_n$ plane onto the Smith Chart.

10.15. Why is a circle with its center at the center of the Smith Chart known as a constant SWR circle? Where on the circle is the corresponding SWR value marked?

10.16. Using the Smith Chart, how do you find the normalized line admittance at a point on the line given the normalized line impedance at that point?

10.17. Briefly discuss the solution of the transmission-line matching problem.

10.18. How is the length of a short-circuited stub for a required input susceptance determined by using the Smith Chart?

PROBLEMS

10.9. With the aid of a computer program, compute values of $\bar{\Gamma}_V$ corresponding to several points along line a in Fig. 10.6(a) and show that the contour a' in Fig. 10.6(b) is a circle of radius $\frac{1}{2}$ and centered at $(1/2, 0)$.

10.10. With the aid of a computer program, compute values of $\bar{\Gamma}_V$ corresponding to several points along line b in Fig. 10.6(a) and show that the contour b' in Fig. 10.6(b) is a portion of a circle of radius 2 and centered at $(1, 2)$.

10.11. For the transmission-line system of Fig. 10.8, and for the values of Z_{01}, Z_{02}, and l specified in the text, find the value of B which minimizes the SWR to the left of jB. What is the minimum value of SWR?

10.12. In Fig. 10.8 assume $Z_{01} = 300$ ohms, $Z_{02} = 75$ ohms, $B = 0.002$ mho, and $l = 0.145\lambda_1$, and find (a) $\bar{Z}_1$, (b) SWR on line 1 to the right of jB, (c) $\bar{Y}_1$, and (d) SWR on line 1 to the left of jB.

10.13. A transmission line of characteristic impedance 50 ohms is terminated by a load impedance of $(73 + j0)$ ohms. Find the location and the length of a short-circuited stub of characteristic impedance 50 ohms for achieving a match between the line and the load.

10.4 REFLECTION AND REFRACTION OF PLANE WAVES

In Sec. 7.6 we considered oblique incidence of uniform plane waves upon an interface between two dielectric media and found the relationships between the angles of incidence, reflection, and transmission. In this section we shall consider the problem in more detail and derive the expressions for the reflection and transmission coefficients at the boundary. To do this, we distinguish between two cases: (a) the electric field vector of the wave linearly polarized parallel to the interface and (b) the magnetic field vector of the wave linearly polarized parallel to the interface. The law of reflection and Snell's law derived in Sec. 7.6 hold for both cases since they result from the fact that the apparent phase velocities of the incident, reflected, and transmitted waves parallel to the boundary must be equal.

The geometry pertinent to the case of the electric field vector parallel to the interface is shown in Fig. 10.13 in which the interface is assumed to be in the $x = 0$ plane, and the subscripts i, r, and t associated with the field symbols denote incident, reflected, and transmitted waves, respectively. The plane of incidence, that is, the plane containing the normal to the interface and the propagation vectors, is assumed to be in the xz plane so that the electric field vectors are entirely in the y direction. The corresponding magnetic field vectors are then as shown in the figure so as to be consistent with the

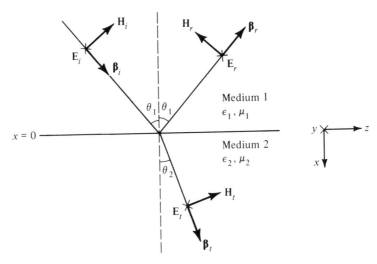

Figure 10.13. For obtaining the reflection and transmission coefficients for an obliquely incident uniform plane wave on a dielectric interface with its electric field perpendicular to the plane of incidence.

condition that $\mathbf{E}$, $\mathbf{H}$, and $\boldsymbol{\beta}$ form a right-handed mutually orthogonal set of vectors. Since the electric field vectors are perpendicular to the plane of incidence, this case is also said to correspond to perpendicular polarization. The angle of incidence is assumed to be θ_1. From the law of reflection (7.69a), the angle of reflection is then also θ_1. The angle of transmission, assumed to be θ_2, is related to θ_1 by Snell's law, given by (7.69b).

The boundary conditions to be satisfied at the interface $x = 0$ are that (a) the tangential component of the electric field intensity be continuous and (b) the tangential component of the magnetic field intensity be continuous. Thus, we have at the interface $x = 0$

$$E_{yi} + E_{yr} = E_{yt} \tag{10.39a}$$

$$H_{zi} + H_{zr} = H_{zt} \tag{10.39b}$$

Expressing the quantities in (10.39a) and (10.39b) in terms of the total fields, we obtain

$$E_i + E_r = E_t \tag{10.40a}$$

$$H_i \cos \theta_1 - H_r \cos \theta_1 = H_t \cos \theta_2 \tag{10.40b}$$

We also know from one of the properties of uniform plane waves that

$$\frac{E_i}{H_i} = \frac{E_r}{H_r} = \eta_1 = \sqrt{\frac{\mu_1}{\epsilon_1}} \tag{10.41a}$$

$$\frac{E_t}{H_t} = \eta_2 = \sqrt{\frac{\mu_2}{\epsilon_2}} \tag{10.41b}$$

Substituting (10.41a) and (10.41b) into (10.40b) and rearranging, we get

$$E_i - E_r = E_t \frac{\eta_1 \cos \theta_2}{\eta_2 \cos \theta_1} \tag{10.42}$$

Solving (10.40a) and (10.42) for E_i and E_r, we have

$$E_i = \frac{E_t}{2}\left(1 + \frac{\eta_1 \cos \theta_2}{\eta_2 \cos \theta_1}\right) \tag{10.43a}$$

$$E_r = \frac{E_t}{2}\left(1 - \frac{\eta_1 \cos \theta_2}{\eta_2 \cos \theta_1}\right) \tag{10.43b}$$

We now define the reflection coefficient $\Gamma_\perp$ and the transmission coefficient $\tau_\perp$ as

$$\Gamma_\perp = \frac{E_{yr}}{E_{yi}} = \frac{E_r}{E_i} \tag{10.44a}$$

$$\tau_\perp = \frac{E_{yt}}{E_{yi}} = \frac{E_t}{E_i} \tag{10.44b}$$

where the subscript $\perp$ refers to perpendicular polarization. From (10.43a) and (10.43b), we then obtain

$$\Gamma_\perp = \frac{\eta_2 \cos \theta_1 - \eta_1 \cos \theta_2}{\eta_2 \cos \theta_1 + \eta_1 \cos \theta_2} \qquad (10.45a)$$

$$\tau_\perp = \frac{2\eta_2 \cos \theta_1}{\eta_2 \cos \theta_1 + \eta_1 \cos \theta_2} \qquad (10.45b)$$

Before we discuss the result given by (10.45a) and (10.45b), we shall derive the corresponding expressions for the case in which the magnetic field of the wave is parallel to the interface. The geometry pertinent to this case is shown in Fig. 10.14. Here again the plane of incidence is chosen to be the xz plane so that the magnetic field vectors are entirely in the y direction. The corresponding electric field vectors are then as shown in the figure so as to be consistent with the condition that $\mathbf{E}$, $\mathbf{H}$, and $\boldsymbol{\beta}$ form a right-handed mutually orthogonal set of vectors. Since the electric field vectors are parallel to the plane of incidence, this case is also said to correspond to parallel polarization.

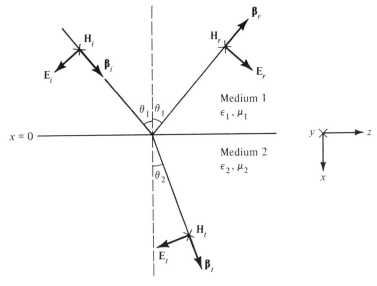

Figure 10.14. For obtaining the reflection and transmission coefficients for an obliquely incident uniform plane wave on a dielectric interface with its electric field parallel to the plane of incidence.

Once again the boundary conditions to be satisfied at the interface $x = 0$ are that (a) the tangential component of the electric field intensity be continous and (b) the tangential component of the magnetic field intensity be continuous. Thus we have at the interface $x = 0$,

$$E_{zi} + E_{zr} = E_{zt} \tag{10.46a}$$

$$H_{yi} + H_{yr} = H_{yt} \tag{10.46b}$$

Expressing the quantities in (10.46a) and (10.46b) in terms of the total fields and also using (10.41a) and (10.41b), we obtain

$$E_i - E_r = E_t \frac{\cos \theta_2}{\cos \theta_1} \tag{10.47a}$$

$$E_i + E_r = E_t \frac{\eta_1}{\eta_2} \tag{10.47b}$$

Solving (10.47a) and (10.47b) for E_i and E_r, we have

$$E_i = \frac{E_t}{2} \left(\frac{\eta_1}{\eta_2} + \frac{\cos \theta_2}{\cos \theta_1} \right) \tag{10.48a}$$

$$E_r = \frac{E_t}{2} \left(\frac{\eta_1}{\eta_2} - \frac{\cos \theta_2}{\cos \theta_1} \right) \tag{10.48b}$$

We now define the reflection coefficient $\Gamma_{||}$ and the transmission coefficient $\tau_{||}$ as

$$\Gamma_{||} = \frac{E_{zr}}{E_{zi}} = \frac{E_r \cos \theta_1}{-E_i \cos \theta_1} = -\frac{E_r}{E_i} \tag{10.49a}$$

$$\tau_{||} = \frac{E_{zt}}{E_{zi}} = \frac{-E_t \cos \theta_2}{-E_i \cos \theta_1} = \frac{E_t \cos \theta_2}{E_i \cos \theta_1} \tag{10.49b}$$

where the subscript $||$ refers to parallel polarization. From (10.48a) and (10.48b), we then obtain

$$\Gamma_{||} = \frac{\eta_2 \cos \theta_2 - \eta_1 \cos \theta_1}{\eta_2 \cos \theta_2 + \eta_1 \cos \theta_1} \tag{10.50a}$$

$$\tau_{||} = \frac{2\eta_2 \cos \theta_2}{\eta_2 \cos \theta_2 + \eta_1 \cos \theta_1} \tag{10.50b}$$

We shall now discuss the results given by (10.45a), (10.45b), (10.50a), and (10.50b) for the reflection and transmission coefficients for the two cases:

1. For $\theta_1 = 0$, that is, for the case of normal incidence of the uniform plane wave upon the interface, $\theta_2 = 0$ and

$$\Gamma_\perp = \frac{\eta_2 - \eta_1}{\eta_2 + \eta_1}, \qquad \Gamma_{||} = \frac{\eta_2 - \eta_1}{\eta_2 + \eta_1}$$

$$\tau_\perp = \frac{2\eta_2}{\eta_2 + \eta_1}, \qquad \tau_{||} = \frac{2\eta_2}{\eta_2 + \eta_1}$$

Thus the reflection coefficients as well as the transmission coefficients for the two cases become equal as they should since for normal incidence, there is no difference between the two polarizations except for rotation by $90°$ parallel to the interface.

2. $\Gamma_\perp = 1$ and $\Gamma_{||} = -1$ if $\cos \theta_2 = 0$, that is,

$$\sqrt{1 - \sin^2 \theta_2} = \sqrt{1 - \frac{\mu_1 \epsilon_1}{\mu_2 \epsilon_2} \sin^2 \theta_1} = 0$$

or

$$\sin \theta_1 = \sqrt{\frac{\mu_2 \epsilon_2}{\mu_1 \epsilon_1}} \qquad (10.51)$$

where we have used Snell's law given by (7.69b) to express $\sin \theta_2$ in terms of $\sin \theta_1$. If we assume $\mu_2 = \mu_1 = \mu_0$ as is usually the case, (10.51) has real solutions for θ_1 for $\epsilon_2 < \epsilon_1$. Thus, for $\epsilon_2 < \epsilon_1$, that is, for transmission from a dielectric medium of higher permittivity into a dielectric medium of lower permittivity, there is a critical angle of incidence θ_c given by

$$\theta_c = \sin^{-1} \sqrt{\frac{\epsilon_2}{\epsilon_1}} \qquad (10.52)$$

for which θ_2 is equal to $90°$, and $|\Gamma_\perp| = |\Gamma_{||}| = 1$. For $\theta_1 > \theta_c$, $\sin \theta_2$ becomes greater than 1, $\cos \theta_2$ becomes imaginary, and $|\Gamma_\perp| = |\Gamma_{||}| = 1$. This is consistent with the phenomenon of "total internal reflection" for $\theta_1 > \theta_c$, which we discussed in Sec. 7.6.

3. $\Gamma_\perp = 0$ for $\eta_2 \cos \theta_1 = \eta_1 \cos \theta_2$, that is

$$\eta_2 \sqrt{1 - \sin^2 \theta_1} = \eta_1 \sqrt{1 - \frac{\mu_1 \epsilon_1}{\mu_2 \epsilon_2} \sin^2 \theta_1}$$

or

$$\sin^2 \theta_1 = \frac{\eta_2^2 - \eta_1^2}{\eta_2^2 - \eta_1^2 (\mu_1 \epsilon_1 / \mu_2 \epsilon_2)} = \mu_2 \frac{\mu_2 - \mu_1 (\epsilon_2 / \epsilon_1)}{\mu_2^2 - \mu_1^2} \qquad (10.53)$$

For the usual case of transmission between two dielectric materials, that is, for $\mu_2 = \mu_1$, and $\epsilon_2 \neq \epsilon_1$, this equation has no real solution for θ_1 and hence there is no angle of incidence for which the reflection coefficient is zero for the case of perpendicular polarization.

4. $\Gamma_{||} = 0$ for $\eta_2 \cos \theta_2 = \eta_1 \cos \theta_1$, that is,

$$\eta_2 \sqrt{1 - \frac{\mu_1 \epsilon_1}{\mu_2 \epsilon_2} \sin^2 \theta_1} = \eta_1 \sqrt{1 - \sin^2 \theta_1}$$

or

$$\sin^2 \theta_1 = \frac{\eta_2^2 - \eta_1^2}{\eta_2^2(\mu_1\epsilon_1/\mu_2\epsilon_2) - \eta_1^2} = \epsilon_2 \frac{(\mu_2/\mu_1)\epsilon_1 - \epsilon_2}{\epsilon_1^2 - \epsilon_2^2} \quad (10.54)$$

If we assume $\mu_2 = \mu_1$, this equation reduces to

$$\sin^2 \theta_1 = \frac{\epsilon_2}{\epsilon_1 + \epsilon_2}$$

which then gives

$$\cos^2 \theta_1 = 1 - \sin^2 \theta_1 = \frac{\epsilon_1}{\epsilon_1 + \epsilon_2}$$

and

$$\tan \theta_1 = \sqrt{\frac{\epsilon_2}{\epsilon_1}}$$

Thus there exists a value of the angle of incidence θ_p, given by

$$\theta_p = \tan^{-1}\sqrt{\frac{\epsilon_2}{\epsilon_1}} \quad (10.55)$$

for which the reflection coefficient is zero and hence there is complete transmission for the case of parallel polarization.

5. In view of (3) and (4) above, for an elliptically polarized wave incident on the interface at the angle θ_p, the reflected wave will be linearly polarized perpendicular to the plane of incidence. For this reason, the angle θ_p is known as the "polarizing angle." It is also known as the "Brewster angle." The phenomenon associated with the Brewster angle has several applications. An example is in gas lasers in which the discharge tube lying between the mirrors of a Fabry Perot resonator is sealed by glass windows placed at the Brewster angle, as shown in Fig. 10.15, to minimize reflections from the ends of the tube so that the laser behavior is governed by the mirrors external to the tube.

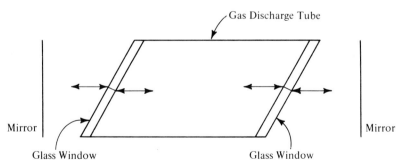

Figure 10.15. For illustrating the application of the Brewster angle effect in gas lasers.

In this section we considered oblique incidence of a uniform plane wave upon the boundary between two perfect dielectric media and derived the expressions for the reflection and transmission coefficients for the cases of perpendicular and parallel polarizations. An examination of these expressions revealed that (a) for incidence from a dielectric medium of higher permittivity onto one of lower permittivity, there is a critical angle of incidence beyond which total internal reflection occurs, as we learned in Sec. 7.6, and (b) for the case of parallel polarization, there is an angle of incidence, known as the Brewster angle, for which the reflection coefficient is zero.

REVIEW QUESTIONS

10.19. What is meant by the plane of incidence? Distinguish between the two different linear polarizations pertinent to the derivation of the reflection and transmission coefficients for oblique incidence on a dielectric interface.

10.20. Briefly discuss the determination of the reflection and transmission coefficients for an obliquely incident wave on a dielectric interface.

10.21. What is the nature of the reflection coefficient for angle of incidence greater than the critical angle for total internal reflection?

10.22. What is the Brewester angle? What is the polarization of the reflected wave for an elliptically polarized wave incident on a dielectric interface at the Brewster angle?

10.23. Discuss an application of the Brewster angle effect.

PROBLEMS

10.14. A uniform plane wave having the electric field given by

$$\mathbf{E} = E_0 \mathbf{i}_y \sin \left[6\pi \times 10^9 t - 10\pi (x + \sqrt{3} z) \right]$$

is incident on the interface between free space and a dielectric of permittivity $1.5\epsilon_0$ as shown in Fig. 10.16. (a) Obtain the expression for the reflected wave field. (b) Obtain the expression for the transmitted wave field.

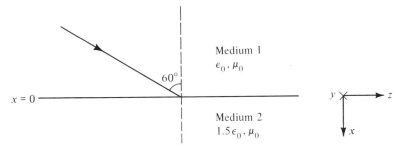

Figure 10.16. For Problem 10.14.

10.15. Repeat Problem 10.14 for the uniform plane wave having the electric field given by

$$\mathbf{E} = E_0 \left(\frac{\sqrt{3}}{2}\mathbf{i}_x - \frac{1}{2}\mathbf{i}_z \right) \cos\left[6\pi \times 10^9 t - 10\pi(x + \sqrt{3}\,z) \right]$$

10.16. Repeat Problem 10.14 for the uniform plane wave having the electric field given by

$$\mathbf{E} = E_0 \left(\frac{\sqrt{3}}{2}\mathbf{i}_x - \frac{1}{2}\mathbf{i}_z \right) \cos\left[6\pi \times 10^9 t - 10\pi(x + \sqrt{3}\,z) \right]$$

$$+ E_0\mathbf{i}_y \sin\left[6\pi \times 10^9 t - 10\pi(x + \sqrt{3}\,z) \right]$$

Also discuss the polarizations of the incident, reflected, and transmitted waves.

10.17. For the dielectric boundary in Fig. 10.16, determine the angle of incidence of an elliptically polarized wave for the reflected wave to be linearly polarized. In which plane is the reflected wave polarized then?

10.5 DESIGN OF A FREQUENCY-INDEPENDENT ANTENNA

In Chap. 8 we studied the directional properties of antennas and antenna arrays. These properties depend on the electrical dimensions of the antenna, that is, the dimensions expressed in terms of the wavelength at the operating frequency. Hence an antenna of fixed physical dimensions exhibits frequency-dependent characteristics. This very fact suggests that for an antenna to be frequency-independent, its electrical size must remain constant with frequency and hence its physical size should increase proportionately to the wavelength. Alternatively, for an antenna of fixed physical dimensions, the active region, that is, the region responsible for the predominant radiation should vary with frequency, that is, scale itself in such a manner that its electrical size remains the same.

A simple illustration of the aforementioned property is provided by the equiangular spiral antenna shown in Fig. 10.17 and so-termed because the angle between the radius vector from the origin and the spiral remains the same for all points on the curve. The equiangular spiral antenna was proposed by Rumsey in 1954 during the early stages of research on frequency-independent antennas at the University of Illinois. When this antenna is excited at the origin, the current flows outward with small attenuation along the spiral until an active region is reached from which essentially all of the incident energy transmitted along the spiral arms is radiated. Since this active region is of constant size in wavelengths, it moves toward the origin as the operating wavelength decreases or the frequency increases. The size of the effective radiating region thus adjusts automatically with the operating frequency such that the antenna behaves the same at all frequencies except for

Figure 10.17. The equiangular spiral antenna.

a rotation of the radiated field about the antenna axis because of the spiraling of the arms.

Another and more conventional example of a frequency-independent antenna is the "log-periodic dipole array," shown in Fig. 10.18. As the name implies, it employs a number of dipoles. The dipole lengths and the spacings between consecutive dipoles increase along the array by a constant scale factor such that

$$\frac{l_{i+1}}{l_i} = \frac{d_{i+1}}{d_i} = \tau \tag{10.56}$$

From the principle of scaling, it is evident that for this structure extending from zero to infinity and energized at the apex, the properties repeat at frequencies given by $\tau^n f$, where n takes integer values. When plotted on a logarithmic scale, these frequencies are equally spaced at intervals of $\log \tau$. It is for this reason that the structure is termed "log-periodic."

The log-periodic dipole array is fed by a transmission line, as shown in Fig. 10.18, such that a 180°-phase shift is introduced between successive elements in addition to that corresponding to the spacing between the elements. The resulting radiation pattern is directed toward the apex, that is toward the source. Almost all of the radiation takes place from those elements which are in the vicinity of a half wavelength long. The operating band of frequencies is therefore bounded on the low side by frequencies at which the largest elements are approximately a half wavelength long and on the high side by frequencies corresponding to the size of the smallest elements. As the frequency is varied, the radiating or active region moves back and forth along

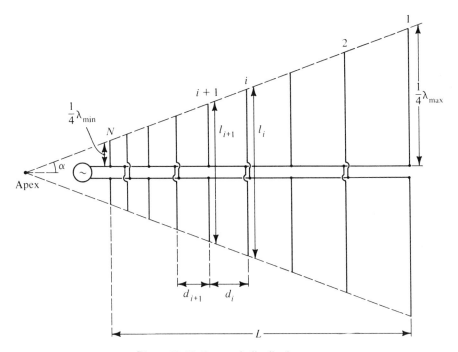

Figure 10.18. Log-periodic dipole array.

the array. Since practically all of the input power is radiated by the active region, the larger elements to the right of it are not excited. Furthermore, because the radiation is toward the apex, these larger elements are essentially in a field-free region and hence do not significantly influence the operation. Although the shorter elements to the left of the active region are in the antenna beam, they have small influence on the pattern because of their short lengths, close spacings, and the 180°-phase shift.

We shall now discuss the design of a log-periodic dipole array. We shall restrict the design to the computation of the lengths and spacings of the elements for a specified bandwidth of operation and directivity of the radiation pattern. The design parameters are the scale factor τ given by (10.56), the half angle α subtended at the apex, and the ratio σ of the element spacing to twice the length of the next larger element. Since the active region is not of negligible length along the structure, the array is designed for a larger bandwidth than the design specification. This larger bandwidth is known as the "bandwidth of the structure," denoted B_s. The ratio of B_s to the design bandwidth B is termed the "bandwidth of the active region," B_{ar} and is related to σ and τ.* We shall first present the relevant definitions and formulas, with

---

*The relationship between B_{ar}, σ and τ and other design curves have been obtained by R. L. Carrell in a Ph.D. dissertation at the University of Illinois.

reference to Fig. 10.18:

$$\tau = \frac{d_{i+1}}{d_i} = \frac{l_{i+1}}{l_i} = \text{scale factor}$$

$$\sigma = \frac{d_i}{2l_i} = \text{relative spacing constant}$$

$\alpha = $ half apex angle

$L = $ boom length, that is, the distance between the shortest and longest element

$N = $ number of elements

$B_s = $ bandwidth of the structure

$$= \frac{\lambda_{max}}{\lambda_{min}} = B_{ar}B \tag{10.57}$$

$B_{ar} = $ bandwidth of the active region

$$= 1.1 + 7.7(1 - \tau)^2 \cot \alpha \tag{10.58}$$

Since

$$\tan \alpha = \frac{l_i - l_{i+1}}{2d_i} = \frac{1 - \tau}{4\sigma}$$

we obtain

$$\alpha = \tan^{-1} \frac{1 - \tau}{4\sigma} \tag{10.59}$$

From

$$\cot \alpha = \frac{L}{(\lambda_{max}/4) - (\lambda_{min}/4)}$$

we have

$$L = \left[\frac{1}{4} \left(1 - \frac{1}{B_s} \right) \cot \alpha \right] \lambda_{max} \tag{10.60}$$

From

$$\frac{\lambda_{min}}{\lambda_{max}} = \frac{l_N}{l_1} = \tau^{N-1}$$

we obtain

$$B_s = \left(\frac{1}{\tau} \right)^{N-1}$$

$$\log B_s = (N - 1) \log \frac{1}{\tau}$$

$$N = 1 + \frac{\log B_s}{\log (1/\tau)} \tag{10.61}$$

Let us now consider an example in which it is desired to design a log-periodic dipole array for which the band of operation is from 12.5 MHz to 30 MHz and the directivity is 9. In terms of decibels, the directivity is $20 \log_{10} 3$, or 9.5 db. The design consists of the following steps:

1. Compute the design bandwidth B.

$$B = \frac{30}{12.5} = 2.4$$

2. Find τ and σ to give the desired directivity. There exists an optimum σ for which the directivity is maximum for each value of τ in the range $0.8 < \tau < 1.0$. Plots of this optimum σ and the corresponding τ versus the directivity are shown in Fig. 10.19. For the desired directivity of 9.5 db, we have from Fig. 10.19,

$$\tau = 0.893$$

$$\sigma = 0.163$$

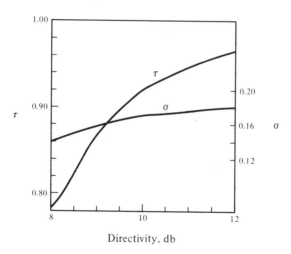

Figure 10.19. Plots of optimum σ and the corresponding τ versus directivity for log-periodic dipole arrays.

3. Determine the half apex angle α from (10.59).

$$\alpha = \tan^{-1} \frac{1 - 0.893}{4 \times 0.163} = 9.32°$$

4. Determine the bandwidth of the active region B_{ar} from (10.58).

$$B_{ar} = 1.1 + 7.7(1 - 0.893)^2 \cot 9.32° = 1.637$$

5. Determine the bandwidth of the structure B_s from (10.57).

$$B_s = 1.637 \times 2.4 = 3.929$$

6. Determine the boom length L from (10.60). Assuming the longest element to be a half wavelength long at the low frequency end of the specified band, $\lambda_{max} = 24$ m and,

$$L = \left[\frac{1}{4} \left(1 - \frac{1}{3.929} \right) \cot 9.32° \right] 24 = 27.25 \text{ m}$$

7. Determine the number of elements N from (10.61).

$$N = 1 + \frac{\log 3.929}{\log (1/0.893)} = 13$$

8. Determine the element lengths and spacings:

$$l_1 = \frac{\lambda_{max}}{2} = \frac{24}{2} = 12 \text{ m}$$

$$l_2 = l_1 \tau = 12 \times 0.893 = 10.72 \text{ m}$$

$$l_3 = l_2 \tau = 10.72 \times 0.893 = 9.57 \text{ m}$$

. . .

$$d_1 = \frac{l_1 - l_2}{2} \cot \alpha = 0.64 \cot 9.32° = 3.91 \text{ m}$$

$$d_2 = d_1 \tau = 3.91 \times 0.893 = 3.49 \text{ m}$$

$$d_3 = d_2 \tau = 3.49 \times 0.893 = 3.12 \text{ m}$$

. . .

Values of the element lengths and spacings and the nearest frequencies at which the elements are a half wavelength long are listed in Table 10.1.

In this section we introduced the concept of frequency-independent antennas based upon the criterion that for the antenna characteristics to be frequency-independent, the active region must vary with frequency such that its electrical size remains approximately constant. We discussed in particular the log-periodic dipole antenna array and illustrated by means of a numerical example the computation of element lengths and spacings for desired operating bandwidth and directivity.

TABLE 10.1.   Computed Values of Log-Periodic Dipole Array Element Parameters

Element number	Length (m)	Spacing (m)	Frequency (MHz)
1	12.00	—	12.50
2	10.72	3.91	14.00
3	9.57	3.49	15.67
4	8.55	3.12	17.55
5	7.63	2.78	19.66
6	6.81	2.49	22.01
7	6.09	2.22	24.65
8	5.43	1.98	27.60
9	4.85	1.77	30.91
10	4.33	1.58	34.61
11	3.87	1.41	38.76
12	3.46	1.26	43.41
13	3.09	1.13	48.61

REVIEW QUESTIONS

10.24. Discuss the criterion for an antenna of fixed physical size to be frequency-independent.

10.25. Describe how the equiangular spiral antenna has frequency-independent characteristics.

10.26. What is a log-periodic dipole array? Briefly discuss its operation.

10.27. Why is a log-periodic dipole array designed for a larger bandwidth than the design specification?

10.28. Outline the steps in the design of a log-periodic dipole array.

PROBLEMS

10.18. Design a log-periodic dipole array for operation over the frequency band from 54 MHz to 108 MHz (VHF TV channels 2 to 6 and FM band) and a directivity of 10 db.

10.19. A log-periodic dipole array is to cover the frequency band from 50 MHz to 250 MHz. Find the boom length and the number of elements for directivity of (a) 9.5 db and (b) 12 db.

10.6   CAPACITANCE OF A PARALLEL-WIRE LINE

In Sec. 9.3 we illustrated the solution of Laplace's equation for the parallel-plate case and discussed the applicability of the static field technique in the determination of transmission-line parameters. In this section we shall use the technique to obtain an analytical expression for the capacitance of a parallel-wire line, consisting of two infinitely long, straight, parallel, cylindrical wires.

Let us first consider an infinitely long, straight, line charge of uniform density ρ_{L0} C/m situated along the z axis, as shown in Fig. 10.20(a), and obtain the electric potential due to the line charge. The symmetry of the problem indicates that the potential is dependent only on the cylindrical coordinate r. Noting then from Appendix B that in cylindrical coordinates,

$$\nabla^2 V = \nabla \cdot \nabla V = \frac{1}{r} \frac{\partial}{\partial r} \left(r \frac{\partial V}{\partial r} \right)$$

we have from Laplace's equation

$$\frac{1}{r} \frac{\partial}{\partial r} \left(r \frac{\partial V}{\partial r} \right) = 0 \qquad \text{for } r \neq 0 \qquad (10.62)$$

Integrating twice, we obtain the solution for (10.62) to be

$$V = A \ln r + B \qquad (10.63)$$

where A and B are arbitrary constants. We can arbitrarily set the potential

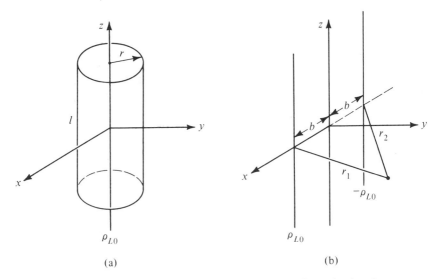

(a)   (b)

Figure 10.20. (a) An infinitely long line charge of uniform density along the z axis. (b) A pair of parallel, infinitely long line charges of equal and opposite uniform densities.

to be zero at a reference value $r = r_0$ giving us

$$0 = A \ln r_0 + B \quad \text{or} \quad B = -A \ln r_0$$

and

$$V = A \ln r - A \ln r_0 = A \ln \frac{r}{r_0} \tag{10.64}$$

To evaluate the arbitrary constant A in (10.64), we find that the electric field intensity due to the line charge is given by

$$\mathbf{E} = -\nabla V = -\frac{\partial V}{\partial r}\mathbf{i}_r = -\frac{A}{r}\mathbf{i}_r$$

The electric field is thus directed radial to the line charge. Let us now consider a cylindrical box of radius r and length l coaxial with the line charge, as shown in Fig. 10.20(a), and apply Gauss' law for the electric field in integral form to the surface of the box. For the cylindrical surface,

$$\int \mathbf{D} \cdot d\mathbf{S} = -\frac{\epsilon A}{r}(2\pi r l)$$

For the top and bottom surfaces, $\int \mathbf{D} \cdot d\mathbf{S} = 0$ since the field is parallel to the surfaces. The charge enclosed by the box is $\rho_{L0}l$. Thus we have

$$-\frac{\epsilon A}{r}(2\pi r l) = \rho_{L0}l \quad \text{or} \quad A = -\frac{\rho_{L0}}{2\pi\epsilon}$$

Substituting this result in (10.64) we obtain the potential field due to the line charge to be

$$V = -\frac{\rho_{L0}}{2\pi\epsilon}\ln\frac{r}{r_0} = \frac{\rho_{L0}}{2\pi\epsilon}\ln\frac{r_0}{r} \tag{10.65}$$

Let us now consider two infinitely long, straight, line charges of equal and opposite uniform charge densities ρ_{L0} C/m and $-\rho_{L0}$ C/m, parallel to the z axis and passing through $x = b$ and $x = -b$, respectively, as shown in Fig. 10.20(b). Applying superposition and using (10.65), we write the potential due to the two line charges as

$$V = \frac{\rho_{L0}}{2\pi\epsilon}\ln\frac{r_{01}}{r_1} - \frac{\rho_{L0}}{2\pi\epsilon}\ln\frac{r_{02}}{r_2} \tag{10.66}$$

where r_1 and r_2 are the distances of the point of interest from the line charges and r_{01} and r_{02} are the distances to the reference point at which the potential is zero. By choosing the reference point to be equidistant from the two line charges, that is, $r_{01} = r_{02}$, we get

$$V = \frac{\rho_{L0}}{2\pi\epsilon} \ln \frac{r_2}{r_1} \qquad (10.67)$$

From (10.67), we note that the equipotential surfaces for the potential field of the line-charge pair are given by

$$\frac{r_2}{r_1} = \text{constant, say, } k \qquad (10.68)$$

where k lies between 0 and ∞. In terms of Cartesian coordinates, (10.68) can be written as

$$\frac{(x + b)^2 + y^2}{(x - b)^2 + y^2} = k^2 \qquad (10.69)$$

Rearranging (10.69), we obtain

$$x^2 - 2b\frac{k^2 + 1}{k^2 - 1}x + y^2 + b^2 = 0$$

or

$$\left(x - b\frac{k^2 + 1}{k^2 - 1}\right)^2 + y^2 = \left(b\frac{2k}{k^2 - 1}\right)^2 \qquad (10.70)$$

Equation (10.70) represents cylinders having their axes along

$$x = b\frac{k^2 + 1}{k^2 - 1}, \qquad y = 0$$

and radii equal to $b\dfrac{2k}{k^2 - 1}$. The corresponding potentials are $(\rho_{L0}/2\pi\epsilon) \ln k$. The cross sections of the equipotential surfaces are shown in Fig. 10.21.

We can now place perfectly conducting cylinders in any two equipotential surfaces without disturbing the field configuration, as shown, for example, by the thick circles in Fig. 10.21, thereby obtaining a parallel-wire line. Letting the distance between their centers be $2d$ and their radii be a, we have

$$\pm d = b\frac{k^2 + 1}{k^2 - 1} \qquad (10.71a)$$

$$a = b\frac{2k}{k^2 - 1} \qquad (10.71b)$$

Solving (10.71a) and (10.71b) for k and accepting only those solutions lying between 0 and ∞, we obtain

$$k = \frac{d \pm \sqrt{d^2 - a^2}}{a} \tag{10.72}$$

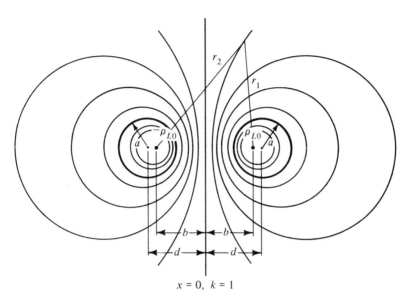

$$x = 0, \quad k = 1$$

Figure 10.21. Cross sections of equipotential surfaces for the line-charge pair of Fig. 10.20(a). Thick circles represent cross section of parallel-wire line.

The potentials of the right ($k > 1$) and left ($k < 1$) conductors are then given, respectively, by

$$V_+ = \frac{\rho_{L0}}{2\pi\epsilon} \ln \frac{d + \sqrt{d^2 - a^2}}{a} \tag{10.73a}$$

$$V_- = \frac{\rho_{L0}}{2\pi\epsilon} \ln \frac{d - \sqrt{d^2 - a^2}}{a}$$

$$= -\frac{\rho_{L0}}{2\pi\epsilon} \ln \frac{d + \sqrt{d^2 - a^2}}{a} \tag{10.73b}$$

The potential difference between the two conductors is

$$V_0 = V_+ - V_- = \frac{\rho_{L0}}{\pi\epsilon} \ln \frac{d + \sqrt{d^2 - a^2}}{a} \tag{10.74}$$

Finally, to find the capacitance, we note that since the electric field lines begin on the positive charge and end on the negative charge orthogonal to the equipotentials, the magnitude of the charge on either conductor, which produces the same field as the line-charge pair, must be the same as the line charge itself. Thus considering unit length of the line, we obtain the capacitance per unit length of the parallel-wire line to be

$$\mathcal{C} = \frac{\rho_{L0}}{V_0} = \frac{\pi\epsilon}{\ln\left[(d + \sqrt{d^2 - a^2})/a\right]}$$

$$= \frac{\pi\epsilon}{\cosh^{-1}(d/a)} \tag{10.75}$$

In this section we obtained the electric potential field of two parallel, infinitely long, straight, line charges of equal and opposite uniform charge densities and we showed that the equipotential surfaces are cylinders having their axes parallel to the line charges. By placing conductors in two equipotential surfaces, thereby forming a parallel-wire line, we obtained the expression for the capacitance per unit length of the line.

REVIEW QUESTIONS

10.29. Discuss the applicability of static field techniques in the determination of transmission-line parameters.

10.30. Briefly discuss the solution for the potential field of the infinitely long, straight, line charge of uniform density.

10.31. Describe the equipotential surfaces for the potential field of two parallel, infinitely long, straight, line charges of equal and opposite uniform densities. What are the shapes of the direction lines of the electric field?

10.32. Briefly discuss the determination of the capacitance of the parallel-wire line from the potential field of the line-charge pair.

PROBLEMS

10.20. For the line-charge pair of Fig. 10.21, show that the direction lines of the electric field are arcs of circles emanating from the positively charged line and terminating on the negatively charged line.

10.21. For the parallel-wire line, show that for $d \gg a$, the capacitance per unit length of the line is $\frac{\pi\epsilon}{\ln(2d/a)}$. Find the value of d/a for which the exact value of the capacitance per unit length is 1.1 times the value given by the approximate expression for $d \gg a$.

10.22. Figure 10.22 shows the cross-sectional view of an arrangement of two infinitely long, parallel, cylindrical conductors of radii a and b and with their axes separated by distance d. Show that the capacitance per unit length of the arrangement is $2\pi\epsilon/\cosh^{-1}[(a^2 + b^2 - d^2)/2ab]$.

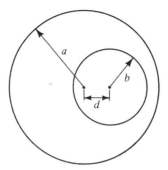

Figure 10.22. For Problem 10.22.

10.7 MAGNETIC VECTOR POTENTIAL

In Sec. 9.1 we learned that since

$$\mathbf{V} \times \mathbf{E} = 0$$

for the static electric field, $\mathbf{E}$ can be expressed as the gradient of a scalar potential in the manner

$$\mathbf{E} = -\mathbf{V}V$$

We then proceeded with the discussion of the electric scalar potential and its application for the computation of static electric fields. In this section we shall introduce a similar tool for the magnetic field computation, namely, the magnetic vector potential. When extended to the time-varying case, the magnetic vector potential has useful application in the determination of fields due to antennas.

To introduce the magnetic vector potential concept, we recall that the divergence of the magnetic flux density vector, whether static or time-varying, is equal to zero, that is,

$$\mathbf{V} \cdot \mathbf{B} = 0 \tag{10.76}$$

If the divergence of a vector is zero, then that vector can be expressed as the curl of another vector since the divergence of the curl of a vector is identically equal to zero, as can be seen by expansion in Cartesian coordinates:

$$\mathbf{V} \cdot \mathbf{V} \times \mathbf{A} = \left(\mathbf{i}_x \frac{\partial}{\partial x} + \mathbf{i}_y \frac{\partial}{\partial y} + \mathbf{i}_z \frac{\partial}{\partial z} \right) \cdot \begin{vmatrix} \mathbf{i}_x & \mathbf{i}_y & \mathbf{i}_z \\ \dfrac{\partial}{\partial x} & \dfrac{\partial}{\partial y} & \dfrac{\partial}{\partial z} \\ A_x & A_y & A_z \end{vmatrix}$$

$$= \begin{vmatrix} \dfrac{\partial}{\partial x} & \dfrac{\partial}{\partial y} & \dfrac{\partial}{\partial z} \\ \dfrac{\partial}{\partial x} & \dfrac{\partial}{\partial y} & \dfrac{\partial}{\partial z} \\ A_x & A_y & A_z \end{vmatrix} = 0$$

Thus the magnetic field vector $\mathbf{B}$ can be expressed as the curl of a vector $\mathbf{A}$, that is,

$$\mathbf{B} = \mathbf{V} \times \mathbf{A} \qquad (10.77)$$

The vector $\mathbf{A}$ is known as the magnetic vector potential in analogy with the electric scalar potential for V.

If we can now find $\mathbf{A}$ due to an infinitesimal current element, we can then find $\mathbf{A}$ for a given current distribution and determine $\mathbf{B}$ by using (10.77). Let us therefore consider an infinitesimal current element of length dl situated at the origin and oriented along the z axis as shown in Fig. 10.23. Assuming first that the current is constant, say, I amperes, we note from (1.68) that the magnetic field at a point P due to the current element is given by

$$\mathbf{B} = \frac{\mu}{4\pi} \frac{I \, d\mathbf{l} \times \mathbf{i}_r}{r^2} \qquad (10.78)$$

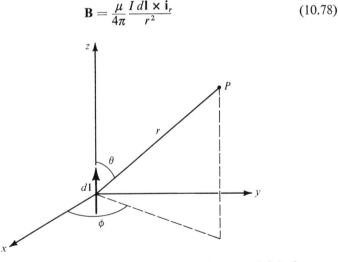

Figure 10.23. For finding the magnetic vector potential due to an infinitesimal current element.

where r is the distance from the current element to the point P and $\mathbf{i}_r$ is the unit vector directed from the element toward P. Expressing $\mathbf{B}$ as

$$\mathbf{B} = \frac{\mu}{4\pi} I \, d\mathbf{l} \times \left(-\nabla \frac{1}{r}\right) \tag{10.79}$$

and using the vector identity

$$\mathbf{A} \times \nabla V = V \nabla \times \mathbf{A} - \nabla \times (V\mathbf{A}) \tag{10.80}$$

we obtain

$$\mathbf{B} = -\frac{\mu I}{4\pi r} \nabla \times d\mathbf{l} + \nabla \times \left(\frac{\mu I \, d\mathbf{l}}{4\pi r}\right) \tag{10.81}$$

Since $d\mathbf{l}$ is a constant, $\nabla \times d\mathbf{l} = 0$, and (10.81) reduces to

$$\mathbf{B} = \nabla \times \left(\frac{\mu I \, d\mathbf{l}}{4\pi r}\right) \tag{10.82}$$

Comparing (10.82) with (10.77), we now see that the vector potential due to the current element situated at the origin is simply given by

$$\mathbf{A} = \frac{\mu I \, d\mathbf{l}}{4\pi r} \tag{10.83}$$

Thus it has a magnitude inversely proportional to the radial distance from the element (similar to the inverse distance dependence of the scalar potential due to a point charge) and direction parallel to the element.

If the current in the element is now assumed to be time-varying in the manner

$$I = I_0 \cos \omega t$$

we would intuitively expect that the corresponding magnetic vector potential would also be time-varying in the same manner but with a time-lag factor included, as discussed in Sec. 8.1 in connection with the determination of the electromagnetic fields due to the time-varying current element (Hertzian dipole). To verify our intuitive expectation, we note from (8.23b) that the magnetic field due to the time-varying current element is given by

$$\begin{aligned}
\mathbf{B} = \mu\mathbf{H} &= \frac{\mu I_0 \, dl \sin\theta}{4\pi}\left[\frac{\cos(\omega t - \beta r)}{r^2} - \frac{\beta \sin(\omega t - \beta r)}{r}\right]\mathbf{i}_\phi \\
&= \frac{\mu I_0 \, d\mathbf{l}}{4\pi} \times \left\{\left[\frac{\cos(\omega t - \beta r)}{r^2} - \frac{\beta \sin(\omega t - \beta r)}{r}\right]\mathbf{i}_r\right\} \\
&= \frac{\mu I_0 \, d\mathbf{l}}{4\pi} \times \left\{-\nabla\left[\frac{\cos(\omega t - \beta r)}{r}\right]\right\}
\end{aligned}$$

and proceed in the same manner as for the constant current case to obtain the vector potential to be

$$\mathbf{A} = \frac{\mu I_0 \, d\mathbf{l}}{4\pi r} \cos(\omega t - \beta r) \tag{10.84}$$

Comparing (10.84) with (10.83), we find that our intuitive expectation is indeed correct for the vector potential case unlike the case of the fields in Sec. 8.1! The result given by (10.84) is familiarly known as the "retarded" vector potential in view of the phase-lag factor βr contained in it.

To illustrate an example of the application of (10.84), we now consider a circular loop antenna having circumference small compared to the wavelength so that it is an electrically small antenna. Under this condition, the current flowing in the loop can be assumed to be uniform around the loop. Let us assume the loop to be in the xy plane with its center at the origin, as shown in Fig. 10.24, and the loop current to be $I = I_0 \cos \omega t$ in the ϕ direction. In view of the circular symmetry around the z axis, we can consider a point P in the xz plane without loss of generality to find the vector potential. To do this, we divide the loop into a series of infinitesimal elements. Considering one such current element $d\mathbf{l} = a \, d\alpha \, (-\sin \alpha \, \mathbf{i}_x + \cos \alpha \, \mathbf{i}_y)$, as shown in Fig. 10.24, and using (10.84) we obtain the vector potential at P due to that current element to be

$$d\mathbf{A} = \frac{\mu I_0 a \, d\alpha \, (-\sin \alpha \, \mathbf{i}_x + \cos \alpha \, \mathbf{i}_y)}{4\pi R} \cos(\omega t - \beta R) \tag{10.85}$$

where

$$R = [(r \sin \theta - a \cos \alpha)^2 + (a \sin \alpha)^2 + (r \cos \theta)^2]^{1/2}$$
$$= [r^2 + a^2 - 2ar \sin \theta \cos \alpha]^{1/2} \tag{10.86}$$

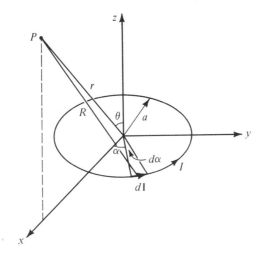

Figure 10.24. For finding the magnetic vector potential due to a small circular loop antenna.

The vector potential at point P due to the entire current loop is then given by

$$
\begin{aligned}
\mathbf{A} &= \int_{\alpha=0}^{2\pi} d\mathbf{A} \\
&= -\left[\int_{\alpha=0}^{2\pi} \frac{\mu I_0 a \sin \alpha \, d\alpha}{4\pi R} \cos(\omega t - \beta R)\right] \mathbf{i}_x \\
&\quad + \left[\int_{\alpha=0}^{2\pi} \frac{\mu I_0 a \cos \alpha \, d\alpha}{4\pi R} \cos(\omega t - \beta R)\right] \mathbf{i}_y
\end{aligned}
\tag{10.87}
$$

The first integral on the right side of (10.87) is, however, zero since the contributions to it due to elements situated symmetrically about the xz plane cancel. Replacing $\mathbf{i}_y$ in the second term by $\mathbf{i}_\phi$ to generalize the result to an arbitrary point $P(r, \theta, \phi)$, we then obtain

$$
\mathbf{A} = \left[\int_{\alpha=0}^{2\pi} \frac{\mu I_0 a \cos \alpha \, d\alpha}{4\pi R} \cos(\omega t - \beta R)\right] \mathbf{i}_\phi
\tag{10.88}
$$

Although the evaluation of the integral in (10.88) is complicated, some approximations can be made for obtaining the "radiation fields." For these fields, we can set the quantity R in the amplitude factor of the integrand equal to r. For R in the phase factor of the integrand, we write

$$
\begin{aligned}
R &= r\left[1 + \frac{a^2}{r^2} - \frac{2a}{r} \sin \theta \cos \alpha\right]^{1/2} \\
&\approx r\left[1 - \frac{a}{r} \sin \theta \cos \alpha\right]
\end{aligned}
\tag{10.89}
$$

Thus for the radiation fields,

$$
\mathbf{A} = \left[\int_{\alpha=0}^{2\pi} \frac{\mu I_0 a \cos \alpha \, d\alpha}{4\pi r} \cos(\omega t - \beta r + \beta a \sin \theta \cos \alpha)\right] \mathbf{i}_\phi
\tag{10.90}
$$

Now, since $2\pi a \ll \lambda$, or $\beta a \ll 1$, we can write

$$
\begin{aligned}
&\cos(\omega t - \beta r + \beta a \sin \theta \cos \alpha) \\
&\approx \cos(\omega t - \beta r) - \beta a \sin \theta \cos \alpha \sin(\omega t - \beta r)
\end{aligned}
\tag{10.91}
$$

Substituting (10.91) into (10.90) and evaluating the integral, we obtain

$$
\mathbf{A} = -\frac{\mu I_0 \pi a^2 \beta \sin \theta}{4\pi r} \sin(\omega t - \beta r) \mathbf{i}_\phi
\tag{10.92}
$$

Having obtained the required magnetic vector potential, we can now determine the radiation fields. Thus from (10.77),

$$\mathbf{H} = \frac{\mathbf{B}}{\mu} = \frac{1}{\mu}\nabla \times \mathbf{A}$$

$$= -\frac{1}{\mu r}\frac{\partial}{\partial r}(rA_\phi)\mathbf{i}_\theta$$

$$= -\frac{I_0\pi a^2\beta^2\sin\theta}{4\pi r}\cos(\omega t - \beta r)\,\mathbf{i}_\theta \tag{10.93}$$

From $\nabla \times \mathbf{H} = \dfrac{\partial \mathbf{D}}{\partial t} = \epsilon\dfrac{\partial \mathbf{E}}{\partial t}$, we have

$$\frac{\partial \mathbf{E}}{\partial t} = \frac{1}{\epsilon}\nabla \times \mathbf{H} = \frac{1}{\epsilon r}\frac{\partial}{\partial r}(rH_\theta)\,\mathbf{i}_\phi$$

$$= -\frac{I_0\pi a^2\beta^3\sin\theta}{4\pi\epsilon r}\sin(\omega t - \beta r)\,\mathbf{i}_\phi$$

$$\mathbf{E} = \frac{I_0\pi a^2\beta^3\sin\theta}{4\pi\omega\epsilon r}\cos(\omega t - \beta r)\,\mathbf{i}_\phi$$

$$= \frac{\eta I_0\pi a^2\beta^2\sin\theta}{4\pi r}\cos(\omega t - \beta r)\,\mathbf{i}_\phi \tag{10.94}$$

Comparing (10.94) and (10.93) with (8.25a) and (8.25b), respectively, we note that a duality exists between the radiation fields of the small current loop and those of the infinitesimal current element aligned along the axis of the current loop.

Proceeding further, we can find the Poynting vector, the instantaneous radiated power and the time-average radiated power due to the loop antenna by following steps similar to those employed for the Hertzian dipole in Sec. 8.2. Thus

$$\mathbf{P} = \mathbf{E} \times \mathbf{H} = E_\phi\mathbf{i}_\phi \times H_\theta\mathbf{i}_\theta = -E_\phi H_\theta\mathbf{i}_r$$

$$= \frac{\eta\beta^4 I_0^2\pi^2 a^4\sin^2\theta}{16\pi^2 r^2}\cos^2(\omega t - \beta r)\,\mathbf{i}_r$$

$$P_{\text{rad}} = \int_{\theta=0}^{\pi}\int_{\phi=0}^{2\pi}\mathbf{P}\cdot r^2\sin\theta\,d\theta\,d\phi\,\mathbf{i}_r$$

$$= \int_{\theta=0}^{\pi}\int_{\phi=0}^{2\pi}\frac{\eta\beta^4 I_0^2\pi^2 a^4\sin^3\theta}{16\pi^2}\cos^2(\omega t - \beta r)\,d\theta\,d\phi$$

$$= \frac{\eta\beta^4 I_0^2\pi^2 a^4}{6\pi}\cos^2(\omega t - \beta r)$$

$$\langle P_{\text{rad}}\rangle = \frac{\eta\beta^4 I_0^2\pi^2 a^4}{6\pi}\langle\cos^2(\omega t - \beta r)\rangle$$

$$= \frac{1}{2}I_0^2\left[\frac{8\pi^5\eta}{3}\left(\frac{a}{\lambda}\right)^4\right]$$

We now identify the radiation resistance of the small loop antenna to be

$$R_{rad} = \frac{8\pi^5 \eta}{3} \left(\frac{a}{\lambda} \right)^4 \tag{10.95}$$

For free space, $\eta = \eta_0 = 120\pi$ ohms, and

$$R_{rad} = 320\pi^6 \left(\frac{a}{\lambda} \right)^4 = 20\pi^2 \left(\frac{2\pi a}{\lambda} \right)^4 \tag{10.96}$$

Comparing this result with the radiation resistance of the Hertzian dipole given by (8.30), we note that the radiation resistance of the small loop antenna is proportional to the fourth power of its electrical size (circumference/wavelength) whereas that of the Hertzian dipole is proportional to the square of its electrical size (length/wavelength). The directivity of the small loop antenna is, however, the same as that of the Hertzian dipole, that is, 1.5, as given by (8.33), in view of the proportionality of the Poynting vectors to $\sin^2 \theta$ in both cases.

In this section we introduced the magnetic vector potential as a tool for computing the magnetic fields due to current distributions. In particular, we derived the expression for the retarded magnetic vector potential for a Hertzian dipole and illustrated its application by considering the case of a small circular loop antenna. We derived the radiation fields for the loop antenna and compared its characteristics with those of the Hertzian dipole.

REVIEW QUESTIONS

10.33. Why can the magnetic flux density vector be expressed as the curl of another vector?

10.34. Discuss the analogy between the magnetic vector potential due to an infinitesimal current element and the electric scalar potential due to a point charge.

10.35. What does the word "retarded" in the terminology "retarded magnetic vector potential" refer to? Explain.

10.36. Discuss the application of the magnetic vector potential in the determination of the electromagnetic fields due to an antenna.

10.37. Discuss the duality between the radiation fields of a small circular loop antenna with those of a Hertzian dipole at the center of the loop and aligned with its axis.

10.38. Compare the radiation resistance and directivity of a small circular loop antenna with those of a Hertzian dipole.

PROBLEMS

10.23. By expansion in Cartesian coordinates, show that

$$\mathbf{A} \times \mathbf{\nabla} V = V\mathbf{\nabla} \times \mathbf{A} - \mathbf{\nabla} \times (V\mathbf{A}).$$

10.24. For the half-wave dipole of Sec. 8.3, determine the magnetic vector potential for the radiation fields. Verify your result by finding the radiation fields and comparing with the results of Sec. 8.3.

10.25. A circular loop antenna of radius 1 m in free space carries a uniform current $10 \cos 4\pi \times 10^6 t$ amp. (a) Calculate the amplitude of the electric field intensity at a distance of 10 km in the plane of the loop. (b) Calculate the radiation resistance and the time-average power radiated by the loop.

10.26. Find the length of a Hertzian dipole that would radiate the same time-average power as the loop antenna of Problem 10.25 for the same current and frequency as in Problem 10.25.

APPENDICES

A. CYLINDRICAL AND SPHERICAL COORDINATE SYSTEMS

In Sec. 1.2 we learned that the Cartesian coordinate system is defined by a set of three mutually orthogonal surfaces, all of which are planes. The cylindrical and spherical coordinate systems also involve sets of three mutually orthogonal surfaces. For the cylindrical coordinate system, the three surfaces are a cylinder and two planes, as shown in Fig. A.1(a). One of these planes is the same as the $z = $ constant plane in the Cartesian coordinate system. The second plane contains the z axis and makes an angle ϕ with a reference plane, conveniently chosen to be the xz plane of the Cartesian coordinate system. This plane is therefore defined by $\phi = $ constant. The cylindrical surface has the z axis as its axis. Since the radial distance r from the z axis to points on the cylindrical surface is a constant, this surface is defined by $r = $ constant. Thus the three orthogonal surfaces defining the cylindrical coordinates of a point are $r = $ constant, $\phi = $ constant, and $z = $ constant. Only two of these coordinates (r and z) are distances; the third coordinate (ϕ) is an angle. We note that the entire space is spanned by varying r from 0 to ∞, ϕ from 0 to 2π, and z from $-\infty$ to ∞.

The origin is given by $r = 0$, $\phi = 0$, and $z = 0$. Any other point in space is given by the intersection of three mutually orthogonal surfaces obtained by incrementing the coordinates by appropriate amounts. For example, the intersection of the three surfaces $r = 2$, $\phi = \pi/4$, and $z = 3$ defines the point $A(2, \pi/4, 3)$, as shown in Fig. A.1(a). These three orthogonal surfaces define three curves that are mutually perpendicular. Two of these are straight lines and the third is a circle. We draw unit vectors, $\mathbf{i}_r$, $\mathbf{i}_\phi$, and $\mathbf{i}_z$ tangential to these

433

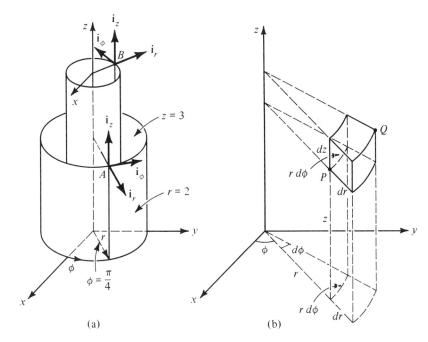

(a)

(b)

Figure A.1. Cylindrical coordinate system. (a) Orthogonal surfaces and unit vectors. (b) Differential volume formed by incrementing the coordinates.

curves at the point A and directed toward increasing values of r, ϕ, and z, respectively. These three unit vectors form a set of mutually orthogonal unit vectors in terms of which vectors drawn at A can be described. In a similar manner, we can draw unit vectors at any other point in the cylindrical coordinate system, as shown, for example, for point $B(1, 3\pi/4, 5)$ in Fig. A.1(a). It can now be seen that the unit vectors $\mathbf{i}_r$ and $\mathbf{i}_\phi$ at point B are not parallel to the corresponding unit vectors at point A. Thus unlike in the Cartesian coordinate system, the unit vectors $\mathbf{i}_r$ and $\mathbf{i}_\phi$ in the cylindrical coordinate system do not have the same directions everywhere, that is, they are not uniform. Only the unit vector $\mathbf{i}_z$, which is the same as in the Cartesian coordinate system, is uniform. Finally, we note that for the choice of ϕ as in Fig. A.1(a), that is, increasing from the positive x axis toward the positive y axis, the coordinate system is right-handed, that is, $\mathbf{i}_r \times \mathbf{i}_\phi = \mathbf{i}_z$.

To obtain expressions for the differential lengths, surfaces, and volume in the cylindrical coordinate system, we now consider two points $P(r, \phi, z)$ and $Q(r + dr, \phi + d\phi, z + dz)$ where Q is obtained by incrementing infinitesimally each coordinate from its value at P, as shown in Fig. A.1(b). The three orthogonal surfaces intersecting at P and the three orthogonal surfaces intersecting at Q define a box which can be considered to be rectangular since

dr, $d\phi$, and dz are infinitesimally small. The three differential length elements forming the contiguous sides of this box are $dr\, \mathbf{i}_r$, $r\, d\phi\, \mathbf{i}_\phi$, and $dz\, \mathbf{i}_z$. The differential length vector $d\mathbf{l}$ from P to Q is thus given by

$$d\mathbf{l} = dr\, \mathbf{i}_r + r\, d\phi\, \mathbf{i}_\phi + dz\, \mathbf{i}_z \qquad (A.1)$$

The differential surfaces formed by pairs of the differential length elements are

$$\pm dS\, \mathbf{i}_z = \pm(dr)\,(r\, d\phi)\, \mathbf{i}_z = \pm dr\, \mathbf{i}_r \times r\, d\phi\, \mathbf{i}_\phi \qquad (A.2a)$$

$$\pm dS\, \mathbf{i}_r = \pm(r\, d\phi)\,(dz)\, \mathbf{i}_r = \pm r\, d\phi\, \mathbf{i}_\phi \times dz\, \mathbf{i}_z \qquad (A.2b)$$

$$\pm dS\, \mathbf{i}_\phi = \pm(dz)\,(dr)\, \mathbf{i}_\phi = \pm dz\, \mathbf{i}_z \times dr\, \mathbf{i}_r \qquad (A.2c)$$

Finally, the differential volume dv formed by the three differential lengths is simply the volume of the box, that is,

$$dv = (dr)\,(r\, d\phi)\,(dz) = r\, dr\, d\phi\, dz \qquad (A.3)$$

For the spherical coordinate system, the three mutually orthogonal surfaces are a sphere, a cone, and a plane, as shown in Fig. A.2(a). The plane is the same as the $\phi = $ constant plane in the cylindrical coordinate system.

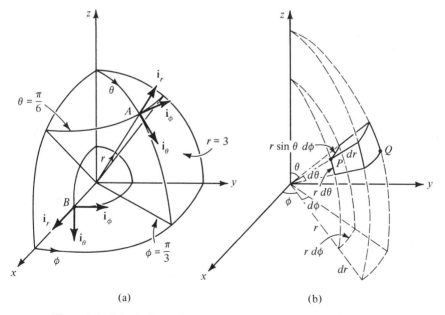

(a)   (b)

Figure A.2. Spherical coordinate system. (a) Orthogonal surfaces and unit vectors. (b) Differential volume formed by incrementing the coordinates.

The sphere has the origin as its center. Since the radial distance r from the origin to points on the spherical surface is a constant, this surface is defined by $r = $ constant. The spherical coordinate r should not be confused with the cylindrical coordinate r. When these two coordinates appear in the same expression, we shall use the subscripts c and s to distinguish between cylindrical and spherical. The cone has its vertex at the origin and its surface is symmetrical about the z axis. Since the angle θ is the angle that the conical surface makes with the z axis, this surface is defined by $\theta = $ constant. Thus the three orthogonal surfaces defining the spherical coordinates of a point are $r = $ constant, $\theta = $ constant, and $\phi = $ constant. Only one of these coordinates (r) is distance; the other two coordinates (θ and ϕ) are angles. We note that the entire space is spanned by varying r from 0 to ∞, θ from 0 to π, and ϕ from 0 to 2π.

The origin is given by $r = 0$, $\theta = 0$, and $\phi = 0$. Any other point in space is given by the intersection of three mutually orthogonal surfaces obtained by incrementing the coordinates by appropriate amounts. For example, the intersection of the three surfaces $r = 3$, $\theta = \pi/6$, and $\phi = \pi/3$ defines the point $A(3, \pi/6, \pi/3)$ as shown in Fig. A.2(a). These three orthogonal surfaces define three curves that are mutually perpendicular. One of these is a straight line and the other two are circles. We draw unit vectors $\mathbf{i}_r$, $\mathbf{i}_\theta$, and $\mathbf{i}_\phi$ tangential to these curves at point A and directed toward increasing values of r, θ, and ϕ, respectively. These three unit vectors form a set of mutually orthogonal unit vectors in terms of which vectors drawn at A can be described. In a similar manner, we can draw unit vectors at any other point in the spherical coordinate system, as shown, for example, for point $B(1, \pi/2, 0)$ in Fig. A.2(a). It can now be seen that these unit vectors at point B are not parallel to the corresponding unit vectors at point A. Thus in the spherical coordinate system all three unit vectors $\mathbf{i}_r$, $\mathbf{i}_\theta$, and $\mathbf{i}_\phi$ do not have the same directions everywhere, that is, they are not uniform. Finally, we note that for the choice of θ as in Fig. A.2(a), that is, increasing from the positive z axis toward the xy plane, the coordinate system is right-handed, that is, $\mathbf{i}_r \times \mathbf{i}_\theta = \mathbf{i}_\phi$.

To obtain expressions for the differential lengths, surfaces, and volume in the spherical coordinate system, we now consider two points $P(r, \theta, \phi)$ and $Q(r + dr, \theta + d\theta, \phi + d\phi)$ where Q is obtained by incrementing infinitesimally each coordinate from its value at P, as shown in Fig. A.2(b). The three orthogonal surfaces intersecting at P and the three orthogonal surfaces intersecting at Q define a box that can be considered to be rectangular since dr, $d\theta$, and $d\phi$ are infinitesimally small. The three differential length elements forming the contiguous sides of this box are $dr\,\mathbf{i}_r$, $r\,d\theta\,\mathbf{i}_\theta$, and $r\sin\theta\,d\phi\,\mathbf{i}_\phi$. The differential length vector $d\mathbf{l}$ from P to Q is thus given by

$$d\mathbf{l} = dr\,\mathbf{i}_r + r\,d\theta\,\mathbf{i}_\theta + r\sin\theta\,d\phi\,\mathbf{i}_\phi \tag{A.4}$$

The differential surfaces formed by pairs of the differential length elements are

$$\pm dS\,\mathbf{i}_\phi = \pm(dr)\,(r\,d\theta)\,\mathbf{i}_\phi = \pm dr\,\mathbf{i}_r \times r\,d\theta\,\mathbf{i}_\theta \tag{A.5a}$$

$$\pm dS\,\mathbf{i}_r = \pm(r\,d\theta)\,(r\sin\theta\,d\phi)\,\mathbf{i}_r = \pm r\,d\theta\,\mathbf{i}_\theta \times r\sin\theta\,d\phi\,\mathbf{i}_\phi \tag{A.5b}$$

$$\pm dS\,\mathbf{i}_\theta = \pm(r\sin\theta\,d\phi)\,(dr)\,\mathbf{i}_\theta = \pm r\sin\theta\,d\phi\,\mathbf{i}_\phi \times dr\,\mathbf{i}_r \tag{A.5c}$$

Finally, the differential volume dv formed by the three differential lengths is simply the volume of the box, that is,

$$dv = (dr)\,(r\,d\theta)\,(r\sin\theta\,d\phi) = r^2\sin\theta\,dr\,d\theta\,d\phi \tag{A.6}$$

In the study of electromagnetics it is sometimes useful to be able to convert the coordinates of a point and vectors drawn at a point from one coordinate system to another, particularly from the Cartesian system to the cylindrical system and vice versa, and from the Cartesian system to the spherical system and vice versa. To derive first the relationships for the conversion of the coordinates, let us consider Fig. A.3(a) which illustrates the geometry pertinent to the coordinates of a point P in the three different coordinate systems. Thus from simple geometrical considerations, we have

$$x = r_c\cos\phi \qquad y = r_c\sin\phi \qquad z = z \tag{A.7}$$

$$x = r_s\sin\theta\cos\phi \qquad y = r_s\sin\theta\sin\phi \qquad z = r_s\cos\theta \tag{A.8}$$

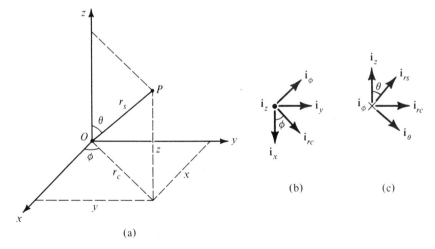

(a)

(b)   (c)

Figure A.3. (a) For conversion of coordinates of a point from one coordinate system to another. (b) and (c) For expressing unit vectors in cylindrical and spherical coordinate systems, respectively, in terms of unit vectors in the Cartesian coordinate system.

Conversely, we have

$$r_c = \sqrt{x^2 + y^2} \qquad \phi = \tan^{-1} \frac{y}{x} \qquad z = z \qquad \text{(A.9)}$$

$$r_s = \sqrt{x^2 + y^2 + z^2} \qquad \theta = \tan^{-1} \frac{\sqrt{x^2 + y^2}}{z} \qquad \phi = \tan^{-1} \frac{y}{x} \qquad \text{(A.10)}$$

Relationships (A.7) and (A.9) correspond to conversion from cylindrical coordinates to Cartesian coordinates and vice versa. Relationships (A.8) and (A.10) correspond to conversion from spherical coordinates to Cartesian coordinates and vice versa.

Considering next the conversion of vectors from one coordinate system to another, we note that in order to do this, we need to express each of the unit vectors of the first coordinate system in terms of its components along the unit vectors in the second coordinate system. From the definition of the dot product of two vectors, the component of a unit vector along another unit vector, that is, the cosine of the angle between the unit vectors, is simply the dot product of the two unit vectors. Thus considering the sets of unit vectors in the cylindrical and Cartesian coordinate systems, we have with the aid of Fig. A.3(b),

$$\mathbf{i}_{r_c} \cdot \mathbf{i}_x = \cos \phi \qquad \mathbf{i}_{r_c} \cdot \mathbf{i}_y = \sin \phi \qquad \mathbf{i}_{r_c} \cdot \mathbf{i}_z = 0 \qquad \text{(A.11a)}$$

$$\mathbf{i}_\phi \cdot \mathbf{i}_x = -\sin \phi \qquad \mathbf{i}_\phi \cdot \mathbf{i}_y = \cos \phi \qquad \mathbf{i}_\phi \cdot \mathbf{i}_z = 0 \qquad \text{(A.11b)}$$

$$\mathbf{i}_z \cdot \mathbf{i}_x = 0 \qquad \mathbf{i}_z \cdot \mathbf{i}_y = 0 \qquad \mathbf{i}_z \cdot \mathbf{i}_z = 1 \qquad \text{(A.11c)}$$

Similarly, for the sets of unit vectors in the spherical and Cartesian coordinate systems, we obtain with the aid of Fig. A.3(c) and Fig. A.3(b),

$$\mathbf{i}_{r_s} \cdot \mathbf{i}_x = \sin \theta \cos \phi \qquad \mathbf{i}_{r_s} \cdot \mathbf{i}_y = \sin \theta \sin \phi \qquad \mathbf{i}_{r_s} \cdot \mathbf{i}_z = \cos \theta \qquad \text{(A.12a)}$$

$$\mathbf{i}_\theta \cdot \mathbf{i}_x = \cos \theta \cos \phi \qquad \mathbf{i}_\theta \cdot \mathbf{i}_y = \cos \theta \sin \phi \qquad \mathbf{i}_\theta \cdot \mathbf{i}_z = -\sin \theta \qquad \text{(A.12b)}$$

$$\mathbf{i}_\phi \cdot \mathbf{i}_x = -\sin \phi \qquad \mathbf{i}_\phi \cdot \mathbf{i}_y = \cos \phi \qquad \mathbf{i}_\phi \cdot \mathbf{i}_z = 0 \qquad \text{(A.12c)}$$

We shall now illustrate the use of these relationships by means of an example.

Example A.1. Let us consider the vector $3\mathbf{i}_x + 4\mathbf{i}_y + 5\mathbf{i}_z$ at the point $(3, 4, 5)$ and convert the vector to one in spherical coordinates.

First, from the relationships (A.10), we obtain the spherical coordinates of the point $(3, 4, 5)$ to be

$$r_s = \sqrt{3^2 + 4^2 + 5^2} = 5\sqrt{2}$$

$$\theta = \tan^{-1} \frac{\sqrt{3^2 + 4^2}}{5} = \tan^{-1} 1 = 45°$$

$$\phi = \tan^{-1} \frac{4}{3} = 53.13°$$

Then noting from the relationships (A.12) that at the point under consideration,

$$i_x = \sin\theta\cos\phi\, i_{rs} + \cos\theta\cos\phi\, i_\theta - \sin\phi\, i_\phi$$
$$= 0.3\sqrt{2}\, i_{rs} + 0.3\sqrt{2}\, i_\theta - 0.8 i_\phi$$
$$i_y = \sin\theta\sin\phi\, i_{rs} + \cos\theta\sin\phi\, i_\theta + \cos\phi\, i_\phi$$
$$= 0.4\sqrt{2}\, i_{rs} + 0.4\sqrt{2}\, i_\theta + 0.6 i_\phi$$
$$i_z = \cos\theta\, i_{rs} - \sin\theta\, i_\theta = 0.5\sqrt{2}\, i_{rs} - 0.5\sqrt{2}\, i_\theta$$

we obtain

$$3i_x + 4i_y + 5i_z = (0.9\sqrt{2} + 1.6\sqrt{2} + 2.5\sqrt{2})i_{rs}$$
$$+ (0.9\sqrt{2} + 1.6\sqrt{2} - 2.5\sqrt{2})i_\theta + (-2.4 + 2.4)i_\phi = 5\sqrt{2}\, i_{rs}$$

This result is to be expected since the given vector has components equal to the coordinates of the point at which it is specified. Hence its magnitude is equal to the distance of the point from the origin, that is, the spherical coordinate r of the point and its direction is along the line drawn from the origin to the point, that is, along the unit vector i_{rs} at that point. In fact, the given vector is a particular case of the vector $xi_x + yi_y + zi_z = r_s i_{rs}$, known as the "position vector," since it is the same as the vector drawn from the origin to the point (x, y, z). ∎

REVIEW QUESTIONS

A.1. What are the three orthogonal surfaces defining the cylindrical coordinate system?

A.2. What are the limits of variation of the cylindrical coordinates?

A.3. Which of the unit vectors in the cylindrical coordinate system are not uniform?

A.4. State whether the vector $3i_r + 4i_\phi + 5i_z$ at the point $(1, 0, 2)$ and the vector $3i_r + 4i_\phi + 5i_z$ at the point $(2, \pi/2, 3)$ are equal or not.

A.5. What are the differential length vectors in cylindrical coordinates?

A.6. What are the three orthogonal surfaces defining the spherical coordinate system?

A.7. What are the limits of variation of the spherical coordinates?

A.8. Which of the unit vectors in the spherical coordinate system are not uniform?

A.9. State if the vector $3i_r + 4i_\theta$ at the point $(1, \pi/2, 0)$ and the vector $3i_r + 4i_\theta$ at the point $(2, 0, \pi/2)$ are equal or not.

A.10. What are the differential length vectors in spherical coordinates?

A.11. Outline the procedure for converting a vector at a point from one coordinate system to another.

A.12. What is the expression for the position vector in the cylindrical coordinate system?

PROBLEMS

A.1. Express in terms of Cartesian coordinates the vector drawn from the point $P(2, \pi/3, 1)$ to the point $Q(4, 2\pi/3, 2)$ in cylindrical coordinates.

A.2. Express in terms of Cartesian coordinates the vector drawn from the point $P(1, \pi/3, \pi/4)$ to the point $Q(2, 2\pi/3, 3\pi/4)$ in spherical coordinates.

A.3. Determine if the vector $i_r + i_\phi + 2i_z$ at the point $(1, \pi/4, 2)$ and the vector $\sqrt{2}\,i_r + 2i_z$ at the point $(2, \pi/2, 3)$ are equal or not.

A.4. Determine if the vector $3i_r + \sqrt{3}\,i_\theta - 2i_\phi$ at the point $(2, \pi/3, \pi/6)$ and the vector $i_r + \sqrt{3}\,i_\theta - 2\sqrt{3}\,i_\phi$ at the point $(1, \pi/6, \pi/3)$ are equal or not.

A.5. Find the dot and cross products of the unit vector i_r at the point $(1, 0, 0)$ and the unit vector i_ϕ at the point $(2, \pi/4, 1)$ in cylindrical coordinates.

A.6. Find the dot and cross products of the unit vector i_r at the point $(1, \pi/4, 0)$ and the unit vector i_θ at the point $(2, \pi/2, \pi/2)$ in spherical coordinates.

A.7. Convert the vector $5i_x + 12i_y + 6i_z$ at the point $(5, 12, 4)$ to one in cylindrical coordinates.

A.8. Convert the vector $3i_x + 4i_y - 5i_z$ at the point $(3, 4, 5)$ to one in spherical coordinates.

B. CURL, DIVERGENCE, AND GRADIENT IN CYLINDRICAL AND SPHERICAL COORDINATE SYSTEMS

In Secs. 3.1, 3.4, and 9.1 we introduced the curl, divergence, and gradient, respectively, and derived the expressions for them in the Cartesian coordinate system. In this appendix we shall derive the corresponding expressions in the cylindrical and spherical coordinate systems. Considering first the cylindrical coordinate system, we recall from Appendix A that the infinitesimal box defined by the three orthogonal surfaces intersecting at point $P(r, \theta, \phi)$ and the three orthogonal surfaces intersecting at point $Q(r + dr, \phi + d\phi, z + dz)$ is as shown in Fig. B.1.

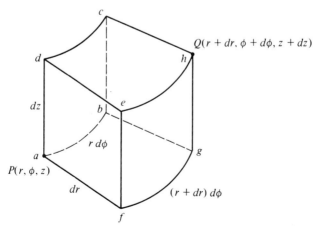

Figure B.1. Infinitesimal box formed by incrementing the coordinates in the cylindrical coordinate system.

From the basic definition of the curl of a vector introduced in Sec. 3.3 and given by

$$\nabla \times \mathbf{A} = \lim_{\Delta S \to 0} \left[\frac{\oint_C \mathbf{A} \cdot d\mathbf{l}}{\Delta S} \right]_{\text{max}} \mathbf{i}_n \tag{B.1}$$

we find the components of $\nabla \times \mathbf{A}$ as follows with the aid of Fig. B.1:

$$(\nabla \times \mathbf{A})_r = \lim_{\substack{d\phi \to 0 \\ dz \to 0}} \frac{\oint_{abcda} \mathbf{A} \cdot d\mathbf{l}}{\text{area } abcd}$$

$$= \lim_{\substack{d\phi \to 0 \\ dz \to 0}} \frac{\left\{ \begin{array}{l} [A_\phi]_{(r,z)} r\, d\phi + [A_z]_{(r,\phi+d\phi)}\, dz \\ - [A_\phi]_{(r,z+dz)} r\, d\phi - [A_z]_{(r,\phi)}\, dz \end{array} \right\}}{r\, d\phi\, dz}$$

$$= \lim_{d\phi \to 0} \frac{[A_z]_{(r,\phi+d\phi)} - [A_z]_{(r,\phi)}}{r\, d\phi} + \lim_{dz \to 0} \frac{[A_\phi]_{(r,z)} - [A_\phi]_{(r,z+dz)}}{dz}$$

$$= \frac{1}{r} \frac{\partial A_z}{\partial \phi} - \frac{\partial A_\phi}{\partial z} \tag{B.2a}$$

$$(\nabla \times \mathbf{A})_\phi = \lim_{\substack{dz \to 0 \\ dr \to 0}} \frac{\oint_{adefa} \mathbf{A} \cdot d\mathbf{l}}{\text{area } adef}$$

$$= \lim_{\substack{dz \to 0 \\ dr \to 0}} \frac{\left\{ \begin{array}{l} [A_z]_{(r,\phi)}\, dz + [A_r]_{(\phi,z+dz)}\, dr \\ - [A_z]_{(r+dr,\phi)}\, dz - [A_r]_{(\phi,z)}\, dr \end{array} \right\}}{dr\, dz}$$

$$= \lim_{dz \to 0} \frac{[A_r]_{(\phi,z+dz)} - [A_r]_{(\phi,z)}}{dz} + \lim_{dr \to 0} \frac{[A_z]_{(r,\phi)} - [A_z]_{(r+dr,\phi)}}{dr}$$

$$= \frac{\partial A_r}{\partial z} - \frac{\partial A_z}{\partial r} \tag{B.2b}$$

$$(\nabla \times \mathbf{A})_z = \lim_{\substack{dr \to 0 \\ d\phi \to 0}} \frac{\oint_{afgba} \mathbf{A} \cdot d\mathbf{l}}{\text{area } afgb}$$

$$= \lim_{\substack{dr \to 0 \\ d\phi \to 0}} \frac{\left\{ \begin{array}{l} [A_r]_{(\phi,z)}\, dr + [A_\phi]_{(r+dr,z)}(r+dr)\, d\phi \\ - [A_r]_{(\phi+d\phi,z)}\, dr - [A_\phi]_{(r,z)} r\, d\phi \end{array} \right\}}{r\, dr\, d\phi}$$

$$= \lim_{dr \to 0} \frac{[rA_\phi]_{(r+dr,z)} - [rA_\phi]_{(r,z)}}{r\, dr} + \lim_{d\phi \to 0} \frac{[A_r]_{(\phi,z)} - [A_r]_{(\phi+d\phi,z)}}{r\, d\phi}$$

$$= \frac{1}{r} \frac{\partial}{\partial r}(rA_\phi) - \frac{1}{r} \frac{\partial A_r}{\partial \phi} \tag{B.2c}$$

Combining (B.2a), (B.2b) and (B.2c), we obtain the expression for the curl of a vector in cylindrical coordinates as

$$\nabla \times \mathbf{A} = \left[\frac{1}{r}\frac{\partial A_z}{\partial \phi} - \frac{\partial A_\phi}{\partial z}\right]\mathbf{i}_r + \left[\frac{\partial A_r}{\partial z} - \frac{\partial A_z}{\partial r}\right]\mathbf{i}_\phi$$

$$+ \frac{1}{r}\left[\frac{\partial}{\partial r}(rA_\phi) - \frac{\partial A_r}{\partial \phi}\right]\mathbf{i}_z$$

$$= \begin{vmatrix} \dfrac{\mathbf{i}_r}{r} & \mathbf{i}_\phi & \dfrac{\mathbf{i}_z}{r} \\ \dfrac{\partial}{\partial r} & \dfrac{\partial}{\partial \phi} & \dfrac{\partial}{\partial z} \\ A_r & rA_\phi & A_z \end{vmatrix} \tag{B.3}$$

To find the expression for the divergence, we make use of the basic definition of the divergence of a vector, introduced in Sec. 3.6 and given by

$$\nabla \cdot \mathbf{A} = \underset{\Delta v \to 0}{\text{Lim}} \frac{\oint_S \mathbf{A} \cdot d\mathbf{S}}{\Delta v} \tag{B.4}$$

Evaluating the right side of (B.4) for the box of Fig. B.1, we obtain

$$\nabla \cdot \mathbf{A} = \underset{\substack{dr\to 0 \\ d\phi\to 0 \\ dz\to 0}}{\text{Lim}} \frac{\left\{\begin{array}{l}[A_r]_{r+dr}(r + dr)\, d\phi\, dz - [A_r]_r r\, d\phi\, dz + [A_\phi]_{\phi+d\phi}\, dr\, dz \\ - [A_\phi]_\phi\, dr\, dz + [A_z]_{z+dz} r\, dr\, d\phi - [A_z]_z r\, dr\, d\phi\end{array}\right\}}{r\, dr\, d\phi\, dz}$$

$$= \underset{dr\to 0}{\text{Lim}} \frac{[rA_r]_{r+dr} - [rA_r]_r}{r\, dr} + \underset{d\phi\to 0}{\text{Lim}} \frac{[A_\phi]_{\phi+d\phi} - [A_\phi]_\phi}{r\, d\phi}$$

$$+ \underset{dz\to 0}{\text{Lim}} \frac{[A_z]_{z+dz} - [A_z]_z}{dz}$$

$$= \frac{1}{r}\frac{\partial}{\partial r}(rA_r) + \frac{1}{r}\frac{\partial A_\phi}{\partial \phi} + \frac{\partial A_z}{\partial z} \tag{B.5}$$

To obtain the expression for the gradient of a scalar, we recall from Appendix A that in cylindrical coordinates,

$$d\mathbf{l} = dr\,\mathbf{i}_r + r\,d\phi\,\mathbf{i}_\phi + dz\,\mathbf{i}_z \tag{B.6}$$

and hence

$$d\Phi = \frac{\partial \Phi}{\partial r}dr + \frac{\partial \Phi}{\partial \phi}d\phi + \frac{\partial \Phi}{\partial z}dz$$

$$= \left(\frac{\partial \Phi}{\partial r}\mathbf{i}_r + \frac{1}{r}\frac{\partial \Phi}{\partial \phi}\mathbf{i}_\phi + \frac{\partial \Phi}{\partial z}\mathbf{i}_z\right) \cdot (dr\,\mathbf{i}_r + r\,d\phi\,\mathbf{i}_\phi + dz\,\mathbf{i}_z)$$

$$= \nabla\Phi \cdot d\mathbf{l} \tag{B.7}$$

Thus

$$\nabla\Phi = \frac{\partial\Phi}{\partial r}\,\mathbf{i}_r + \frac{1}{r}\frac{\partial\Phi}{\partial\phi}\,\mathbf{i}_\phi + \frac{\partial\Phi}{\partial z}\,\mathbf{i}_z \qquad (B.8)$$

Turning now to the spherical coordinate system, we recall from Appendix A that the infinitesimal box defined by the three orthogonal surfaces intersecting at $P(r, \theta, \phi)$ and the three orthogonal surfaces intersecting at $Q(r + dr, \theta + d\theta, \phi + d\phi)$ is as shown in Fig. B.2. From the basic definition of the curl of a vector given by (B.1), we then find the components of $\nabla \times \mathbf{A}$ as follows with the aid of Fig. B.2:

$$(\nabla \times \mathbf{A})_r = \lim_{\substack{d\theta\to0\\d\phi\to0}} \frac{\oint_{abcda} \mathbf{A}\cdot d\mathbf{l}}{\text{area } abcd}$$

$$= \lim_{\substack{d\theta\to0\\d\phi\to0}} \frac{\left\{\begin{array}{l}[A_\theta]_{(r,\phi)}r\,d\theta + [A_\phi]_{(r,\theta+d\theta)}r\sin(\theta+d\theta)\,d\phi\\ - [A_\theta]_{(r,\phi+d\phi)}r\,d\theta - [A_\phi]_{(r,\theta)}r\sin\theta\,d\phi\end{array}\right\}}{r^2 \sin\theta\,d\theta\,d\phi}$$

$$= \lim_{d\theta\to0} \frac{[A_\phi\sin\theta]_{(r,\theta+d\theta)} - [A_\phi\sin\theta]_{(r,\theta)}}{r\sin\theta\,d\theta}$$

$$+ \lim_{d\phi\to0} \frac{[A_\theta]_{(r,\phi)} - [A_\theta]_{(r,\phi+d\phi)}}{r\sin\theta\,d\phi}$$

$$= \frac{1}{r\sin\theta}\frac{\partial}{\partial\theta}(A_\phi\sin\theta) - \frac{1}{r\sin\theta}\frac{\partial A_\theta}{\partial\phi} \qquad (B.9a)$$

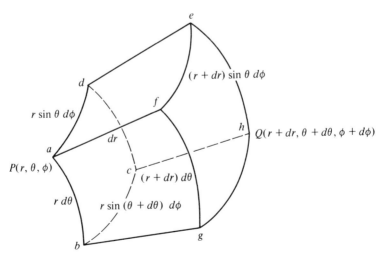

Figure B.2. Infinitesimal box formed by incrementing the coordinates in the spherical coordinate system.

$$(\boldsymbol{\nabla} \times \mathbf{A})_\theta = \underset{\substack{d\phi \to 0 \\ dr \to 0}}{\text{Lim}} \frac{\oint_{adefa} \mathbf{A} \cdot d\mathbf{l}}{\text{area } adef}$$

$$= \underset{\substack{d\phi \to 0 \\ dr \to 0}}{\text{Lim}} \frac{\left\{ \begin{array}{l} [A_\phi]_{(r,\theta)} r \sin\theta\, d\phi + [A_r]_{(\theta,\phi+d\phi)}\, dr \\ - [A_\phi]_{(r+dr,\theta)}(r+dr)\sin\theta\, d\phi - [A_r]_{(\theta,\phi)}\, dr \end{array} \right\}}{r \sin\theta\, dr\, d\phi}$$

$$= \underset{d\phi \to 0}{\text{Lim}} \frac{[A_r]_{(\theta,\phi+d\phi)} - [A_r]_{(\theta,\phi)}}{r \sin\theta\, d\phi}$$

$$+ \underset{dr \to 0}{\text{Lim}} \frac{[rA_\phi]_{(r,\theta)} - [rA_\phi]_{(r+dr,\theta)}}{r\, dr}$$

$$= \frac{1}{r\sin\theta}\frac{\partial A_r}{\partial\phi} - \frac{1}{r}\frac{\partial}{\partial r}(rA_\phi) \tag{B.9b}$$

$$(\boldsymbol{\nabla} \times \mathbf{A})_\phi = \underset{\substack{dr \to 0 \\ d\theta \to 0}}{\text{Lim}} \frac{\oint_{afgba} \mathbf{A} \cdot d\mathbf{l}}{\text{area } afgb}$$

$$= \underset{\substack{dr \to 0 \\ d\theta \to 0}}{\text{Lim}} \frac{\left\{ \begin{array}{l} [A_r]_{(\theta,\phi)}\, dr + [A_\theta]_{(r+dr,\phi)}(r+dr)\, d\theta \\ - [A_r]_{(\theta+d\theta,\phi)}\, dr - [A_\theta]_{(r,\phi)} r\, d\theta \end{array} \right\}}{r\, dr\, d\theta}$$

$$= \underset{dr \to 0}{\text{Lim}} \frac{[rA_\theta]_{(r+dr,\phi)} - [rA_\theta]_{(r,\phi)}}{r\, dr}$$

$$+ \underset{d\theta \to 0}{\text{Lim}} \frac{[A_r]_{(\theta,\phi)}\, dr - [A_r]_{(\theta+d\theta,\phi)}\, dr}{r\, d\theta}$$

$$= \frac{1}{r}\frac{\partial}{\partial r}(rA_\theta) - \frac{1}{r}\frac{\partial A_r}{\partial\theta} \tag{B.9c}$$

Combining (B.9a), (B.9b), and (B.9c), we obtain the expression for the curl of a vector in spherical coordinates as

$$\boldsymbol{\nabla} \times \mathbf{A} = \frac{1}{r\sin\theta}\left[\frac{\partial}{\partial\theta}(A_\phi \sin\theta) - \frac{\partial A_\theta}{\partial\phi}\right]\mathbf{i}_r$$

$$+ \frac{1}{r}\left[\frac{1}{\sin\theta}\frac{\partial A_r}{\partial\phi} - \frac{\partial}{\partial r}(rA_\phi)\right]\mathbf{i}_\theta + \frac{1}{r}\left[\frac{\partial}{\partial r}(rA_\theta) - \frac{\partial A_r}{\partial\theta}\right]\mathbf{i}_\phi$$

$$= \begin{vmatrix} \dfrac{\mathbf{i}_r}{r^2 \sin\theta} & \dfrac{\mathbf{i}_\theta}{r\sin\theta} & \dfrac{\mathbf{i}_\phi}{r} \\[2mm] \dfrac{\partial}{\partial r} & \dfrac{\partial}{\partial\theta} & \dfrac{\partial}{\partial\phi} \\[2mm] A_r & rA_\theta & r\sin\theta\, A_\phi \end{vmatrix} \tag{B.10}$$

To find the expression for the divergence, we make use of the basic definition of the divergence of a vector given by (B.4) and by evaluating its right side for the box of Fig. B.2, we obtain

$$\mathbf{V} \cdot \mathbf{A} = \lim_{\substack{dr \to 0 \\ d\theta \to 0 \\ d\phi \to 0}} \frac{\left\{ \begin{array}{l} [A_r]_{r+dr}(r+dr)^2 \sin\theta \, d\theta \, d\phi - [A_r]_r r^2 \sin\theta \, d\theta \, d\phi \\ + [A_\theta]_{\theta+d\theta} r \sin(\theta+d\theta) \, dr \, d\phi - [A_\theta]_\theta r \sin\theta \, dr \, d\phi \\ + [A_\phi]_{\phi+d\phi} r \, dr \, d\theta - [A_\phi]_\phi r \, dr \, d\theta \end{array} \right\}}{r^2 \sin\theta \, dr \, d\theta \, d\phi}$$

$$= \lim_{dr \to 0} \frac{[r^2 A_r]_{r+dr} - [r^2 A_r]_r}{r^2 \, dr} + \lim_{d\theta \to 0} \frac{[A_\theta \sin\theta]_{\theta+d\theta} - [A_\theta \sin\theta]_\theta}{r \sin\theta \, d\theta}$$

$$+ \lim_{d\phi \to 0} \frac{[A_\phi]_{\phi+d\phi} - [A_\phi]_\phi}{r \sin\theta \, d\phi}$$

$$= \frac{1}{r^2} \frac{\partial}{\partial r}(r^2 A_r) + \frac{1}{r \sin\theta} \frac{\partial}{\partial \theta}(A_\theta \sin\theta) + \frac{1}{r \sin\theta} \frac{\partial A_\phi}{\partial \phi} \qquad (B.11)$$

To obtain the expression for the gradient of a scalar, we recall from Appendix A that in spherical coordinates,

$$dl = dr \, \mathbf{i}_r + r \, d\theta \, \mathbf{i}_\theta + r \sin\theta \, d\phi \, \mathbf{i}_\phi \qquad (B.12)$$

and hence

$$d\Phi = \frac{\partial \Phi}{\partial r} dr + \frac{\partial \Phi}{\partial \theta} d\theta + \frac{\partial \Phi}{\partial \phi} d\phi$$

$$= \left(\frac{\partial \Phi}{\partial r} \mathbf{i}_r + \frac{1}{r} \frac{\partial \Phi}{\partial \theta} \mathbf{i}_\theta + \frac{1}{r \sin\theta} \frac{\partial \Phi}{\partial \phi} \mathbf{i}_\phi \right) \cdot (dr \, \mathbf{i}_r + r \, d\theta \, \mathbf{i}_\theta + r \sin\theta \, d\phi \, \mathbf{i}_\phi)$$

$$= \mathbf{V}\Phi \cdot dl \qquad (B.13)$$

Thus

$$\mathbf{V}\Phi = \frac{\partial \Phi}{\partial r} \mathbf{i}_r + \frac{1}{r} \frac{\partial \Phi}{\partial \theta} \mathbf{i}_\theta + \frac{1}{r \sin\theta} \frac{\partial \Phi}{\partial \phi} \mathbf{i}_\phi \qquad (B.14)$$

REVIEW QUESTIONS

B.1. Briefly discuss the basic definition of the curl of a vector.

B.2. Justify the application of the basic definition of the curl of a vector to determine separately the individual components of the curl.

B.3. How would you generalize the interpretations for the components of the curl of a vector in terms of the lateral derivatives involving the components of the vector to hold in cylindrical and spherical coordinate systems?

B.4. Briefly discuss the basic definition of the divergence of a vector.

B.5. How would you generalize the interpretation for the divergence of a vector in terms of the longitudinal derivatives involving the components of the vector to hold in cylindrical and spherical coordinate systems?

B.6. Provide general interpretation for the components of the gradient of a scalar.

PROBLEMS

B.1. Find the curl and the divergence for each of the following vectors in cylindrical coordinates: (a) $r \cos \phi \, \mathbf{i}_r - r \sin \phi \, \mathbf{i}_\phi$; (b) $\dfrac{1}{r} \mathbf{i}_r$; (c) $\dfrac{1}{r} \mathbf{i}_\phi$.

B.2. Find the gradient for each of the following scalar functions in cylindrical coordinates: (a) $\dfrac{1}{r} \cos \phi$; (b) $r \sin \phi$.

B.3. Find the expansion for the Laplacian, that is, the divergence of the gradient, of a scalar in cylindrical coordinates.

B.4. Find the curl and the divergence for each of the following vectors in spherical coordinates: (a) $r^2 \mathbf{i}_r + r \sin \theta \, \mathbf{i}_\theta$; (b) $\dfrac{e^{-r}}{r} \mathbf{i}_\theta$; (c) $\dfrac{1}{r^2} \mathbf{i}_r$.

B.5. Find the gradient for each of the following scalar functions in spherical coordinates: (a) $\dfrac{\sin \theta}{r}$; (b) $r \cos \theta$.

B.6. Find the expansion for the Laplacian, that is, the divergence of the gradient, of a scalar in spherical coordinates.

C. UNITS AND DIMENSIONS

In 1960 the International System of Units was given official status at the Eleventh General Conference on weights and measures held in Paris, France. This system of units is an expanded version of the rationalized meter-kilogram-second-ampere (MKSA) system of units and is based on six fundamental or basic units. The six basic units are the units of length, mass, time, current, temperature, and luminous intensity.

The international unit of length is the meter. It is exactly 1,650,763.73 times the wavelength in vacuum of the radiation corresponding to the unperturbed transition between the levels $2p_{10}$ and $5d_5$ of the atom of krypton-86, the orange-red line. The international unit of mass is the kilogram. It is the mass of the International Prototype Kilogram which is a particular cylinder of platinum-iridium alloy preserved in a vault at Sevres, France, by the International Bureau of Weights and Measures. The international unit of time is the second. It is equal to 9,192,631,770 times the period corresponding to the frequency of the transition between the hyperfine levels $F = 4$, $M = 0$ and $F = 3$, $M = 0$ of the fundamental state $^2S_{1/2}$ of the cesium–133 atom unperturbed by external fields.

To present the definition for the international unit of current, we first define the newton, which is the unit of force, derived from the fundamental units meter, kilogram, and second in the following manner. Since velocity is rate of change of distance with time, its unit is meter per second. Since acceleration is rate of change of velocity with time, its unit is meter per second per second or meter per second squared. Since force is mass times acceleration, its unit is kilogram-meter per second squared, also known as the newton. Thus, the newton is that force which imparts an acceleration of 1 meter per second squared to a mass of 1 kilogram. The international unit of current, which is the ampere, can now be defined. It is the constant current which when maintained in two straight, infinitely long, parallel conductors of negligible cross section and placed one meter apart in vacuum produces a force of 2×10^{-7} newtons per meter length of the conductors.

The international unit of temperature is the Kelvin degree. It is based on the definition of the thermodynamic scale of temperature by means of the triple-point of water as a fixed fundamental point to which a temperature of exactly 273.16 degrees Kelvin is attributed. The international unit of luminous intensity is the candela. It is defined such that the luminance of a blackbody radiator at the freezing temperature of platinum is 60 candelas per square centimeter.

We have just defined the six basic units of the International System of Units. Two supplementary units are the radian and the steradian for plane angle and solid angle, respectively. All other units are derived units. For example, the unit of charge which is the coulomb is the amount of charge transported in 1 second by a current of 1 ampere; the unit of energy which is the joule is the work done when the point of application of a force of 1 newton is displaced a distance of 1 meter in the direction of the force; the unit of power which is the watt is the power which gives rise to the production of energy at the rate of 1 joule per second; the unit of electric potential difference which is the volt is the difference of electric potential between two points of a conducting wire carrying constant current of 1 ampere when the power dissipated between these points is equal to 1 watt; and so on. The units for the various quantities used in this book are listed in Table C.1, together with the symbols of the quantities and their dimensions.

Dimensions are a convenient means of checking the possible validity of a derived equation. The dimension of a given quantity can be expressed as some combination of a set of fundamental dimensions. These fundamental dimensions are mass (M), length (L) and time (T). In electromagnetics, it is the usual practice to consider the charge (Q), instead of the current, as the additional fundamental dimension. For the quantities listed in Table C.1, these four dimensions are sufficient. Thus, for example, the dimension of velocity is length (L) divided by time (T), that is LT^{-1}; the dimension of acceleration is length (L) divided by time squared (T^2), that is, LT^{-2}; the dimension of force is mass (M) times acceleration (LT^{-2}), that is, MLT^{-2}; the dimension of ampere is charge (Q) divided by time (T), that is, QT^{-1}; and so on.

To illustrate the application of dimensions for checking the possible validity of a derived equation, let us consider the equation for the phase velocity of an electromagnetic wave in free space, given by

$$v_p = \frac{1}{\sqrt{\mu_0 \epsilon_0}}$$

We know that the dimension of v_p is LT^{-1}. Hence we have to show that the dimension of $1/\sqrt{\mu_0 \epsilon_0}$ is also LT^{-1}. To do this, we note from Coulomb's law that

$$\epsilon_0 = \frac{Q_1 Q_2}{4\pi F R^2}$$

Hence, the dimension of ϵ_0 is $Q^2/[(MLT^{-2})(L^2)]$ or $M^{-1}L^{-3}T^2Q^2$. We note from Ampere's law of force applied to two infinitesimal current elements parallel to each other and normal to the line joining them that

$$\mu_0 = \frac{4\pi F R^2}{(I_1 \, dl_1)(I_2 \, dl_2)}$$

Hence the dimension of μ_0 is $[(MLT^{-2})(L^2)]/(QT^{-1}L)^2$ or MLQ^{-2}. Now we obtain the dimension of $1/\sqrt{\mu_0 \epsilon_0}$ as $1/\sqrt{(M^{-1}L^{-3}T^2Q^2)(MLQ^{-2})}$ or LT^{-1}, which is the same as the dimension of v_p. It should, however, be noted that the test for the equality of the dimensions of the two sides of a derived equation is not a sufficient test to establish the equality of the two sides since any dimensionless constants associated with the equation may be in error.

It is not always necessary to refer to the table of dimensions for checking the possible validity of a derived equation. For example, let us assume that we have derived the expression for the characteristic impedance of a transmission line, i.e., $\sqrt{\mathcal{L}/\mathcal{C}}$ and we wish to verify that $\sqrt{\mathcal{L}/\mathcal{C}}$ does indeed have the dimension of impedance. To do this, we write

$$\sqrt{\frac{\mathcal{L}}{\mathcal{C}}} = \sqrt{\frac{\omega \mathcal{L} 1}{\omega \mathcal{C} 1}} = \sqrt{\frac{\omega L}{\omega C}} = \sqrt{(\omega L)\left(\frac{1}{\omega C}\right)}$$

We now recognize from our knowledge of circuit theory that both ωL and $1/\omega C$, being the reactances of L and C, respectively, have the dimension of impedance. Hence we conclude that $\sqrt{\mathcal{L}/\mathcal{C}}$ has the dimension of $\sqrt{(\text{impedance})^2}$ or impedance.

TABLE C.1.   Symbols, Units, and Dimensions of Various Quantities

Quantity	Symbol	Unit	Dimensions
Admittance	$\bar{Y}$	mho	$M^{-1}L^{-2}TQ^2$
Area	A	square meter	L^2
Attenuation constant	α	neper/meter	L^{-1}
Capacitance	C	farad	$M^{-1}L^{-2}T^2Q^2$
Capacitance per unit length	$\mathcal{C}$	farad/meter	$M^{-1}L^{-3}T^2Q^2$

TABLE C.1. Continued

Quantity	Symbol	Unit	Dimensions
Cartesian coordinates	$\begin{cases} x \\ y \\ z \end{cases}$	meter meter meter	L L L
Characteristic admittance	Y_0	mho	$M^{-1}L^{-2}TQ^2$
Characteristic impedance	Z_0	ohm	$ML^2T^{-1}Q^{-2}$
Charge	Q, q	coulomb	Q
Conductance	G	mho	$M^{-1}L^{-2}TQ^2$
Conductance per unit length	$\mathcal{G}$	mho/meter	$M^{-1}L^{-3}TQ^2$
Conduction current density	$\mathbf{J}_c$	ampere/square meter	$L^{-2}T^{-1}Q$
Conductivity	σ	mho/meter	$M^{-1}L^{-3}TQ^2$
Current	I	ampere	$T^{-1}Q$
Cutoff frequency	f_c	hertz	T^{-1}
Cutoff wavelength	λ_c	meter	L
Cylindrical coordinates	$\begin{cases} r, r_c \\ \phi \\ z \end{cases}$	meter radian meter	L — L
Differential length element	$d\mathbf{l}$	meter	L
Differential surface element	$d\mathbf{S}$	square meter	L^2
Differential volume element	dv	cubic meter	L^3
Directivity	D	—	—
Displacement flux density	$\mathbf{D}$	coulomb/square meter	$L^{-2}Q$
Electric dipole moment	$\mathbf{p}$	coulomb-meter	LQ
Electric field intensity	$\mathbf{E}$	volt/meter	$MLT^{-2}Q^{-1}$
Electric potential	V	volt	$ML^2T^{-2}Q^{-1}$
Electric susceptibility	χ_e	—	—
Electron density	N	(meter)$^{-3}$	L^{-3}
Electronic charge	e	coulomb	Q
Energy	W	joule	ML^2T^{-2}
Energy density	w	joule/cubic meter	$ML^{-1}T^{-2}$
Force	$\mathbf{F}$	newton	MLT^{-2}
Frequency	f	hertz	T^{-1}
Group velocity	v_g	meter/second	LT^{-1}
Guide impedance	η_g	ohm	$ML^2T^{-1}Q^{-2}$
Guide wavelength	λ_g	meter	L
Impedance	$\bar{Z}$	ohm	$ML^2T^{-1}Q^{-2}$
Inductance	L	henry	ML^2Q^{-2}
Inductance per unit length	$\mathcal{L}$	henry/meter	MLQ^{-2}
Intrinsic impedance	η	ohm	$ML^2T^{-1}Q^{-2}$
Length	l	meter	L
Line charge density	ρ_L	coulomb/meter	$L^{-1}Q$
Magnetic dipole moment	$\mathbf{m}$	ampere-square meter	$L^2T^{-1}Q$
Magnetic field intensity	$\mathbf{H}$	ampere/meter	$L^{-1}T^{-1}Q$
Magnetic flux	ψ	weber	$ML^2T^{-1}Q^{-1}$
Magnetic flux density	$\mathbf{B}$	tesla or weber/square meter	$MT^{-1}Q^{-1}$
Magnetic susceptibility	χ_m	—	—
Magnetic vector potential	$\mathbf{A}$	weber/meter	$MLT^{-1}Q^{-1}$

Table C.1.   Continued

Quantity	Symbol	Unit	Dimensions
Magnetization current density	$\mathbf{J}_m$	ampere/square meter	$L^{-2}T^{-1}Q$
Magnetization vector	$\mathbf{M}$	ampere/meter	$L^{-1}T^{-1}Q$
Mass	m	kilogram	M
Mobility	μ	square meter/volt-second	$M^{-1}TQ$
Permeability	μ	henry/meter	MLQ^{-2}
Permeability of free space	μ_0	henry/meter	MLQ^{-2}
Permittivity	ϵ	farad/meter	$M^{-1}L^{-3}T^2Q^2$
Permittivity of free space	ϵ_0	farad/meter	$M^{-1}L^{-3}T^2Q^2$
Phase constant	β	radian/meter	L^{-1}
Phase velocity	v_p	meter/second	LT^{-1}
Plasma frequency	f_N	hertz	T^{-1}
Polarization current density	$\mathbf{J}_p$	ampere/square meter	$L^{-2}T^{-1}Q$
Polarization vector	$\mathbf{P}$	coulomb/square meter	$L^{-2}Q$
Power	P	watt	ML^2T^{-3}
Power density	p	watt/square meter	MT^{-3}
Poynting vector	$\mathbf{P}$	watt/square meter	MT^{-3}
Propagation constant	$\bar{\gamma}$	complex neper/meter	L^{-1}
Propagation vector	$\boldsymbol{\beta}$	radian/meter	L^{-1}
Radian frequency	ω	radian/second	T^{-1}
Radiation resistance	R_{rad}	ohm	$ML^2T^{-1}Q^{-2}$
Reactance	X	ohm	$ML^2T^{-1}Q^{-2}$
Reflection coefficient	Γ	—	—
Refractive index	n	—	—
Relative permeability	μ_r	—	—
Relative permittivity	ϵ_r	—	—
Reluctance	$\mathcal{R}$	ampere (turn)/weber	$M^{-1}L^{-2}Q^2$
Resistance	R	ohm	$ML^2T^{-1}Q^{-2}$
Skin depth	δ	meter	L
Spherical coordinates	$\begin{cases} r, r_s \\ \theta \\ \phi \end{cases}$	meter / radian / radian	L / — / —
Standing wave ratio	SWR	—	—
Surface charge density	ρ_S	coulomb/square meter	$L^{-2}Q$
Surface current density	$\mathbf{J}_S$	ampere/meter	$L^{-1}T^{-1}Q$
Susceptance	B	mho	$M^{-1}L^{-2}TQ^2$
Time	t	second	T
Transmission coefficient	τ	—	—
Unit normal vector	$\mathbf{i}_n$	—	—
Velocity	v	meter/second	LT^{-1}
Velocity of light in free space	c	meter/second	LT^{-1}
Voltage	V	volt	$ML^2T^{-2}Q^{-1}$
Volume	V	cubic meter	L^3
Volume charge density	ρ	coulomb/cubic meter	$L^{-3}Q$
Volume current density	$\mathbf{J}$	ampere/square meter	$L^{-2}T^{-1}Q$
Wavelength	λ	meter	L
Work	W	joule	ML^2T^{-2}

SUGGESTED COLLATERAL AND FURTHER READING

ADLER, R. B., L. J. CHU, AND R. M. FANO, *Electromagnetic Energy Transmission and Radiation*, John Wiley & Sons, Inc., New York, 1960.

HAYT, W. H., JR., *Engineering Electromagnetics* (3rd ed.), McGraw-Hill Book Company, Inc., New York, 1974.

JORDAN, E. C., AND K. G. BALMAIN, *Electromagnetic Waves and Radiating Systems* (2nd ed.), Prentice-Hall, Inc., Englewood Cliffs, N.J., 1968.

KRAUS, J. D., AND K. R. CARVER, *Electromagnetics* (2nd ed.), McGraw-Hill Book Company, Inc., New York, 1973.

MAGID, L. M., *Electromagnetic Fields, Energy, and Waves*, John Wiley & Sons, Inc., New York, 1972.

PARIS, D. T., AND F. K. HURD, *Basic Electromagnetic Theory*, McGraw-Hill Book Company, Inc., New York, 1969.

RAMO, S., J. R. WHINNERY, AND T. VAN DUZER, *Fields and Waves in Communication Electronics*, John Wiley & Sons, Inc., New York, 1965.

RAO, N. N., *Basic Electromagnetics with Applications*, Prentice-Hall, Inc., Englewood Cliffs, N.J., 1972.

SESHADRI, S. R., *Fundamentals of Transmission Lines and Electromagnetic Fields*, Addison-Wesley Publishing Company, Reading, Mass., 1971.

THOMAS, D. T., *Engineering Electromagnetics*, Pergamon Press, Inc., New York, 1972.

ANSWERS TO ODD-NUMBERED PROBLEMS

CHAPTER 1

1.1. (a) 2 m; (b) 0.8 m northward and 0.4 m eastward; (c) 0.8944 m

1.5. 21

1.7. $2\mathbf{i}_x + 2\mathbf{i}_y + \mathbf{i}_z$

1.9. $(4\mathbf{i}_x - 5\mathbf{i}_y + 3\mathbf{i}_z)/5\sqrt{2}$; $6\sqrt{2}$

1.11. $(4\mathbf{i}_x + 4\mathbf{i}_y + \mathbf{i}_z)\,dz$

1.13. $(4\mathbf{i}_x - \mathbf{i}_y)/\sqrt{17}$

1.15. $x + y + z = \text{constant}$

1.17. $\omega(-y\mathbf{i}_x + x\mathbf{i}_y)$

1.19. Traveling wave progressing in the negative z direction

1.21. (a) Linear; (b) circular; (c) elliptical

1.23. Elliptical polarization

1.25. $5\cos(\omega t + 6.87°)$

1.27. $\sqrt{8\pi\epsilon_0 l^2\, mg}$

1.29. $\dfrac{0.0555Q}{\epsilon_0}(\mathbf{i}_x + \mathbf{i}_y + \mathbf{i}_z)$ N/C

1.31. $\dfrac{10^{-7}}{\pi\epsilon_0}\displaystyle\sum_{i=1}^{50}(2i - 1)[10^{-4}(2i - 1)^2 + 1]^{-3/2}\mathbf{i}_y$

1.33. $\dfrac{4 \times 10^{-7}}{\pi \epsilon_0} \sum\limits_{i=1}^{50} \sum\limits_{j=1}^{50} [10^{-4}(2i - 1)^2 + 10^{-4}(2j - 1)^2 + 1]^{-3/2} \mathbf{i}_z$

1.35. (a) $0.4485 \times 10^{-6} \sin 2\pi \times 10^7 t \, \mathbf{i}_x$ amp/m²

(b) $0.4485 \times 10^{-8} \sin 2\pi \times 10^7 t$ amp

1.37. $d\mathbf{F}_1 = 0; \, d\mathbf{F}_2 = \dfrac{\mu_0}{4\pi} I_1 I_2 \, dx \, dy \, \mathbf{i}_x$

1.39. (a) $(5 \times 10^{-5} \mu_0/\pi)\mathbf{i}_z$; (b) $-(10^{-4}\mu_0/4\pi)\mathbf{i}_z$

1.41. $0.179\mu_0\mathbf{i}_z$

1.43. $-v_0 B_0 (14\mathbf{i}_y + 7\mathbf{i}_z)$

CHAPTER 2

2.1. 0.855

2.3. 1

2.7. $1/6$

2.9. $\dfrac{(4n^2 - 1)(1 - e^{-1})}{12n^3(1 - e^{-1/n})} e^{-1/2n}$; $0.20825, 0.21009, 0.21070, 0.21071$

2.11. 16π

2.13. 30 amp

2.15. $-B_0 b v_0 \left(\dfrac{1}{x_0 + a} - \dfrac{1}{x_0} \right)$

2.17. $B_0 b \omega \ln \dfrac{x_0 + a}{x_0} \sin \omega t - B_0 b v_0 \left(\dfrac{1}{x_0 + a} - \dfrac{1}{x_0} \right) \cos \omega t$

2.19. $2B_0 \omega \sin \omega t$

2.21. 0

2.23. (a) 0; (b) $I_1 - I_2$

2.25. $\dfrac{J_0 r}{2}$ for $r < a$ and $\dfrac{J_0 a^2}{2r}$ for $r > a$, direction circular to the axis of the wire

2.27. (a) $I/4$; (b) $I/4$

2.29. $1/2$

2.31. $\dfrac{\rho_{L0}}{2\pi\epsilon_0 r}$, direction radially away from the line charge

2.33. (a) $Q/8$; (b) $Q/8$

2.35. $-1/2$ amp

CHAPTER 3

3.1. $\omega B_0 \dfrac{z^2}{2} \sin \omega t \, \mathbf{i}_x$

3.3. (a) $z\mathbf{i}_x + x\mathbf{i}_y + y\mathbf{i}_z$; (b) 0

3.5. $\frac{1}{3} \times 10^{-7} \cos(6\pi \times 10^8 t - 2\pi z) \, \mathbf{i}_y$ Wb/m^2

3.7. $\mathbf{B} = -\omega \mu_0 \epsilon_0 E_0 \frac{z^3}{3} \cos \omega t \, \mathbf{i}_y$

$\mathbf{E} = -\omega^2 \mu_0 \epsilon_0 E_0 \frac{z^4}{12} \sin \omega t \, \mathbf{i}_x$

3.9. $\mathbf{E} = 10 \cos(6\pi \times 10^8 t - 2\pi z) \, \mathbf{i}_x$

$\mathbf{B} = \frac{10^{-7}}{3} \cos(6\pi \times 10^8 t - 2\pi z) \, \mathbf{i}_y$

3.11. $J_0(a + z)\mathbf{i}_y$ for $-a < z < 0$, $J_0(a - z) \, \mathbf{i}_y$ for $0 < z < a$, 0 otherwise

3.13. Curl will have a component in the y direction in addition to the x component

3.15. Curl has only a z component

3.17. $\oint_C \mathbf{A} \cdot d\mathbf{l} = 0$ for any C

3.19. (a) $3(x^2 + y^2 + z^2)$; (b) 0

3.21. (a) $-x\mathbf{i}_z, y$; (b) $-\mathbf{i}_z, 0$; (c) 0, 1; (d) 0, 0

3.23. $\frac{\rho_0}{2a\epsilon_0}(x^2 - a^2)\mathbf{i}_x$ for $-a < x < a$, 0 otherwise

3.25. (a) and (c)

3.27. $\nabla \cdot \mathbf{r} = 3$

3.29. $\oint_S \mathbf{A} \cdot d\mathbf{S} = 2\pi$, $\nabla \cdot \mathbf{A} = 3$

3.31. 0

CHAPTER 4

4.1. (a) 0.2 amp; (b) 0; (c) 0.2 amp

4.3. (a) 0.2 cos ωt amp; (b) 0.2 sin ωt amp; (c) 0.2828 sin $(\omega t + 45°)$ amp

4.5. (a) $\pm 0.0368 \cos \omega t \, \mathbf{i}_y$; (b) $\pm 0.0135 \cos \omega t \, \mathbf{i}_y$

4.7. $J_0 \frac{a}{2} \mathbf{i}_y$ for $z < -a$, $-J_0\left(z + \frac{z^2}{2a}\right)\mathbf{i}_y$ for $-a < z < 0$, $-J_0\left(z - \frac{z^2}{2a}\right)\mathbf{i}_y$ for $0 < z < a$, $-J_0 \frac{a}{2} \mathbf{i}_y$ for $z > a$

4.9. $-(\rho_0 a/\epsilon_0)\mathbf{i}_x$ for $x < -a$, $(\rho_0 x/\epsilon_0)\mathbf{i}_x$ for $-a < x < a$, $(\rho_0 a/\epsilon_0)\mathbf{i}_x$ for $x > a$

4.15. $(t - z\sqrt{\mu_0\epsilon_0})^2$ corresponds to a $(+)$ wave; $(t + z\sqrt{\mu_0\epsilon_0})^2$ corresponds to a $(-)$ wave

4.17. $C = \frac{\eta_0 J_{s0}}{2}$

For Problem 4.13, $E_x = \frac{\eta_0 J_{s0}}{2}(t \mp z\sqrt{\mu_0\epsilon_0})^2$ for $z \gtrless 0$ and

$H_y = \pm\frac{E_x}{\eta_0}$ for $z \gtrless 0$

4.19. $\mathbf{E} = [0.1\eta_0 \cos(6\pi \times 10^8 t \mp 2\pi z) + 0.05\eta_0 \cos(12\pi \times 10^8 t \mp 4\pi z)]\mathbf{i}_x$
for $z \gtrless 0$

$\mathbf{H} = \pm \dfrac{E_x}{\eta_0}\mathbf{i}_y$ for $z \gtrless 0$

4.21. (a) Same as in Fig. 4.17, except displaced to the left by $1/3$ μs; (b) 75.4 V/m for $300(n - 1/3) < |z| < 300n$ and -37.7 V/m for $300(n - 1) < |z| < 300(n - 1/3)$, $n = 1, 2, 3, \ldots$; (c) $0.2z/|z|$ amp/m for $300(n - 1) < |z| < 300(n - 2/3)$ and $-0.1z/|z|$ amp/m for $300(n - 2/3) < |z| < 300n$, $n = 1, 2, 3, \ldots$

4.23. (a) 0; (b) $\eta_0 J_{S0} \sin \omega t \sin \beta z \, \mathbf{i}_x$; (c) 0

4.25. (a) $\dfrac{\eta_0 J_{S0}}{2} [\cos(\omega t + \beta z)\mathbf{i}_x - \cos(\omega t + \beta z)\mathbf{i}_y]$, linear;

(b) $\dfrac{\eta_0 J_{S0}}{2} [\cos(\omega t - \beta z)\mathbf{i}_x - \cos(\omega t + \beta z)\mathbf{i}_y]$, elliptical except at $z = 0$, $\lambda/8$, $\lambda/4$, $3\lambda/8$, and $\lambda/2$;

(c) $\dfrac{\eta_0 J_{S0}}{2} [\cos(\omega t - \beta z)\mathbf{i}_x - \cos(\omega t - \beta z)\mathbf{i}_y]$, linear

4.27. (a) 0; (b) -3.00 kHz; (c) 1.732 kHz

4.31. (a) $-\dfrac{E_0}{\eta_0} \sin(\omega t - \beta z)\mathbf{i}_x + \dfrac{E_0}{\eta_0} \cos(\omega t - \beta z)\mathbf{i}_y$; (b) $\dfrac{E_0^2}{\eta_0}\mathbf{i}_z$

CHAPTER 5

5.1. (a) 0.1724×10^{-4} V/m, 0.1724×10^{-6} V, 0.1724×10^{-5} ohms;
(b) 0.2857×10^{-4} V/m, 0.2857×10^{-6} V, 0.2857×10^{-5} ohms;
(c) 250 V/m, 2.5 V, 25 ohms

5.3. 1.5245×10^{-19} s

5.5. (a) $-8.667 \times 10^{-7} \sin 2\pi \times 10^9 t$ amp; (b) $-2.778 \times 10^{-6} \sin 2\pi \times 10^9 t$ amp; (c) $-4.444 \times 10^{-5} \sin 2\pi \times 10^9 t$ amp

5.7. (a) $\epsilon_0 E_0(4\mathbf{i}_x + 2\mathbf{i}_y + 2\mathbf{i}_z)$; (b) $8\epsilon_0 E_0(\mathbf{i}_x + \mathbf{i}_y + \mathbf{i}_z)$; (c) $0.5E_0(3\mathbf{i}_x - \mathbf{i}_y - \mathbf{i}_z)$

5.9. $|e|^2 B_0 a^2/2m$, 0.7035×10^{-18} amp-m^2

5.11. $0.5 \times 10^{-6} \sin 2\pi z$ amp

5.13. $\dfrac{\partial^2 \bar{H}_y}{\partial z^2} = \bar{\gamma}^2 \bar{H}_y$

5.15. 0.00083 nepers/m, 4.7562×10^{-3} rad/m, 1.32105×10^8 m/s, 1321.05 m, $(161.102 + j28.115)$ ohms

5.17. $\mathbf{E} = 3.736 e^{\mp 0.0404 z} \cos\left(2\pi \times 10^6 t \mp 0.0976 z + \dfrac{\pi}{8}\right)\mathbf{i}_x$ for $z \gtrless 0$

$\mathbf{H} = \pm 0.05 \, e^{\mp 0.0404 z} \cos(2\pi \times 10^6 t \mp 0.0976 z)\mathbf{i}_y$ for $z \gtrless 0$

5.19. 16.09 m, 1.917:1, $90°$ out of phase

5.21. (a) 30 MHz; (b) 5 m; (c) 1.5×10^8 m/s; (d) $4\epsilon_0$;

(e) $\dfrac{1}{6\pi} \cos (6\pi \times 10^7 t - 0.4\pi z)\, \mathbf{i}_y$ amp/m

5.23. (a) $\mathbf{H} = 0.1 \cos (3\pi \times 10^7 t - 0.4\pi z)\, \mathbf{i}_y$, $\mathbf{B} = 2\mu_0\mathbf{H}$, $\mathbf{M} = \mathbf{H}$, $\mathbf{J}_m = -0.04\pi \cdot$
$\sin (3\pi \times 10^7 t - 0.4\pi z)\, \mathbf{i}_x$;   (b) $\mathbf{E} = 6\pi \cos (3\pi \times 10^7 t - 0.4\pi z)\, \mathbf{i}_x$, $\mathbf{D} = 8\epsilon_0\mathbf{E}$, $\mathbf{P} = 7\epsilon_0\mathbf{E}$, $\mathbf{J}_p = -0.035\pi \sin (3\pi \times 10^7 t - 0.4\pi z)\, \mathbf{i}_x$

5.25. (a) 0.0211 nepers/m, 18.73 rad/m, 0.3354×10^8 m/s, 0.3354 m, 42.15 ohms;
(b) $2\pi \times 10^{-3}$ nepers/m, $2\pi \times 10^{-3}$ rad/m, 10^7 m/s, 1000 m, $2\pi(1 + j)$ ohms

5.27. 1 Hz

CHAPTER 6

6.1. $\pm(\mathbf{i}_x + 2\mathbf{i}_y + 3\mathbf{i}_z)/\sqrt{14}$

6.3. $H_0(2\mathbf{i}_x - \mathbf{i}_y)/\sqrt{5}$

6.5. $\rho_S = 3\,|D_0|$, $\mathbf{J}_S = H_0(2\mathbf{i}_x + \mathbf{i}_y - 2\mathbf{i}_z)\,D_0/|D_0|$

6.7. (a) $2\pi \cos (2\pi \times 10^6 t - 0.02\pi z)$ V; (b) $0.25 \cos (2\pi \times 10^6 t - 0.02\,\pi z)$ amp;
(c) $0.5\pi \cos^2 (2\pi \times 10^6 t - 0.02\pi z)$ W

6.11. Exact values are $\mathcal{L} = 0.1994\mu$, $\mathcal{C} = 5.0155\epsilon$, and $\mathcal{G} = 5.0155\sigma$

6.13. (a) $\mathcal{L} = 0.278 \times 10^{-6}$ H/m, $\mathcal{G} = 4.524 \times 10^{-16}$ mho/m;
(b) $(52.73 + j0)$ ohms

6.17. (a) $\dfrac{\partial \bar{V}}{\partial z} = -\mathbf{Z}\bar{I}$, $\dfrac{\partial \bar{I}}{\partial z} = -\mathcal{Y}\bar{V}$; (c) $\bar{\gamma} = \sqrt{\dfrac{\mathcal{L}_1}{\mathcal{L}_2} - \omega^2\mathcal{L}_1\mathcal{C}}$

6.19. $1667\dfrac{\lambda}{4}$ at $f = 50.01$ MHz

6.21. $2.25\epsilon_0$

6.23. 4.7746 cm

6.25. $\mathbf{E}_2 = E_0(4\mathbf{i}_x + 2\mathbf{i}_y - 6\mathbf{i}_z)$, $\mathbf{H}_2 = H_0(4\mathbf{i}_x - 3\mathbf{i}_y)$

6.27. All boundary conditions are satisfied

6.29. (a) $\dfrac{1}{16}P_i$; (b) $\dfrac{9}{16}P_i$; (c) $\dfrac{3}{8}P_i$

6.31. 150 ohms

CHAPTER 7

7.1. $0.05\pi(\sqrt{3}\,\mathbf{i}_x + \mathbf{i}_y)$

7.3. $\dfrac{1}{2\sqrt{2}}\mathbf{i}_x + \dfrac{\sqrt{3}}{2}\mathbf{i}_y + \dfrac{1}{2\sqrt{2}}\mathbf{i}_z$

7.5. (a) Yes; (b) $\dfrac{1}{24\pi}(\sqrt{3}\,\mathbf{i}_y - \mathbf{i}_z)\cos[6\pi \times 10^7 t - 0.1\pi(y + \sqrt{3}\,z)]$

7.7. (a) $\dfrac{1}{2}(\mathbf{i}_x + \sqrt{3}\,\mathbf{i}_z)$; (b) $8\sqrt{3}$ m, 24 m

7.9. 1 cm

7.11. 3600 MHz, 5400 MHz

7.13. $TE_{1,0}$ mode; $10\sin 20\pi x \sin\left(10^{10}\pi t - \dfrac{80\pi}{3}z\right)\mathbf{i}_y$

7.15. $\Gamma = -0.3252$, $\tau = 0.6748$

7.17. (a) 0; (b) -5 m/s

7.19. 2.4×10^8 m/s

7.23. $TE_{1,0}$, $TF_{0,1}$, $TE_{2,0}$, $TE_{1,1}$, and $TM_{1,1}$

7.25. 6.5 cm, 3.5 cm

7.27. 3535.5 MHz ($TE_{1,0,1}$, $TE_{0,1,1}$), 4330.1 MHz ($TE_{1,1,1}$, $TM_{1,1,1}$), 5590.2 MHz ($TE_{2,0,1}$, $TE_{0,2,1}$, $TE_{1,0,2}$, $TE_{0,1,2}$)

7.29. (a) $41.81°$; (b) $48.6°$

CHAPTER 8

8.1. $0.2\pi\cos 2\pi \times 10^7 t$ amp

8.5. 0.2λ

8.7. (a) 1.257×10^{-3} V/m; (b) $R_{rad} = 0.0351$ ohm, $\langle P_{rad}\rangle = 1.7546$ W

8.9. 1.111 W

8.11. $\sqrt{(D_2 R_{rad2})/(D_1 R_{rad1})}$

8.13. $1\frac{7}{8}$

8.15. 0.60943

8.17. 1.015 W

8.19. (a) $E_\theta = -\dfrac{\eta\beta L I_0 \sin\theta}{8\pi r}\sin(\omega t - \beta r)$, $H_\phi = \dfrac{E_\theta}{\eta}$;

(b) $R_{rad} = 20\pi^2(L/\lambda)^2$, $D = 1.5$

8.21. $-\dfrac{\pi}{4}$, $\cos\left(\dfrac{\pi}{4}\cos\psi - \dfrac{\pi}{8}\right)$

8.23. $\cos^2\left(\dfrac{\pi}{2}\cos\psi\right)$

8.25. $\left|\cos\psi\cos\left(\dfrac{\pi}{4}\cos\psi - \dfrac{\pi}{4}\right)\right|$

8.27. $\left[\cos\left(\dfrac{\pi}{2}\cos\theta\right)\right]\Big/\sin\theta$, where θ is the angle from the vertical, $D = 3.284$

8.29. 4

8.31. 0.00587 V

CHAPTER 9

9.1. (a) $\dfrac{x\mathbf{i}_x + y\mathbf{i}_y + z\mathbf{i}_z}{\sqrt{x^2 + y^2 + z^2}}$; (b) $yz\mathbf{i}_x + zx\mathbf{i}_y + xy\mathbf{i}_z$

9.3. $\dfrac{1}{3\sqrt{5}}(5\mathbf{i}_x + 2\mathbf{i}_y + 4\mathbf{i}_z)$

9.5. 2.121

9.7. $Q/30\pi\epsilon_0$

9.9. $V = \dfrac{10^{-5}}{\pi\epsilon_0} \sum\limits_{i=1}^{50} [10^{-4}(2i-1)^2 + y^2]^{-1/2}$

$\mathbf{E} = \dfrac{10^{-5}}{\pi\epsilon_0} \sum\limits_{i=1}^{50} [10^{-4}(2i-1)^2 + 1]^{-3/2}\mathbf{i}_y$

9.11. (a) $-\dfrac{4\epsilon_0 V_0}{9d^2}\left(\dfrac{x}{d}\right)^{-2/3}$; (b) $[\rho_s]_{x=0} = 0$, $[\rho_s]_{x=d} = \dfrac{4\epsilon_0 V_0}{3d}$

9.13. $V = -\dfrac{kx^3}{6\epsilon} + \dfrac{kd^2 x}{8\epsilon}$ for $-\dfrac{d}{2} < x < \dfrac{d}{2}$

9.15. (a) $\dfrac{\epsilon_2 x}{\epsilon_2 t + \epsilon_1(d-t)} V_0$ for $0 < x < t$, $\dfrac{\epsilon_2 t + \epsilon_1(x-t)}{\epsilon_2 t + \epsilon_1(d-t)} V_0$ for $t < x < d$

(b) $\dfrac{\epsilon_2 t}{\epsilon_2 t + \epsilon_1(d-t)} V_0$

9.17. (a) $V_0 \dfrac{\sinh(\pi x/b)}{\sinh(\pi a/b)} \sin\dfrac{\pi y}{b} + \dfrac{V_0}{3} \dfrac{\sinh(3\pi x/b)}{\sinh(3\pi a/b)} \sin\dfrac{3\pi y}{b}$
(b) $0.1963 V_0$

9.19. $\sum\limits_{n=1,3,5,\ldots}^{\infty} \dfrac{4V_0}{n\pi} \cos\dfrac{n\pi}{4} \dfrac{\sinh(n\pi x/b)}{\sinh(n\pi a/b)} \sin\dfrac{n\pi y}{b}$

9.21. (a) 16.91 V; (b) 16.92 V; (c) 15.53 V. Exact value $= 15.17$ V

9.23. 90.886 V

9.25. (b) $\bar{Z}_{\text{in}} = \dfrac{j\omega\mu dl}{w}\left(1 - \dfrac{j\omega\mu\sigma l^2}{3}\right)$; equivalent circuit consists of an inductor L in parallel with a resistor $3R$ where $L = \mu dl/w$ and $R = d/\sigma lw$

9.27. $\bar{Y}_{\text{in}} = \dfrac{j\omega\epsilon wl}{d}\left(1 + \dfrac{\omega^2\mu\epsilon l^2}{3}\right)$; equivalent circuit consists of a capacitor C in series with an inductor $(1/3)L$ where $C = \epsilon wl/d$ and $L = \mu dl/w$

9.29. 5×10^5 amp-turns/Wb

9.31. 8.4×10^{-4} Wb

CHAPTER 10

10.1. (a) 240 km; (b) 6 MHz; (c) 10 MHz

10.3. (a) 0.9475 Hz; (b) 0.0347 Hz

10.5. 15 cm

10.7. 45.39°

10.11. -0.00533 mho, 1.667

10.13. 0.14λ, 0.192λ

10.15. (a) $0.0359E_0(-\sqrt{3}\,\mathbf{i}_x - \mathbf{i}_z)\cos[6\pi \times 10^9 t - 10\pi(-x + \sqrt{3}\,z)]$; (b) $0.5359E_0(\mathbf{i}_x - \mathbf{i}_z)\cos[6\pi \times 10^9 t - 17.32\pi(x + z)]$

10.17. 50.77°; perpendicular to the plane of incidence

10.19. (a) 8.02 m, 20; (b) 25.38 m, 52

10.21. 1.645

10.25. (a) 0.1654×10^{-3} V/m; (b) $R_{\mathrm{rad}} = 0.6077 \times 10^{-3}$ ohms, $\langle P_{\mathrm{rad}} \rangle = 0.0304$ W

APPENDIX A

A.1. $-3\,\mathbf{i}_x + \sqrt{3}\,\mathbf{i}_y + \mathbf{i}_z$

A.3. Equal

A.5. $-\dfrac{1}{\sqrt{2}}$, $\dfrac{1}{\sqrt{2}}\mathbf{i}_z$

A.7. $13\mathbf{i}_r + 6\mathbf{i}_z$

APPENDIX B

B.1. (a) $-\sin\phi\,\mathbf{i}_z$, $\cos\phi$; (b) 0, 0 except at $r = 0$; (c) 0 except at $r = 0, 0$

B.3. $\dfrac{1}{r}\dfrac{\partial}{\partial r}\left(r\dfrac{\partial\Phi}{\partial r}\right) + \dfrac{1}{r^2}\dfrac{\partial^2\Phi}{\partial\phi^2} + \dfrac{\partial^2\Phi}{\partial z^2}$

B.5. (a) $-\dfrac{1}{r^2}(\sin\theta\,\mathbf{i}_r - \cos\theta\,\mathbf{i}_\theta)$; (b) $\cos\theta\,\mathbf{i}_r - \sin\theta\,\mathbf{i}_\theta$

INDEX